Human Genetics: A Comprehensive Approach

Human Genetics:
A Comprehensive Approach

Editor: Lesley Easton

R CALLISTO REFERENCE

www.callistoreference.com

Callisto Reference,
118-35 Queens Blvd., Suite 400,
Forest Hills, NY 11375, USA

Visit us on the World Wide Web at:
www.callistoreference.com

ISBN: 978-1-64116-160-2 (Hardback)

Cataloging-in-Publication Data

Human genetics : a comprehensive approach / edited by Lesley Easton.
    p. cm.
Includes bibliographical references and index.
ISBN 978-1-64116-160-2
1. Human genetics. 2. Human genome. 3. Human biology. I. Easton, Lesley.
QH431 .H86 2019
599.935--dc23

Table of Contents

Preface

The main aim of this book is to educate learners and enhance their research focus by presenting diverse topics covering this vast field. This is an advanced book which compiles significant studies by distinguished experts in the area of analysis. This book addresses successive solutions to the challenges arising in the area of application, along with it; the book provides scope for future developments.

Human genetics is the study of inheritance in human beings. It is an interdisciplinary science that encompasses the fields of classical genetics, molecular genetics, clinical genetics, cytogenetics, genomics and developmental genetics. The study of human genetics aids in the understanding of genetic diseases and their potential treatments. The concepts of autosomal dominant and recessive inheritance, X-linked and Y-linked inheritance, pedigree analysis and karyotype are significant for understanding genetic differences and inheritance patterns. This book discusses the fundamentals as well as the modern approaches to human genetics. It includes some of the vital pieces of work being conducted across the world, on various topics related to human genetics. It attempts to assist those with a goal of delving into this field.

It was a great honour to edit this book, though there were challenges, as it involved a lot of communication and networking between me and the editorial team. However, the end result was this all-inclusive book covering diverse themes in the field.

Finally, it is important to acknowledge the efforts of the contributors for their excellent chapters, through which a wide variety of issues have been addressed. I would also like to thank my colleagues for their valuable feedback during the making of this book.

Editor

A droplet digital PCR detection method for rare L1 insertions in tumors

Travis B White[1†], Adam M McCoy[2,3†], Vincent A Streva[1,4†], Joshua Fenrich[2] and Prescott L Deininger[1*]

Abstract

Background: The active human mobile element, long interspersed element 1 (L1) currently populates human genomes in excess of 500,000 copies per haploid genome. Through its mobility via a process called target primed reverse transcription (TPRT), L1 mobilization has resulted in over 100 *de novo* cases of human disease and has recently been associated with various cancer types. Large advances in high-throughput sequencing (HTS) technology have allowed for an increased understanding of the role of L1 in human cancer; however, researchers are still limited by the ability to validate potentially rare L1 insertion events detected by HTS that may occur in only a small fraction of tumor cells. Additionally, HTS detection of rare events varies greatly as a function of read depth, and new tools for *de novo* element discovery are needed to fill in gaps created by HTS.

Results: We have employed droplet digital PCR (ddPCR) to detect rare L1 loci in mosaic human genomes. Our assay allows for the detection of L1 insertions as rare as one cell in every 10,000.

Conclusions: ddPCR represents a robust method to be used alongside HTS techniques for detecting, validating and quantitating rare L1 insertion events in tumors and other tissues.

Keywords: L1, retrotransposon, droplet digital PCR, tumor

Background

The human retrotransposon, long interspersed element 1 (L1) exists in over half a million copies per genome and constitutes 17% of genomic content [1]. The majority of these copies are nonfunctional relics that litter the genome; however, on average, approximately 100 L1 elements remain active in any given individual [1,2]. These active L1 elements mobilize in both germline and somatic tissues [3-11]. *De novo* L1 retrotransposition has been responsible for numerous germline diseases, as well as being implicated in tumorigenesis [8,10,12]. Notably, *de novo* L1 insertions have been identified in numerous cancer types including lung, colon, prostate, ovarian, and hepatocellular carcinoma through the use of high-throughput sequencing (HTS) technology [3-11].

Because tumors are often heterogeneous in genomic content, discovery and validation of *de novo* L1 insertion events detected by HTS in tumors can be problematic [13]. Validation statistics for HTS hits of *de novo* L1

somatic insertions have been reported to be as low as 67% [11]. One explanation for this fairly low rate of validation is tumor heterogeneity. Somatic L1 insertion events that occur late in tumorigenesis may represent a small minority of cells, and even insertion events that occur early in tumorigenesis may not be present in all tissue derived from that tumor. Some studies have had significantly higher validations, [3,7,10] but as methods develop to detect insertions present in smaller proportions of tumor cells, we can expect validation to become progressively more difficult.

Droplet digital PCR (ddPCR) has recently emerged as a robust tool to provide precise measurements of nucleic acid target concentrations [14,15]. In ddPCR, input DNA is partitioned, along with PCR reagents, into approximately 20,000 droplets as a water-in-oil emulsion within a single thermocycled reaction well [16]. Detection of target DNA relies on fluorogenic probes in a 5'-nuclease assay (TaqMan™) [17,18]. Briefly, an oligonucleotide probe, which anneals specifically to a target DNA within the primer binding sites, is included in the PCR with the primers. The probe is modified at the 5' end with a fluorescent moiety, which is quenched in the intact probe by a

* Correspondence: pdeinin@tulane.edu
†Equal contributors
[1]Tulane Cancer Center, 1430 Tulane Avenue, New Orleans, LA 70112, USA
Full list of author information is available at the end of the article

modification at the 3' end with a quencher moiety. The probe anneals to the target DNA during the annealing/extension step of the PCR. During extension of the primer that anneals to the same DNA strand as the probe, the 5' to 3' nuclease activity of *Taq* polymerase cleaves the probe, which separates the 5'-fluorescent nucleotide of the probe from the 3' quencher, generating a fluorescent signal.

Sequestration of template DNA occurs in ddPCR, such that some droplets contain no copies and others one or more copies of the template target DNA [14,16]. Identification of template target DNA-containing droplets is achieved through fluorescence analysis of the droplets according to the 5'-fluorogenic probes used in the ddPCR. Droplets containing one or more target templates generate increased fluorescence compared to droplets containing non-target DNA. Thus, the quantification comes from the ability to essentially detect a single DNA template sequestered into a droplet through PCR amplification of the templates followed by counting of fluorescent droplets. The concentration of the input target DNA is calculated according to a Poisson distribution of template DNA molecules partitioned into the fluorescence-positive droplets [16]. Recent reports use ddPCR to successfully identify very rare alleles (that is, <1%) in heterogeneous tumor samples, making ddPCR an ideal method to apply for detection of rare *de novo* L1 insertion events [16]. Additionally, the utility of ddPCR over traditional qPCR methods has recently been examined [19].

Due to the high copy number of L1 sequence in the human genome, detection of specific polymorphic loci in a heterogeneous sample by traditional qPCR approaches is particularly difficult due to the high background signal created from nonspecific amplification from templates that do not contain the polymorphic L1.

Partitioning of template DNA in ddPCR not only affords a reduction of this nonspecific background due to template dilution, it also allows an accurate determination of the concentration of the polymorphic L1 of interest in the input DNA. In this report, we apply ddPCR technology to the detection of rare L1 elements, allowing detection levels as low as one in every 10,000 cells. Our ddPCR assays incorporate L1 primers and probes that are common to each 5' or 3' junction ddPCR and specifically detect the youngest, actively mobile, L1Hs subfamiliy. By using universal L1 5'- and 3'-end primers and probes, paired with locus-specific flanking primers, this L1 detection method will prove useful as a way to rapidly identify *de novo* L1 insertion events in a heterogeneous tumor sample and to quantitate their frequency within an individual tumor sample. Additionally, L1 ddPCR allows heterozygote and homozygote loci to be easily distinguished through parallel detection of a second genomic locus.

Results

For validation or discovery of *de novo* L1 insertion events, we designed assays to detect either the 5'- or 3'-insertion junctions at specific genomic loci. The core of each assay is a single primer and probe specific to the youngest L1 subfamily, L1Hs [2]. One primer and probe set is located at the 3' end of L1Hs (Table 1; 3' L1Hs primer, 3' L1Hs probe), which may be used to detect both full-length and truncated L1Hs elements when paired with an appropriate locus-specific primer (Figure 1). The other primer and probe set is located at the 5' end of L1Hs (Table 1; 5' L1Hs primer, 5' L1Hs probe) to detect full-length L1Hs 5'-insertion junctions when paired with an appropriate locus-specific primer (Figure 1). Amplification

Table 1 Primers and probes used in this study

Primer/Probe Name	Sequence
5' L1Hs Primer	5' GGAAATGCAGAAATCACCGTCTTC 3'
5' L1Hs Probe (Chr 15)	5' FAM AGGAACAGCTCCGGTCTACAGCTC BHQ1 3'
5' L1Hs Probe (other loci tested)	5' FAM AGGAACAGC/ZEN/TCCGGTCTACAGCTC IABkFQ 3'
5' Chr15 AC216176 Locus Primer	5' GTGGACAAAGAAAAGCATCCTTGAT 3'
RPP30 Forward Primer	5' GATTTGGACCTGCGAGCG 3'
RPP30 Reverse Primer	5' GCGGCTGTCTCCACAAGT 3'
RPP30 Probe	5' VIC CTGACCTGAAGGCTCT MGB 3'
5' Chr4 esv3475 Locus Primer	5' CCACATGGTATAAGATAAAAACACGAG 3'
5' Chr4 esv4912 Locus Primer	5' CTAAGCAATGGAGGAAAATATCG 3'
3' L1Hs Primer	5' GGGAGATATACCTAATGCTAGATGACAC 3'
3' L1Hs Probe	5' FAM ATTATACTCTAAGTTTTAGGGTACATGTGCACATTGTGC BHQ1 3'
3' Chr15 AC216176 Locus Primer	5' TCTATAAGCAGTGGAAGCACATG 3'
3' Chr4 esv3475 Locus Primer	5' GACCAATTTTTTTTTTGCCTGTACTGAC 3'
3' Chr4 esv4912 Locus Primer	5' TGCATTTTGAGTTAATTTTTGTACATGGTG 3'

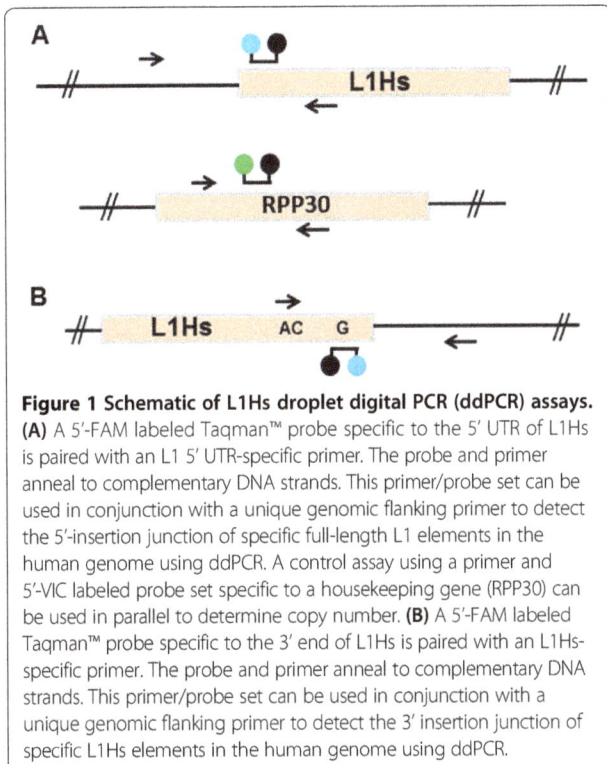

Figure 1 Schematic of L1Hs droplet digital PCR (ddPCR) assays.
(A) A 5'-FAM labeled Taqman™ probe specific to the 5' UTR of L1Hs is paired with an L1 5' UTR-specific primer. The probe and primer anneal to complementary DNA strands. This primer/probe set can be used in conjunction with a unique genomic flanking primer to detect the 5'-insertion junction of specific full-length L1 elements in the human genome using ddPCR. A control assay using a primer and 5'-VIC labeled probe set specific to a housekeeping gene (RPP30) can be used in parallel to determine copy number. **(B)** A 5'-FAM labeled Taqman™ probe specific to the 3' end of L1Hs is paired with an L1Hs-specific primer. The probe and primer anneal to complementary DNA strands. This primer/probe set can be used in conjunction with a unique genomic flanking primer to detect the 3' insertion junction of specific L1Hs elements in the human genome using ddPCR.

of the locus-specific L1Hs 5'- or 3'-insertion junction generates FAM fluorescence through nucleolytic cleavage of the annealed L1Hs-specific probe by *Taq* polymerase. For each experiment, a threshold of fluorescence is set relative to negative controls to measure the quantity of droplets that do or do not contain template target DNA. Through separation of DNA templates for PCR in up to 20,000 droplets and measurement of fluorescence for each droplet at the terminal plateau phase of PCR, ddPCR-based L1 detection is capable of a high degree of discrimination that is not possible with standard TaqMan™ assays [14]. Additionally, L1 ddPCR assays can be multiplexed with control ddPCR assays for housekeeping genes such as RPP30 to allow for accurate copy number determination [20].

Because they exist in high genomic copy number, L1 elements can contribute to significant background signal in PCR-based assays. In our assay designs, the L1Hs 5' and 3' end probes anneal to the same DNA strand as the locus-specific primer in each assay to ensure fluorescent signal is generated by extension of the primer at the L1-occupied chromosomal locus (Figure 1). This minimizes fluorescent signal arising from linear extensions off primers annealed at the numerous other genomic L1 loci. There is, however, still the possibility of amplification of two inverted L1Hs elements by two L1-specific primers resulting in background fluorescence in negative samples.

We were able to generate L1-specific primers and probes that target young L1 insertions and result in

only a minimal degree of non-specific background (Table 1, Figures 2 and 3). We developed a ddPCR assay for the L1 5' end to detect *de novo* full-length L1 insertion events. Using a known polymorphic full-length L1Hs on Chromosome 15 (AC216176; [21]) as a model for our assay, we were able to successfully design a ddPCR assay that is able to robustly detect a specific L1Hs 5'-insertion junction known to be homozygous for the polymorphic L1 element in the cell line tested (Figure 2). To determine the limit of sensitivity of our ddPCR assay, we performed ten-fold dilutions of this sample as a mixture with DNA from a sample known to be negative for the insertion, thus keeping the total input genomic DNA constant for each ddPCR. Detection of RPP30 by VIC fluorescence is consistent in each dilution experiment. This analysis allowed us to determine that the limit of sensitivity of our assay is as low as one positive cell in 10,000 total cells (0.01%) (Figure 2).

To assay specific 3'-insertion junctions of polymorphic L1 elements by ddPCR, we designed primers and probes unique to the 3' end of the youngest L1Hs subfamily, which constitute the vast majority of L1 elements capable of retrotransposition [2]. The 3' end of the 3' L1Hs primer makes use of an AC dinucleotide at position 5926 of L1Hs, which gives the primer specificity to only these youngest L1 elements. Thus, although the primer can probably anneal to a significant portion of genomic L1 elements, it will only be able to prime DNA synthesis from these actively mobile, and therefore most interesting, L1 elements. Additionally, the L1 3'-end probe makes use of a G nucleotide at position 6011 of L1Hs, making it also specific for only the youngest L1 elements [2,22,23].

For the 3' junction ddPCR experiment, we investigated the same known polymorphic full-length L1Hs on Chromosome 15 used as a model for our 5'-junction assay (AC216176; [21]). Using this primer and probe set, we were able to robustly detect a single, specific L1 3'-insertion junction with only minimal background (Figure 3). We additionally performed a dilution experiment as described above and were able to detect the L1 3'-insertion junction to one positive cell in every 1,000 total cells (0.1%) (Figures 3 and 4). Because establishment of the polymorphic L1 detection limit was our goal in these experiments, this ddPCR did not include RPP30 detection. Additionally, we showed that the 3' end L1 primer and probe sets are specific to young L1 elements, as they do not amplify a known older (L1PA4) genomic L1 (data not shown).

To show that the L1-specific primers and probes may be used to detect the 5'- and 3'-insertion junctions of multiple polymorphic loci, locus specific primers were designed to detect the 5' and 3' junctions of two other known polymorphic full-length L1Hs elements (Additional file 1: Figure S1, Additional file 2: Figure S2, Additional file 3: Figure S3, Additional file 4: Figure S4, Additional file 5:

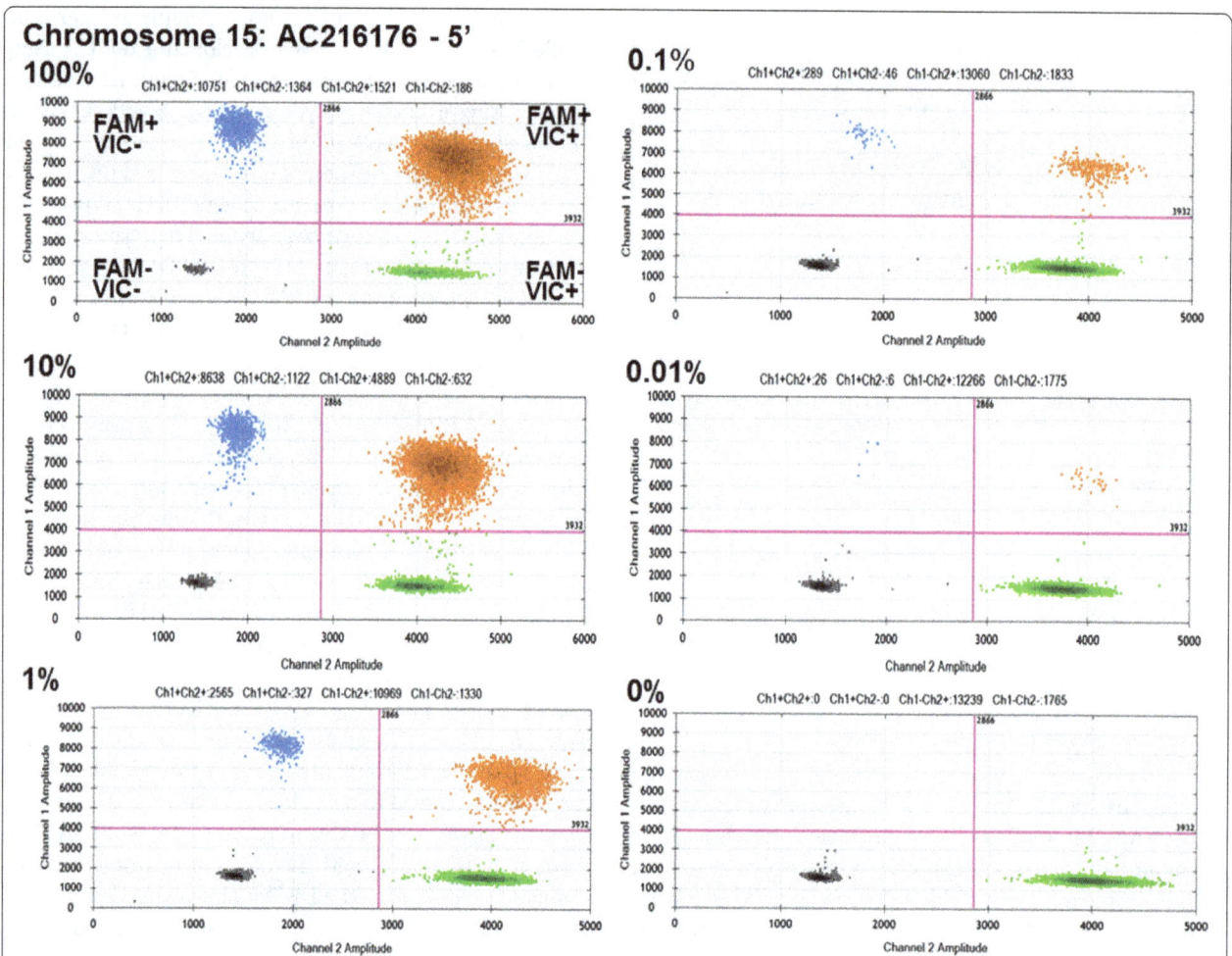

Figure 2 Detection of chromosome 15 AC216176 L1Hs by the 5′ junction droplet digital PCR (ddPCR) assay. Each panel represents a single ddPCR experiment whereby a DNA sample (defined below) is segregated into individual droplets and assessed for the presence of the L1 locus (FAM) and RPP30 locus (VIC) using two different fluorophores in Taqman™ assays (see Figure 1). The FAM and VIC fluorescence for each droplet is plotted as a data point on each graph. FAM fluorescent signal (Channel 1) is plotted on the y-axis and VIC fluorescent signal (Channel 2) is plotted on the x-axis. The droplet threshold for each fluorophore used is indicated by the magenta lines, determining whether a droplet is considered positive or negative for either FAM or VIC fluorescence. The positive or negative fluorescence assessment for each quadrant is labeled accordingly for the plot describing the experiment with 100% GM01632 DNA. The blue dots represent individual droplets that contain at least one copy of the L1 locus tested but not the RPP30 locus (FAM positive, VIC negative), the green dots represent droplets that contain at least one copy of the RPP30 gene and not the L1 locus (VIC positive, FAM negative), and the orange dots represent droplets that contain at least one copy of both the RPP30 gene DNA and the L1 locus tested (positive for both FAM and VIC). We tested 160 ng of *BsaJI*-digested genomic DNA from GM01632 cells, which are homozygous for the polymorphic L1 element (100%), and tenfold dilutions of this same sample as a mixture with *BsaJI*-digested genomic DNA from GM01631 cells, which do not have this polymorphic L1 insertion (10%-0.01%), thus keeping the total input genomic DNA constant for each ddPCR. Additionally, as a negative control, 160 ng of *BsaJI*-digested genomic DNA from GM01631 cells was tested (0%).

Figure S5, Additional file 6: Figure S6, Additional file 7: Figure S7, Additional file 8: Figure S8). Detection of both 5′- and 3′-insertion junctions for a polymorphic element on Chromosome 4 (Database of Genomic Variants ID: esv 3475, [24,25]) was sensitive to one positive cell in 1,000 total cells (0.1%) (Additional file 1: Figure S1, Additional file 2: Figure S2, Additional file 3: Figure S3, Additional file 4: Figure S4). Likewise, detection of both 5′- and 3′-insertion junctions of another polymorphic element on Chromosome 4 (Database of Genomic Variants ID: esv4912, [24,25]) was sensitive to one positive cell in 1,000 total cells (0.1%)

(Additional file 5: Figure S5, Additional file 6: Figure S6, Additional file 7: Figure S7, Additional file 8: Figure S8).

Discussion

Recent advances in detection of *de novo* L1 integration events by HTS have resulted in an increased understanding of the potential role L1 elements might play in the development of tumors. To date, L1 insertions have been detected by HTS in five different cancer types, and many of these insertions have been fully validated by traditional PCR-based strategies [6-11]. There are many

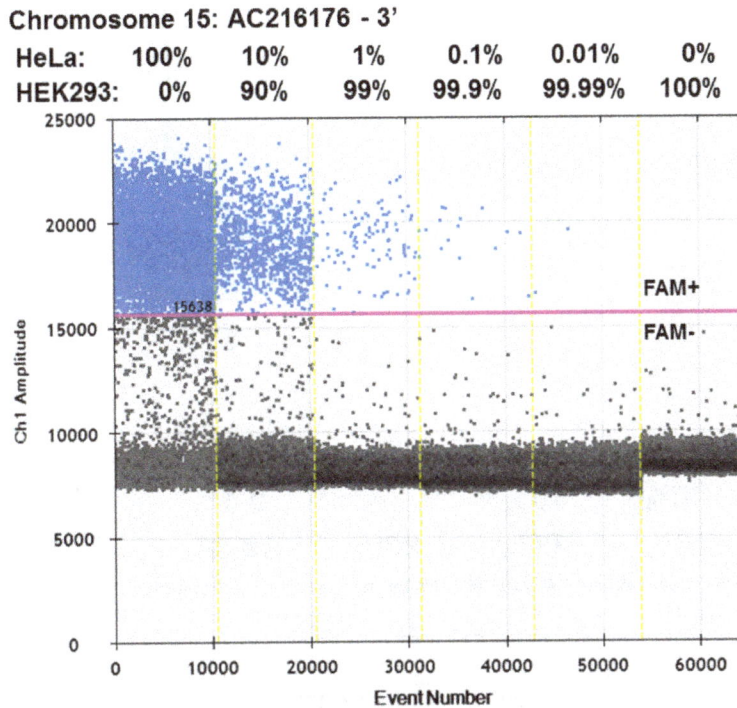

Figure 3 Detection of chromosome 15 AC216176 L1Hs by the 3' junction ddPCR assay. The L1Hs 3' junction ddPCR assay uses a L1-specific primer, L1-specific 5'-FAM labeled Taqman™ probe, and a locus-specific primer near the Chromosome 15 AC216176 3'-insertion junction, as shown in Figure 1B. The FAM fluorescent signal (Ch 1) for each droplet is plotted on the y-axis for each of the ddPCR experiments, which are separated by a dotted yellow line, with input DNA indicated above each experiment. Each droplet is cumulatively counted as an 'Event Number' for the ddPCR experiments analyzed in tandem, and plotted along the x-axis. The positive droplet fluorescence threshold is indicated by the magenta line, which determines whether a droplet is considered positive or negative for FAM fluorescence. Thus, the blue dots represent individual droplets that contain at least one copy of the L1 locus tested. We tested 200 ng of *BamHI*-digested genomic DNA from HeLa cells, which contain the polymorphic L1 element, and tenfold dilutions of this same sample as a mixture with *BamHI*-digested genomic DNA from HEK293 cells, which do not have this polymorphic L1 insertion. Percentages given reflect the amount of input DNA with 100% corresponding to 200 ng of DNA. This assay robustly detects the 3'-insertion junction of the polymorphic full-length AC216176 L1Hs element when present in the genomic DNA from a cell line positive for that polymorphism (HeLa 100%), but not in a cell line negative for that polymorphism (HEK293 100%). L1-positive droplets are observed at dilutions as low as 0.01% of the DNA with this assay.

more *de novo* L1 insertions, however, that have been detected through the use of HTS, but have not been able to be successfully validated. One likely explanation for this discrepancy is the genomic heterogeneity associated with tumors.

HTS technology has afforded researchers the ability to identify extremely low-frequency events that are difficult to validate by traditional PCR-based methods due to a high rate of background signal. *De novo* L1 insertions in tumors can often be classified as low-frequency events for a number of reasons. First, it is often difficult to fully separate normal adjacent tissue from tumors, with tissue dissected from certain tumor types sometimes containing a greater fraction normal than cancerous tissue [26]. Second, the timing of L1 mobilization in tumors has not been fully established. If L1 insertions occur at late stages in the development of a tumor, they will only be represented in a small fraction of the cells that compose

the tumor. In this case, it remains very likely that such *de novo* L1 insertion events would be detected by some HTS studies, but would not necessarily be detectable by traditional PCR.

Droplet digital PCR (ddPCR) has proven itself capable of detecting extremely low frequency events [27]. In this study, we report the ability of a ddPCR assay to detect an L1 insertion event in as few as 0.01% to 0.1% of cells. This assay has a minimal level of background signal, which is surprising given the prolific nature of the L1 template in the human genome. The most likely source of background signal is the high level of L1Hs 3' ends (approximately 5000 matching the 3' L1Hs primer, but 500,000 with a partial match), resulting in off-target amplification between two L1-specific primers. Regardless of this, we are able to robustly detect an L1-positive signal in a low fraction of cells. Our ddPCR assay is not only a robust, straightforward tool to validate L1 insertion events

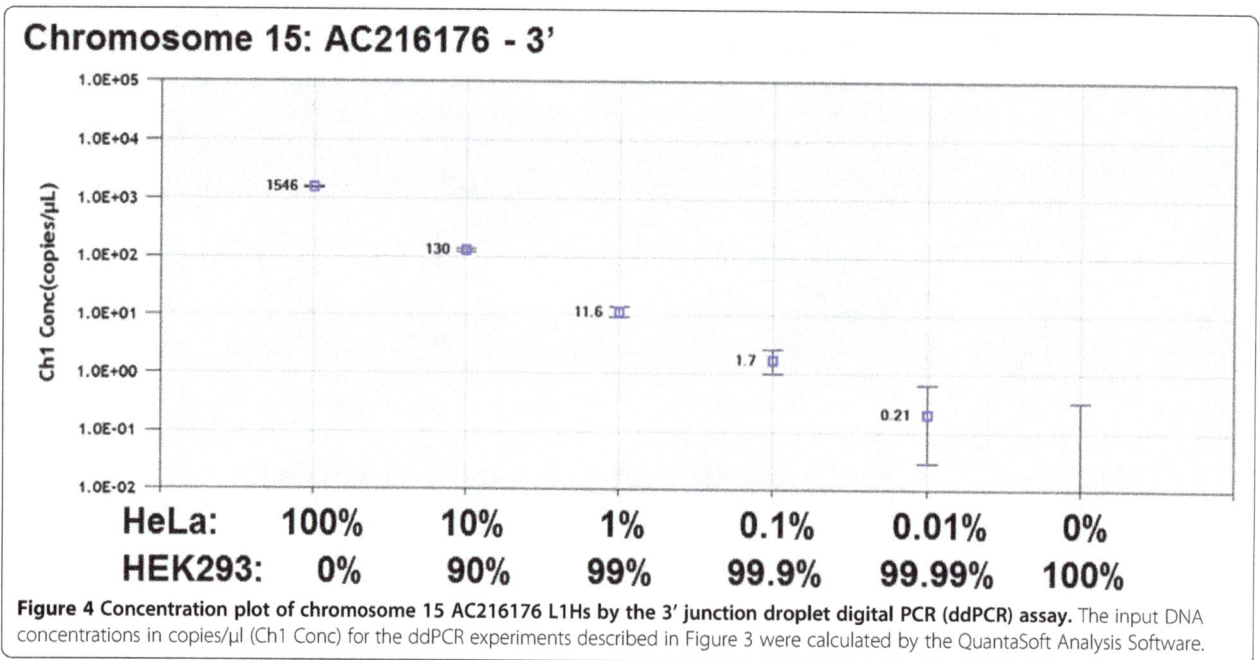

Figure 4 Concentration plot of chromosome 15 AC216176 L1Hs by the 3' junction droplet digital PCR (ddPCR) assay. The input DNA concentrations in copies/μl (Ch1 Conc) for the ddPCR experiments described in Figure 3 were calculated by the QuantaSoft Analysis Software.

detected by HTS from tumors, but is also capable of quantifying the fraction of cells in the tumor, or other material, that have that particular insertion.

Tumor cells undergo constant evolution and produce subclonal populations of cells, each containing different signatures of genomic rearrangements [28]. These chromosomal aberrations may serve as biomarkers for minor subclonal populations that harbor the capacity for relapse [28]. Indeed, there is a major effort to use HTS data to describe the subclonal genomic constituency of tumors and identify biomarkers for invasive subclonal populations of cells [29,30]. In addition to the validation of unique L1 insertions identified by HTS of tumors, the assays described here may be used to track and quantify mosaic L1 loci used as biomarkers for subclonal populations of cancer cells and, if an L1 insertion is unique to an individual's identified cancer, establish a minimum level of residual disease detection.

Detection of rare alleles in the human population such as single-nucleotide polymorphisms, small insertions or deletions, or mobile element polymorphisms allows determination of disease-causing candidate genomic loci through association studies and shows regions of our genome that have been subject to selective pressures [31,32]. Establishment of rare allele frequencies through individual genotyping is a laborious and expensive process that can be overcome through methods that interrogate pools of human genomic DNA [33,34]. Our assay may be used as a means of establishing rare allele frequencies, in the range of 0.01%, in pools of human genomic DNA. Detection of the rare *BRAF* V600E allele has been previously demonstrated by ddPCR [14].

Conclusions

Retrotransposition of long interspersed element 1 (L1) in human germline and somatic cells contributes to genomic variation in human populations and is implicated in tumorigenesis. In this study, we designed droplet digital PCR (ddPCR) assays to detect rare L1 insertion events in heterogeneous human genomic DNA samples. Traditional qPCR methods are unable to confidently discern rare target DNA sequences among input DNA as complex as a human genome due to low-chance priming events that cause background signal and lead to false-positive determinations. This effect is exacerbated when the target DNA involves L1 sequence, which occupies approximately 17% of the human genome. Using universal 5' and 3' L1 primers and probes in ddPCR, paired with a locus-specific primer near the assayed insertion site, we detected polymorphic L1 5' and 3' junctions in genomic DNA from a heterogeneous sample when as few as 0.01% of the cells contained the polymorphic L1. The ability to confidently detect and simultaneously quantify the level of a L1 insertion locus in a mosaic sample, such as tumor biopsy genomic DNA, will allow rapid validation of high-throughput sequencing data on *de novo* L1 insertions for a given sample, establishment of a minimum of residual disease detection for a cancer cell-specific L1 insertion, or sampling of pools of human genomic DNA for rare L1 allele detection.

Methods

Selection of L1 loci and primer/probe design

Polymorphic L1 elements were detected from genomic DNA from fibroblast cell lines GM01630, GM01631 and

GM01632 (Coriell Institute; Camden, NJ, USA) using Sequencing Identification and Mapping of Primed L1 Elements (SIMPLE) (VAS, unpublished data). These polymorphic elements were previously confirmed by PCR. An identified polymorphic L1Hs locus was selected on the basis of identity to the L1Hs consensus sequence: a Chromosome 15 full-length L1Hs locus (AC216176; [21]). Additional polymorphic L1 loci tested were chosen among previously characterized polymorphic full-length L1 elements on Chromosome 4 (Database of Genomic Variants ID: esv4912, esv3475 [24,25]), and were assayed on the basis of identity to the L1Hs consensus sequence. Probes and primers were designed to match either the 5' or 3' end of the L1Hs consensus sequence (Table 1). The L1Hs 3' primer and probe incorporate diagnostic nucleotides specific to L1Hs that are not present in older L1 elements. These L1Hs-specific primer/probe sets can be paired with a unique flanking primer specific to the genomic region of interest for detection of the 5'- and 3'-insertions junctions. Primers and probes were synthesized by Integrated DNA Technologies (Coralville, IA, USA), with the exception of those used for detection of RPP30 (Applied Biosystems now Life Technologies; Grand Island, NY, USA) (Table 1).

Droplet digital PCR reaction conditions

Genomic DNA from fibroblast cell lines was extracted using the DNEasy Blood and Tissue Kit (Qiagen; Germantown, MD, USA). The 5' junction ddPCR assays were performed in 20-µL reactions using ddPCR Supermix for Probes (Bio-Rad; Hercules, CA, USA) and 150 to 200 ng of *Bsa*JI- or *Bam*HI-digested input DNA. Restriction enzyme digestions were done according to manufacturers' protocol (New England BioLabs; Ipswitch, MA, USA). In the 5' junction ddPCR assays, 900 nM 5' L1Hs primer, 900 nM locus-specific primer, and 250 nM 5' L1Hs probe was used. The 5' junction ddPCR assays for the Chromosome 15 AC216176 locus included detection of the housekeeping gene RPP30 with 900 nM of each RPP30-specific primer and 250 nM RPP30 probe. Because the two loci are not linked, each droplet has a probability of being positive for either one of the loci, and some droplets will be either negative or positive for both. The relationship between presence/absence of each locus in a droplet is defined by the Poisson distribution and allows robust, digital quantification of the two loci relative to one another. Droplet generation was performed as per the manufacturer's instructions. Cycling conditions were 95°C for 10 min, followed by 40 cycles of 94°C for 30 seconds and 64°C for two minutes, and then a final 10-min incubation at 98°C. Droplet reading was performed on a QX100 ddPCR droplet reader (Bio-Rad; Hercules, CA, USA) for the Chromosome 15 AC216176 locus, a QX200 ddPCR droplet reader (Bio-Rad; Hercules, CA, USA) for the other loci tested

and analysis was done using QuantaSoft Analysis software (Bio-Rad; Hercules, CA, USA).

The 3' junction ddPCR assays were performed in 20-µL reactions using ddPCR Supermix for Probes (No dUTP) (Bio-Rad; Hercules, CA, USA) and 200 ng *Bam*HI-digested input DNA. In the 3' junction ddPCR assay of the Chromosome 4 esv4912 polymorphic L1, 900nM of 3' L1Hs primer, 900nM of locus-specific primer, and 200nM of 3' L1Hs probe was used. In all other 3' junction ddPCR assays, 900nM of 3' L1Hs primer, 4.5 µM of locus-specific primer, and 200nM of 3' L1Hs probe was used. Droplet generation was performed as per the manufacturer's instructions. Cycling conditions were 95°C for 10 min, followed by 40 cycles of 94°C for 30 seconds and 64°C for one minute, and then a final 10-min incubation at 98°C. Droplet reading was performed on a QX200 ddPCR droplet reader (Bio-Rad; Hercules, CA, USA), and analysis was done using QuantaSoft Analysis software (Bio-Rad; Hercules, CA, USA).

Genomic DNA from the following cell lines was used: GM01630, GM01631 and GM01632 (Coriell Institute; Camden, NJ, USA), Flp-In-293 (denoted HEK293 in figures and figure legends, the parental line of these cells; Invitrogen now Life Technologies; Grand Island, NY, USA), HeLa (American Type Culture Collection; Manassas, VA, USA, item number: CCL-2), LoVo (American Type Culture Collection; Manassas, VA, USA, item number: CCL-229), HCT116D (HCT116 derivative with a Flp-In site integrated kindly provided by J. Issa [35]; denoted HCT116 in figures and figure legends).

Droplet digital PCR mixing experiments

For mixing experiments, cell line genomic DNA positive for a particular L1 insertion was mixed via tenfold dilutions with cell line genomic DNA negative for that particular L1 insertion. Following mixing, dilutions were added at 150 ng to 200 ng per ddPCR reaction as described above.

Additional files

Additional file 1: Figure S1. Detection of Chromosome 4 esv3475 L1Hs by the 5' junction ddPCR assay. The L1Hs 5' junction ddPCR assay uses a L1-specific primer, L1-specific 5'-FAM labeled Taqman™ probe, and a locus-specific primer near the Chromosome 4 esv3475 5'-insertion junction, as shown in Figure 1A. The FAM fluorescent signal (Ch 1) for each droplet is plotted on the y-axis for each ddPCR experiment, which are separated by a dotted yellow line and indicated above each experiment with the input DNA. Each droplet is cumulatively counted as an "Event Number" for the ddPCR experiments analyzed in tandem, and plotted along the x-axis. The positive droplet fluorescence threshold for each fluorophore used is indicated by the magenta line, which determines whether a droplet is considered positive or negative for FAM fluorescence. Thus, the blue dots represent individual droplets that contain at least one copy of the L1 locus tested. We tested 200 ng of *Bam*HI-digested genomic DNA from LoVo cells, which contain the polymorphic L1 element, and tenfold dilutions of this same sample as a mixture with *BamHI*-digested genomic DNA from HEK293 cells, which do not have this polymorphic L1 insertion, thus keeping the input genomic DNA constant for each ddPCR. Percentages given reflect the amount of input DNA with 100% corresponding to 200 ng of DNA. This assay robustly detects the 5′-insertion junction of the polymorphic full-length esv3475 L1Hs element when present in the genomic DNA from a cell line positive for that polymorphism (LoVo 100%), but not in a cell line negative for that polymorphism (HEK293 100%). L1-positive droplets are observed at dilutions as low as 0.1% of the DNA for this locus.

Additional file 2: Figure S2. Concentration plot of Chromosome 4 esv3475 L1Hs by the 5′ junction ddPCR assay. The input DNA concentrations in copies/μl (Ch1 Conc) for the ddPCR experiments described in Additional file 1: Figure S1 were calculated by the QuantaSoft Analysis Software.

Additional file 3: Figure S3. Detection of Chromosome 4 esv3475 L1Hs by the 3′ junction ddPCR assay. The L1Hs 3′ junction ddPCR assay used a L1-specific primer, L1-specific 5′-FAM labeled Taqman™ probe, and a locus-specific primer near the Chromosome 4 esv3475 3′-insertion junction, as shown in Figure 1B. The FAM fluorescent signal (Ch 1) for each droplet is plotted on the y-axis for each ddPCR experiment, which are separated by a dotted yellow line and indicated above each experiment with the input DNA. Each droplet is cumulatively counted as an "Event Number" for the ddPCR experiments analyzed in tandem, and plotted along the x-axis. The positive droplet fluorescence threshold is indicated by the magenta line, which determines whether a droplet is considered positive or negative for FAM fluorescence. Thus, the blue dots represent individual droplets that contain at least one copy of the L1 locus tested. We tested 200 ng of *BamHI*-digested genomic DNA from LoVo cells, which contain the polymorphic L1 element, and ten-fold dilutions of this same sample as a mixture with *BamHI*-digested genomic DNA from HEK293 cells, which do not have this polymorphic L1 insertion. Percentages given reflect the amount of input DNA with 100% corresponding to 200 ng of DNA. The positive droplet fluorescence threshold is indicated by the magenta line. This assay robustly detects the 3′-insertion junction of the polymorphic full-length esv3475 L1Hs element when present in the genomic DNA from a cell line positive for that polymorphism (LoVo 100%), but not in a cell line negative for that polymorphism (HEK293 100%). L1-positive droplets are observed at dilutions as low as 0.1% of the DNA for this locus.

Additional file 4: Figure S4. Concentration plot of Chromosome 4 esv3475 L1Hs by the 3′ junction ddPCR assay. The input DNA concentrations in copies/ μl (Ch1 Conc) for the ddPCR experiments described in Additional file 3: Figure S3 were calculated by the QuantaSoft Analysis Software.

Additional file 5: Figure S5. Detection of Chromosome 4 esv4912 L1Hs by the 5′ junction ddPCR assay. Experiments were performed as in Additional file 1: Figure S1 to determine the limit of detection for the 5′ junction of the polymorphic full-length esv4912 L1Hs element on Chromosome 4 in a cell line positive for that polymorphism (HCT116 100%), in indicated experiments as an input mixture with genomic DNA from a cell line negative for that polymorphism (HeLa).

Additional file 6: Figure S6. Concentration plot of Chromosome 4 esv4912 L1Hs by the 5′ junction ddPCR assay. The input DNA concentrations in copies/μl (Ch1 Conc) for the ddPCR experiments described in Additional file 5: Figure S5 were calculated by the QuantaSoft Analysis Software.

Additional file 7: Figure S7. Detection of Chromosome 4 esv4912 L1Hs by the 3′ junction ddPCR assay. Experiments were performed as in Additional file 3: Figure S3 to determine the limit of detection for the 3′ junction of the polymorphic full-length esv4912 L1Hs element on Chromosome 4 in a cell line positive for that polymorphism (HCT116 100%), in indicated experiments as an input mixture with genomic DNA from a cell line negative for that polymorphism (HeLa).

Additional file 8: Figure S8. Concentration plot of Chromosome 4 esv4912 L1Hs by the 3′ junction ddPCR assay. The input DNA concentrations in copies/μl (Ch1 Conc) for the ddPCR experiments described in Additional file 7: Figure S7 were calculated by the QuantaSoft Analysis Software.

Abbreviations
ddPCR: droplet digital PCR; HTS: high-throughput sequencing; L1: long interspersed element 1; qPCR: quantitative PCR; TPRT: target primed reverse transcription.

Competing interests
AMM and JF are employed by Bio-Rad Laboratories. TBW, VAS, and PLD have no competing interests.

Authors' contributions
TBW, AMM, VAS, and PLD conceived and designed the experiments. TBW, AMM, VAS, and JF performed the experiments. TBW and VAS wrote the manuscript. AMM and PLD edited the manuscript. All authors read and approved the final manuscript before submission.

Acknowledgements
This work was funded by grants from the National Institutes of Health to PLD (R01GM045668, P20RR020152 and P20GM103518). VAS was supported by a Louisiana State Board of Regents Fellowship.

Author details
[1]Tulane Cancer Center, 1430 Tulane Avenue, New Orleans, LA 70112, USA. [2]Bio-Rad Laboratories, 750 Alfred Nobel Drive, Hercules, CA, 94547, USA. [3]Present Address: Eureka Genomics, 2000 Alfred Nobel Drive, Hercules, CA, 94547, USA. [4]Present Address: Division of Infectious Diseases, Boston Children's Hospital and Harvard Medical School, 300 Longwood Avenue, Boston, MA 02115, USA.

References
1. Lander ES, Linton LM, Birren B, Nusbaum C, Zody MC, Baldwin J, Devon K, Dewar K, Doyle M, FitzHugh W, Funke R, Gage D, Harris K, Heaford A, Howland J, Kann L, Lehoczky J, LeVine R, McEwan P, McKernan K, Meldrim J, Mesirov JP, Miranda C, Morris W, Naylor J, Raymond C, Rosetti M, Santos R, Sheridan A, Sougnez C, *et al*: **Initial sequencing and analysis of the human genome.** *Nature* 2001, **409**:860–921.
2. Brouha B, Schustak J, Badge RM, Lutz-Prigge S, Farley AH, Moran JV, Kazazian HH Jr: **Hot L1s account for the bulk of retrotransposition in the human population.** *Proc Natl Acad Sci U S A* 2003, **100**:5280–5285.
3. Baillie JK, Barnett MW, Upton KR, Gerhardt DJ, Richmond TA, De Sapio F, Brennan PM, Rizzu P, Smith S, Fell M, Talbot RT, Gustincich S, Freeman TC, Mattick JS, Hume DA, Heutink P, Carninci P, Jeddeloh JA, Faulkner GJ: **Somatic retrotransposition alters the genetic landscape of the human brain.** *Nature* 2011, **479**:534–537.
4. Ewing AD, Kazazian HH Jr: **High-throughput sequencing reveals extensive variation in human-specific L1 content in individual human genomes.** *Genome Res* 2010, **20**:1262–1270.
5. Ewing AD, Kazazian HH Jr: **Whole-genome resequencing allows detection of many rare LINE-1 insertion alleles in humans.** *Genome Res* 2011, **21**:985–990.
6. Helman E, Lawrence ML, Stewart C, Sougnez C, Getz G, Meyerson M: **Somatic retrotransposition in human cancer revealed by whole-genome and exome sequencing.** *Genome Res* 2014, **24**:1053–1063.
7. Iskow RC, McCabe MT, Mills RE, Torene S, Pittard WS, Neuwald AF, Van Meir EG, Vertino PM, Devine SE: **Natural mutagenesis of human genomes by endogenous retrotransposons.** *Cell* 2010, **141**:1253–1261.
8. Lee E, Iskow R, Yang L, Gokcumen O, Haseley P, Luquette LJ 3rd, Lohr JG, Harris CC, Ding L, Wilson RK, Wheeler DA, Gibbs RA, Kucherlapati R, Lee C, Kharchenko PV, Park PJ: **Landscape of somatic retrotransposition in human cancers.** *Science* 2012, **337**:967–971.
9. Pitkanen E, Cajuso T, Katainen R, Kaasinen E, Valimaki N, Palin K, Taipale J, Aaltonen LA, Kilpivaara O: **Frequent L1 retrotranspositions originating from TTC28 in colorectal cancer.** *Oncotarget* 2014, **5**:853–859.
10. Shukla R, Upton KR, Muñoz-Lopez M, Gerhardt DJ, Fisher ME, Nguyen T, Brennan PM, Baillie JK, Collino A, Ghisletti S, Sinha S, Iannelli F, Radaelli E, Dos Santos A, Rapoud D, Guettier C, Samuel D, Natoli G, Carninci P, Ciccarelli FD, Garcia-Perez JL, Faivre J, Faulkner GJ: **Endogenous retrotransposition activates oncogenic pathways in hepatocellular carcinoma.** *Cell* 2013, **153**:101–111.

11. Solyom S, Ewing AD, Rahrmann EP, Doucet T, Nelson HH, Burns MB, Harris RS, Sigmon DF, Casella A, Erlanger B, Wheelan S, Upton KR, Shukla R, Faulkner GJ, Largaespada DA, Kazazian HH Jr: **Extensive somatic L1 retrotransposition in colorectal tumors.** *Genome Res* 2012, **22:**2328–2338.

12. Harris CR, Normart R, Yang Q, Stevenson E, Haffty BG, Ganesan S, Cordon-Cardo C, Levine AJ, Tang LH: **Association of nuclear localization of a long interspersed nuclear element-1 protein in breast tumors with poor prognostic outcomes.** *Genes Cancer* 2010, **1:**115–124.

13. Marusyk A, Polyak K: **Tumor heterogeneity: causes and consequences.** *Biochim Biophys Acta* 2010, **1805:**105–117.

14. Hindson BJ, Ness KD, Masquelier DA, Belgrader P, Heredia NJ, Makarewicz AJ, Bright IJ, Lucero MY, Hiddessen AL, Legler TC, Kitano TK, Hodel MR, Petersen JF, Wyatt PW, Steenblock ER, Shah PH, Bousse LJ, Troup CB, Mellen JC, Wittmann DK, Erndt NG, Cauley TH, Koehler RT, So AP, Dube S, Rose KA, Montesclaros L, Wang S, Stumbo DP, Hodges SP, *et al*: **High-throughput droplet digital PCR system for absolute quantitation of DNA copy number.** *Anal Chem* 2011, **83:**8604–8610.

15. Sykes PJ, Neoh SH, Brisco MJ, Hughes E, Condon J, Morley AA: **Quantitation of targets for PCR by use of limiting dilution.** *Biotechniques* 1992, **13:**444–449.

16. Pekin D, Skhiri Y, Baret JC, Le Corre D, Mazutis L, Salem CB, Millot F, El Harrak A, Hutchison JB, Larson JW, Link DR, Laurent-Puig P, Griffiths AD, Taly V: **Quantitative and sensitive detection of rare mutations using droplet-based microfluidics.** *Lab Chip* 2011, **11:**2156–2166.

17. Holland PM, Abramson RD, Watson R, Gelfand DH: **Detection of specific polymerase chain reaction product by utilizing the 5′–3′ exonuclease activity of Thermus aquaticus DNA polymerase.** *Proc Natl Acad Sci U S A* 1991, **88:**7276–7280.

18. Livak KJ: **Allelic discrimination using fluorogenic probes and the 5′ nuclease assay.** *Genet Anal* 1999, **14:**143–149.

19. Whale AS, Huggett JF, Cowen S, Speirs V, Shaw J, Ellison S, Foy CA, Scott DJ: **Comparison of microfluidic digital PCR and conventional quantitative PCR for measuring copy number variation.** *Nucleic Acids Res* 2012, **40:**e82.

20. Roberts CH, Jiang W, Jayaraman J, Trowsdale J, Holland MJ, Traherne JA: **Killer-cell immunoglobulin-like receptor gene linkage and copy number variation analysis by droplet digital PCR.** *Genome Med* 2014, **6:**20.

21. Kidd JM, Graves T, Newman TL, Fulton R, Hayden HS, Malig M, Kallicki J, Kaul R, Wilson RK, Eichler EE: **A human genome structural variation sequencing resource reveals insights into mutational mechanisms.** *Cell* 2010, **143:**837–847.

22. Boissinot S, Chevret P, Furano AV: **L1 (LINE-1) retrotransposon evolution and amplification in recent human history.** *Mol Biol Evol* 2000, **17:**915–928.

23. Ovchinnikov I, Rubin A, Swergold GD: **Tracing the LINEs of human evolution.** *Proc Natl Acad Sci U S A* 2002, **99:**10522–10527.

24. MacDonald JR, Ziman R, Yuen RK, Feuk L, Scherer SW: **The database of genomic variants: a curated collection of structural variation in the human genome.** *Nucleic Acids Res* 2014, **42:**D986–D992.

25. Wang J, Wang W, Li R, Li Y, Tian G, Goodman L, Fan W, Zhang J, Li J, Zhang J, Guo Y, Feng B, Li H, Lu Y, Fang X, Liang H, Du Z, Li D, Zhao Y, Hu Y, Yang Z, Zheng H, Hellmann I, Inouye M, Pool J, Yi X, Zhao J, Duan J, Zhou Y, Qin J, *et al*: **The diploid genome sequence of an Asian individual.** *Nature* 2008, **456:**60–65.

26. Stjernqvist S, Ryden T, Greenman CD: **Model-integrated estimation of normal tissue contamination for cancer SNP allelic copy number data.** *Cancer Inform* 2011, **10:**159–173.

27. Abyzov A, Mariani J, Palejev D, Zhang Y, Haney MS, Tomasini L, Ferrandino AF, Rosenberg Belmaker LA, Szekely A, Wilson M, Kocabas A, Calixto NE, Grigorenko EL, Huttner A, Chawarska K, Weissman S, Urban AE, Gerstein M, Vaccarino FM: **Somatic copy number mosaicism in human skin revealed by induced pluripotent stem cells.** *Nature* 2012, **492:**438–442.

28. Mullighan CG, Phillips LA, Su X, Ma J, Miller CB, Shurtleff SA, Downing JR: **Genomic analysis of the clonal origins of relapsed acute lymphoblastic leukemia.** *Science* 2008, **322:**1377–1380.

29. Navin N, Krasnitz A, Rodgers L, Cook K, Meth J, Kendall J, Riggs M, Eberling Y, Troge J, Grubor V, Levy D, Lundin P, Månér S, Zetterberg A, Hicks J, Wigler M: **Inferring tumor progression from genomic heterogeneity.** *Genome Res* 2010, **20:**68–80.

30. Parisi F, Ariyan S, Narayan D, Bacchiocchi A, Hoyt K, Cheng E, Xu F, Li P, Halaban R, Kluger Y: **Detecting copy number status and uncovering subclonal markers in heterogeneous tumor biopsies.** *BMC Genomics* 2011, **12:**230.

31. Sabeti PC, Varilly P, Fry B, Lohmueller J, Hostetter E, Cotsapas C, Xie X, Byrne EH, McCarroll SA, Gaudet R, Schaffner SF, Lander ES, International HapMap Consortium, Frazer KA, Ballinger DG, Cox DR, Hinds DA, Stuve LL, Gibbs RA, Belmont JW, Boudreau A, Hardenbol P, Leal SM, Pasternak S, Wheeler DA, Willis TD, Yu F, Yang H, Zeng C, Gao Y, *et al*: **Genome-wide detection and characterization of positive selection in human populations.** *Nature* 2007, **449:**913–918.

32. International HapMap 3 Consortium: **Integrating common and rare genetic variation in diverse human populations.** *Nature* 2010, **467:**52–58.

33. Neve B, Froguel P, Corset L, Vaillant E, Vatin V, Boutin P: **Rapid SNP allele frequency determination in genomic DNA pools by pyrosequencing.** *Biotechniques* 2002, **32:**1138–1142.

34. Li-Sucholeiki XC, Tomita-Mitchell A, Arnold K, Glassner BJ, Thompson T, Murthy JV, Berk L, Lange C, Leong-Morgenthaler PM, MacDougall D, Munro J, Cannon D, Mistry T, Miller A, Deka C, Karger B, Gillespie KM, Ekström PO, Todd JA, Thilly WG: **Detection and frequency estimation of rare variants in pools of genomic DNA from large populations using mutational spectrometry.** *Mutat Res* 2005, **570:**267–280.

35. Zhang Y, Shu J, Si J, Shen L, Estecio MR, Issa JP: **Repetitive elements and enforced transcriptional repression co-operate to enhance DNA methylation spreading into a promoter CpG-island.** *Nucleic Acids Res* 2012, **40:**7257–7268.

Identification of polymorphic SVA retrotransposons using a mobile element scanning method for SVA (ME-Scan-SVA)

Hongseok Ha[1,2], Jui Wan Loh[1] and Jinchuan Xing[1,2*] (iD)

Abstract

Background: Mobile element insertions are a major source of human genomic variation. SVA (SINE-R/VNTR/Alu) is the youngest retrotransposon family in the human genome and a number of diseases are known to be caused by SVA insertions. However, inter-individual genomic variations generated by SVA insertions and their impacts have not been studied extensively due to the difficulty in identifying polymorphic SVA insertions.

Results: To systematically identify SVA insertions at the population level and assess their genomic impact, we developed a mobile element scanning (ME-Scan) protocol we called ME-Scan-SVA. Using a nested SVA-specific PCR enrichment method, ME-Scan-SVA selectively amplify the 5′ end of SVA elements and their flanking genomic regions. To demonstrate the utility of the protocol, we constructed and sequenced a ME-Scan-SVA library of 21 individuals and analyzed the data using a new analysis pipeline designed for the protocol. Overall, the method achieved high SVA-specificity and over >90 % of the sequenced reads are from SVA insertions. The method also had high sensitivity (>90 %) for fixed SVA insertions that contain the SVA-specific primer-binding sites in the reference genome. Using candidate locus selection criteria that are expected to have a 90 % sensitivity, we identified 151 and 29 novel polymorphic SVA candidates under relaxed and stringent cutoffs, respectively (average 12 and 2 per individual). For six polymorphic SVAs that we were able to validate by PCR, the average individual genotype accuracy is 92 %, demonstrating a high accuracy of the computational genotype calling pipeline.

Conclusions: The new approach allows identifying novel SVA insertions using high-throughput sequencing. It is cost-effective and can be applied in large-scale population study. It also can be applied for detecting potential active SVA elements, and somatic SVA retrotransposition events in different tissues or developmental stages.

Keywords: SVA, Retrotransposon, High-throughput sequencing, ME-Scan

Background

Mobile elements are discrete DNA fragments that can move and integrate into other locations in a genome. More than two-thirds of human genome are occupied by repetitive or repeat-derived sequences, including active mobile elements that are still capable of transposition [1]. Mobile elements can insert and disrupt host genes or participate in genomic rearrangement, resulting in diseases (for review, see [2–4]). In humans, some mobile element insertions (MEIs) are polymorphic across individuals [4, 5]. Besides their functional and structural genomic impact, these polymorphic MEIs (pMEIs) are also important markers for ascertaining human population relationships and evolutionary history [6–8]. Therefore, it is of great interest to identify pMEIs in human populations. In general, there are two high-throughput sequencing based strategies for identifying pMEIs; whole genome and MEI-targeted sequencing. Compared with whole genome sequencing, MEI-targeted high-throughput sequencing methods are more cost-effective [9]. Although a number of targeted high-throughput sequencing methods have been developed for *Alu* and L1 elements [10–14], to date the only targeted sequencing method for SVA (SINE-R/VNTR/Alu) elements is retrotransposon capture sequencing (RC-seq) [10, 15–17].

* Correspondence: xing@biology.rutgers.edu
[1]Department of Genetics, The State University of New Jersey, Piscataway 08854, NJ, USA
[2]Human Genetic Institute of New Jersey, Rutgers, The State University of New Jersey, Piscataway 08854, NJ, USA

SVA is a composite element consisting of a $(CCCTCT)_n$ hexamer simple repeat region at the 5′ end, an *Alu*-like region, a variable number of tandem repeats (VNTR) region, a short interspersed element of retroviral origin (SINE-R) region, and a poly-A tail after the putative polyadenylation signal (Fig. 1a). SVA insertions have all the hallmarks of L1-mediated target primed reverse transcription, such as poly(A) tail, target-site duplications (TSDs), 5′ truncation, and have been shown to mobilize by hijacking the L1-encoded protein machinery [18–21]. SVA elements represent the youngest retrotransposon family in the human genome and many insertions are polymorphic among human populations [5,

18, 22]. The polymorphism rates of members of the youngest subfamilies SVA_E and SVA_F were estimated as 37.5 and 27.6 %, respectively [18].

Although SVA elements only constitute approximately 0.1 % of the human genome, they have substantial biological impact in human. Insertion of SVA elements can trigger exonization, polyadenylation, enhancer and alternative promoter events, which lead to the formation of various transcript isoforms and evolutionary dynamics that contributes to the differences in gene expression level [19, 23–28]. Several human diseases have been attributed to SVA insertions or SVA-associated deletions, including Fukuyama congenital muscular dystrophy,

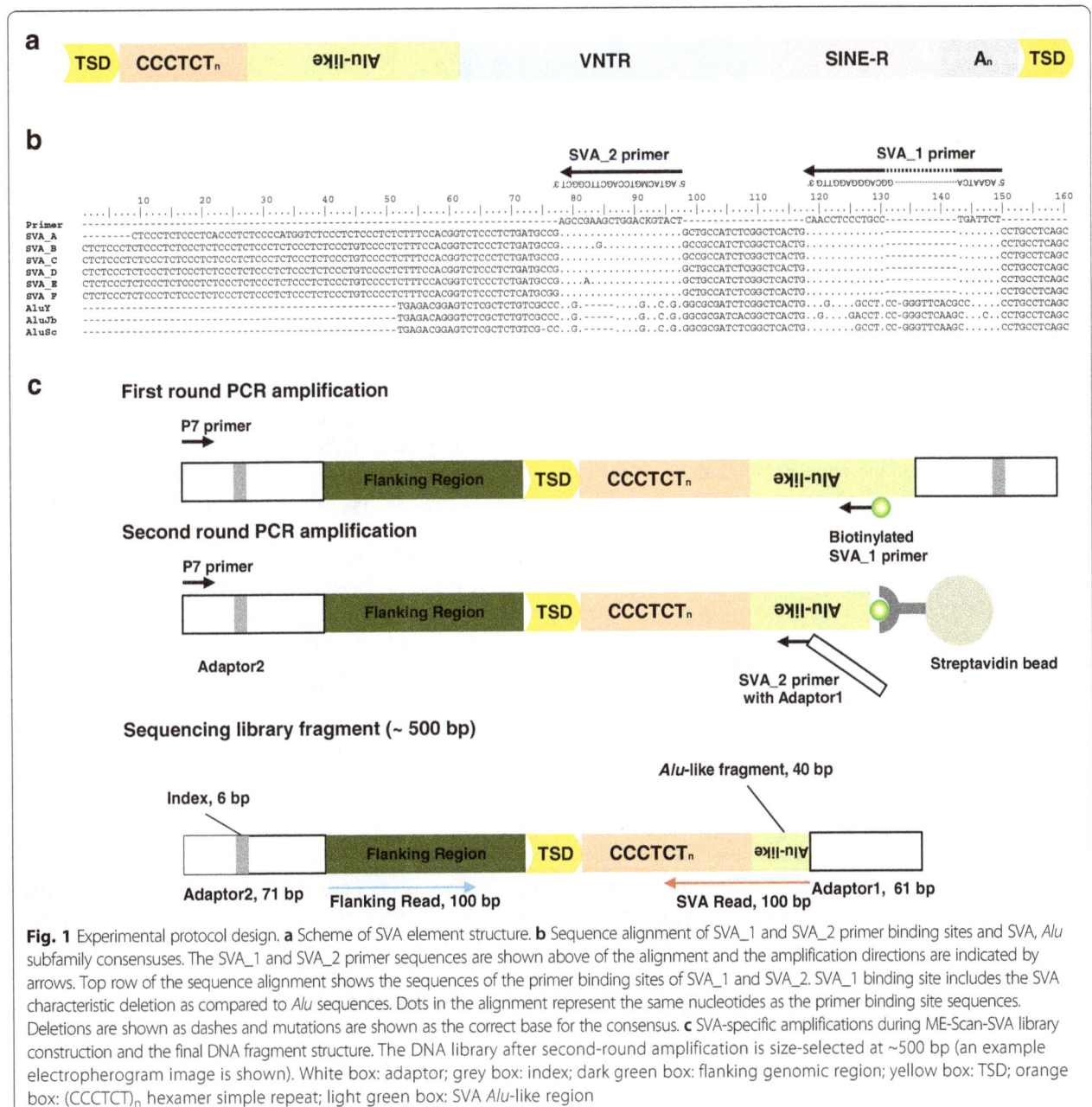

Fig. 1 Experimental protocol design. **a** Scheme of SVA element structure. **b** Sequence alignment of SVA_1 and SVA_2 primer binding sites and SVA, *Alu* subfamily consensuses. The SVA_1 and SVA_2 primer sequences are shown above of the alignment and the amplification directions are indicated by arrows. Top row of the sequence alignment shows the sequences of the primer binding sites of SVA_1 and SVA_2. SVA_1 binding site includes the SVA characteristic deletion as compared to *Alu* sequences. Dots in the alignment represent the same nucleotides as the primer binding site sequences. Deletions are shown as dashes and mutations are shown as the correct base for the consensus. **c** SVA-specific amplifications during ME-Scan-SVA library construction and the final DNA fragment structure. The DNA library after second-round amplification is size-selected at ~500 bp (an example electropherogram image is shown). White box: adaptor; grey box: index; dark green box: flanking genomic region; yellow box: TSD; orange box: $(CCCTCT)_n$ hexamer simple repeat; light green box: SVA *Alu*-like region

Lynch syndrome, X-linked agammaglobulinemia, autosomal recessive hypercholesterolemia, hemophilia B, and neurofibromatosis type 1 [29–33]. Therefore, it is important to systematically analyze polymorphic SVA insertions in human populations.

Mobile element scanning (ME-Scan) is a targeted high-throughput sequencing strategy for MEIs. In previous studies, the technique was applied for identifying AluYb8/9 insertion polymorphisms in human genomes [11, 14], and Ves SINE insertions in bat genomes [34]. In this study, we developed a ME-Scan method and an associated data analysis pipeline for SVA elements, which we termed ME-Scan-SVA. We then demonstrated the method by examining SVA insertions in 21 individuals.

Results

ME-Scan-SVA overview

Experimental protocol design

We designed a two-round nested PCR amplification protocol for SVA following the existing ME-Scan method [35]. We targeted the 5′ Alu-like region of the SVA elements to selectively enrich for SVA elements. Despite the high similarity between the SVA Alu-like region and Alu subfamily consensus sequences, one insertion and one deletion are shared by all SVA sequences (Fig. 1b). Therefore we designed SVA-specific primers in these regions. A biotinylated primer (SVA_1) was used for the first round PCR reaction and the second-round nested primer (SVA_2) was used to further improve specificity and add Illumina sequencing adaptors. Because typical SVA truncations happen at the 5′ of the insertion, this nested-PCR design at the 5′ end of the SVA element allows us to selectively enrich full-length SVA elements. In addition, 5′ or 3′ truncated SVA elements that contains both SVA_1 and SVA_2 primer binding sites (Fig. 1b, SVA consensus position 78 - 137) will also be amplified. Based on the human reference genome (hg19), we estimate that this method can amplify 65 % of SVA_D (828/1274), 27 % of SVA_E (52/192), and 24 % of SVA_F elements (198/821), respectively.

A DNA fragment in the final sequencing library contains a variable-length 5′ flanking genomic sequence, the 5′ terminus of an SVA element ends at the primer binding site of SVA_2, and 132 base pair (bp) of sequencing adapters that flank either end of the fragment (Fig. 1c bottom). The expected SVA fragment size is the size of the $(CCCTCT)_n$ hexamer simple repeats plus 40 bp in the Alu-like region. Because of the variable size of the simple repeat and possible the 5′ truncation, the size of an SVA fragment could vary between 20 bp (SVA_2 primer binding site only) to several hundred bps. We aim to minimize the library size for sequencing efficiency while maintaining sufficient flanking sequence for

identifying the genomic location of the SVA insertions. Therefore, we first fragment the genomic DNA to about 1,000 bp in size. After library construction, we select DNA fragments around 500 bp for sequencing (~130 bp adaptor sequence + ~370 bp SVA sequences and genomic flanking sequence).

Computational analysis pipeline

We designed a pipeline for ME-Scan-SVA analysis based on the general ME-Scan workflow [35]. Figure 2 shows an outline for the analysis pipeline. Using the Illumina 100 bp pair-end sequencing format, two sequencing reads are generated from each DNA fragment (Fig. 2a). We use the 40 bp Alu-like region in the first read (referred as the SVA Read in the following text) to determine if a read-pair is derived from an SVA locus (Fig. 2a). For each SVA Read, the Alu-like region is compared with the SVA consensus sequence [36] using BLAST [37] and the resulted bit-scores are recorded. The BLAST bit-score is a normalized measurement of the similarity between the SVA Read and the corresponding SVA consensus sequence. To choose a suitable cutoff for the BLAST bit-score, we determined the BLAST score distribution of SVA sequences in the human reference genome (Fig. 3). As expected, almost all SVAs from SVA_F, the youngest SVA subfamily, are present in the highest BLAST bit-score bins (>65). The majority of SVAs in the subfamilies SVA_D, SVA_E, and SVA_F have BLAST bit-scores higher than 48. Because these three subfamilies contain all known polymorphic SVA insertions, we selected BLAST bit-score 48 as a relaxed cutoff and 65 as a stringent cutoff. The relaxed cutoff is expected to capture more candidate loci. The stringent cutoff will enrich for the youngest subfamily SVA_F, which is expected to contain higher proportion of very recent insertions (Fig. 3). We then filter SVA Read based on selected bit-score cutoffs. A typical 100 bp SVA Read contains 40 bp SVA Alu-like region, and the variable $(CCCTCT)_n$ hexamer simple repeats region. Because the simple repeat region are often longer than 50 bp in size, most of the SVA Reads are expected to contain little or no flanking genomic sequences. Therefore, we use the second read in the read pair (referred as the Flanking Read in the following text) to identify the genomic location of an SVA insertion. Flanking Read sequences are aligned to the reference genome using the program BWA-MEM (Burrows-Wheeler Alignment Tool- maximal exact matches) [38]. The mapped Flanking Reads are then filtered based on their mapping quality scores to ensure the high-confidence mapping of the read. After mapping, the end positions of the mapped Flanking Reads are sorted, and then clustered within a sliding window of 500 bp in size. Within each cluster, the Flanking Read mapping position that is

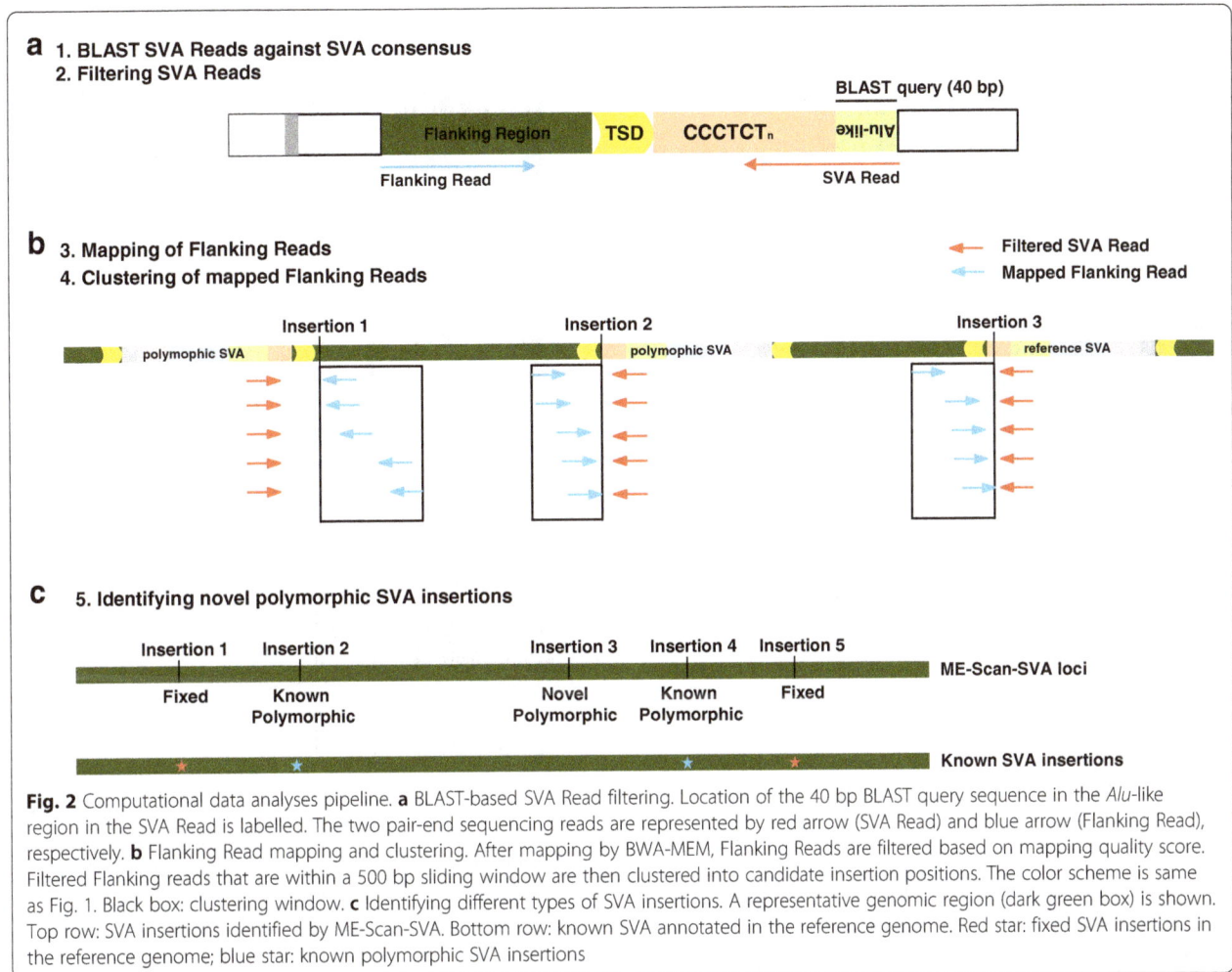

Fig. 2 Computational data analyses pipeline. **a** BLAST-based SVA Read filtering. Location of the 40 bp BLAST query sequence in the *Alu*-like region in the SVA Read is labelled. The two pair-end sequencing reads are represented by red arrow (SVA Read) and blue arrow (Flanking Read), respectively. **b** Flanking Read mapping and clustering. After mapping by BWA-MEM, Flanking Reads are filtered based on mapping quality score. Filtered Flanking reads that are within a 500 bp sliding window are then clustered into candidate insertion positions. The color scheme is same as Fig. 1. Black box: clustering window. **c** Identifying different types of SVA insertions. A representative genomic region (dark green box) is shown. Top row: SVA insertions identified by ME-Scan-SVA. Bottom row: known SVA annotated in the reference genome. Red star: fixed SVA insertions in the reference genome; blue star: known polymorphic SVA insertions

closest to the SVA insertion site is chosen as the insertion position for that locus (Fig. 2b). Depending on the length of the SVA element in the DNA fragment, the Flanking Reads might not cover the exact SVA insertion site. The candidate SVA insertion loci are then separated into several types (Fig. 2c). Reference SVAs are loci that are annotated by RepeatMasker in the human reference genome and passed the BLAST score cutoff. Fixed SVAs are reference SVA loci that are not known to be polymorphic. Known polymorphic SVAs are loci reported in previous studies [5, 22, 39, 40]. Finally, novel polymorphic SVA insertions are loci that do not overlap reference and known polymorphic SVAs.

Applying ME-Scan-SVA to 21 human samples
Data generation
To demonstrate the feasibility of our protocol, we constructed a ME-Scan-SVA library using 21 individuals from two HapMap populations, including six parent-offspring trios (Table 1). All samples were pooled after indexing and the pooled library was used to construct a

ME-Scan-SVA sequencing library. The library was sequenced using the Illumina Hiseq 2000 with 100 bp paired-end format. We obtained 152.9 million total read pairs from the library, and the average and median of individual read number is 7.3 and 6.3 million, respectively (Additional file 1: Table S1).

Read filtering and candidate loci identification
As described in the "Computational pipeline" section, we filtered SVA Read based on BLAST bit-score cutoffs. We used BLAST bit-score 48 as a relaxed cutoff and 65 as a stringent cutoff. Using the relaxed and stringent cutoffs, 93.8 and 17.6 % of the SVA Read passed the cutoff, respectively (Additional file 1: Table S1).

The vast majority (99.2 %) of Flanking Reads was mapped to the reference genome. More than 82 % of the reads in each individual passed a BWA-MEM mapping quality score cutoff of 29. We used this mapping quality cutoff to exclude low-quality reads and reads that mapped to multiple genomic locations. Overall, 78.1 and 14.5 % of the read-pairs passed both SVA Read and

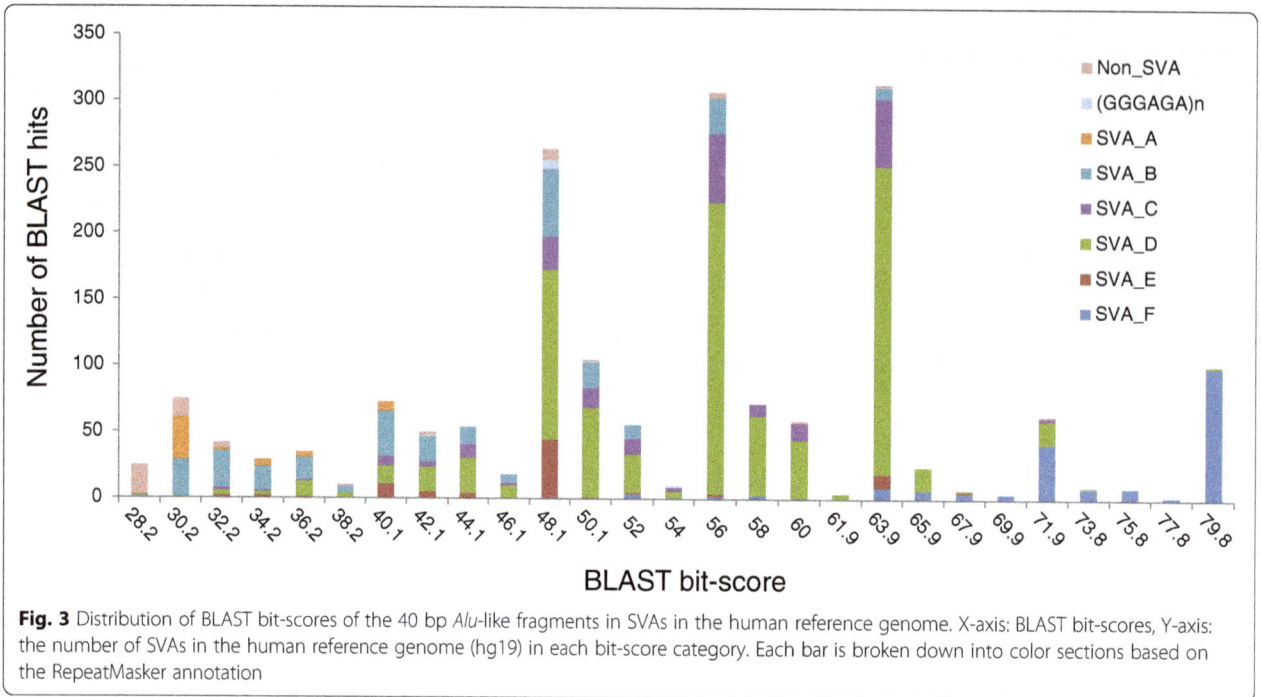

Fig. 3 Distribution of BLAST bit-scores of the 40 bp *Alu*-like fragments in SVAs in the human reference genome. X-axis: BLAST bit-scores, Y-axis: the number of SVAs in the human reference genome (hg19) in each bit-score category. Each bar is broken down into color sections based on the RepeatMasker annotation

Table 1 The cutoffs used and the number of SVA loci identified in each individual

Individual	Population	Family	Relation	Relaxed				Stringent			
				Cutoff (TPM,UR)	All	Poly-morphic	Novel	Cutoff (TPM,UR)	All	Poly-morphic	Novel
NA12872	CEPH	1459	paternal grandfather	(5,10)	1388	157	15	(16,10)	254	68	1
NA12873	CEPH	1459	paternal grandmother	(5,10)	1383	159	6	(14,10)	252	64	1
NA12864	CEPH	1459	father	(5,10)	1407	178	15	(12,10)	263	76	2
NA12874	CEPH	1459	maternal grandfather	(3,4)	1394	169	16	(4,6)	339	156	0
NA12875	CEPH	1459	maternal grandmother	(4,10)	1407	174	12	(13,10)	263	73	4
NA12865	CEPH	1459	mother	(4,10)	1399	171	13	(9,10)	270	80	1
NA12891	CEPH	1463	maternal grandfather	(4,10)	1394	164	10	(11,10)	266	76	1
NA12892	CEPH	1463	maternal grandmother	(5,10)	1387	158	6	(11,10)	265	75	0
NA12878	CEPH	1463	mother	(4,10)	1397	167	13	(13,10)	262	73	1
NA18501	YRI	Y004	father	(3,10)	1399	178	12	(4,10)	390	202	4
NA18502	YRI	Y004	mother	(5,10)	1398	173	14	(11,10)	271	83	2
NA18500	YRI	Y004	child	(4,9)	1401	167	13	(9,10)	285	96	4
NA18504	YRI	Y005	father	(3,10)	1398	168	7	(9,10)	283	92	3
NA18505	YRI	Y005	mother	(4,10)	1408	175	9	(15,10)	268	80	2
NA18503	YRI	Y005	child	(4,10)	1393	167	10	(11,10)	277	87	3
NA18507	YRI	Y009	father	(3,10)	1408	176	11	(11,10)	268	80	3
NA18508	YRI	Y009	mother	(5,10)	1408	175	15	(10,10)	276	87	2
NA18506	YRI	Y009	child	(3,7)	1404	177	13	(10,10)	270	84	2
NA18517	YRI	Y013	mother	(4,10)	1420	185	23	(10,10)	268	82	5
NA18515	YRI	Y013	child	(5,10)	1394	162	13	(17,10)	261	73	2
NA18521	YRI	Y016	child	(6,10)	1388	161	14	(9,10)	280	91	3
Total					1722	428	151		521	310	29

Flanking Read filtering under the relaxed and stringent SVA Read cutoffs, respectively (Additional file 1: Table S1).

To obtain candidate SVA insertion loci, the mapping positions of mapped Flanking Reads were sorted and then clustered within a sliding window of 500 bp in size (Fig. 2b). A total of 28,130 and 7,972 insertion positions were generated from the 21 individuals under relaxed and stringent SVA Read cutoffs, respectively.

Sensitivity analysis

To estimate the sensitivity of ME-Scan-SVA, we first identified presumed fixed SVA insertion loci in the human reference genome. The presumed fixed SVA insertion loci are defined as SVA insertions that are present in the reference genome and are known to be not polymorphic in previous studies [5, 22, 39, 40]. Using the relaxed and stringent SVA Read cutoffs, we identified 1,343 and 200 loci as presumed fixed SVAs, respectively. Using this set of SVA insertion loci, we calculated the depth of coverage and the number of unique reads (URs) for each locus. To account for inter-library variation, we normalized the depth of coverage at each locus by the total number of mapped reads in each individual as TPM (tags per million).

Using the TPM and UR info for each locus, we calculated the sensitivity for identifying fixed loci under different TPM and UR cutoffs (Fig. 4). Overall, we achieve high sensitivity: even at a stringent TPM/UR cutoff 15/15, the pooled data has 89 and 96 % sensitivity, for the relaxed and stringent conditions, respectively (Fig. 4). Among individuals, the sensitivities are similar but lower than pooled data at high cutoffs (Additional file 2: Figure S1).

SVA candidate loci identification and validation

To identify SVA insertion candidates, we started from the list of candidate insertion positions and used TPM/UR cutoffs that achieve 90 % sensitivity in each

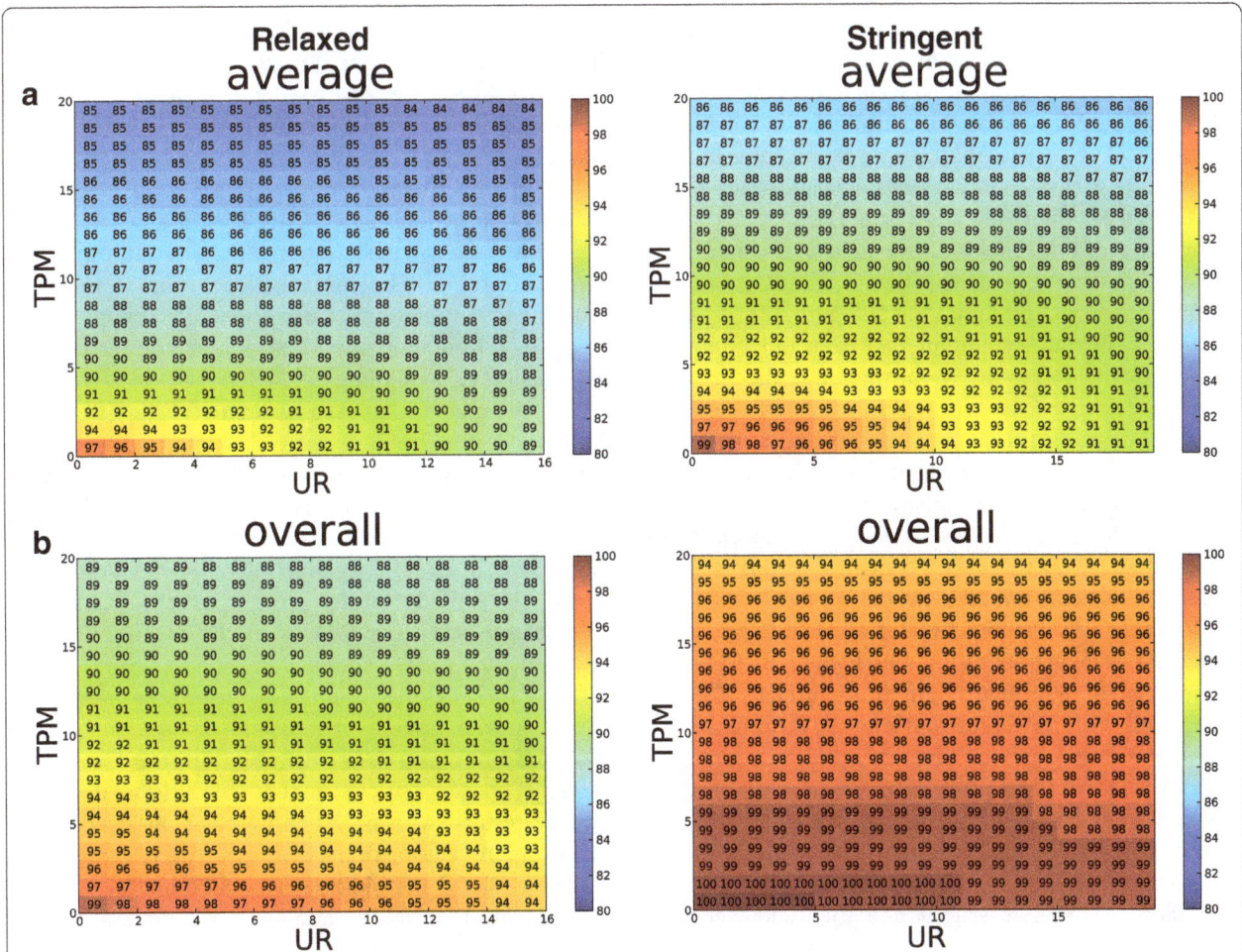

Fig. 4 Sensitivity analysis. The sensitivity for identifying fixed SVA insertions under different TPM and UR cutoffs. **a** average individual sensitivity; **b** overall sensitivity. The sensitivity is shown as the percentage of fixed insertions identified. Results under relaxed and stringent SVA Read cutoffs are shown in the left and right panel, respectively

individual based on the presumably fixed SVA insertions (Table 1). In each individual, ~1,400/~300 SVA insertion loci were selected under the relaxed/stringent conditions. Among them, ~200/~100 loci are polymorphic, and ~10/2 loci are novel (Table 1). In total, 428 polymorphic SVAs were identified among the 21 individuals under relaxed condition, and 151 of them are novel. As expected, the vast majority of novel insertions are rare, and ~80 % of the loci are only present in one sample. In comparison, some of the known polymorphic loci are more common and are present in all individuals in our dataset (Fig. 5a). Candidate loci from the stringent cutoff exhibit similar allele frequency pattern (Fig. 5b). The final relaxed and stringent call sets are available in Additional files 3 and 4.

To validate polymorphic SVA insertions, we performed PCR validation on 11 candidates (Additional files 5 and 6: Figure S2, Table S2). We used a combination of internal and external PCR for validation, similar to the protocol in the 1000 Genomes Project [22]. Out of the 11 loci, six showed clear and distinct bands for SVA insertions. We did not achieve specific amplification for SVA internal products for the remaining loci despite multiple attempts with different PCR conditions (see Method section for detail). This result might partially due to the difficulty in amplifying the complex SVA 5′ region. Although we expect some of these loci are true positives, our current validation results give a minimum true positive rate of 55 % (6/11).

For the six confirmed loci, we then performed individual genotyping to assess the individual genotype calling accuracy (Additional file 5: Figure S2). We consider an individual's genotype call from our computational pipeline correct if: 1) our pipeline called an SVA insertion and the PCR genotyping validated the insertion (either homozygous or heterozygous); or 2) our pipeline did not call an insertion and the genotyping result is no insertion. In general the individual genotypes are in agreement with computational calls: we achieved 93 % accuracy for individual genotype calls under the relaxed condition for the six loci (Additional file 7: Table S3). For the five loci that are also called under the stringent condition, one locus (Loc 5) has an accuracy of 17 %, primarily due to the under-calling of individuals with the SVA insertion (i.e., false-negative). The remaining four loci have an average accuracy of 96 % (Additional file 6: Table S2).

Next we compared our results with the 1000 Genomes Project phase 3 dataset [22], where 12 samples in our dataset are included. For these 12 overlapping samples, we called 363 SVA insertions and the 1000 Genomes Project called 223 insertions. Based on the primer-binding site position (78–137 in the SVA consensus sequence), 67 SVA insertions in the 1000 Genomes dataset are expected to be amplified by ME-Scan-SVA. Among these 67, 39 loci (58.2 %) were called in our data set. The individual genotype concordance rate for the 39 loci is 78 % (366/468 genotypes). The high genotype concordance rate suggests both datasets have high quality genotype calls for the shared loci.

Because our DNA samples include six parent-offspring trios, we can investigate the inheritance pattern and identify potential *de novo* SVA insertions in the offspring of each trio. To identify *de novo* SVA insertions, SVA insertions in each offspring that are found in parents or shared with unrelated individuals in the dataset (background) were removed. In total, 10 and 3 de novo insertion candidates were identified in the six offspring under the relaxed and stringent cutoffs, respectively. A close inspection showed that all candidate insertion loci are within old retrotransposons or simple repeats in the reference genome. The supporting flanking reads have low mapping quality in general because of the repetitive nature of these regions. Therefore these loci are unlikely to be authentic insertions. Consistent with this observation, two de novo insertion candidates failed validation (Additional file 6: Table S2). Given the SVA retrotransposition rate is estimated to be one in 916 births [39], in six trios the expected chance of identifying a de novo SVA insertion is < 0.01. Therefore, it is not surprising that we did not identify *de novo* SVA insertion in our dataset.

Potential functional impact of SVA insertions

Next we assessed the potential biological impact of SVA insertions. The insertion loci were intersected with gene annotations from the GENCODE project (Fig. 6). Given less than 5 % of the human genome are annotated as coding sequences (CDS, GENCODE v19), we expect the vast majority of insertions are located in intergenic or intronic regions, assuming a random insertion pattern. As expected, more than 93 % of SVA insertions are located in intergenic or intronic regions and only a small number of insertions overlap exonic regions: polymorphic SVA insertions identified under the relaxed condition intersected with four CDSs, six UTRs (untranslated regions), and one undefined exonic region (Fig. 6a, left). Three of the four CDS insertions were also found in the novel polymorphic dataset, suggesting most exonic insertions identified in this study are novel (Fig. 6b, left). Stringent conditions produced similar results, with only one insertion intersected the CDS region (Fig. 6, right). SVA insertions overlapping CDSs are listed in Additional file 8: Table S5.

Given most polymorphic SVA insertions are in noncoding regions, we investigated the relationship between SVA insertions and epigenetic modifications. Using the 15 chromatin state profile from nine cell lines as defined by ChromHMM [41], we calculated the normalized

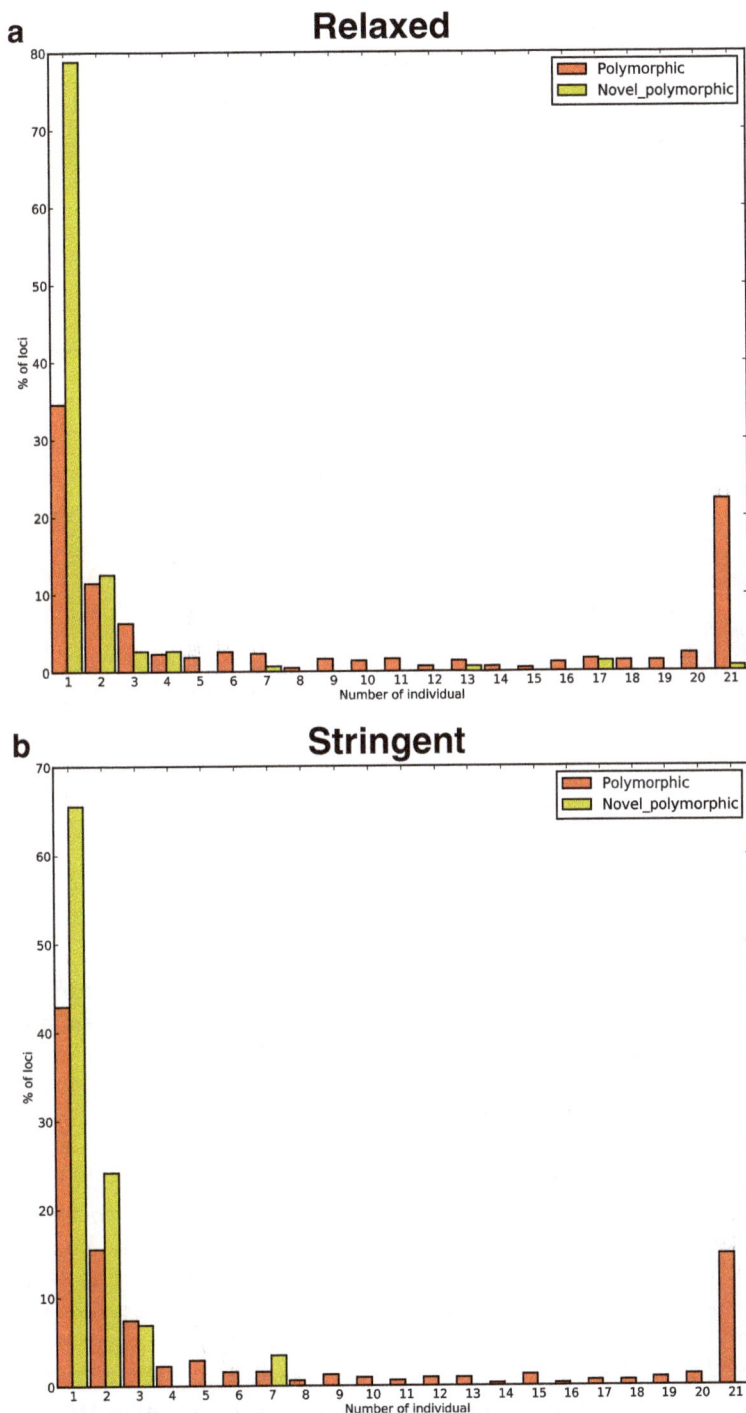

Fig. 5 Allele frequency distribution of polymorphic SVA insertions. The number of individuals having an SVA insertion is shown on the X-axis. The percentage of polymorphic or novel polymorphic SVAs in each individual bin is shown on the Y-axis. **a** relaxed SVA Read cutoff; **b** stringent SVA Read cutoff

number of SVA insertions in each state. The majority of polymorphic SVA insertions are enriched in non- or less- functional genomic regions, especially state 13 (heterochromatin, low signal), suggesting most of these insertions will not affect gene expression (Fig. 7).

Discussion

As the youngest retrotransposon family in the human genome, SVA insertions are highly polymorphic among human populations and play an important role in gene regulation and contribute to human

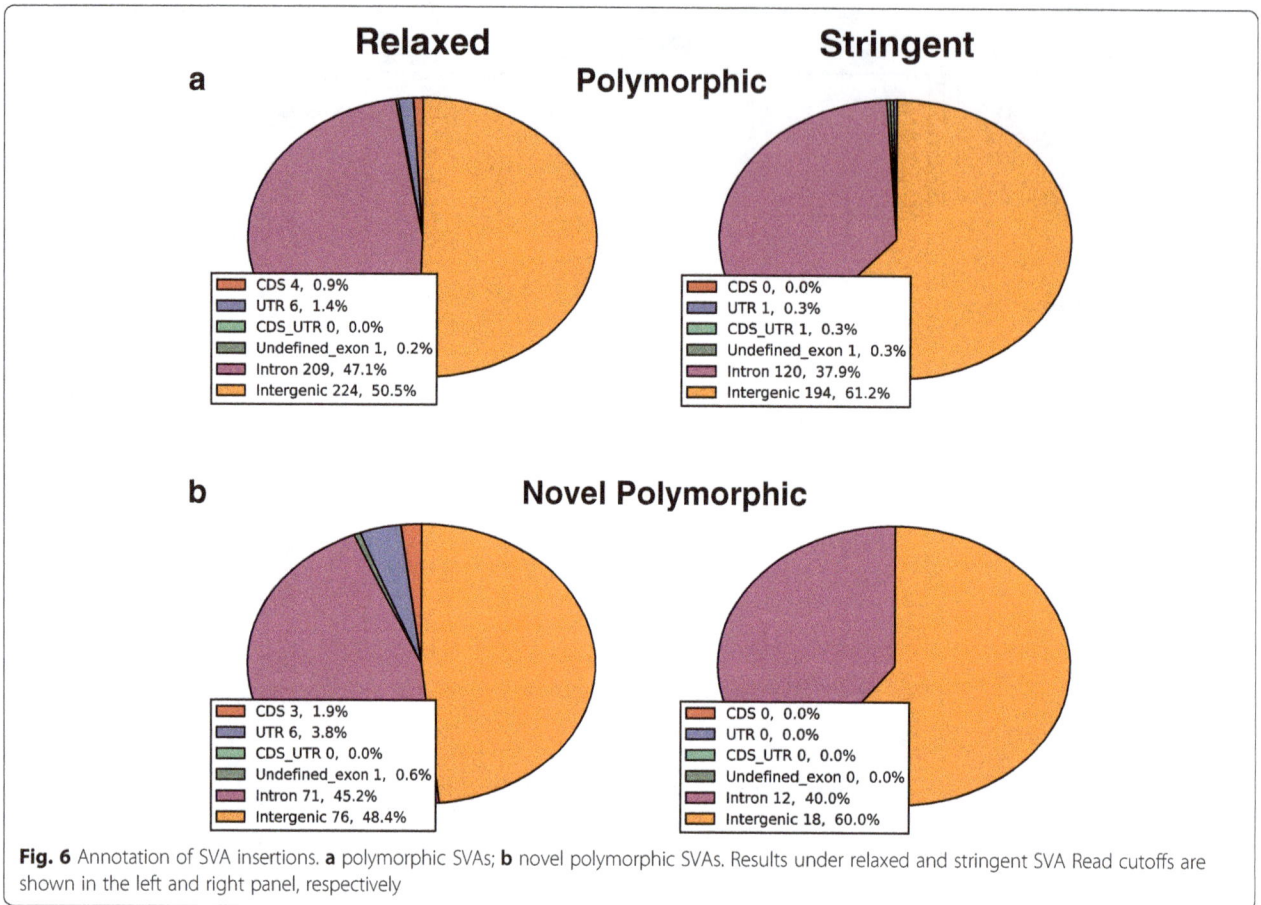

Fig. 6 Annotation of SVA insertions. **a** polymorphic SVAs; **b** novel polymorphic SVAs. Results under relaxed and stringent SVA Read cutoffs are shown in the left and right panel, respectively

diseases [19, 23–28]. However, the composite and complex structure of the SVA element has made it difficult to study the insertions using high-throughput sequencing. Here we described ME-Scan-SVA, a protocol for identifying polymorphic SVA insertions in a large number of samples.

Compare to RC-seq [10, 15–17], which uses a probe-based enrichment protocol to selectively enrich for SVAs, ME-Scan-SVA uses a two-round, nested SVA-specific PCR enrichment method. Unlike RC-seq which enriches for both ends of SVA insertions, ME-Scan-SVA only identify the flanking genomic region on the 5′ end of an SVA insertion. This design prevents us from identifying the TSDs of an SVA insertion without follow-up locus-specific sequencing. In addition, because ME-Scan-SVA is designed to preferentially amplify full-length insertions, we will not identify 5′ truncated SVAs that do not have the primer binding sites. Despite of these limitations, this PCR enrichment method has a high specificity: ~94 % of the DNA fragments in the sequencing library passed the SVA Read filtering and are derived from SVA loci. An average 78 % of the total read-pairs passed both SVA Read and Flanking Read filters and we can determine the genomic locations of these potential SVA insertions (Additional file 1: Table S1). This high-specificity for SVA insertions allows us to pool a large number of individuals (e.g., 48) in one sequencing library to save the sequencing cost. Therefore, ME-Scan-SVA is particularly useful in projects that require cost-effective discovery of SVA insertions in a large number of samples.

Another potential future application of the ME-Scan-SVA method is to identify active SVA elements. SVA insertions can carry both 5′ and 3′ flanking sequences during their retrotransposition, in a process known as transduction [18, 26]. The unique genomic sequence carried by the transduction event can be used to trace a new SVA insertion to the active SVA element where the insertion was generated [26]. With the current sequencing length (100 bps), we do not have sufficient flanking sequence to identify most transduction events. In the future, with long read sequencing technology we will be able to identify the transduction events using the ME-Scan-SVA protocol.

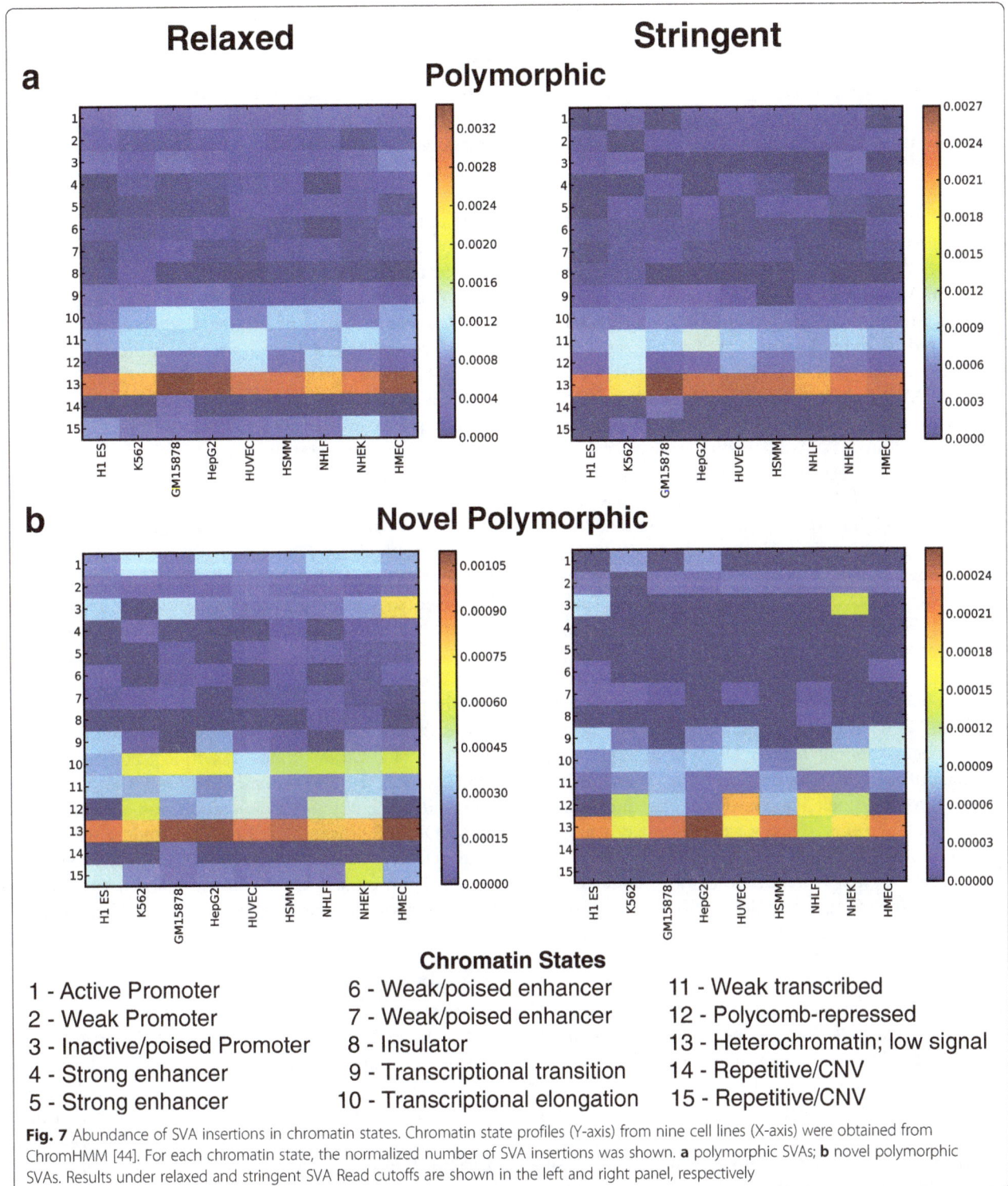

Fig. 7 Abundance of SVA insertions in chromatin states. Chromatin state profiles (Y-axis) from nine cell lines (X-axis) were obtained from ChromHMM [44]. For each chromatin state, the normalized number of SVA insertions was shown. **a** polymorphic SVAs; **b** novel polymorphic SVAs. Results under relaxed and stringent SVA Read cutoffs are shown in the left and right panel, respectively

Conclusions

ME-Scan-SVA allows accurate and cost-effective SVA insertions discovery and genotyping. It can be applied in large-scale population studies. It also can be used to study endogenous somatic SVA retrotransposition events in different tissues or developmental stages.

Methods

Genomic DNA samples

Genomic DNA samples from 21 individuals were obtained from Coriell Cell Repositories (https://coriell.org/). The samples contain three parent-offspring trios with northern and western European ancestry from the CEPH collection

(CEU), three parent-offspring trios from Yoruba in Ibadan, Nigeria (YRI), and three additional YRI individuals. Information including population, family and individual relationships is shown in Table 1.

Library construction and sequencing

The ME-Scan-SVA libraries were prepared following the ME-Scan protocol described previously [35] with SVA-specific modifications. All the adaptor and primer sequences used in this study were synthesized by Integrated DNA Technologies (Coralville, IA, USA) and are shown in Additional file 9: Table S4.

For each sample, 5 µg genomic DNA was randomly fragmented to about 1 kb in size using Covaris system (Covaris, Woburn, MA, USA) and concentrated using AMPure XP beads (cat. no. A63881, Beckman Coulter, Brea, CA, USA), following the manufacturer's protocol. The concentrated DNA fragments were then used to construct the sequencing library using KAPA Library Preparation Kits with SPRI solution for Illumina (KAPA Biosystems, Wilmington, MA, USA, cat. no KK8201).

DNA fragments were end-repaired, A-tailed on both ends following the kit protocol. The concentration of the A-tailed DNA was determined using a Nanodrop (Thermo Fisher Scientific, Wilmington, DE, USA). A-tailed DNA fragments were then ligated with adaptors following the protocol of adaptor ligation of KAPA Library Preparation Kit. Each individual was characterized by a unique 6 bp index for downstream identification. The concentration of ligated DNA from each sample was quantified using Nanodrop and the 21 libraries were pooled into one single library with equal concentration. All of the following steps were performed using the pooled library.

SVA-specific first amplification was conducted for 10 cycles with 200 ng of template DNA and 2.5 µl of primer, following the library prep kit amplification protocol (initial denaturation at 98 °C for 45 s, followed by the thermocycling conditions of 98 °C for 15 s, 65 °C for 30 s, and 72 °C for 30 s, and a final extension at 72 °C for 1 min). Size selection was performed on the amplified PCR product using 0.5X of PEG/NaCl SPRI Solution. After size selection, biotinylated SVA-enriched DNA fragments were magnetically separated from other genomic DNA fragments using 5 µl Dynabeads[R] M-270 Streptavidin (cat. no. 65305, Invitrogen, Life Technologies, Oslo, Norway) following the manufacturer's protocol. Second amplification was conducted for 12 cycles under the same condition as first amplification, with 24 µl of biotinylated SVA-enriched DNA as template in a 75 µl reaction. The amplified PCR product was electrophoresed at 120 volts for 90 min on a 2 % NuSieve[R] GTG[R] Agarose gel (cat. no. 50080, Lonza, Rockland, Maine, USA). Fragments around 500 bp were size

selected and purified using Wizard SV Gel and PCR Clean-up system (cat. no. A9281, Promega, Madison, WI, USA).

Before the library was sequenced, its fragment size and concentration was determined using Bioanalyzer and quantitative PCR by the RUCDR Infinite Biologics (Piscataway, NJ, USA). The library was sequenced using the Illumina Hiseq 2000 with 100PE format at RUCDR Infinite Biologics.

Computational analysis

The computational analysis pipeline was constructed using a combination of bash and python codes. The codes are available at https://github.com/JXing-Lab/ME-SCAN-SVA/.

Briefly, ncbi-blast-2.2.28+ [37] was used to compare SVA sequence in the SVA Read to the SVA consensus sequence to generate BLAST bit-scores. BWA-MEM (ver. 0.7.5a) [38] was used to map Flanking Read against the human reference genome (hg19). Samtools-1.1 [42] were used to count the number of Flanking Read mapped to the human reference genome in each individual for TPM calculation. BEDTools (Ver. 2.16.2) [43] was used to cluster all mapped reads in a region and generate a list of candidate insertion loci for downstream analyses. Using customized python and bash codes, results from all applications were integrated into the current pipeline.

Known polymorphic loci were obtained from the Database of Retrotransposon Insertion Polymorphisms (dbRIP, [40]), HuRef genome [39], and the 1000 Genomes data [5, 22]. Gene annotation was obtained from GENCODE (Release v19). Chromatin state profiles from nine cell lines were obtained from ChromHMM [44]. For each chromatin state, the normalized number of SVA insertions (number of insertions divided by total number of locations in each state) was calculated.

Genotyping PCR for validation

Three separate PCR reactions were performed for each of the 13 loci (11 polymorphic and 2 de novo candidates): one outside primer with two different internal primers (SVA_1 internal, and SVA_2 internal, Additional file 9: Table S4) in two reactions and external primer pair in one reaction (Additional file 5: Figure S2B). Because the 5′ end of an SVA element contains a (CTCCCT)$_n$ simple repeat region and an *Alu* region that shares homology with *Alu* elements, non-specific amplifications occurred at many loci. In these cases different DNA polymerases, annealing temperatures, PCR buffers (standard and high GC buffer), PCR additive betaine, and primer locations were attempted. However, for 7 loci (5 polymorphic, 2 de novo) no specific internal/external amplification was achieved. The PCRs were

Identification of polymorphic SVA retrotransposons using a mobile element scanning method for SVA...

21

performed using One Taq hot start DNA polymerase with GC buffer (cat. no. M0481, New England Biolabs, Ipswich, MA, USA). The thermocycling condition is: an initial denaturation at 94 °C for 30 s, followed by 30 cycles of 94 °C for 30 s, a locus-specific annealing temperature (Additional file 6: Table S2) for 1 min, and 68 °C for 3 min, followed by a final extension at 68 °C for 3 min The PCR products were electrophoresed at 300 volts for 25 min on a 1.5 % GenePure LE Agarose gel (cat. no. E-3120-500, BioExpress, Kaysville, UT, USA). For loci that showed clear and distinct bands, individual genotyping was performed. The DNA fragments of all these loci from at least one individual were validated by Sanger sequencing.

Additional files

Additional file 1: Table S1. Number of passed filter reads in each sample. (XLSX 16 kb)

Additional file 2: Figure S1. Sensitivity Analysis. The sensitivity for identifying fixed SVA insertions under different TPM and UR cutoffs in each individual. The sensitivity is shown as the percentage of fixed insertions identified. (A) Relaxed SVA Read cutoff; (B) stringent SVA Read cutoff. (PDF 2419 kb)

Additional file 3: Polymorphic SVA candidate loci with relaxed SVA Read cutoff. (TXT 89 kb)

Additional file 4: Polymorphic SVA candidate loci with stringent SVA Read cutoff. (TXT 60 kb)

Additional file 5: Figure S2. Individual genotypes of polymorphic SVA insertions. For each individual, three PCR reactions were performed: SVA_1 + outside primer; SVA_2 + outside primer; and outside primer pairs. (A) Genotyping results of Locus 3. Each individual ID is labelled on the top of the lane. For a sample with a homozygous no insertion genotype (e.g., NA12873), the two internal-external primer pairs (SVA_1 + 3R; SVA_2 + 3R) are expected to have no PCR product, and the outside primer pairs (3F + 3R) is expected to amplify the genomic region without SVA insertion. The expected empty (i.e., no insertion) product size for the outside primer pairs is 566 bps. For a sample with a heterozygous insertion genotype (e.g., NA12872), all three reactions will have PCR products. The expected PCR product sizes for the internal-external primer pairs are uncertain because of the unknown size of the SVA 5′ (CCCTCT)$_n$ hexamer simple repeat region. For a sample with a homozygous insertion genotype (e.g., NA18504), the two internal-external primer pairs are expected to have PCR products, and the outside primer pairs is expected to either have no amplification or a large PCR product (SVA + flanking sequence). (B) PCR primer location diagram for Locus 3. The primers are represented by arrows. The color scheme is same as Additional file 2: Figure S1. (C-G) Individual genotyping results of Locus 1, 4, 5, 6, and 9. The expected empty product sizes are shown in Additional file 6: Table S2. (PDF 5287 kb)

Additional file 6: Table S2. Candidate SVA insertion loci subjected to PCR validation. (XLSX 12 kb)

Additional file 7: Table S3. Individual genotypes of validated loci. (XLSX 16 kb)

Additional file 8: Table S5. SVA insertions overlapping protein coding regions. (XLSX 11 kb)

Additional file 9: Table S4. Oligo and primers used in this study. (XLSX 10 kb)

Abbreviations
CDS, coding sequence; MEIs, Mobile element insertions; ME-Scan, mobile element scanning; pMEIs, polymorphic mobile element insertions; TPM, tags per million; UR, unique reads; UTR, untranslated region; VNTR, variable number of tandem repeats

Acknowledgements
We thank Drs David Ray and Roy Platt, and the two anonymous reviewers for their helpful comments.

Funding
This study was supported by the National Institutes of Health (R00HG005846).

Authors' contributions
JX designed the overall strategy. HH designed SVA-specific primers, optimized the protocol, and analyzed the data. JWL constructed the ME-Scan libraries, optimized the protocol, and analyzed the data. All authors wrote the paper. All authors read and approved the final manuscript.

Competing interests
The authors declare that they have no competing interests.

References
1. de Koning AP, Gu W, Castoe TA, Batzer MA, Pollock DD. Repetitive elements may comprise over two-thirds of the human genome. PLoS Genet. 2011;7(12):e1002384.
2. Cordaux R, Batzer MA. The impact of retrotransposons on human genome evolution. Nat Rev Genet. 2009;10(10):691–703.
3. Beck CR, Garcia-Perez JL, Badge RM, Moran JV. LINE-1 elements in structural variation and disease. Annu Rev Genomics Hum Genet. 2011;12:187–215.
4. Hancks DC, Kazazian Jr HH. Active human retrotransposons: variation and disease. Curr Opin Genet Dev. 2012;22(3):191–203.
5. Stewart C, Kural D, Stromberg MP, Walker JA, Konkel MK, Stutz AM, Urban AE, Grubert F, Lam HY, Lee WP, et al. A comprehensive map of mobile element insertion polymorphisms in humans. PLoS Genet. 2011;7(8): e1002236.
6. Nishihara H, Okada N. Retroposons: genetic footprints on the evolutionary paths of life. Methods Mol Biol. 2008;422:201–25.
7. Ray DA, Xing J, Salem AH, Batzer MA. SINEs of a nearly perfect character. Syst Biol. 2006;55(6):928–35.
8. Xing J, Witherspoon DJ, Ray DA, Batzer MA, Jorde LB. Mobile DNA elements in primate and human evolution. Am J Phys Anthropol. 2007;134(S45):2–19.
9. Xing J, Witherspoon DJ, Jorde LB. Mobile element biology: new possibilities with high-throughput sequencing. Trends Genet. 2013;29(5):280–9.
10. Baillie JK, Barnett MW, Upton KR, Gerhardt DJ, Richmond TA, De Sapio F, Brennan PM, Rizzu P, Smith S, Fell M, et al. Somatic retrotransposition alters the genetic landscape of the human brain. Nature. 2011;479(7374):534–7.
11. Witherspoon DJ, Xing J, Zhang Y, Watkins WS, Batzer MA, Jorde LB. Mobile element scanning (ME-Scan) by targeted high-throughput sequencing. BMC Genomics. 2010;11:410.
12. Iskow RC, McCabe MT, Mills RE, Torene S, Pittard WS, Neuwald AF, Van Meir EG, Vertino PM, Devine SE. Natural mutagenesis of human genomes by endogenous retrotransposons. Cell. 2010;141(7):1253–61.
13. Ewing AD, Kazazian Jr HH. High-throughput sequencing reveals extensive variation in human-specific L1 content in individual human genomes. Genome Res. 2010;20(9):1262–70.
14. Witherspoon DJ, Zhang Y, Xing J, Watkins WS, Ha H, Batzer MA, Jorde LB. Mobile element scanning (ME-Scan) identifies thousands of novel Alu insertions in diverse human populations. Genome Res. 2013;23(7):1170–81.
15. Sanchez-Luque FJ, Richardson SR, Faulkner GJ. Retrotransposon Capture Sequencing (RC-Seq): A Targeted, High-Throughput Approach to Resolve Somatic L1 Retrotransposition in Humans. Methods Mol Biol. 2016;1400:47–77.
16. Klawitter S, Fuchs NV, Upton KR, Munoz-Lopez M, Shukla R, Wang J, Garcia-Canadas M, Lopez-Ruiz C, Gerhardt DJ, Sebe A, et al. Reprogramming triggers endogenous L1 and Alu retrotransposition in human induced pluripotent stem cells. Nat Commun. 2016;7:10286.
17. Shukla R, Upton KR, Munoz-Lopez M, Gerhardt DJ, Fisher ME, Nguyen T, Brennan PM, Baillie JK, Collino A, Ghisletti S, et al. Endogenous retrotransposition activates oncogenic pathways in hepatocellular carcinoma. Cell. 2013;153(1):101–11.
18. Wang H, Xing J, Grover D, Hedges DJ, Han K, Walker JA, Batzer MA. SVA elements: a hominid-specific retroposon family. J Mol Biol. 2005;354(4):994–1007.

19. Ostertag EM, Goodier JL, Zhang Y, Kazazian Jr HH. SVA elements are nonautonomous retrotransposons that cause disease in humans. Am J Hum Genet. 2003;73(6):1444–51.

20. Raiz J, Damert A, Chira S, Held U, Klawitter S, Hamdorf M, Lower J, Stratling WH, Lower R, Schumann GG. The non-autonomous retrotransposon SVA is trans-mobilized by the human LINE-1 protein machinery. Nucleic Acids Res. 2012;40(4):1666–83.

21. Hancks DC, Goodier JL, Mandal PK, Cheung LE, Kazazian Jr HH. Retrotransposition of marked SVA elements by human L1s in cultured cells. Hum Mol Genet. 2011;20(17):3386–400.

22. Sudmant PH, Rausch T, Gardner EJ, Handsaker RE, Abyzov A, Huddleston J, Zhang Y, Ye K, Jun G, Hsi-Yang Fritz M, et al. An integrated map of structural variation in 2,504 human genomes. Nature. 2015;526(7571):75–81.

23. Kwon YJ, Choi Y, Eo J, Noh YN, Gim JA, Jung YD, Lee JR, Kim HS. Structure and Expression Analyses of SVA Elements in Relation to Functional Genes. Genome Inform. 2013;11(3):142–8.

24. Xing J, Wang H, Belancio VP, Cordaux R, Deininger PL, Batzer MA. Emergence of primate genes by retrotransposon-mediated sequence transduction. Proc Natl Acad Sci U S A. 2006;103(47):17608–13.

25. Hancks DC, Kazazian Jr HH. SVA retrotransposons: Evolution and genetic instability. Semin Cancer Biol. 2010;20(4):234–45.

26. Damert A, Raiz J, Horn AV, Lower J, Wang H, Xing J, Batzer MA, Lower R, Schumann GG. 5′-Transducing SVA retrotransposon groups spread efficiently throughout the human genome. Genome Res. 2009;19(11):1992–2008.

27. Hancks DC, Ewing AD, Chen JE, Tokunaga K, Kazazian Jr HH. Exon-trapping mediated by the human retrotransposon SVA. Genome Res. 2009;19(11):1983–91.

28. Quinn JP, Bubb VJ. SVA retrotransposons as modulators of gene expression. Mobile Genet Elem. e32102;4.

29. van der Klift HM, Tops CM, Hes FJ, Devilee P, Wijnen JT. Insertion of an SVA element, a nonautonomous retrotransposon, in PMS2 intron 7 as a novel cause of Lynch syndrome. Hum Mutat. 2012;33(7):1051–5.

30. Conley ME, Partain JD, Norland SM, Shurtleff SA, Kazazian Jr HH. Two independent retrotransposon insertions at the same site within the coding region of BTK. Hum Mutat. 2005;25(3):324–5.

31. Wilund KR, Yi M, Campagna F, Arca M, Zuliani G, Fellin R, Ho YK, Garcia JV, Hobbs HH, Cohen JC. Molecular mechanisms of autosomal recessive hypercholesterolemia. Hum Mol Genet. 2002;11(24):3019–30.

32. Nakamura Y, Murata M, Takagi Y, Kozuka T, Nakata Y, Hasebe R, Takagi A, Kitazawa J, Shima M, Kojima T. SVA retrotransposition in exon 6 of the coagulation factor IX gene causing severe hemophilia B. Int J Hematol. 2015;102(1):134–9.

33. Vogt J, Bengesser K, Claes KB, Wimmer K, Mautner VF, van Minkelen R, Legius E, Brems H, Upadhyaya M, Hogel J, et al. SVA retrotransposon insertion-associated deletion represents a novel mutational mechanism underlying large genomic copy number changes with non-recurrent breakpoints. Genome Biol. 2014;15(6):R80.

34. Platt 2nd RN, Zhang Y, Witherspoon DJ, Xing J, Suh A, Keith MS, Jorde LB, Stevens RD, Ray DA. Targeted Capture of Phylogenetically Informative Ves SINE Insertions in Genus Myotis. Genome Biol Evol. 2015;7(6):1664–75.

35. Ha H, Wang N, Xing J. Library construction for high-throughput mobile element identification and genotyping. Methods Mol Biol. 2015. [Epub ahead of print].

36. Jurka J, Kapitonov V, Pavlicek A, Klonowski P, Kohany O, Walichiewicz J. Repbase Update, a database of eukaryotic repetitive elements. Cytogenet Genome Res. 2005;110(1–4):462–7.

37. Altschul SF, Gish W, Miller W, Myers EW, Lipman DJ. Basic local alignment search tool. J Mol Biol. 1990;215(3):403–10.

38. Li H, Durbin R. Fast and accurate short read alignment with Burrows-Wheeler transform. Bioinformatics. 2009;25(14):1754–60.

39. Xing J, Zhang Y, Han K, Salem AH, Sen SK, Huff CD, Zhou Q, Kirkness EF, Levy S, Batzer MA, et al. Mobile elements create structural variation: analysis of a complete human genome. Genome Res. 2009;19(9):1516–26.

40. Wang J, Song L, Grover D, Azrak S, Batzer MA, Liang P. dbRIP: a highly integrated database of retrotransposon insertion polymorphisms in humans. Hum Mutat. 2006;27(4):323–9.

41. Ernst J, Kheradpour P, Mikkelsen TS, Shoresh N, Ward LD, Epstein CB, Zhang X, Wang L, Issner R, Coyne M, et al. Mapping and analysis of chromatin state dynamics in nine human cell types. Nature. 2011;473(7345):43–9.

42. Li H, Handsaker B, Wysoker A, Fennell T, Ruan J, Homer N, Marth G, Abecasis G, Durbin R, Genome Project Data Processing S. The Sequence Alignment/Map format and SAMtools. Bioinformatics. 2009;25(16):2078–9.

43. Quinlan AR, Hall IM. BEDTools: a flexible suite of utilities for comparing genomic features. Bioinformatics. 2010;26(6):841–2.

44. Ernst J, Kellis M. ChromHMM: automating chromatin-state discovery and characterization. Nat Methods. 2012;9(3):215–6.

3

Evidence for L1-associated DNA rearrangements and negligible L1 retrotransposition in glioblastoma multiforme

Patricia E. Carreira[1†], Adam D. Ewing[1†], Guibo Li[2,3†], Stephanie N. Schauer[1], Kyle R. Upton[1,4], Allister C. Fagg[1], Santiago Morell[1], Michaela Kindlova[1], Patricia Gerdes[1], Sandra R. Richardson[1], Bo Li[2], Daniel J. Gerhardt[1], Jun Wang[2,3], Paul M. Brennan[5*] and Geoffrey J. Faulkner[1,6*]

Abstract

Background: LINE-1 (L1) retrotransposons are a notable endogenous source of mutagenesis in mammals. Notably, cancer cells can support unusual L1 retrotransposition and L1-associated sequence rearrangement mechanisms following DNA damage. Recent reports suggest that L1 is mobile in epithelial tumours and neural cells but, paradoxically, not in brain cancers.

Results: Here, using retrotransposon capture sequencing (RC-seq), we surveyed L1 mutations in 14 tumours classified as glioblastoma multiforme (GBM) or as a lower grade glioma. In four GBM tumours, we characterised one probable endonuclease-independent L1 insertion, two L1-associated rearrangements and one likely *Alu-Alu* recombination event adjacent to an L1. These mutations included PCR validated intronic events in MeCP2 and EGFR. Despite sequencing L1 integration sites at up to 250× depth by RC-seq, we found no tumour-specific, endonuclease-dependent L1 insertions. Whole genome sequencing analysis of the tumours carrying the MeCP2 and EGFR L1 mutations also revealed no endonuclease-dependent L1 insertions. In a complementary in vitro assay, wild-type and endonuclease mutant L1 reporter constructs each mobilised very inefficiently in four cultured GBM cell lines.

Conclusions: These experiments altogether highlight the consistent absence of canonical L1 retrotransposition in GBM tumours and cultured cell lines, as well as atypical L1-associated sequence rearrangements following DNA damage in vivo.

Background

Glioblastoma multiforme (GBM) is the most common and aggressive brain tumour in adults [1]. Ninety-five percent of diagnosed GBM tumours originate *de novo* (primary GBM), while the remainder progress from a lower grade glioma (secondary GBM) [2]. Primary and secondary GBM tumours are histologically indistinguishable [3]. To date, genomic analyses have elucidated somatic mutations and intra-tumoural heterogeneity governing GBM progression and resistance to therapy [4–6]. Defects in several DNA repair mechanisms, especially in the repair of DNA double strand breaks (DSBs), are known to enable genomic aberrations, such as deletions and amplifications, in GBM [7, 8]. Despite the extensive genomic analyses performed thus far, the GBM genome may yet harbour additional etiological clues that could improve treatment and patient outcomes.

L1 retrotransposons are endogenous mutagens known to cause sporadic disease, including cancer [9]. A full-length human L1 is ~6 kb long [10, 11] and contains a 5′ untranslated region (UTR), two non-overlapping open reading frames that encode respectively for a 40KDa

* Correspondence: paul.brennan@ed.ac.uk; faulknergj@gmail.com
†Equal contributors
5Edinburgh Cancer Research Centre, IGMM, University of Edinburgh, Edinburgh EH42XR, UK
1Mater Research Institute - University of Queensland, TRI Building, Woolloongabba, QLD 4102, Australia
Full list of author information is available at the end of the article

RNA binding protein (ORF1p) [12, 13] and a 150KDa protein with both endonuclease (EN) and reverse transcriptase (RT) activities (ORF2p) [14, 15], and a 3′UTR. The L1 5′UTR bears an internal promoter with sense and antisense activity [16, 17] and a recently described antisense open reading frame (ORF0) [18]. Canonical L1 mobilisation depends on the transcription and translation of L1 and the formation of a ribonucleoprotein particle (RNP) consisting of ORF1p and ORF2p, and their encoding mRNA. Once the RNP enters the nucleus, the L1-encoded EN can cleave genomic DNA [15] and, typically, generate a new L1 insertion via target-primed reverse transcription (TPRT) [19]. Hallmarks of L1 integration by TPRT include use of an L1 EN recognition motif (5′-TT/AAAA), target site duplications (TSDs), and an L1 poly-A tail [20]. Endonuclease-independent (ENi) L1 mobilisation can also occur into pre-existing DNA double strand breaks, producing insertions that lack TPRT hallmarks [21–24]. Notably, L1 can mobilise other polyadenylated RNAs, such as *Alu* retrotransposons, in *trans* [25–27]. L1 and *Alu* elements can also participate in DNA rearrangements driven by recombination [28, 29]. Although TPRT-mediated L1 mobilisation occurs in many cancers [30–38] and neural cells [39–42], several recent studies employing high-throughput sequencing have reported a surprising absence of somatic L1 insertions in brain tumours [6, 30–32, 35].

We hypothesised that L1-associated DNA rearrangements in GBM might occur via recombination or an atypical retrotransposition mechanism and therefore may lack the TPRT hallmarks required for L1 insertion recognition by previous genomic analyses. Alternatively, we considered that L1 insertions in GBM could be restricted to sub-clonal and highly heterogeneous events. We therefore applied deep retrotransposon capture sequencing (RC-seq) [34, 42] to 14 brain tumour patients (9 GBM and 5 lower grade glioma) and detected tumour-specific L1-associated mutations lacking TPRT hallmarks in 4 GBM tumour samples, and also found no examples of TPRT-driven L1 mobilisation. Complementary assays using an engineered L1 reporter assay [43] revealed negligible in vitro L1 activity in all tested GBM cell lines. These experiments confirm that L1 mobilisation is absent or very rare in GBM tumours and cell lines. Unusual endonuclease-independent L1 retrotransposition or L1-associated recombination events can however occasionally occur, and may impact the expression of genes relevant to GBM aetiology and neural cell morphology.

Methods

Patient samples

Tissues were obtained from 14 patients undergoing surgical removal of a brain tumour at the Department of Clinical Neurosciences, Western General Hospital,

Edinburgh, UK. All patients gave informed consent for tumour and peri-tumoural tissue removed in the normal course of surgery, and blood obtained intra-operatively, to be used for research. Ethical approval for the study was granted to P.M.B. by the East of Scotland Research Ethics Service (SR018). Tissue designation as 'tumour' or 'adjacent brain' was determined at the time of sampling based on pre-operative imaging, intra-operative image guidance and macroscopic inspection. Blood was sampled from 9 patients (Additional file 1: Table S1) intra-operatively and stored in lithium/heparin tubes. Tissue and blood samples were snap frozen on dry ice within 30 min of sampling and stored at −80 °C. Ethical approval for subsequent experiments performed at the Mater Research Institute – University of Queensland was granted to G.J.F. by the Mater Health Services Human Research Ethics Committee (Reference: 1915A) and the University of Queensland Human Research Ethics Committee (Reference: 2014000221). From all samples, genomic DNA was isolated by standard phenol-chloroform extraction.

RC-seq libraries and sequencing

Paired-end 150mer multiplexed Illumina libraries were constructed from genomic DNA samples as described previously [34], with the following minor modifications: sonicated DNA was size selected for fragments of 230–260 bp by gel purification and used as template for 10 - cycles of ligation-mediated PCR (LM-PCR). Libraries were quantified and insert size confirmed using an Agilent Bioanalyzer 2100 with a DNA1000 chip (Agilent Technologies, USA). L1 enrichment was achieved via two different RC-seq capture designs. Equimolar quantities of tumour and adjacent brain libraries from patients #1-#5 were pooled and hybridised to a second generation (V2) RC-seq capture pool [34] composed of 80 biotinylated oligonucleotide probes tiled across the L1-Ta consensus sequence L1.4 [44] 5′ and 3′ ends (Additional file 2: Figure S1A). Hybridisation and library processing were performed as described previously [34]. L1 enriched libraries were sequenced on Illumina HiSeq2000 platform (BGI-Shenzhen, China).

Three additional library pools comprising i) tumour and adjacent brain from patients #1-#7, ii) tumour and adjacent brain from patients #8-#14 and iii) blood from patients #6-#12 and #14 were hybridised using a third generation (V3) RC-seq capture protocol involving only two optimised, custom locked nucleic acid (LNA) oligonucleotide probes (Exiqon Vedbaek, Denmark) respectively targeting the 5′ and 3′ ends of L1.4 [42] (Fig. 1a). LNA probe LNA-D/5Biosg/CTCCGGT + C + T + ACAG CTC + C + C + AGC targeted the 5′ end and LNA-B/ 5Biosg/AG + A + TGAC + A + C + ATTAGTGGGTGC + A + GCG targeted the 3′ end (+ denotes LNA positions within each probe). Pools i) and ii) were sequenced on

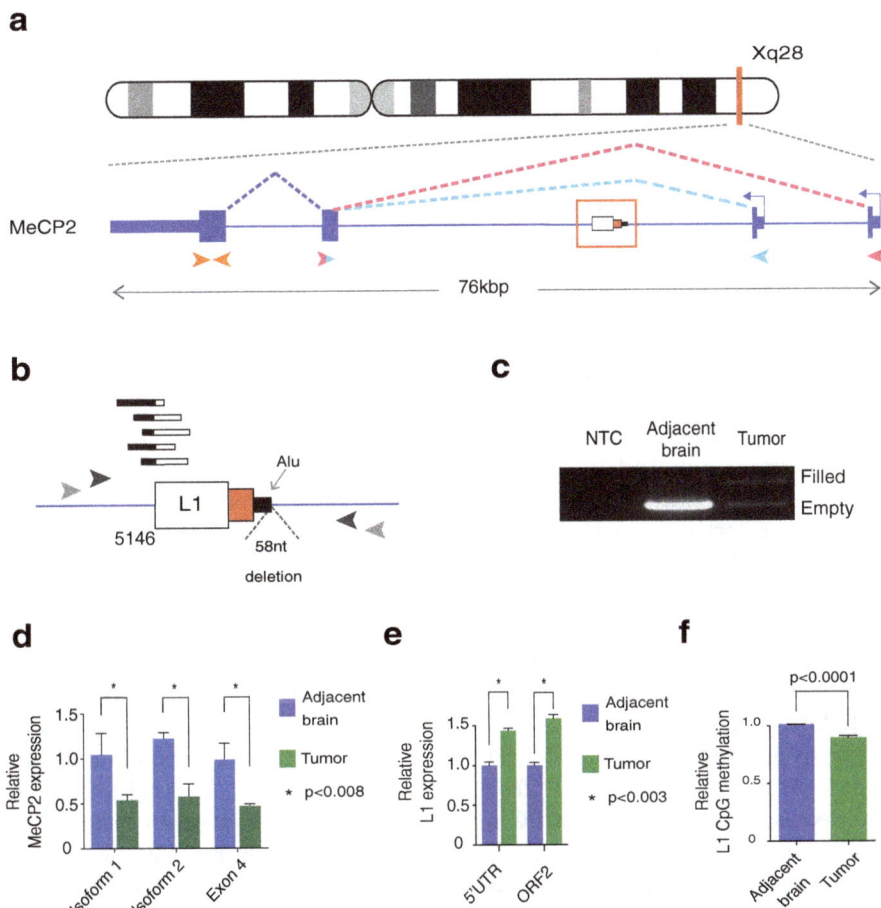

Fig. 1 Characterisation of a somatic L1-associated DNA rearrangement within MeCP2. **a** Patient #2 MeCP2 mutant allele: a 0.9 kb L1PA2 sequence antisense to MeCP2. Direction of transcription (*blue arrows*), transcript isoforms (*purple/pink lines*) and qRT-PCR primers for MeCP2 expression assays (*arrowheads*) are indicated. **b** L1 mutation magnified view: RC-seq reads detected at the L1 5′ terminus (*black/white bars*). The L1 sequence comprises a truncated fragment of L1 ORF2 (*white box*), the 3′UTR without a poly-A tail (*red box*) and 37 nt from an Alu (*black box*). A 58 nucleotide deletion was also identified (*triangle*). Primers used for PCR validation are indicated as grey arrows. **c** Mutation site PCR validation: the mutant MeCP2 allele carrying L1 (filled) was only detected in patient #2 tumour whilst the empty site was found in both tumour and adjacent brain samples. No amplification was detected when water was used as template (NTC). **d** qRT-PCR measurement of MeCP2 transcript isoforms: The relative levels of RNA from both isoforms were significantly reduced in tumour (*blue*) versus adjacent brain (*green*) samples. Data for each group were normalised to non-tumour values, pooled and presented as mean +/− SEM (*$p < 0.008$, two tailed t-test, df = 6). Text colour relates with the primer pair used as represented in (a). **e** qRT-PCR measurement of L1 transcript abundance measured at the L1 5′UTR and ORF2 regions: The relative levels of RNA from both regions were significantly increased in tumour (*blue*) versus adjacent brain (*green*) samples. Data for each group were normalised to adjacent brain values, pooled and presented as mean +/− SEM (*$p < 0.001$, two tailed t-test, df = 10). **f** L1 promoter methylation: CpG methylation was measured across the L1 promoter CpG-island sequence. Tumour samples (*blue*) showed reduced methylation when compared to adjacent brain samples (*green*). Data for each group were normalised to non-tumour values, pooled and presented as mean +/− SEM (*$p < 0.001$, paired t-test, df = 18)

an Illumina HiSeq2500 (Ambry Genetics, USA). Pool iii) was sequenced by multiple Illumina MiSeq runs. A total of 3,252,752,806 2x150mer RC-seq reads were generated (Additional file 1: Table S1). RC-seq FASTQ files are available from the European Nucleotide Archive (ENA) under the identifier PRJEB1785.

RC-seq bioinformatic analysis
RC-seq read data were analysed using TEBreak (https://github.com/adamewing/tebreak/tree/f7f01c1) with settings −mincluster 2, −minclip 30, −minq 1. Briefly,

TEBreak aligned RC-seq reads against the human reference genome (hg19) using BWA-MEM [45] with settings −Y and −M to output soft-clipped secondary alignments. PCR duplicates were marked with MarkDuplicates from the Picard Tools library (http://broadinstitute.github.io/picard/). Non-duplicate reads that aligned partially to the reference genome but had ≥30 nt soft-clipped from either end were retained; clipped ends were then aligned against L1.4 using the same BWA-MEM settings. Split-read mappings joining an L1 to the reference genome were then clustered and annotated for the presence of TSDs or deletions. Clusters with at least two reads supporting a consistent breakpoint were output in VCF format and further post-processed using the summary.py script included in the TEBreak distribution. This further filtered candidates by ensuring a consistent breakpoint between BWA-MEM and BLAT [46] alignments, excluded clusters mapping to locations in the reference genome occupied by other L1s and required that the consensus sequence of each cluster matched L1.4 by at least 90 % over ≥30 nt and the reference genome by at least 95 % based on the BLAT alignment.

RC-seq sensitivity for each library was assessed based on a cohort [41] of 960 reference genome L1-Ta and L1 pre-Ta insertions with intact 3′ ends (Evrony et al. Table S5, "Category 4") detected by ≥20 and ≥8 reads for patients #1-#5 and #6-#14, respectively (Additional file 1: Table S1). Sensitivity was further assessed in terms of polymorphic L1 insertions detected in each individual, with ≥10 and ≥4 reads required in both the tumour and other (adjacent brain and blood) libraries for patients #1-#5 and #6-#14, respectively, to report an L1 insertion (Additional file 1: Table S1). These thresholds exceed those used in another recent work, where we reported a 98.5 % validation rate for polymorphic L1 insertions detected by 2 RC-seq reads and tested by PCR [34]. RC-seq coverage statistics presented in Additional file 1: Table S1 were calculated as the total number of RC-seq reads in a given library that spanned a 5′ or 3′ L1-genome junction of the abovementioned cohort of 960 reference genome L1 insertions, divided by 960 to generate an average value. Non-reference L1 insertions were annotated as tumour-specific if they were: i) found in only one tumour sample with ≥8 RC-seq reads and >10× more RC-seq reads than all other samples combined, ii) absent from published L1 polymorphism databases [30, 32–35, 47–51], iii) likely to be found by the corresponding RC-seq capture design (a 5′ L1-genome junction for an L1 > 6000 nt, <1000 nt in length or a 3′ L1-genome junction), iv) with TEBreak 'strand confidence', 'family confidence' and 'position confidence' scores of >0.9, >0.9 and >0.3, respectively, and v) presented microhomology of <10 nt between the integration site and L1.4 (to exclude possible molecular chimeras). Four putative tumour-specific L1 mutations were reported at these thresholds (Additional file 1: Table S2).

L1 mutation PCR validation

Empty/filled site PCR assays were used to validate tumour-specific L1 mutations detected by RC-seq. Primers flanking either side of each insertion were designed using Primer3 [52] (Additional file 1: Table S3). PCR reactions involved the following reagents: 1U MyTaq DNA polymerase (Bioline, Australia, #BIO-21106), 1× MyTaq Reaction Buffer, 2 µM primers and

20 ng template DNA in a 25 µL reaction volume with the following cycling conditions for the MeCP2 L1 mutation: 3 min at 92 °C, then 10 cycles of 30s at 92 °C, 30s at 60 °C and 6 min 30s at 68 °C, followed by 20 cycles of 30s at 92 °C, 30s at 58 °C and 6 min 30s at 68 °C (increasing by 20s per cycle), followed by a single extension step at 68 °C for 10 min. Products were treated with ExoSap-IT (Affymetrix, USA), with 2 µL of product then used for a second PCR reaction with the same conditions as the first round, except with 30 cycles in the second phase. For the EGFR L1 mutation, PCR was performed using primers targeting the 5′ L1-genome junction with the following cycling conditions: 2 min at 95 °C, then 20 cycles of 15 s at 95 °C, 30s at 59 °C and 30s at 72 °C, followed by a single extension step at 72 °C for 10 min. Products were again treated with ExoSap-IT, with 2 µL of product used for a second PCR reaction with the same conditions as the first round, except with 30 cycles. PCR products were capillary sequenced using an ABI3730 (AGRF, Brisbane, AUS) and the results are provided in Additional file 1: Table S3.

qRT-PCR analyses

Snap frozen tumour and adjacent brain tissues from patient #2 were shaved with a scalpel on dry ice and re-suspended in Trizol Reagent (Invitrogen, Life Technologies, USA, #15596-026) following manufacturer's instructions for total RNA isolation. Quantification was performed using Nano-Drop 1000 (Thermo Fisher Scientific, USA). 2 µg total RNA was treated with DNase I (Ambion, Life Technologies, USA, #AM1906) and used as template for cDNA synthesis with SuperScript III Reverse Transcriptase (Invitrogen, Life Technologies, USA, #18080-093) following manufacturer's instructions. 2 µg total RNA was processed as described with no reverse transcriptase added to the cDNA synthesis reaction for further use as negative control (RT-). cDNA from adjacent brain tissue was diluted to final concentrations of 1:5, 1:10, 1:20, 1:40 and 1:80 and used to generate a standard curve for each primer set. Samples were diluted to 1:20 final concentration for qRT-PCR. Real time PCRs were performed using SensiFast SYBR Lo-ROX kit (Bioline, Australia, #BIO-94005) and run on a ViiA 7 Real-Time PCR System (Life Technologies, USA) with standard curve experiment analysis settings. Negative control qRT-PCRs were performed using water as template (no template control, NTC) and 2 µl of RT- reaction; no amplification was detected. MeCP2 isoforms (NM_004992) were assessed using 5′-GAGGCGAGGAGGAGAGAC and 5′-TGGTAGCTGGGATGTTAGGG as forward primers for isoforms 1 and 2, respectively, and 5′-GCAGAGTGGT GGGCTGAT as a common reverse primer to amplify 154 nt of isoform 1 and 161 nt of isoform 2. Additionally, 143 nt of the MeCP2 exon 4, present in both isoforms, were amplified using 5′-CAGAGGAGGCTCACTGGAGA as forward primer and 5′-GGCATGGAGGATGAA ACAAT as reverse primer. EGFR (NM_005228) was amplified on the 5′UTR (amplicon of 156 nt) and the junction between exons 11 and 12 (176 nt) with the following primers:

EGFR 5′ UTR, 5′ -CCAGTATTGATCGGGAGAGC C, 5′ -CTCGTGCCTTGGCAAACTTTC
EGFR exon 11–12 junction, 5′ -GACCAAGCAACAT GGTCAGT, 5′ -TTTTCTGACCGGAGGTCCCA

L1 (L1.4) expression was assessed by targeting 61 nt of the 5′UTR and 85 nt of the ORF2 with the primers: L1 5′UTR:

5′ -ACAGCTTTGAAGAGAGCAGTGGTT, 5′ -AG TCTGCCCGTTCTCAGATCT
L1 ORF2: 5′ -TGCGGAGAAATAGGAACACTTTT, 5′ -TGAGGAATCGCCACACTGACT

156nt of TATA-binding protein mRNA (TBP, NM_003194) and 173 nt of beta actin mRNA (ACTB, NM_001101) were amplified with the following primers:

TBP, 5′ -GCAAGGGTTTCTGGTTTGCC, 5′ -GGG TCAGTCCAGTGCCATAA
ACTB, 5′ -AGAAAATCTGGCACCACACC, 5′ -TA GCACAGCCTGGATAGCAA

Standard curve parameters for qRT-PCR (slope, y-intercept, r^2) are as follows: MeCP2 isoform 1: −3.312, 29.329, 0.95; MeCP2 isoform 2: −2.946, 29.578, 0.943; MeCP2 exon 4: −3.796, 27.991, 0.897; EGFR 5′UTR: −4.057, 30.028, 0.931; EGFR exon 11–12 junction: −4.660, 28.317, 0.985; TBP: −3.754, 27.208, 0.973; L1 5′ UTR: −3.094, 21.209, 0.996; L1 ORF2: −3.11, 18.836, 0.9992; ACTB: −3.954, 22.606, 0.999.

Relative expression levels were calculated using five technical replicates and normalised to TBP (for MeCP2 and EGFR) or ACTB (for L1). Statistical analysis was performed with Prism5 (GraphPad Software), applying a t-test with a 99 % confidence interval.

cDNA from patient #2 tumour was used as template for RT-PCR to amplify possible L1-MeCP2 chimeric transcripts. The following chimeric variants were tested; Exon 1–L1, using a forward primer within exon 1 5′-GAGGCGAGGAGGAGAGAC and a reverse primer within the new L1, 5′-CACCAGCATGGCACATGTAT. Exon 2–L1, forward primer (5′-TGGTAGCTGGGAT GTTAGGG) in exon 2 and reverse primer within L1 (5′-CACCAGCATGGCACATGTAT). L1-Exon 3, with the L1 primer as a reverse primer (5′-GCACATTGTGC AGGTTAGTTAC) and a forward primer within exon 3, (5′-GCAGAGTGGTGGGCTGAT).

MeCP2 deletion quantification

5 ng of genomic DNA from patient #2 adjacent brain and tumour were used as a template for a PCR to amplify the deleted region within MeCp2 using 5′-AAATTA GCCAGGCGTGGTG as forward primer within the deleted region and 5′-TCCTGTTTTGTCTTACGTCTTG A as reverse primer downstream of the deleted region. The PCR conditions were as follow; 1U MyTaq DNA polymerase (Bioline, Australia, #BIO-21106), 1× MyTaq Reaction Buffer, 2 µM primers and 5 ng template DNA in a 25 µL reaction volume with the following cycling conditions: 3 min at 92 °C, then 35 cycles of 15 s at 92 °C, 15 s at 56 °C and 15 s at 72 °C, followed by a single extension step at 72 °C for 10 min. PCR amplicons were resolved in a 2 % agarose gel and scanned (Typhoon FLA 9500, GE Healthcare life science, US). Amplicons present on the scanned picture was quantified using Image Studio Lite version 4 software (LI-COR Biosciences). Values were corrected for background and normalised to adjacent brain value.

L1 promoter methylation

200 ng of genomic DNA extracted from tumour and adjacent brain tissues was bisulfite converted using EZ DNA Methylation-Lightning Kit (Zymo Research, CA, USA) following manufacturer's instructions. After purification, 2 µL was used as template for a PCR reaction using L1-Bis-F and L1-Bis-R primers as described by Shukla et al. [34]. PCR reactions were performed using MyTaq DNA polymerase (Bioline, Australia, #BIO-21106) in a 25 µL volume with the following cycling conditions: 2 min at 95 °C, then 25 cycles of 15 s at 95 °C, 60s at 55 °C and 60s at 72 °C, followed by a single extension step at 72 °C for 10 min. The ~350 bp PCR product was gel purified using QIAquick gel extraction kit (QIAGEN, NLD). Illumina libraries were generated for each purified PCR product using the NEBNext Ultra DNA Library Prep Kit (New England Biolabs Inc., USA) following manufacturer's instructions and sequenced on an Illumina MiSeq. 250mer paired-end reads were assembled into single contigs using FLASH [53] (−m 15 -M 150 -× 0.3). Contigs were then aligned to the mock bisulfite converted L1-Ta consensus L1.4 using blastn (−dust no -penalty −1 -gapopen 2 -gapextend 1 -max_target_seqs 1). The methylation status of CpG sites in the L1.4 promoter CpG island was used to compare tumour and adjacent brain L1 promoter methylation, as performed previously [34]. Mutated CpG dinucleotides were excluded from analysis, as were reads with less than 95 % conversion of non-CpG cytosines.

Patient #2 and #8 whole genome sequencing and analysis

Illumina libraries (TruSeq Nano DNA sample preparation kit) were generated from patient #2 tumour and adjacent brain and, for patient #8, tumour and blood genomic DNA. Libraries had an insert size of ~300 nt and were sequenced on an Illumina HiSeq X Ten platform (Kinghorn Centre for Clinical Genomics, Garvan Institute of Medical Research, Australia). Reads were aligned to hg19 using BWA-MEM (parameters −Y −M −R < read group name>) and sorted using SAMtools [45]. PCR duplicates were marked using MarkDuplicates and local INDEL realignment was carried out with GATK 3.3. Patient #2 adjacent brain and tumour libraries were sequenced to 44.3× and 84.6×, respectively. Patient #8 blood and tumour libraries were sequenced to 59.8× and 124.2× aligned read depth, respectively. Point mutations and short insertions/deletions were detected using Strelka [54] and Platypus [55]. Structural rearrangements were detected using Delly [56] and Manta [57] and somatic CNVs were detected using cn.MOPS [58], cross-referenced with SV calls, and manually inspected (Additional file 3: Figure S3, Additional file 4: Figure S4, Additional file 5: Figure S5 and Additional file 6: Figure S6). WGS FASTQ files are also available from the ENA under the identifier PRJEB1785.

Cell culture

GBM cell lines were purchased from American Type Culture Collection (ATCC, USA) and grown in a humidified, 5 % CO_2 incubator at 37 °C in the complete media as described by the provider. DBTRG-05MG (ATCC, USA, #CRL-2020) cells were grown in ATCC-formulated RMPI-1640 medium (Gibco, Life Technologies, USA, #A10491-01) supplemented with 10 % foetal bovine calf serum (Gibco, Life Technologies, USA, #16000044), non-essential amino acids (Gibco, Life Technologies, USA, #11140050) and 100 U/mL penicillin, 0.1 mg/mL streptomycin (Gibco, Life Technologies, USA, #15140122) to generate complete media. M059J (ATCC, USA, #CRL-2366) cells were grown in media containing a 1:1 mixture of Dulbecco's Modified Eagle's Medium and Ham's F12 Medium (DMEM-F12, ATCC, USA, #30-2006) supplemented with 10 % foetal bovine calf serum, non-essential amino acids and 100 U/mL penicillin, 0.1 mg/mL streptomycin. LN18 (ATCC, USA, #CRL-2610) and LN229 (ATCC, USA, #CRL-2611) were grown in Dulbecco's Modified Eagle's Medium (DMEM, ATCC, USA, #30-2002) supplemented with 5 % foetal bovine calf serum and 100 U/mL penicillin, 0.1 mg/mL streptomycin. HeLa cells were grown in a humidified, 5 % CO_2 incubator at 37 °C in the complete media, Dulbecco's Modified Eagle's Medium (DMEM, ATCC, USA, #30-2002) supplemented with 10 % foetal bovine calf serum and 100 U/mL penicillin, 0.1 mg/mL streptomycin.

Generation of L1 retrotransposition assay plasmids

Plasmids carrying i) an L1 EN mutant (pCEP4-L1.3D205A) [23] and ii) an L1 reverse transcriptase

mutant (JJ105-L1.3-D702A) [59, 60] were digested with NotI-Hf and BstZ17I restriction enzymes (New England Biolabs, USA, #R3189 and #R0594) at 37 °C for 2 h to obtain L1.3-D205A and the JJ-NotI-Hf/BstZ17I backbone. This backbone was also treated with alkaline phosphatase (New England Biolabs, USA, #M0290) for 30 min at 37 °C to use for subsequent cloning. Purified L1.3-D205A was ligated to the JJ-NotI-Hf/BstZ17I backbone using T4 ligase (New England Biolabs, USA, #M0202) for 2 h at room temperature generating JJ-L1.3-D205A. To obtain wild-type L1.3 (L1.3 WT) [61] and L1.3-D205A/D702A (EN and RT double mutant) constructs, plasmids carrying the individual mutations were digested with EcoRI (New England Biolabs USA, #R3101), NotI and BstZ17I simultaneously (2 h at 37 °C). Fragments corresponding to L1.3-NotI/EcoRI, L1.3-D205A-NotI/EcoRI, L1.3-EcoRI/BstZ17I and L1.3-D702A-EcoRI/BstZ17I were gel purified. L1.3-NotI/EcoRI and L1.3-EcoRI/BstZ17I were ligated to JJ-NotI-Hf/BstZ17I backbone to obtain JJ-L1.3 plasmid. JJ-L1.3-D205A-D702A was generated by ligation of L1.3-D205A-NotI/EcoRI and L1.3-D702A-EcoRI/BstZ17I to the JJ-NotI-Hf/BstZ17I backbone. Ligations were transformed in One Shot TOP10 Chemically Competent E. coli bacteria (Invitrogen, Life Technologies, USA, #C4040), and plated in LB-agar (Sigma-Aldrich, USA, #L2897) and 100 μg/ml ampicillin (Sigma-Aldrich, USA, #A9518). All plasmid sequences were confirmed by capillary sequencing.

L1 retrotransposition assay

Plasmid DNA was purified on maxiprep columns (Qiagen, NED, #12143) and diluted in sterile dH_2O to 0.5 μg/μL. GBM cells were seeded in 6-well dishes in their respective complete media to ~25 % confluence. Cells were transfected at the time of seeding with FuGENE HD transfection reagent (Promega, USA, #E2312) following the manufacturer's protocol using 1:4 DNA: FuGENE ratio. Each transfection well received 1 μg plasmid DNA, 4 μL FuGENE reagent and 2 mL of complete media. Media was changed 24 h post transfection and selection with blasticidin S HCl (Life technologies, USA, #A11139) began 4 days post transfection. Cells were selected for antibiotic resistance for 10 days using a final concentration of 5 μg/ml for HeLa, DBTRG-05MG and LN229 and 2 μg/mL for LN18 and M059J.13-14 days post transfection, cells were washed twice with 1xPBS and fixed and stained as described by Moran et al. [43].

Transfection efficiency for each plasmid was calculated by flow cytometry (BD FACSCanto II, BD Bioscience, USA). GBM cell lines were co-transfected as described above with 0.5 μg of each JJ construct and 0.5 μg of pCAG-GFP (plasmid that constitutively expresses GFP). 72 h post-transfection, cells were harvested and re-

Evidence for L1-associated DNA rearrangements and negligible L1 retrotransposition in glioblastoma...

29

suspended in 1× PBS. Propidium iodide (Thermo Fisher, Life Technologies, USA, #P3566) was added to the samples for identification of dead cells. GFP positive cells were gated based on fluorescence of untransfected cells and transfection efficiency calculated as the percentage of GFP positive cells using FlowJo 10.0.8 software (FlowJo LLc., USA) (Additional file 7: Figure S2, Additional file 1: Table S4).

Results

L1 mutations identified in 4 GBM tumours

We applied RC-seq to 9 GBM and 5 lower grade glioma sample sets, including tumour and matched adjacent brain or blood (Additional file 2: Figure S1A, Additional file 1: Table S1). Tumour and non-tumour samples from five of these patients were sequenced to ~250× coverage of targeted L1-genome junctions by RC-seq, while samples from the remaining individuals were sequenced to ~55× coverage. Overall, we detected 93.6 % of 960 reference genome copies [41] of the most active human L1 subfamily, L1-Ta, as well as an average of 208 polymorphic L1-Ta insertions per sample (Additional file 1: Table S1), despite stringent RC-seq reporting thresholds (see Materials and Methods).

We identified four putative tumour-specific L1 mutations in four different GBM patients (Additional file 1: Table S2). Three of these mutations were located within genes known to be active in the brain (MeCP2, EGFR and CEP112) while the other was intergenic and was not associated with a known regulatory element, such as a promoter region or annotated enhancer (Additional file 1: Table S2). The putative tumour-specific L1 mutations in MeCP2 and EGFR were validated via PCR (see below). The remaining two events identified by RC-seq could not be confirmed by PCR and hence their structures could not be fully elucidated. RC-seq read information however indicated that the putative L1 insertion in CEP112 involved an L1-Ta donor sequence, a long (102 nt) poly-A tail and a degenerate L1 EN recognition motif, suggesting potential TPRT-mediated L1 mobilisation, albeit without corroboration via PCR. The remaining putative L1 mutation was annotated as an older L1PA2 element, which are usually not capable of autonomous retrotransposition in humans [62], and also lacked an L1 EN motif. These features are consistent with a DNA rearrangement rather than a retrotransposition event. No somatic L1 sequence variants were detected in lower grade glioma samples.

Structure and impact of a *de novo* L1-associated DNA rearrangement within MeCP2

In patient #2, a female, we identified a putative intronic L1 mutation in the methyl CpG binding protein 2 (MeCP2) gene by RC-seq that was confirmed by PCR

(Fig. 1a-c, Additional file 2: Figure S1B-C, Additional file 1: Table S2). MeCP2 is an X-chromosome linked transcription factor necessary for neural differentiation and is defective in the neurodevelopmental disorder Rett syndrome [63]. MeCP2 binds and generally represses methylated DNA genome-wide, including the CpG-island present in the canonical L1 5′ promoter [64]. Sequence characterisation revealed that the L1 sequence belonged to the L1PA2 subfamily, was 5′ truncated and carried a 49 nt 3′ flanking region from its L1 donor element on chromosome 9. The L1 mutation site lacked an L1 EN recognition motif and TSDs, and incorporated a 58 nt genomic deletion (Fig. 1b, Additional file 2: Figure S1B), features strongly inconsistent with retrotransposition through TPRT. Further analysis revealed that the 3′ flanking region carried with the L1 comprised an *Alu* retrotransposon and that integration occurred into another *Alu* element. These features led us to conclude that this L1-associated event was probably driven by recombination of the *Alu* adjacent to the L1PA2 donor sequence with the *Alu* present in MeCP2.

Quantitative RT-PCR revealed significant ($p < 0.008$, t-test) reductions in tumoural expression of both of the two main MeCP2 transcript isoforms (Fig. 1a, d). Using additional primers specific to the L1 mutation, we performed qRT-PCR to evaluate whether the L1 generated a chimeric transcript with upstream or downstream MeCP2 exons. However, chimeric L1-MeCP2 RNA species were not detected by this assay (data not shown). WGS applied to patient #2 tumour and adjacent brain revealed tumour-specific copy number gain at the MeCP2 locus and an absence of single nucleotide variants or DNA rearrangements, other than the L1 mutation. Quantitative PCR measuring copy number of the genomic region deleted 3′ of the L1 mutation confirmed that the L1-mutant MeCP2 allele was amplified in the tumour as, despite overall amplification of the MeCP2 locus as detected by WGS, we identified copy number loss of this deleted 3′ sequence in the tumour. MeCP2 is known to regulate L1 activity by binding the methylated CpG-island present in the canonical L1 5′ promoter [64]. Therefore, as an evidence of a reduction in MeCP2 activity, we detected significantly higher ($p < 0.0001$, t-test) L1 mRNA abundance in patient #2 tumour versus adjacent brain (Fig. 1e), as well as significant tumour-restricted hypomethylation of the canonical L1-Ta promoter ($p < 0.0001$, t-test) (Fig. 1f). Replicate PCR performed on seven spatially disparate tumour foci detected the L1 mutation in all locations (Additional file 2: Figure S1C). These data suggest that the MeCP2 L1 mutation underwent copy number gain, was present in clonally amplified cells, and may have impacted MeCP2 expression and function throughout the tumour mass. Given these results, we propose transcriptional disruption by the L1 [65] as a plausible cause for reduced MeCP2 expression, although we cannot rule out the involvement of another mechanism.

Structure of a tumour-specific L1 mutation within EGFR

In patient #8, RC-seq detected a putative tumour-specific L1 mutation in the first intron of the epidermal growth factor receptor gene (EGFR) (Fig. 2a), a major oncogene amplified or otherwise altered in >60 % of GBM cases [6]. PCR validation and capillary sequencing showed a 5′ truncated L1 sequence of the L1-Ta subfamily and lacking a 3′ poly-A tail and TSDs, and here incorporating a 550 nt genomic deletion (Fig. 2b-c, Additional file 2: Figure S1D). Although these features

are consistent with ENi L1 retrotransposition, we cannot fully exclude the possibility that this event arose via L1-associated DNA recombination. PCR upon multiple tumour foci suggested that the L1 mutation was clonally amplified (Additional file 2: Figure S1E). Unlike the MeCP2 L1 mutation, the EGFR L1 mutation did not appear to impact host gene expression. Indeed, EGFR expression was significantly up-regulated in patient #8 tumour versus adjacent brain tissue ($p < 0.0001$, t-test) (Fig. 2d). As EGFR structural and copy number variation is a common feature of GBM [6], we performed WGS on patient #8 tumour and blood samples, elucidating major (>50×) copy number amplification of EGFR and the surrounding genomic locus (Fig. 2e). Given the

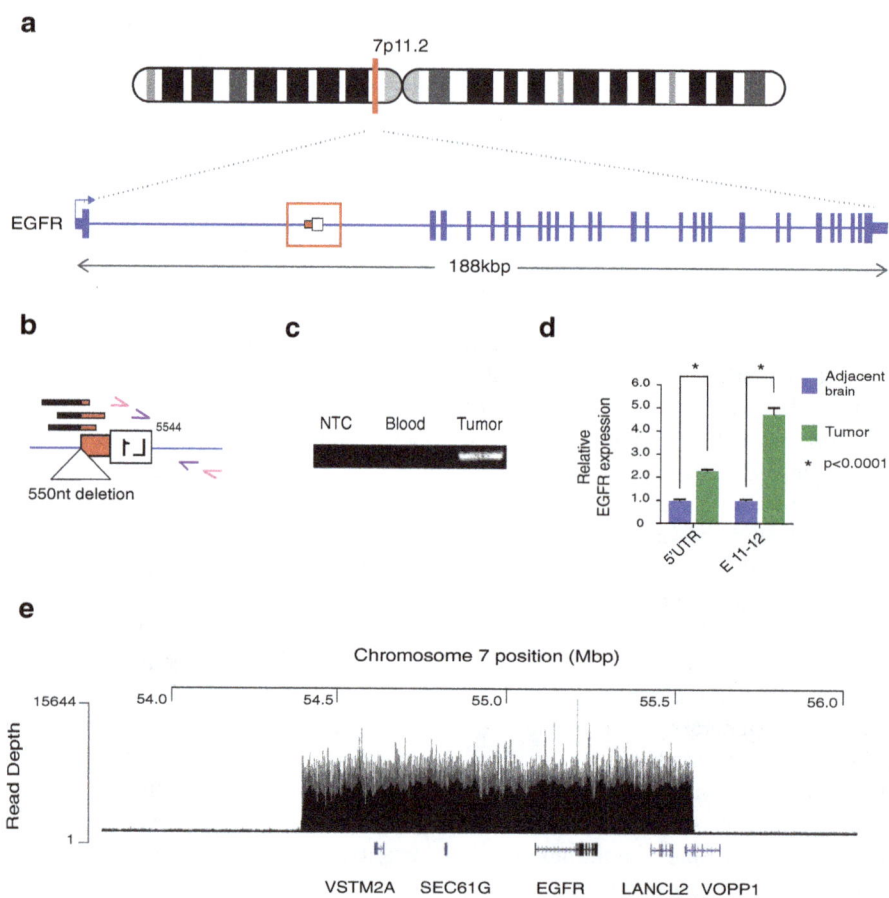

Fig. 2 Characterisation of a somatic L1 mutation within EGFR. **a** Patient #8 EGFR mutant allele: a 0.5 kb L1-Ta sequence antisense to EGFR. Direction of transcription is indicated with *blue arrow*. **b** L1 mutation magnified view: RC-seq reads detected at the L1 3′ terminus (*black/red bars*). The L1 mutation comprised a truncated fragment of L1 ORF2 (*white box*) and the 3′UTR without a poly-A tail (*red box*). A 550 nucleotide deletion at the integration site was also identified (triangle). Primers used for PCR validation are indicated as pink and purple arrows. **c** Mutation site PCR validation: Region comprising the EGFR-L1 5′ junction was detected in patient #8 tumour sample. No amplification was detected when water (NTC) or genomic DNA from blood were used as template. **d** qRT-PCR measurement of EGFR transcription at its 5′UTR and exon 11-to-12 junction (E 11–12): The relative levels of RNA from both regions were significantly increased in tumour (*blue*) versus adjacent brain (*green*) samples. Data for each group were normalised to adjacent brain values, pooled and presented as mean +/− SEM (*$p < 0.001$, two tailed t-test, df = 10). **e** Amplified chromosome 7 region including EGFR: mapped read depth in EGFR region. Positions in Mbp are marked across the top horizontal axis. Read depth is reflected by the height of vertical lines as indicated on the vertical axis. Genes present in the amplified region are placed based on the locations of representative transcripts from UCSC Genes (hg19)

extreme copy number gain and the presence of discordant read pairs supporting a junction between the 5′ and 3′ ends of the amplified genomic segment, this event was likely to represent a double minute chromosome [66] containing EGFR. Further WGS data analysis suggested that the majority of additional EGFR alleles did not incorporate the L1 mutation, indicating that copy number amplification primarily drove EGFR induction.

Whole genome analysis of patients #2 and #8

To place the PCR validated L1-associated mutations in MeCP2 and EGFR into a broader context of genomic abnormality, we analysed WGS data from patients #2 and #8 (Additional file 1: Table S5). For patient #2, we compared tumour tissue and adjacent tissue that appeared pathologically normal. Here we found that a significant portion of cells from the normal tissue did in fact contain tumour cells based on the presence of mutations at low variant allele fraction (VAF) in the normal tissue and at an increased VAF in the tumour tissue. For example, we identified a TP53 mutation (c.421C > T/p.Arg141Cys) present in 18.5 % of sequencing reads from adjacent tissue and 90.9 % of tumour reads. This mutation corresponded to rs121913343/COSM3719990 (dbSNP/COSMIC) but may have been somatic in this instance given the low VAF in normal tissue. Other potentially pathogenic point mutations in patient #2 included an established GBM mutation in the IDH1 gene at p.Arg132His [6]. We observe LOH over APC leading to at least two non-synonymous changes increasing in VAF to >95 % (rs139196838, rs459552). The tumour from patient #2 presented variable 5–10 fold amplification across 4q12 (Additional file 3: Figure S3A), which included the tyrosine kinase KIT, and tyrosine kinase receptors PDGFRA and KDR (VEGFR2) (Additional file 3: Figure S3B). Amplifications of this region occur frequently in GBM [67]. We also detected a focal ~2.7kbp deletion removing both copies of CDKN2A exon 2 inside of a larger single-copy region encompassing CDKN2A/B (Additional file 4: Figure S4). Finally, we detected copy number amplifications at the end of the q arm on chromosome X, which indicated one additional copy in two regions, one of which included MeCP2 (Additional file 5: Figure S5). This copy number increase provides a reasonable explanation for the 50 % decrease in MeCP2 expression in this tumour relative to the adjacent tissue, due to the aforementioned intronic L1 mutation in the amplified MeCP2 allele (Fig. 1).

As noted above, the tumour sample from patient #8 showed a remarkable amplification of EGFR (>50 fold, Fig. 2e, Additional file 6: Figure S6A), likely due to a double minute chromosome as observed in many other GBM cases [66]. Additionally this tumour contained a deletion surrounding CDKN2A/B, and an amplification of the RAS-related oncogene RAB14 on chr9q33.2

(Additional file 6: Figure S6B). There were few point mutations affecting known cancer or GBM-associated genes detected in this sample at appreciable (>10 %) VAF. The detected examples include a frameshift mutation of the histone methyltransferase and known tumour suppressor SETD2 (p.Asp14fs) that can activate TP53 and is necessary for DSB repair via homologous recombination [68, 69], and a putative splice donor site mutation affecting CIITA (class II MHC transactivator). Thus, the L1 mutation observed here in EGFR was likely a passenger to the main oncogenic transformation enabling tumorigenesis in patient #8 and occurred in an environment of impinged DNA repair.

GBM cell lines rarely support L1 retrotransposition

To assess whether GBM cell lines support canonical or ENi L1 retrotransposition in vitro we performed an established cultured cell retrotransposition assay [43] on HeLa cells and four GBM cell lines (DBTRG-05MG, M059J, LN 18 and LN 229). This assay relies on the expression of a blasticidin resistance gene carried by an L1 reporter construct, where blasticidin is only expressed and confers resistance after L1 retrotransposition (Fig. 3a). Each cell line was transfected in triplicate with a set of 4 plasmids bearing different L1 sequences upstream of the antisense orientated blasticidin-resistance gene [60] (Fig. 3b); a wild type full-length L1 (JJ L1.3 WT) [61], an L1 with an EN domain missense mutation that abolishes L1 ORF2p EN activity (JJ L1.3 D205A) [23], an RT mutated L1 with no reverse-transcriptase activity (JJ L1.3 D702A) [59] and a double mutant L1 bearing both EN and reverse-transcriptase mutations (JJ L1.3 D205A D702A). No L1 retrotransposition events were detected for DBTRG-05MG, M059J or LN18 cell lines and very few events (<4 events per well) were detected for the L1.3 WT construct in LN 229 cells (Fig. 3c). Mobilisation of the RT mutant and double mutant L1 reporter was not observed in any of the GBM cell lines. By contrast, the positive control HeLa cells supported the expected "hot" L1.3 WT activity [43, 70, 71] as well as mobilisation of each L1.3 mutant to lesser extents. These data indicate that GBM cell lines typically only support very low or negligible L1 retrotransposition, in line with our RC-seq data obtained from patient tumour samples.

Discussion

These experiments reveal rare L1-associated mutations caused by recombination or L1 ENi retrotransposition in GBM tumours, accompanied by an absence of TPRT-driven L1 insertions. Endonuclease-independent L1 insertions have been reported by several prior studies employing engineered L1 reporter constructs in cultured cancer cells or cells otherwise deficient for DNA damage

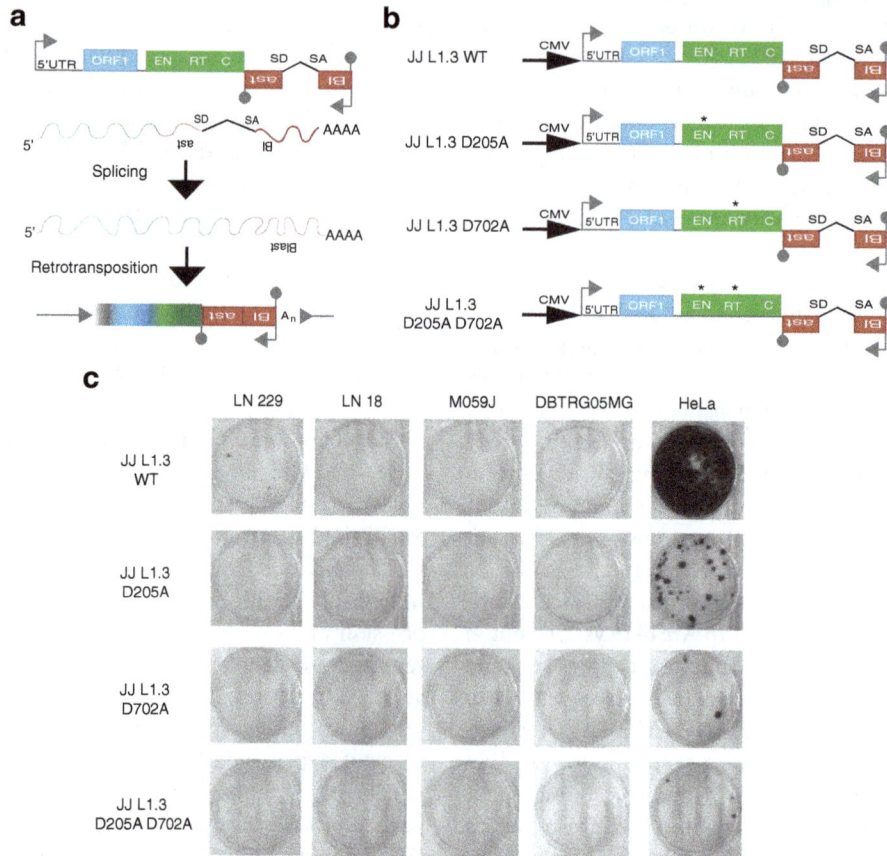

Fig. 3 L1 retrotransposition rarely occurs in GBM cell lines. **a** Schematic representing L1 retrotransposition assay. A full-length L1 (L1.3) [61] is located upstream of the antisense oriented blasticidin resistance gene (*red boxes*). The L1 internal promoter is represented by an arrow on the 5'UTR region. Two L1 open reading frames (ORF1 and ORF2) are indicated by *blue* and *green boxes*, respectively. Functional domains of ORF2, endonuclease (EN), reverse transcriptase (RT) and cysteine rich domain (C) are also indicated. The blasticidin resistance gene is interrupted by an intron in the same orientation as the L1. Splice donor (SD) and splice acceptor (SA) sites are indicated. Polyadenylation signals are denoted by grey lollipops. **b** Schematic representation of retrotransposition assay constructs. JJ L1.3 WT contains an external promoter (cytomegalovirus promoter, CMV) upstream of a full length retrotransposition-competent L1.3 element [61]. *Asterisk* indicates missense mutation to abolish endonuclease activity (JJ L1.3 D205A), reverse-transcriptase activity (JJ L1.3 D702A) or both (JJ L1.3 D205A D702A). **c** Results of cell culture-based L1 retrotransposition assay. Each stained colony represents a cell where a retrotransposition event took place allowing the expression of the blasticidin resistance gene

repair factors [21–23, 72, 73], or through bioinformatic analysis of the human reference genome [24]. Unusual L1 integration sites identified by these studies incorporated, amongst other features, genomic deletions, deletions of the L1 3′ end and poly-A tail, absence of TSDs and absence of an L1 EN recognition motif. Here, by fully resolving the structures of GBM tumour-specific L1 mutations through RC-seq and capillary sequencing, we confirmed they lacked a recognisable L1 poly-A tail or TSDs, and incorporated genomic deletions. In the case of the EGFR L1 mutation, these features are suggestive of L1 ENi mobilisation, as primarily reported by others using engineered L1 systems in vitro [21–23, 72, 73] or, potentially, DNA recombination. Genomic abnormality at L1 mutation sites may also explain failure to PCR validate 2/4 observed putative tumour-specific L1

mutations. Notably, our WGS analyses elucidated mutations in key DSB repair factors, such as SETD2 in patient #8, as well as TP53 deficiency in patient #2. Thus, rare tumour-specific L1-associated mutations occur in GBM in a milieu of deficient DSB repair previously encountered for similar events in vitro [21, 23].

Although identified at first by RC-seq as a potential L1 mobilisation event, further characterisation of the MeCP2 L1 mutation indicated a probable DNA rearrangement event mediated by *Alu:Alu* recombination. Notably, the *Alu*-flanked L1PA2 donor sequence, located on chromosome 9, was 5′ truncated and did not encode a viable L1 ORF2p. The involvement of an older L1 family in a tumour-specific DNA rearrangement event is reminiscent of one of the earliest L1 mutations detected in cancer, in that case affecting the myc locus in breast

Table 1 Published analyses of L1-associated mutations in brain tumours

Study	Brain tumour types (sample count)	Sequencing method	L1 coverage (tumour)	Potential somatic L1-associated mutations (PCR validated)	Ref.
Iskow et al.	GBM (5), medulloblastoma (5)	L1-seq	3.6× - 6.4×	74 (0)	[30]
Lee et al.	GBM (16)	WGS	39.2×	16 (0)	[32]
Brennan et al.	GBM (42)	WGS	35.3×	0 (0)	[6]
Helman et al.	GBM (20)	WGS	40.8×	0 (0)	[35]
Tubio et al.	Glioma cell line (1)	WGS	42.6×	1 (0)	[31]
Carreira et al.	GBM (9), glioma (5)	RC-seq	43.2×–231.0×	4 (2)	–

tumour [74]. A lesser possibility is that the L1PA2 transduced its flanking 3′ *Alu* during ENi L1 mobilisation in *trans*, in agreement with a previous observation where 71 % of detected ENi L1 mobilisation events involved non-L1 DNA [24], and involved single strand annealing [72, 73, 75]. The tumour-specific L1 mutation in MeCP2 appeared to reduce the expression of this gene concomitant with copy number gain for the L1 mutant MeCP2 allele and, as suggested by increased L1 expression and reduced L1 5′UTR methylation, may have reduced MeCP2 function throughout the tumour mass, causing a molecular phenotype.

Notably, the EGFR L1 mutation provides an interesting example because, despite not having a direct effect on gene activity, it is one of very few tumour-specific L1 mutations noted in a major oncogene or tumour suppressor since the first such example was discovered more than 20 years ago by Miki et al. [9].

Previous studies employing high-throughput sequencing reported no tumour-specific L1 insertions in brain tumours [30–32, 35]. Iskow et al. did observe putative tumour-specific L1 insertions, though subsequent PCR validation experiments were unsuccessful [30]. Thus, although the atypical L1-associated mutations reported here represent the first PCR validated variants of this type in in GBM, our in vivo and in vitro results agree with prior reports of a lack of TPRT-driven L1 insertions in brain cancers (Table 1). At the same time, the absence of *de novo* L1 insertions in this context is intriguing, given frequent somatic L1 mobilisation via TPRT in neural cells [39–42], including glia [42]. To speculate, one explanation for consistently limited L1 activity in GBM could be that the relevant tumour initiating cells only support L1 retrotransposition in the context of a deficient DSB repair, perhaps due to the presence of L1 inhibiting host factors, including those affecting subcellular localisation of the L1 RNP [76]. This conclusion would disagree with GBM typically arising from dedifferentiated neural cells. Still, given the high RC-seq coverage employed here, and evidence obtained using the L1 reporter system in vitro, we consider this explanation more appealing than tumoural heterogeneity obscuring TPRT-driven L1 mobilisation.

Conclusions

Although we conclusively find that L1 mobilisation is a rare event in GBM, our discovery of atypical L1-associated mutations in MeCP2 and EGFR demonstrates that L1 can otherwise contribute to GBM genome abnormality in key loci regulating neural cell differentiation and proliferation. Future experiments are required to ascertain whether this phenomena correlates with patient prognosis and whether potential DNA damage caused by chemotherapy or radiotherapy [77] activates L1 in recurrent GBM.

Additional files

Additional file 1: Table S1. RC-seq output summary. **Table S2.** Polymorphic and somatic L1 mutations called by TEBreak. **Table S3.** List of somatic L1 mutations and PCR validation information. **Table S4.** GBM cell line transfection efficiencies. **Table S5.** Summary of point mutations and CNVs detected from WGS (XLSX 335 kb)

Additional file 2: Figure S1. Structure and detection of L1 mutations in MeCP2 and EGFR. (a) RC-seq versions: Capture probes from version 2 (V2, marked in blue) correspond to two sets of DNA oligonucleotides mapping within the 5′UTR and the 3′UTR of the L1 sequence (UTRs represented as red boxes). Version 3 (V3, marked in green) includes a single LNA probe for each UTR. The 5′ end probes capture only the 5′ junction of full or nearly full length insertions. The 3′ end probes capture both 3′ junctions of insertions of all lengths and 5′ junctions of heavily truncated insertions. (b) Pre- and post- mutation site sequences for MeCP2: Pre-integration panel shows the sequence at the L1 mutation site as obtained from the human reference genome (GRCh37/hg19). A 58 nt deletion was detected after the L1 mutation (marked in green within box). Post-integration panel shows the 5′ and 3′ termini sequences of the L1 mutation and flanking genomic region. L1PA2 sequence is marked in red. The 49 additional nucleotides from an *Alu* are marked in grey. Microhomology with the pre-integration genomic region is denoted (underlined). Slashes represent L1PA2 sequence not shown. Lowercase indicate mutations versus reference sequences. (c) MeCP2 L1 mutation PCR validation: DNA extracted from different regions of the tumour (denoted from A to G) was used as template for the amplification of MeCP2 5′end L1 junction. No amplification was detected when water was used as template (NTC). (d) Pre- and post- mutation site sequences for EGFR: Pre-integration panel shows the sequence at the L1 insertion site as obtained from the human reference genome (GRCh37/hg19). A 550 nt deletion was detected after the L1 mutation (marked in green within box). Post-integration panel shows the 5′ and 3′ termini sequences of the L1 mutation and flanking genomic region. L1-Ta sequence is marked in red. Slashes represent genomic or L1-Ta sequence not shown. Lowercase indicate mutations relative to reference sequences (L1-Ta, human reference genome). Untemplated nucleotides are represented in grey. (e) EGFR mutation site PCR validation: DNA extracted from different regions of the tumour (denoted from A to G) was used as template for the amplification of the EGFR 5' end L1 junction. No amplification was detected when water was used as template (NTC) (EPS 2005 kb)

Additional file 3: Figure S3. Copy number aberrations in the tumour sample of patient #2 on chr4q12. (a) Detailed view of the 4q12 region with aberrant and highly variable copy number changes in oncogenes PDGFRA, KIT, and KDR (VEGFR2). (b) altered region in the context of chromosome 4. Y-axis of both panels indicates the log10 fold-change in counts per million (CPM), X-axis indicates position on the indicated chromosome. Shaded regions indicate gaps in the chromosome reference sequence. (PDF 133 kb)

Additional file 4: Figure S4. Copy number aberration of CDKN2A on chromosome 9 in patient 2. (a) Detailed view of the CDKN2A region with the regional single-copy deletion and focal deletion of both copies i ndicated. (b) Altered region in the context of chromosome 9. Axes and shading are as described for Additional file 3: Figure S3. (PDF 130 kb)

Additional file 5: Figure S5. Copy number aberration of MECP2 on chromosome X in patient 2. (a) Detailed view of the MECP2 region with the regional single-copy amplification indicated. (b) Altered region in the context of chromosome X. Axes and shading are as described for Additional file 3: Figure S3. (PDF 135 kb)

Additional file 6: Figure S6. Copy number aberration of EGFR, CDKN2A, and RAB14 in patient 8. (a) Amplification of EGFR on chromosome 7 (see also Fig. 2e). (b) Amplifications on chromosome 9 including CDKN2A and RAB14. (PDF 151 kb)

Additional file 7: Figure S2. Retrotransposition assay flow cytometry gating parameters. Flow cytometry gating strategy for LN 229 cell line is shown as an example. Cells were distinguished from debris based on internal complexity (side scatter light area, SSC-A) and size (forward scatter light area, FSC-A). To identify single cells only, the original population was gated based on the size (FSC-A) and height (forward scatter light height, FSC-H) of the cells. Propidium iodide staining was used to distinguish dead and alive cells. GFP positive events were gated based on the absence of GFP signal from the untransfected live cell population. Flow cytometry analysis settings were established for untransfected cells and applied to transfected cells. (EPS 2683 kb)

Abbreviations

APC: Adenomatous polyposis coli gene; CDKN 2a/2b: Cyclin-Dependent Kinase Inhibitor 2a/2b genes; CEP112: Centrosomal protein 112 gene; CIITA: Class II, Major Histocompatibility Complex, Transactivator gene; DSBs: Double strand breaks; EGFR: Epidermal growth factor receptor; EN: Endonuclease; ENi: Endonuclease independent; GBM: Glioblastoma multiforme; IDH1: Isocitrate Dehydrogenase 1 gene; KDR: Kinase Insert Domain Receptor; L1 or LINE1: Long interspersed nuclear element 1; LM-PCR: Ligation-mediated PCR; LOH: Loss of heterozigosity; MeCP2: Methyl CpG binding protein 2 gene; ORF: Open reading frame; PDGFRA: Platelet-derived growth factor receptor alpha gene; RC-seq: Retrotransposon capture sequencing; RNP: Ribonucleoprotein particle; RT: Reverse transcriptase; TP53: Tumour protein p53 gene; TPRT: Target-primed reverse transcription; TSDs: Target site duplications; UTR: Untranslated region; VAF: Variant allele fraction

Acknowledgements

We thank J. L. Garcia-Perez for kindly providing plasmid constructs for the retrotransposition assay and F.J. Sanchez-Luque and G.O. Bodea for helpful discussion. We also thank J.M.C. Tubio for providing GBM genome sequencing depth statistics for Tubio et al. [31].

Funding

G.J.F. acknowledges the support of an Australian National Health and Medical Research Council Senior Research Fellowship [GNT1106214], National Health and Medical Research Council Project Grants [GNT1042449, GNT1045991, GNT1067983, GNT1068789, GNT1106206] and the European Union's Seventh Framework Programme (FP7/2007-2013) under grant agreement No. 259743 underpinning the MODHEP consortium. A.D.E. acknowledges the support of an

Australian Research Council Discovery Early Career Researcher Award [DE150101117]. G.J.F. and A.D.E. were supported by the Mater Foundation. P.M.B. was supported by a Welcome Trust Clinical Fellowship [090386/Z/09/Z]. The Translational Research Institute (TRI) is supported by a grant from the Australian Government.

Authors' contributions

PEC, KRU, DJG and PMB performed RC-seq experiments. ADE, GL and GJF analysed RC-seq data. ADE analysed WGS data. PEC and KRU performed L1 promoter methylation analyses. PEC. and SNS performed PCR validations. PEC, PG, SRR and SM generated the retrotransposition assay plasmids. PEC, ACF, SM and MK optimised and performed the retrotransposition assay. GL, BL and JW provided resources. PMB performed tissue sampling and annotation. PMB and GJF conceived the study. GJF directed the study. PEC and GJF wrote the manuscript with comments from the other authors. All authors read and approved the final manuscript.

Competing interests

The authors declare that they have no competing interests.

Author details

[1]Mater Research Institute - University of Queensland, TRI Building, Woolloongabba, QLD 4102, Australia. [2]BGI-Shenzhen, Shenzhen 518083, China. [3]Department of Biology and the Novo Nordisk Foundation Center for Basic Metabolic Research, University of Copenhagen, Copenhagen 1599, Denmark. [4]School of Chemistry and Molecular Biosciences, University of Queensland, Brisbane, QLD 4072, Australia. [5]Edinburgh Cancer Research Centre, IGMM, University of Edinburgh, Edinburgh EH42XR, UK. [6]Queensland Brain Institute, University of Queensland, Brisbane, QLD 4072, Australia.

References

1. Wen PY, Kesari S. Malignant gliomas in adults. N Engl J Med. 2008;359:492–507.
2. Bleeker FE, Molenaar RJ, Leenstra S. Recent advances in the molecular understanding of glioblastoma. J Neuro-Oncol. 2012;108:11–27.
3. Ohgaki H, Kleihues P. Genetic pathways to primary and secondary glioblastoma. Am J Pathol. 2007;170:1445–53.
4. Parsons DW, Jones S, Zhang X, Lin JC, Leary RJ, Angenendt P, Mankoo P, Carter H, Siu IM, Gallia GL, et al. An integrated genomic analysis of human glioblastoma multiforme. Science. 2008;321:1807–12.
5. Sottoriva A, Spiteri I, Piccirillo SG, Touloumis A, Collins VP, Marioni JC, Curtis C, Watts C, Tavare S. Intratumor heterogeneity in human glioblastoma reflects cancer evolutionary dynamics. Proc Natl Acad Sci U S A. 2013;110: 4009–14.
6. Brennan Cameron W, Verhaak Roel GW, McKenna A, Campos B, Noushmehr H, Salama Sofie R, Zheng S, Chakravarty D, Sanborn JZ, Berman Samuel H, et al. The somatic genomic landscape of glioblastoma. Cell. 2013;155:462–77.
7. Maher EA, Brennan C, Wen PY, Durso L, Ligon KL, Richardson A, Khatry D, Feng B, Sinha R, Louis DN, et al. Marked genomic differences characterize primary and secondary glioblastoma subtypes and identify two distinct molecular and clinical secondary glioblastoma entities. Cancer Res. 2006;66:11502–13.
8. Fischer U, Meese E. Glioblastoma multiforme: the role of DSB repair between genotype and phenotype. Oncogene. 2007;26:7809–15.
9. Miki Y, Nishisho I, Horii A, Miyoshi Y, Utsunomiya J, Kinzler KW, Vogelstein B, Nakamura Y. Disruption of the APC gene by a retrotransposal insertion of L1 sequence in a colon cancer. Cancer Res. 1992;52:643–5.
10. Grimaldi G, Skowronski J, Singer MF. Defining the beginning and end of KpnI family segments. EMBO J. 1984;3:1753–9.
11. Scott AF, Schmeckpeper BJ, Abdelrazik M, Comey CT, O'Hara B, Rossiter JP, Cooley T, Heath P, Smith KD, Margolet L. Origin of the human L1 elements: proposed progenitor genes deduced from a consensus DNA sequence. Genomics. 1987;1:113–25.
12. Hohjoh H, Singer MF. Cytoplasmic ribonucleoprotein complexes containing human LINE-1 protein and RNA. EMBO J. 1996;15:630–9.

13. Kolosha VO, Martin SL. In vitro properties of the first ORF protein from mouse LINE-1 support its role in ribonucleoprotein particle formation during retrotransposition. Proc Natl Acad Sci U S A. 1997;94:10155–60.

14. Mathias SL, Scott AF, Kazazian Jr HH, Boeke JD, Gabriel A. Reverse transcriptase encoded by a human transposable element. Science. 1991;254:1808–10.

15. Feng Q, Moran JV, Kazazian Jr HH, Boeke JD. Human L1 retrotransposon encodes a conserved endonuclease required for retrotransposition. Cell. 1996;87:905–16.

16. Swergold GD. Identification, characterization, and cell specificity of a human LINE-1 promoter. Mol Cell Biol. 1990;10:6718–29.

17. Speek M. Antisense promoter of human L1 retrotransposon drives transcription of adjacent cellular genes. Mol Cell Biol. 2001;21:1973–85.

18. Denli AM, Narvaiza I, Kerman BE, Pena M, Benner C, Marchetto MC, Diedrich JK, Aslanian A, Ma J, Moresco JJ, et al. Primate-specific ORF0 contributes to retrotransposon-mediated diversity. Cell. 2015;163:583–93.

19. Luan DD, Korman MH, Jakubczak JL, Eickbush TH. Reverse transcription of R2Bm RNA is primed by a nick at the chromosomal target site: a mechanism for non-LTR retrotransposition. Cell. 1993;72:595–605.

20. Jurka J. Sequence patterns indicate an enzymatic involvement in integration of mammalian retroposons. Proc Natl Acad Sci U S A. 1997;94:1872–7.

21. Morrish TA, Garcia-Perez JL, Stamato TD, Taccioli GE, Sekiguchi J, Moran JV. Endonuclease-independent LINE-1 retrotransposition at mammalian telomeres. Nature. 2007;446:208–12.

22. Symer DE, Connelly C, Szak ST, Caputo EM, Cost GJ, Parmigiani G, Boeke JD. Human l1 retrotransposition is associated with genetic instability in vivo. Cell. 2002;110:327–38.

23. Morrish TA, Gilbert N, Myers JS, Vincent BJ, Stamato TD, Taccioli GE, Batzer MA, Moran JV. DNA repair mediated by endonuclease-independent LINE-1 retrotransposition. Nat Genet. 2002;31:159–65.

24. Sen SK, Huang CT, Han K, Batzer MA. Endonuclease-independent insertion provides an alternative pathway for L1 retrotransposition in the human genome. Nucleic Acids Res. 2007;35:3741–51.

25. Ahl V, Keller H, Schmidt S, Weichenrieder O. Retrotransposition and crystal structure of an Alu RNP in the ribosome-stalling conformation. Mol Cell. 2015;60:715–27.

26. Dewannieux M, Esnault C, Heidmann T. LINE-mediated retrotransposition of marked Alu sequences. Nat Genet. 2003;35:41–8.

27. Doucet AJ, Wilusz JE, Miyoshi T, Liu Y, Moran JV. A 3' Poly(A) tract is required for LINE-1 retrotransposition. Mol Cell. 2015;60:728–41.

28. Han K, Lee J, Meyer TJ, Remedios P, Goodwin L, Batzer MA. L1 recombination-associated deletions generate human genomic variation. Proc Natl Acad Sci U S A. 2008;105:19366–71.

29. Sen SK, Han K, Wang J, Lee J, Wang H, Callinan PA, Dyer M, Cordaux R, Liang P, Batzer MA. Human genomic deletions mediated by recombination between Alu elements. Am J Hum Genet. 2006;79:41–53.

30. Iskow RC, McCabe MT, Mills RE, Torene S, Pittard WS, Neuwald AF, Van Meir EG, Vertino PM, Devine SE. Natural mutagenesis of human genomes by endogenous retrotransposons. Cell. 2010;141:1253–61.

31. Tubio JM, Li Y, Ju YS, Martincorena I, Cooke SL, Tojo M, Gundem G, Pipinikas CP, Zamora J, Raine K, et al. Mobile DNA in cancer. Extensive transduction of nonrepetitive DNA mediated by L1 retrotransposition in cancer genomes. Science. 2014;345:1251343.

32. Lee E, Iskow R, Yang L, Gokcumen O, Haseley P, Luquette 3rd LJ, Lohr JG, Harris CC, Ding L, Wilson RK, et al. Landscape of somatic retrotransposition in human cancers. Science. 2012;337:967–71.

33. Solyom S, Ewing AD, Rahrmann EP, Doucet T, Nelson HH, Burns MB, Harris RS, Sigmon DF, Casella A, Erlanger B, et al. Extensive somatic L1 retrotransposition in colorectal tumors. Genome Res. 2012;22:2328–38.

34. Shukla R, Upton KR, Munoz-Lopez M, Gerhardt DJ, Fisher ME, Nguyen T, Brennan PM, Baillie JK, Collino A, Ghisletti S, et al. Endogenous retrotransposition activates oncogenic pathways in hepatocellular carcinoma. Cell. 2013;153:101–11.

35. Helman E, Lawrence ML, Stewart C, Sougnez C, Getz G, Meyerson M. Somatic retrotransposition in human cancer revealed by whole-genome and exome sequencing. Genome Res. 2014;24(7):1053–63.

36. Ewing AD, Gacita A, Wood LD, Ma F, Xing D, Kim MS, Manda SS, Abril G, Pereira G, Makohon-Moore A, et al. Widespread somatic L1 retrotransposition occurs early during gastrointestinal cancer evolution. Genome Res. 2015;25:1536–45.

37. Rodic N, Steranka JP, Makohon-Moore A, Moyer A, Shen P, Sharma R, Kohutek ZA, Huang CR, Ahn D, Mita P, et al. Retrotransposon insertions in the clonal evolution of pancreatic ductal adenocarcinoma. Nat Med. 2015;21:1060–4.

38. Doucet-O'Hare TT, Rodić N, Sharma R, Darbari I, Abril G, Choi JA, Young Ahn J, Cheng Y, Anders RA, Burns KH, et al. LINE-1 expression and retrotransposition in Barrett's esophagus and esophageal carcinoma. Proc Natl Acad Sci. 2015;112:E4894–900.

39. Baillie JK, Barnett MW, Upton KR, Gerhardt DJ, Richmond TA, De Sapio F, Brennan PM, Rizzu P, Smith S, Fell M, et al. Somatic retrotransposition alters the genetic landscape of the human brain. Nature. 2011;479:534–7.

40. Coufal NG, Garcia-Perez JL, Peng GE, Yeo GW, Mu Y, Lovci MT, Morell M, O'Shea KS, Moran JV, Gage FH. L1 retrotransposition in human neural progenitor cells. Nature. 2009;460:1127–31.

41. Evrony GD, Cai X, Lee E, Hills LB, Elhosary PC, Lehmann HS, Parker JJ, Atabay KD, Gilmore EC, Poduri A, et al. Single-neuron sequencing analysis of L1 retrotransposition and somatic mutation in the human brain. Cell. 2012;151:483–96.

42. Upton KR, Gerhardt DJ, Jesuadian JS, Richardson SR, Sanchez-Luque FJ, Bodea GO, Ewing AD, Salvador-Palomeque C, van der Knaap MS, Brennan PM, et al. Ubiquitous L1 mosaicism in hippocampal neurons. Cell. 2015;161: 228–39.

43. Moran JV, Holmes SE, Naas TP, DeBerardinis RJ, Boeke JD, Kazazian Jr HH. High frequency retrotransposition in cultured mammalian cells. Cell. 1996; 87:917–27.

44. Dombroski BA, Mathias SL, Nanthakumar E, Scott AF, Kazazian Jr HH. Isolation of an active human transposable element. Science. 1991;254:1805–8.

45. Li H, Handsaker B, Wysoker A, Fennell T, Ruan J, Homer N, Marth G, Abecasis G, Durbin R. The Sequence Alignment/Map format and SAMtools. Bioinformatics. 2009;25:2078–9.

46. Kent WJ. BLAT–the BLAST-like alignment tool. Genome Res. 2002;12:656–64.

47. Beck CR, Collier P, Macfarlane C, Malig M, Kidd JM, Eichler EE, Badge RM, Moran JV. LINE-1 retrotransposition activity in human genomes. Cell. 2010; 141:1159–70.

48. Ewing AD, Kazazian Jr HH. High-throughput sequencing reveals extensive variation in human-specific L1 content in individual human genomes. Genome Res. 2010;20:1262–70.

49. Ewing AD, Kazazian Jr HH. Whole-genome resequencing allows detection of many rare LINE-1 insertion alleles in humans. Genome Res. 2011;21:985–90.

50. Kuhn A, Ong YM, Cheng CY, Wong TY, Quake SR, Burkholder WF. Linkage disequilibrium and signatures of positive selection around LINE-1 retrotransposons in the human genome. Proc Natl Acad Sci U S A. 2014;111:8131–6.

51. Wang J, Song L, Grover D, Azrak S, Batzer MA, Liang P. dbRIP: a highly integrated database of retrotransposon insertion polymorphisms in humans. Hum Mutat. 2006;27:323–9.

52. Rozen S, Skaletsky H. Primer3 on the WWW for general users and for biologist programmers. Methods Mol Biol. 2000;132:365–86.

53. Magoc T, Salzberg SL. FLASH: fast length adjustment of short reads to improve genome assemblies. Bioinformatics. 2011;27:2957–63.

54. Saunders CT, Wong WS, Swamy S, Becq J, Murray LJ, Cheetham RK. Strelka: accurate somatic small-variant calling from sequenced tumor-normal sample pairs. Bioinformatics. 2012;28:1811–7.

55. Rimmer A, Phan H, Mathieson I, Iqbal Z, Twigg SR, Wilkie AO, McVean G, Lunter G. Integrating mapping-, assembly- and haplotype-based approaches for calling variants in clinical sequencing applications. Nat Genet. 2014;46: 912–8.

56. Rausch T, Zichner T, Schlattl A, Stutz AM, Benes V, Korbel JO. DELLY: structural variant discovery by integrated paired-end and split-read analysis. Bioinformatics. 2012;28:i333–9.

57. Chen X, Schulz-Trieglaff O, Shaw R, Barnes B, Schlesinger F, Kallberg M, Cox AJ, Kruglyak S, Saunders CT. Manta: Rapid detection of structural variants and indels for germline and cancer sequencing applications. Bioinformatics. 2016;32(8):1220–2.

58. Klambauer G, Schwarzbauer K, Mayr A, Clevert DA, Mitterecker A, Bodenhofer U, Hochreiter S. cn.MOPS: mixture of Poissons for discovering copy number variations in next-generation sequencing data with a low false discovery rate. Nucleic Acids Res. 2012;40:e69.

59. Wei W, Morrish TA, Alisch RS, Moran JV. A transient assay reveals that cultured human cells can accommodate multiple LINE-1 retrotransposition events. Anal Biochem. 2000;284:435–8.

60. Kopera HC, Moldovan JB, Morrish TA, Garcia-Perez JL, Moran JV. Similarities between long interspersed element-1 (LINE-1) reverse transcriptase and telomerase. Proc Natl Acad Sci U S A. 2011;108:20345–50.

61. Dombroski BA, Scott AF, Kazazian Jr HH. Two additional potential

retrotransposons isolated from a human L1 subfamily that contains an active retrotransposable element. Proc Natl Acad Sci U S A. 1993;90:6513–7.

62. Mills RE, Bennett EA, Iskow RC, Devine SE. Which transposable elements are active in the human genome? Trends Genet. 2007;23:183–91.

63. Guy J, Hendrich B, Holmes M, Martin JE, Bird A. A mouse Mecp2-null mutation causes neurological symptoms that mimic Rett syndrome. Nat Genet. 2001;27:322–6.

64. Muotri AR, Marchetto MC, Coufal NG, Oefner R, Yeo G, Nakashima K, Gage FH. L1 retrotransposition in neurons is modulated by MeCP2. Nature. 2010; 468:443–6.

65. Han JS, Szak ST, Boeke JD. Transcriptional disruption by the L1 retrotransposon and implications for mammalian transcriptomes. Nature. 2004;429:268–74.

66. Sanborn JZ, Salama SR, Grifford M, Brennan CW, Mikkelsen T, Jhanwar S, Katzman S, Chin L, Haussler D. Double minute chromosomes in glioblastoma multiforme are revealed by precise reconstruction of oncogenic amplicons. Cancer Res. 2013;73:6036–45.

67. Joensuu H, Puputti M, Sihto H, Tynninen O, Nupponen NN. Amplification of genes encoding KIT, PDGFRalpha and VEGFR2 receptor tyrosine kinases is frequent in glioblastoma multiforme. J Pathol. 2005;207:224–31.

68. Carvalho S, Vitor AC, Sridhara SC, Martins FB, Raposo AC, Desterro JM, Ferreira J, de Almeida SF. SETD2 is required for DNA double-strand break repair and activation of the p53-mediated checkpoint. Elife. 2014;3:e02482.

69. Pfister SX, Ahrabi S, Zalmas LP, Sarkar S, Aymard F, Bachrati CZ, Helleday T, Legube G, La Thangue NB, Porter AC, Humphrey TC. SETD2-dependent histone H3K36 trimethylation is required for homologous recombination repair and genome stability. Cell Rep. 2014;7:2006–18.

70. Sassaman DM, Dombroski BA, Moran JV, Kimberland ML, Naas TP, DeBerardinis RJ, Gabriel A, Swergold GD, Kazazian Jr HH. Many human L1 elements are capable of retrotransposition. Nat Genet. 1997;16:37–43.

71. Brouha B, Schustak J, Badge RM, Lutz-Prigge S, Farley AH, Moran JV, Kazazian Jr HH. Hot L1s account for the bulk of retrotransposition in the human population. Proc Natl Acad Sci U S A. 2003;100:5280–5.

72. Gilbert N, Lutz S, Morrish TA, Moran JV. Multiple fates of L1 retrotransposition intermediates in cultured human cells. Mol Cell Biol. 2005; 25:7780–95.

73. Gilbert N, Lutz-Prigge S, Moran JV. Genomic deletions created upon LINE-1 retrotransposition. Cell. 2002;110:315–25.

74. Morse B, Rotherg PG, South VJ, Spandorfer JM, Astrin SM. Insertional mutagenesis of the myc locus by a LINE-1 sequence in a human breast carcinoma. Nature. 1988;333:87–90.

75. McVey M, Lee SE. MMEJ repair of double-strand breaks (director's cut): deleted sequences and alternative endings. Trends Genet. 2008;24:529–38.

76. Goodier JL, Cheung LE, Kazazian Jr HH. MOV10 RNA helicase is a potent inhibitor of retrotransposition in cells. PLoS Genet. 2012;8:e1002941.

77. Short SC, Martindale C, Bourne S, Brand G, Woodcock M, Johnston P. DNA repair after irradiation in glioma cells and normal human astrocytes. Neuro Oncol. 2007;9:404–11.

4

Somatic retrotransposition is infrequent in glioblastomas

Pragathi Achanta[1†], Jared P. Steranka[2,3†], Zuojian Tang[4,5], Nemanja Rodić[2,7], Reema Sharma[2], Wan Rou Yang[2], Sisi Ma[4], Mark Grivainis[4,5], Cheng Ran Lisa Huang[2,8], Anna M. Schneider[2,9], Gary L. Gallia[6], Gregory J. Riggins[6], Alfredo Quinones-Hinojosa[6,10], David Fenyö[4,5], Jef D. Boeke[5*] and Kathleen H. Burns[2,3*]

Abstract

Background: Gliomas are the most common primary brain tumors in adults. We sought to understand the roles of endogenous transposable elements in these malignancies by identifying evidence of somatic retrotransposition in glioblastomas (GBM). We performed transposon insertion profiling of the active subfamily of Long INterspersed Element-1 (LINE-1) elements by deep sequencing (TIPseq) on genomic DNA of low passage oncosphere cell lines derived from 7 primary GBM biopsies, 3 secondary GBM tissue samples, and matched normal intravenous blood samples from the same individuals.

Results: We found and PCR validated one somatically acquired tumor-specific insertion in a case of secondary GBM. No LINE-1 insertions present in primary GBM oncosphere cultures were missing from corresponding blood samples. However, several copies of the element (11) were found in genomic DNA from blood and not in the oncosphere cultures. SNP 6.0 microarray analysis revealed deletions or loss of heterozygosity in the tumor genomes over the intervals corresponding to these LINE-1 insertions.

Conclusions: These findings indicate that LINE-1 retrotransposon can act as an infrequent insertional mutagen in secondary GBM, but that retrotransposition is uncommon in these central nervous system tumors as compared to other neoplasias.

Keywords: LINE-1, Retrotransposition, Cancer, Glioblastoma

Background

Glioblastomas (GBMs) are the most common malignant form of primary brain tumor in adults, and are typically fatal. These are histologically aggressive gliomas, categorized by the World Health Organization (WHO) as grade IV astrocytomas; they are hypercellular with frequent mitotic figures, vascular proliferation and pseudopalisading necrosis. Although morphologically indistinguishable, distinct primary and secondary types of GBM are recognized clinically. Primary GBMs arise *de novo*, and usually present as advanced cancers in patients over 50 years old.

These are characterized genetically by amplification of epidermal growth factor receptor (*EGFR*), loss of heterozygosity (LOH) on chromosomes 10q and 17p, and phosphatase and tensin homologue (*PTEN*) mutation. Secondary GBMs arise from preexisting low-grade tumors over a period of a few years and are more common among younger patients. This class of tumors is characterized by mutations in isocitrate dehydrogenase 1 (*IDH1*) and p53 tumor suppressor genes as well *PDGFA* amplification [1, 2].

Activation of endogenous transposable elements as a mechanism of mutagenesis is being increasingly recognized in human tumors. Retrotransposons are a class of mobile genetic elements that use a 'copy and paste' mechanism to replicate in the genome through RNA intermediates. Among these, the autonomous Long INterspersed Element-1 (LINE-1 or L1s) are the most active elements in humans [3]. Recently, methods have

* Correspondence: jef.boeke@nyumc.org; kburns@jhmi.edu
†Equal contributors
5Institute for Systems Genetics, New York University Langone Medical Center, ACLSW Room 503, 430 East 29th Street, New York, NY 10016, USA
2Department of Pathology, Johns Hopkins University School of Medicine, Miller Research Building (MRB) Room 447, 733 North Broadway, Baltimore, MD 21205, USA
Full list of author information is available at the end of the article

been developed to identify LINE-1 sequences in human genomes that collectively underscore their ongoing potential for retrotransposition in the germline [4–9] and in malignancy [6, 10–18]. Recent studies have also implicated LINE-1 expression and activity in normal brain and in brain malformations and disease [19–22].

In this study, we mapped LINE-1 insertion sites in GBMs and matched blood samples using a targeted sequencing approach, Transposon Insertion Profiling (TIPseq). We profiled oncosphere cell lines derived from primary GBMs as compared to matched normal genomic DNA from the same patients [23]. We also used TIPseq to compare genomic DNA isolates from primary and secondary GBMs and from normal blood DNA from the same patients.

Methods

Consent statement

Blood and brain tumor tissue samples were obtained from glioma patients who underwent surgery at the Johns Hopkins Hospital under the approval of the Institutional Review Board (IRB) and with consent. This study included 7 primary GBM and 3 secondary GBM patients.

Oncosphere cell cultures from primary glioblastoma tissue

Fresh primary glioblastoma tissue was dissociated enzymatically using TrypLE (Gibco). The homogenized tissue was passed through a narrow fire-polished Pasteur pipette and 40 μm cell strainer to obtain single cell suspension. Primary cells were then plated at a density of 1×10^5 viable cells in 25-cm^2 non-adherent flasks in DMEM/F12 medium supplemented with 20 ng/mL of human epidermal growth factor (EGF), and 10 ng/mL of human fibroblast growth factor (FGF). Oncospheres of approximately 100 μm were passaged and replated.

Genomic DNA preparation and Vectorette PCRs

Genomic DNA from peripheral blood samples was isolated using the QIAamp DNA blood mini kit (Qiagen). Genomic DNA from tumor tissue and oncospheres was isolated by Trizol homogenization, phenol-chloroform-iso amyl alcohol extraction and ethanol precipitation. Aliquots of ~0.5–2 μg of genomic DNA from each sample were digested individually with six different restriction enzymes (*AseI, BspHI, BstYI, HindIII, NcoI, PstI*) generating fragments averaging 1–3 kb in length. Vectorette matched with restriction enzyme sticky-end sites were designed and ligated to the digested DNA fragments. Vectorette PCR was performed using *ExTaq* HS polymerase (Takara Bio) and a touch-down PCR program to generate amplicons spanning the transposon insertion end and the flanking unique genomic sequences [24, 25].

Deep sequencing DNA libraries and quality control

Vectorette PCR products from each patient sample were pooled, purified and fragmented to an average length of 300 bp using a Covaris E210. TruSeq DNA Sample Preparation kit v2 (Illumina) was used for end-repair, A-tailing, index-specific adapter ligation and PCR enrichment. We size-selected our DNA fragments at ~450 bp using 2 % Size-Select E-gels (Life Technologies) prior to PCR. The enriched PCR products were purified and checked for quality control using an Agilent Bioanalyzer. The DNA libraries were pooled and submitted for single-end or paired-end deep sequencing with Illumina HiSeq 2000 platform either at Johns Hopkins high-throughput sequencing center or the HudsonAlpha Institute for Biotechnology (HudsonAlpha, Huntsville, AL). The sequencing batch, facility, indexes, barcodes and read lengths for each sample are provided in Additional file 1: Table S1.

Computational analysis

Two analytical approaches were used. In the first, all trimmed reads (75–100 bp) were first aligned to the human reference genome (hg18) using Bowtie [26], and cisGenome was used for identifying peaks. The peaks were ranked based on the maximum base pair read coverage. Unmappable reads from the Bowtie alignment were used to identify the junctional reads. 35 bp from each of the 5′ and 3′ end of the unmappable reads were trimmed and aligned with the human reference genome. Reads aligning uniquely with only one end to the genome were extracted and grouped together according to the peak list using SAMtools. Those with at least six consecutive As or Ts were used for further analysis to enrich for transposon junctions. A maximum of 200 such junction reads per peak were used to generate the consensus sequences using multiple sequence alignment (MSA) and the bioperl AlignIO module. BLAT was used to compare each consensus sequence to both the hg18 reference genome and a 3′ LINE-1 sequence with polyA tail. Galaxy was used to identify lists of putative insertions occurring in *either* a blood or tumor sample for an individual and not both. The Integrative Genomics Viewer (IGV) was used to visualize the read alignments to the reference genome.

For secondary GBM TIPseq samples, a second machine learning algorithm analysis was also conducted using paired-end read samples. Low quality sequences, base pairs, and vectorette sequences were trimmed using Trimmomatic software [27]. Qualified read pairs were aligned to an L1Hs-masked reference genome (hg19) and the L1Hs consensus sequence using Bowtie2 software. Candidate insertion sites were identified as peaks with at least one junction-containing read pair. The machine learning model was trained on known LINE-1 insertions using five sequencing features, namely the peak width and depth; variant index for reads mapping in the peak interval; the polyA tail purity; and the number of junction reads. The trained model was used to predict probabilities of the candidate insertions being the true insertion sites. This pipeline, TIPseqHunter, will be reported in more detail elsewhere (Tang, Z., et al. in review).

PCR validations and Sanger sequencing
Primers were designed to flank putative LINE-1 insertion sites using Primer3 software. PCR products with insertions were cut out of the gel and DNA was extracted using a QIAquick Gel extraction kit (Qiagen). The purified PCR products were then sent for Sanger sequencing to obtain the 5′ and 3′ junction coordinates, length and orientation of the inserted L1, and target site duplication.

Copy number variation (CNV) and loss of heterozygosity (LOH)
Genomic DNA samples extracted from primary glioblastoma oncosphere lines were run on an Affymetrix Genome-wide Human Single Nucleotide Polymorphism (SNP) 6.0 Array in the Johns Hopkins University School of Medicine High Throughput Biology Center Microarray Core facility. The fluorescent intensity values were used by CNViewer and Partek software for copy number variant (CNV) and loss of heterozygosity (LOH) analysis.

Immunohistochemistry
Formalin-fixed, paraffin-embedded primary glioblastoma tissue samples were analyzed for endogenous L1 ORF1p expression as previously described [28, 29]. Briefly, 5-μm thick sections were deparaffinized and hydrated by baking at 65 °C for 20 min and then with xylene and ethanol washes. Sections were heated at 98 °C in citrate buffer for 20 min for antigen retrieval. Sections were blocked at room temperature for 10 min and then incubated with primary monoclonal mouse L1-ORF1p antibody (1:1000 dilution) in Tris-buffered saline (TBS), pH 7.2 with Tween 20 and 1%BSA overnight at 4 °C. Sections were washed with TBS and incubated with biotin-conjugated anti-mouse IgG for 10 min at room temperature. Sections were developed with 3,3′-Diaminobenzidine (DAB) chromagen mix, counter stained with hematoxylin, dehydrated and coverslipped.

Results

Transposon insertion profiling
We obtained seven primary GBM samples from patients 50–65 years old (average age 57) as well as peripheral blood draws from the same individuals. We established oncosphere cultures to expand the tumor cells and

extracted high molecular weight genomic DNA from both the oncospheres and peripheral blood mononuclear cells. Primary tissue directly from these resections was available in sufficient quantity to assay directly for four cases. We digested these samples with restriction enzymes, ligated vectorette oligonucleotides to their ends, and selectively amplified LINE-1 genomic insertion sites using vectorette PCR as previously described.

Vectorette PCR is a ligation mediated PCR that allows for the amplification of unknown sequence downstream of a sequence of interest. In this case, the PCR recovers L1Hs or L1(Ta) insertions and the genomic sequence immediately downstream. L1Hs are known to be the most active LINE-1 in modern humans [4]; they are responsible for the most variation in human populations as well as the largest proportion of *de novo* and somatic LINE-1 insertions [3]. The specificity of this PCR for L1Hs is imposed by the position of one of the amplification primers located in the 3′UTR of the LINE-1 elements. The primer positioning is also advantageous because it allows for the recovery of insertions that are severely 5′ truncated, a common feature of LINE-1 insertions.

In addition to the primary GBM oncosphere cultures, we also acquired three secondary GBM samples and matching blood samples from patients 33–53 years old (average age 44). These tumors are less readily expanded in vitro, so tissue from the resected tumors was used directly to make genomic DNA for TIPseq in each case. Figure 1 illustrates the TIPseq workflow and Additional file 1: Table S1 summarizes the sequencing batches, TIPseq library indices, and total reads obtained from each sample for this study.

Germline LINE-1 insertions not found in oncosphere cultures
We sequenced 14 TIPseq amplicons preparations from primary GBM oncosphere cultures and matching peripheral blood samples from 7 patients in 3 different sequencing batches. We compared TIPseq profiles from blood and oncosphere samples for each individual, and identified numerous putative LINE-1 insertions in the blood samples that were absent from the corresponding oncosphere cultures. We validated 11 of these by PCR; in 8 we were also able to Sanger sequence the LINE-1 insertion and report both the 5′ and 3′ ends and the target site duplication (Tables 1 and 2, Fig. 2a). Instances of this occurred in six of our seven cases. The size of these LINE-1 elements ranged from full-length insertions of 6059 bp to 5′ truncated insertions as short as 684 bp. Several of these were in gene introns, namely guanylate cyclase 1, soluble, beta 2 pseudogene (*GUCY1B2*), neuronal PAS domain 3 (*NPAS3*), the uncharacterized KIAA159-like gene (KIAA1549L), sterile alpha motif

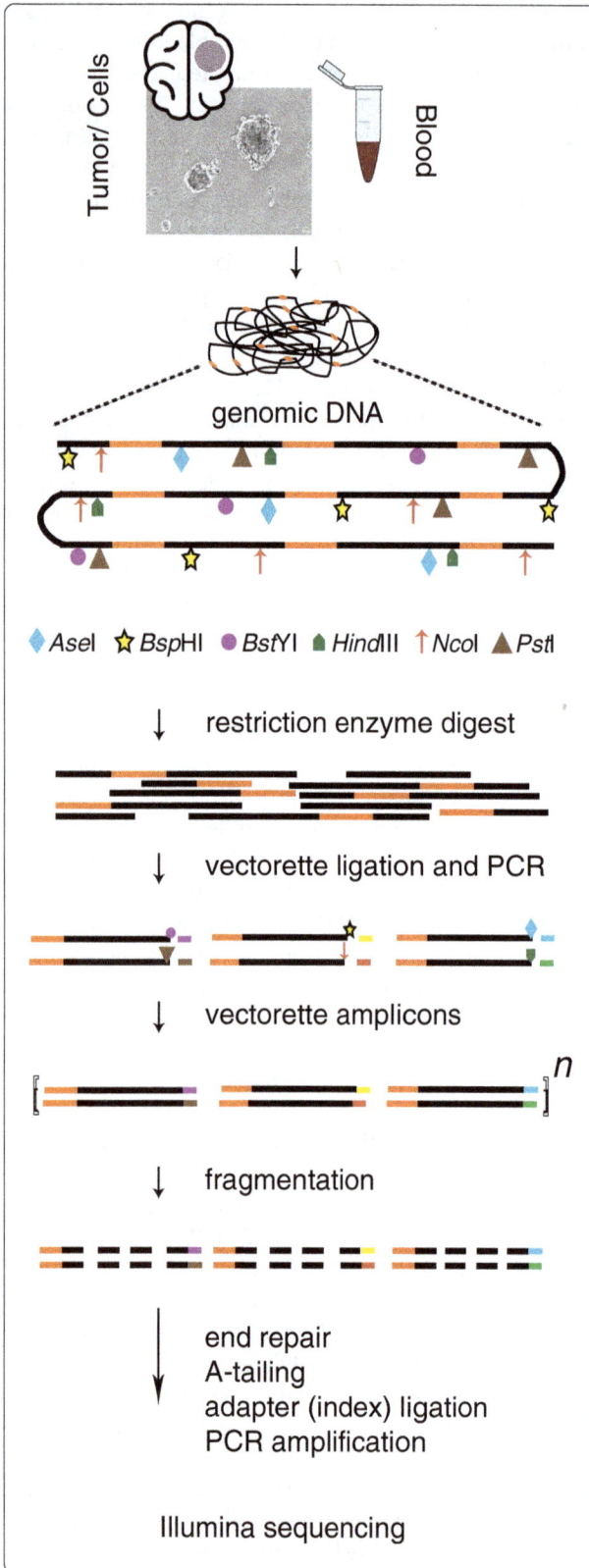

Fig. 1 Transposon Insertion Profiling by sequencing (TIPseq) workflow. High molecular weight genomic DNA was extracted from primary and secondary glioblastoma (GBM) tumors, oncosphere cultures expanded from primary GBM, and matched blood samples from the same patients. Genomic DNA was then digested in six parallel reactions each using one of a panel of restriction enzymes. In the diagram, LINE-1 insertions are depicted as orange segments of the genomic DNA; restriction enzyme cuts sites are illustrated with different symbols. Vectorette oligonucleotides designed to match each restriction enzyme sticky end were ligated to the DNA fragments, and the 3′ ends of LINE-1 sequences and downstream DNA were amplified. Genomic DNA fragments without binding sites for the LINE-1 amplification primer are not enriched in this PCR. Amplified DNA was then randomly sheared and prepared for Illumina sequencing

domain 12 antisense RNA-1 (*SAMD12-AS1*), and sec1 family domain 1 (*SCFD1*). None were in exons.

One of these insertions, the LINE-1 at the KIAA1549L locus, was found in two unrelated patient blood samples, although it was absent from the corresponding oncosphere cell lines. Several others, including the insertions at *GUCY1B2*, *NPAS3*, *SAMD12-AS1* and *SCFD1* loci had been previously reported to be polymorphic LINE-1 insertions [30]. Knowing that these LINE-1 insertions are segregating in human populations, we could not attribute the discordance between blood and oncosphere cell culture genotypes to somatically acquired insertions resulting in mosaicism in these patients. Rather, it seemed likely that these represented polymorphic and heterozygous germline copies that were lost in the glioblastoma, e.g. by loss of heterozygosity.

Table 1 Candidate L1 insertions tested

Sample ID	Tumor	Blood
Primary glioblastoma		
714	0/19	0/6
750	0/18	**2**/9
772	0/24	**1**/11
832	0/9	**1**/19
847	0/8	**2**/6
897	0/2	**4**/23
922	0/11	**1**/7
Secondary glioblastoma		
007	0/93 + 0/13	0/0
023	0/25	0/0
083	**1**/44	0/0

Number of candidate L1 insertions detected in either tumor or peripheral blood DNA and tested by PCR from putative insertions in primary glioblastoma cell cultures, secondary glioblastoma tissue and matched blood samples
Numerators indicate numbers of productive PCR reactions. Denominators indicate candidate loci tested. Non-zero numerators are in bold

Table 2 PCR validated L1 insertions

Sample ID	Chromosome	5′ junction coordinate	3′ junction coordinate	L1 strand	Length of inserted L1	Length of TSD	Gene name	CN	LOH
Blood specific insertions									
750	5	.	34495746	+	.	.	.	1	Y
750	13	50488000	50487984	+	4162 bp	17 bp	GUCY1B2	1	Y
772	14	33209876	33209888	-	5422 bp	13 bp	NPAS3	1	Y
832	10	31557469	31557476	-	6059 bp	8 bp	.	1	Y
847	11	33626714	33626699	+	1281 bp	16 bp	KIAA1549L	1	Y
847	11	.	94317493	-	.	.	.	1	Y
897	8	24982467	24982461	-	2845 bp	7 bp	.	2	Y
897	8	119790870	119790884	-	965 bp	15 bp	SAMD12AS1	2	Y
897	13	85238154	85238150	-	684 bp	5 bp	.	1	Y
897	14	.	30220577	-	.	.	SCFD1	2	Y
922	11	33626714	33626699	+	1281 bp	16 bp	KIAA1549L	1	Y
Tumor specific insertions									
083	17	47881841	47881832	+	1839 bp	10 bp	.	NT	NT

List of L1 insertions validated by PCR in glioblastoma patients' samples

TSD target site duplication, *CN* copy number, *LOH* loss of heterozygosity, *NT* not tested

To distinguish between sample contamination and the possibility that these germline insertions had been lost in the GBM cell lines, we assessed copy number and heterozygosity in the oncosphere cell lines using Affymetrix SNP 6.0 microarrays. All LINE-1 insertions that we detected uniquely in blood were indeed located in regions where the corresponding oncosphere cell lines showed loss of heterozygosity (LOH). LOH was seen at some loci with maintenance of copy number (2n) and was seen at others where deletions resulted in reduced copy number (1n) (Table 2, Fig. 2b). We conclude that our findings reflect deletions of genomic LINE-1 associated with LOH events that had occurred either in the primary GBM or *in vitro* as oncosphere cultures of these tumors were established.

No tumor-specific LINE-1 insertions in primary GBM

We found no evidence of somatically acquired, tumor-specific insertions in primary GBM samples. We compared the seven TIPseq profiles from oncosphere cultures to those from matched peripheral blood gDNA. Additionally, for four of these samples, we had sufficient primary resected tumor to conduct tumor:normal comparisons without *in vitro* GBM expansion. We used an analytical pipeline that ranks peaks by numbers of contributing reads and manually reviewed hundreds of candidates. Although most lacked evidence of junction reads (*i.e.*, reads spanning the 3′ end of the LINE-1 and adjacent, unique genomic sequence), we tested a total of 91 putative tumor-specific insertions in spanning PCR assays and could not detect any with a LINE-1 insertion (Table 1). This

is the same approach we used to identify somatically acquired insertions in pancreatic ductal adenocarcinomas and malignancies of the nearby tubular gastrointestinal tract [13].

Infrequent tumor-specific LINE-1 insertions in secondary GBM

LINE-1 sequences code for two proteins essential for their retrotransposition; these are termed open reading frame 1 and open reading frame 2 proteins (ORF1p and ORF2p). We previously reported that about 33 % of cases in a tissue microarray collection of glioblastomas (GBM) express LINE-1 ORF1p [28]. This was higher than for low grade gliomas. When we distinguished between primary and secondary GBM cases in this study, it was clear that this frequency reflected the numbers of secondary GBM cases included in our survey. These secondary GBMs showed the greatest proportion of LINE-1 ORF1p positivity by immunohistochemistry (74 %, $n = 39$). Although many cases showed weak immunoreactivity, we viewed this as evidence that perhaps these secondary GBM tumors would be relatively permissive for retrotransposition (Fig. 3a-b).

To test secondary GBM for somatically acquired LINE-1 integrations, we obtained 3 tumor samples from neurosurgical resections and matched peripheral blood DNA from the same individuals. We analyzed these sequencing data using two approaches. For the first, we used the same strategy described above which ranks peaks on the basis of the numbers of contributing reads. We manually reviewed these and identified a total of 162 putative tumor specific somatic insertions carried

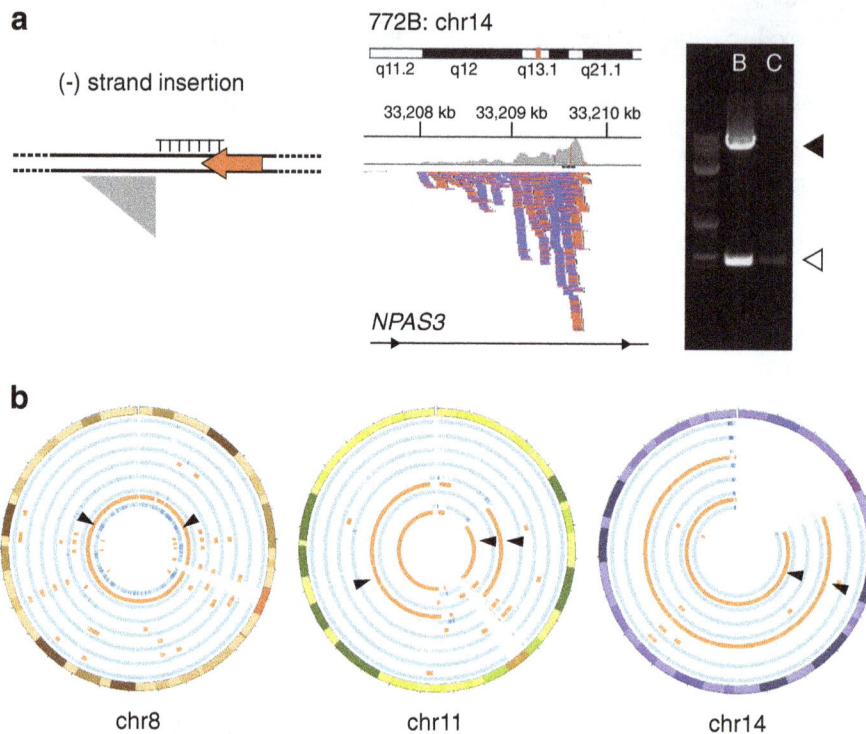

Fig. 2 TIPseq in primary glioblastoma GBM oncosphere lines and corresponding blood samples. **a**. TIPseq data. (Leftmost panel) The schematic depicts a minus (-) strand L1 as a leftward facing orange arrow. The LINE-1 sequence ends with a 3′ polyA tail, shown as a homopolymer of thymine (T) on the complementary strand. The gray right triangle illustrates the shape of sequencing reads piling up (*vertically, downward*) when mapped against in the reference genome (i.e., with genome coordinates depicted on the horizontal axis). (*Central panel*) TIPseq read alignments corresponding to an insertion detected in blood and not the patient's oncosphere cell line. The insertion is in an intron of the *NPAS3* gene (14q13.1). Read depth is illustrated on the top (*gray*) and individual reads are represented as blue and red bars denoting orientation. (*Rightmost panel*) An agarose gel electrophoresis of a validation PCR. The open arrowhead (*lower*) marks the pre-insertion allele and the solid arrowhead (upper) marks the amplicon spanning the LINE-1 insertion. The insertion is detected in the blood (B) sample for this patient and not the corresponding tumor cells (C). The LINE-1 is 5.4 kb. **b**. Copy number and loss of heterozygosity (LOH) studies on the oncosphere cell lines. Results for chromosomes 8, 11 and 14 are shown in Circos plots. The seven samples are each depicted as two circular tracks of data. The blue track indicates copy number; medium blue is diploid, darker blue shows amplifications, and lighter blue shows deletions. The orange track highlights regions with LOH. (Leftmost circle) Two insertions on chromosome 8 are marked with arrowheads at 25 and 120 MB; both were identified in a blood sample and not the corresponding patient's oncosphere cell line, which showed a copy neutral LOH of the entire chromosome (847). (*Central circle*) Three LINE-1 insertions on chromosome 11 found in blood only are marked; two are the same insertion at 33.6 MB found in two different patient samples (847, 922). Both tumor cell lines had deletions with copy number decreases and LOH at this site. One of these cases (847) also had loss of material near 94.3 MB associated with deletion of a second LINE-1. (*Rightmost circle*) Two LINE-1 are marked with arrowheads at 30.2 and 33.2 MB on acrocentric chromosome 14. These were found in genomic DNA from blood, but were lost owing to LOH in the corresponding oncosphere lines (897, 772)

forward to validation studies (Table 1). Although most lacked evidence of junction reads (*i.e.*, reads spanning the 3′ end of the LINE-1 and adjacent, unique genomic sequence), we tested these in spanning PCR assays.

We validated a single tumor-specific insertion at 17q22. Sanger sequencing showed that this insertion had all of the features of a LINE-1 retrotransposed by target primed reverse transcription (TPRT). The element has an intact 3′ polyA tail and 3′ LINE-1 sequence and is flanked by a 10 bp target site duplication. The insertion is 5′ truncated to a length of 1839 bp, which includes a 662 bp inversion of its 5′ end (Fig. 3c-f). These features are characteristic of

many somatically acquired LINE-1 insertions [13]. The insertion is intergenic, and to our knowledge, no heritable or somatically acquired element variants have been reported at this position.

To confirm this finding and more thoroughly review the entirety of these data, we also analyzed these sequences using a more advanced machine learning based approach. This algorithm combines five types of information at each locus to identify insertions. The pipeline imposes requirements for 3′ LINE-1 sequence and incorporates metrics to reflect the quantity, quality, and distribution of read alignments to the reference genome as well as measures of polyA tail purity and the numbers

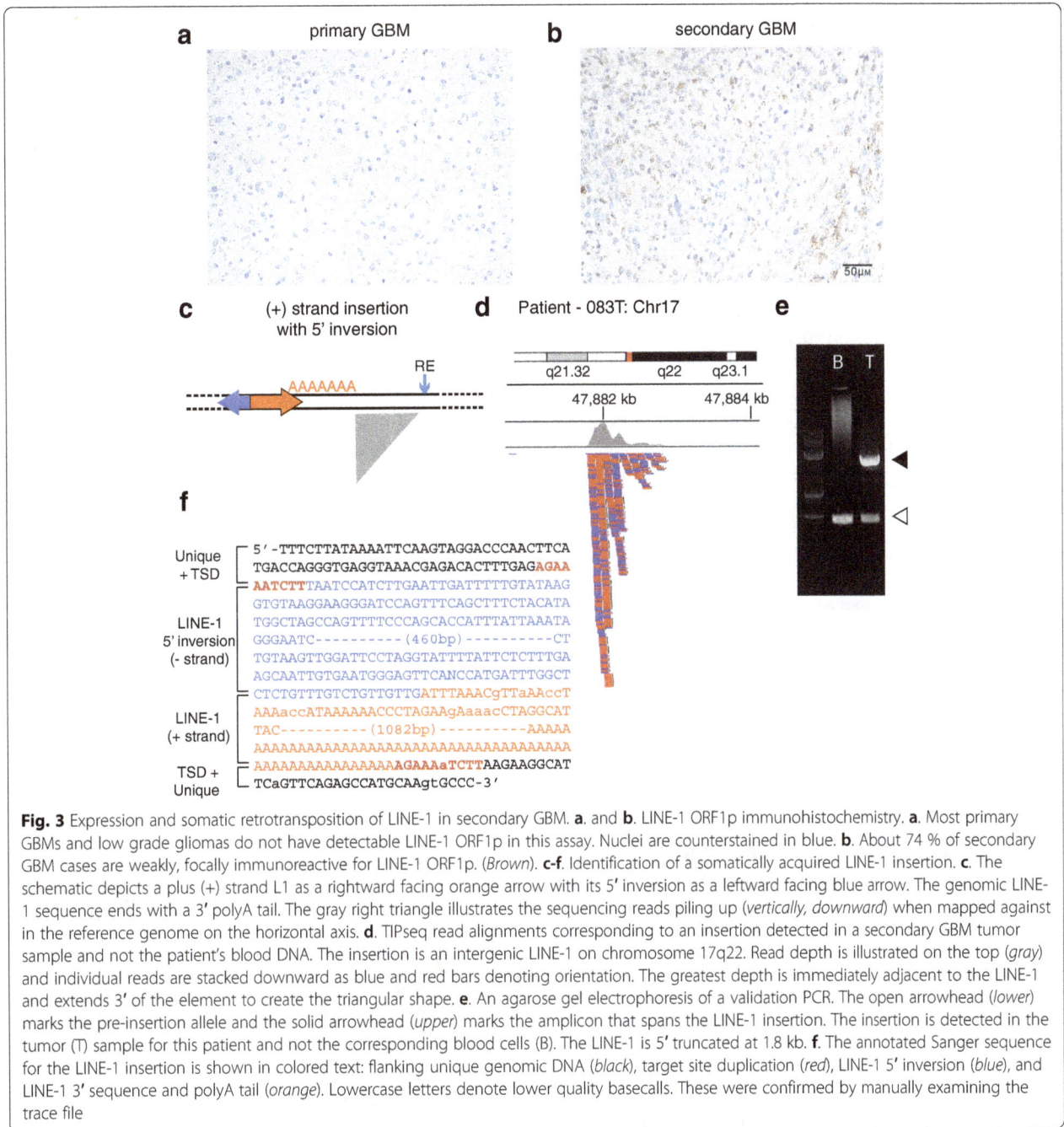

Fig. 3 Expression and somatic retrotransposition of LINE-1 in secondary GBM. **a.** and **b.** LINE-1 ORF1p immunohistochemistry. **a.** Most primary GBMs and low grade gliomas do not have detectable LINE-1 ORF1p in this assay. Nuclei are counterstained in blue. **b.** About 74 % of secondary GBM cases are weakly, focally immunoreactive for LINE-1 ORF1p. (*Brown*). **c-f.** Identification of a somatically acquired LINE-1 insertion. **c.** The schematic depicts a plus (+) strand L1 as a rightward facing orange arrow with its 5' inversion as a leftward facing blue arrow. The genomic LINE-1 sequence ends with a 3' polyA tail. The gray right triangle illustrates the sequencing reads piling up (*vertically, downward*) when mapped against in the reference genome on the horizontal axis. **d.** TIPseq read alignments corresponding to an insertion detected in a secondary GBM tumor sample and not the patient's blood DNA. The insertion is an intergenic LINE-1 on chromosome 17q22. Read depth is illustrated on the top (*gray*) and individual reads are stacked downward as blue and red bars denoting orientation. The greatest depth is immediately adjacent to the LINE-1 and extends 3' of the element to create the triangular shape. **e.** An agarose gel electrophoresis of a validation PCR. The open arrowhead (*lower*) marks the pre-insertion allele and the solid arrowhead (*upper*) marks the amplicon that spans the LINE-1 insertion. The insertion is detected in the tumor (T) sample for this patient and not the corresponding blood cells (B). The LINE-1 is 5' truncated at 1.8 kb. **f.** The annotated Sanger sequence for the LINE-1 insertion is shown in colored text: flanking unique genomic DNA (*black*), target site duplication (*red*), LINE-1 5' inversion (*blue*), and LINE-1 3' sequence and polyA tail (*orange*). Lowercase letters denote lower quality basecalls. These were confirmed by manually examining the trace file

of junction reads. It is trained on known LINE-1 insertions recovered within the same run of the same sample, and then used to predict other insertions. The outcome was 26 low probability insertions called in one sample (007); no predicted somatic insertions in the second sample (029); and only the chr17q22 insertion in the third sample (083). For samples 029 and 083, this outcome agreed perfectly with our previous PCR validations. In light of these new predictions for sample 007, we designed 13 additional pairs of spanning PCRs to test half of the 26 putative insertions. Gel electrophoresis of these PCR products provided no support for somatically acquired transposition events. All amplicons matched the predicted sizes for an empty site.

Thus, although we examined only a few cases of secondary GBM, our data suggest that somatic LINE-1 retrotransposition is not prominent in these malignancies.

Discussion

Somatic retrotransposition of LINE-1 leading to cancer was first described by Miki *et al.* in 1992 [31]. It would take more than 20 years to demonstrate the potential of next generation sequencing technologies to find such events [6]. Since that time, progress has been quick, with both targeted and whole genome sequencing demonstrating LINE-1 instability in a large number of human cancers. Chief among these has been the gastrointestinal tract tumors, including colon [10, 12, 15], esophagus [32, 33], hepatocellular carcinomas [11], and pancreatic ductal adenocarcinomas [13]. Lung and ovarian cancers also demonstrate LINE-1 retrotransposition [6, 14, 17, 18]. In contrast, surveys of selected hematolymphoid tumors and glioblastomas have indicated that these malignancies are not as prone to somatic LINE-1 reintegration [14].

To test this conclusion using a targeted sequencing approach, we profiled LINE-1 insertion sites in 10 cases of GBM. Our samples included 7 primary GBM cases; each was represented by an oncosphere cell line, and 4 primary GBM tissue biopsies were also assayed. We also profiled LINE-1 insertions in 3 secondary GBM tissue samples. In each case, normal brain tissue was not available for comparison, and peripheral blood draws were used to infer the normal genetic make-up of LINE-1 in each of these individuals.

We found no evidence of somatically acquired LINE-1 insertions in primary GBM cases in this study. We do not think that this is attributable to limitations of our assay. In previous studies, the same assay and analyses in our hands have been effective in detecting somatic LINE-1 insertion [13] (Zuojian Tang, et al. in review). Indeed, in this study, our approach was effective in identifying LINE-1 insertions deleted in loss-of-heterozygosity events, which effectively shows that we can detect elements present in one sample and absent from the other. Similar targeted sequencing studies, performed by an orthogonal method, reported in *Mobile DNA* by the Faulkner laboratory also reveal no canonical LINE-1 retrotransposition events (Carreira, et al. MDNA-D-16-00017).

Our work does suggest an interesting distinction between primary and secondary GBM. Unlike primary GBM, a majority of secondary GBM cases show some immunoreactivity for the LINE-1 encoded RNA binding protein ORF1p. Here, we also report finding a single, apparently somatically-acquired LINE-1 insertion in a case of secondary GBM. This insertion has several sequence features to indicate it resulted from a canonical, LINE-1 retrotransposition event. It is a 5′ truncated LINE-1 insertion, with a 5′ inversion and a 3′ polyA tail; the insertion is flanked by a short target site duplication.

When this LINE-1 was acquired is an open question. Our finding of increased ORF1p in secondary GBM implies that these tumors may provide a cellular context permissive for LINE-1 expression and retrotransposition that is unlike the normal adult brain or primary high grade gliomas. Although we favor this possibility, there is also evidence for somatic retrotransposition in the central nervous system as well as genetic variation within the brain reflecting retrotransposition events in early development. Since in all cases, we used blood as the germline comparison, genetic mosaicism antedating tumor initiation cannot be excluded. In either scenario, we presume that this LINE-1 integration has had no direct role in promoting tumor development in this case; it is an intergenic insertion several tens of kilobases away from the nearest gene and in a location with no recognized significance for the development of brain cancer.

Conclusions

Our findings indicate that LINE-1 retrotransposon events are infrequent in glioblastomas. While examples of driving mutations mediated by target-primed reverse transcription (TPRT) are being recognized in some types of malignancies, we expect these to be relatively uncommon in glioblastoma.

Abbreviations
CNV: Copy number variant; GBM: Glioblastoma; LINE-1 or L1: Long INterspersed element-1; LOH: Loss of heterozygosity; ORF1p: Open reading frame 1 protein; SNP: Single nucleotide polymorphism; TIPseq: Transposon insertion profiling by sequencing; TPRT: Target primed reverse transcription; UTR: Untranslated region

Acknowledgements
We thank Peilin Shen for technical assistance.

Funding
This work was supported by the American Brain Tumor Association (ABTA, K.H.B.); a Burroughs Wellcome Fund Career Award for Biomedical Scientists Program (K.H.B.); US National Institutes of Health awards R01CA161210 (J.D.B.), R01CA163705 (K.H.B.) and R01GM103999 (K.H.B.), R01NS070024 (A.Q.-H.), as well as the National Institute of General Medical Sciences Center for Systems Biology of Retrotransposition P50GM107632 (K.H.B. and J.D.B.).

Authors' contributions
PA and JPS conducted molecular biology experiments, interpreted results, and wrote the paper; ZT, CRLH, SM, MG and DF conducted analyses; NR and RS acquired tissues and conducted immunostaining; WRY edited the manuscript; GLG and GJR provided tissues and expertise on GBM; AMS, AQ-H, JDB and KHB conceived of the study, oversaw the project, wrote and edited the manuscript. All authors read and approved the final manuscript.

Competing interests
The authors declare that they have no competing interests.

Author details
[1]Molecular Biology and Genetics, Johns Hopkins University School of Medicine, Baltimore, MD, USA. [2]Department of Pathology, Johns Hopkins University School of Medicine, Miller Research Building (MRB) Room 447, 733 North Broadway, Baltimore, MD 21205, USA. [3]McKusick-Nathans Institute of Genetic Medicine, Johns Hopkins University School of Medicine, Miller Research Building (MRB) Room 447, 733 North Broadway, Baltimore, MD 21205, USA. [4]Center for Health Informatics and Bioinformatics, New York University Langone Medical Center, New York, NY, USA. [5]Institute for Systems Genetics, New York University Langone Medical Center, ACLSW Room 503, 430 East 29th Street, New York, NY 10016, USA. [6]Department of

Neurosurgery, Johns Hopkins University School of Medicine, Baltimore, MD, USA. [7]Present address: Yale University, New Haven, CT, USA. [8]Present address: L.E.K. Consulting, Boston, MA, USA. [9]Present address: BioNTech AG, Mainz, Germany. [10]Present address: Mayo Clinic, Jacksonville, FL, USA.

References

1. Cancer Genome Atlas Research N. Comprehensive genomic characterization defines human glioblastoma genes and core pathways. Nature. 2008;455(7216):1061–8.
2. Parsons DW, Jones S, Zhang X, Lin JC, Leary RJ, Angenendt P, Mankoo P, Carter H, Siu IM, Gallia GL, et al. An integrated genomic analysis of human glioblastoma multiforme. Science. 2008;321(5897):1807–12.
3. Burns KH, Boeke JD. Human transposon tectonics. Cell. 2012;149(4):740–52.
4. Huang CR, Schneider AM, Lu Y, Niranjan T, Shen P, Robinson MA, Steranka JP, Valle D, Civin CI, Wang T, et al. Mobile interspersed repeats are major structural variants in the human genome. Cell. 2010;141(7):1171–82.
5. Beck CR, Collier P, Macfarlane C, Malig M, Kidd JM, Eichler EE, Badge RM, Moran JV. LINE-1 retrotransposition activity in human genomes. Cell. 2010;141(7):1159–70.
6. Iskow RC, McCabe MT, Mills RE, Torene S, Pittard WS, Neuwald AF, Van Meir EG, Vertino PM, Devine SE. Natural mutagenesis of human genomes by endogenous retrotransposons. Cell. 2010;141(7):1253–61.
7. Ewing AD, Kazazian Jr HH. High-throughput sequencing reveals extensive variation in human-specific L1 content in individual human genomes. Genome Res. 2010;20(9):1262–70.
8. Stewart C, Kural D, Stromberg MP, Walker JA, Konkel MK, Stutz AM, Urban AE, Grubert F, Lam HY, Lee WP, et al. A comprehensive map of mobile element insertion polymorphisms in humans. PLoS Genet. 2011;7(8):e1002236.
9. Sudmant PH, Rausch T, Gardner EJ, Handsaker RE, Abyzov A, Huddleston J, Zhang Y, Ye K, Jun G, Hsi-Yang Fritz M, et al. An integrated map of structural variation in 2,504 human genomes. Nature. 2015;526(7571):75–81.
10. Solyom S, Ewing AD, Rahrmann EP, Doucet T, Nelson HH, Burns MB, Harris RS, Sigmon DF, Casella A, Erlanger B, et al. Extensive somatic L1 retrotransposition in colorectal tumors. Genome Res. 2012;22(12):2328–38.
11. Shukla R, Upton KR, Munoz-Lopez M, Gerhardt DJ, Fisher ME, Nguyen T, Brennan PM, Baillie JK, Collino A, Ghisletti S, et al. Endogenous retrotransposition activates oncogenic pathways in hepatocellular carcinoma. Cell. 2013;153(1):101–11.
12. Scott EC, Gardner EJ, Masood A, Chuang NT, Vertino PM, Devine SE. A hot L1 retrotransposon evades somatic repression and initiates human colorectal cancer. Genome Res. 2016;26(6):745-55.
13. Rodic N, Steranka JP, Makohon-Moore A, Moyer A, Shen P, Sharma R, Kohutek ZA, Huang CR, Ahn D, Mita P, et al. Retrotransposon insertions in the clonal evolution of pancreatic ductal adenocarcinoma. Nat Med. 2015;21(9):1060–4.
14. Lee E, Iskow R, Yang L, Gokcumen O, Haseley P, Luquette 3rd LJ, Lohr JG, Harris CC, Ding L, Wilson RK, et al. Landscape of somatic retrotransposition in human cancers. Science. 2012;337(6097):967–71.
15. Ewing AD, Gacita A, Wood LD, Ma F, Xing D, Kim MS, Manda SS, Abril G, Pereira G, Makohon-Moore A, et al. Widespread somatic L1 retrotransposition occurs early during gastrointestinal cancer evolution. Genome Res. 2015;25(10):1536–45.
16. Doucet TT, Kazazian Jr HH. Long Interspersed Element Sequencing (L1-Seq): A Method to Identify Somatic LINE-1 Insertions in the Human Genome. Methods Mol Biol. 2016;1400:79–93.
17. Tubio JM, Li Y, Ju YS, Martincorena I, Cooke SL, Tojo M, Gundem G, Pipinikas CP, Zamora J, Raine K, et al. Mobile DNA in cancer. Extensive transduction of nonrepetitive DNA mediated by L1 retrotransposition in cancer genomes. Science. 2014;345(6196):1251343.
18. Helman E, Lawrence MS, Stewart C, Sougnez C, Getz G, Meyerson M. Somatic retrotransposition in human cancer revealed by whole-genome and exome sequencing. Genome Res. 2014;24(7):1053–63.
19. Evrony GD, Cai X, Lee E, Hills LB, Elhosary PC, Lehmann HS, Parker JJ, Atabay KD, Gilmore EC, Poduri A, et al. Single-neuron sequencing analysis of L1 retrotransposition and somatic mutation in the human brain. Cell. 2012;151(3):483–96.
20. Evrony GD, Lee E, Mehta BK, Benjamini Y, Johnson RM, Cai X, Yang L, Haseley P, Lehmann HS, Park PJ, et al. Cell lineage analysis in human brain using endogenous retroelements. Neuron. 2015;85(1):49–59.
21. Upton KR, Gerhardt DJ, Jesuadian JS, Richardson SR, Sanchez-Luque FJ, Bodea GO, Ewing AD, Salvador-Palomeque C, van der Knaap MS, Brennan PM, et al. Ubiquitous L1 mosaicism in hippocampal neurons. Cell. 2015;161(2):228–39.
22. Erwin JA, Marchetto MC, Gage FH. Mobile DNA elements in the generation of diversity and complexity in the brain. Nat Rev Neurosci. 2014;15(8):497–506.
23. Guerrero-Cazares H, Chaichana KL, Quinones-Hinojosa A. Neurosphere culture and human organotypic model to evaluate brain tumor stem cells. Methods Mol Biol. 2009;568:73–83.
24. Arnold C, Hodgson IJ. Vectorette PCR: a novel approach to genomic walking. PCR Methods Appl. 1991;1(1):39–42.
25. Eggert H, Bergemann K, Saumweber H. Molecular screening for P-element insertions in a large genomic region of Drosophila melanogaster using polymerase chain reaction mediated by the vectorette. Genetics. 1998;149(3):1427–34.
26. Langmead B, Trapnell C, Pop M, Salzberg SL. Ultrafast and memory-efficient alignment of short DNA sequences to the human genome. Genome Biol. 2009;10(3):R25.
27. Bolger AM, Lohse M, Usadel B. Trimmomatic: a flexible trimmer for Illumina sequence data. Bioinformatics. 2014;30(15):2114–20.
28. Rodic N, Sharma R, Sharma R, Zampella J, Dai L, Taylor MS, Hruban RH, Iacobuzio-Donahue CA, Maitra A, Torbenson MS, et al. Long interspersed element-1 protein expression is a hallmark of many human cancers. Am J Pathol. 2014;184(5):1280–6.
29. Sharma R, Rodic N, Burns KH, Taylor MS. Immunodetection of Human LINE-1 Expression in Cultured Cells and Human Tissues. Methods Mol Biol. 2016;1400:261–80.
30. Mir AA, Philippe C, Cristofari G. euL1db: the European database of L1HS retrotransposon insertions in humans. Nucleic Acids Res. 2015;43(Database issue):D43–47.
31. Miki Y, Nishisho I, Horii A, Miyoshi Y, Utsunomiya J, Kinzler KW, Vogelstein B, Nakamura Y. Disruption of the APC gene by a retrotransposal insertion of L1 sequence in a colon cancer. Cancer Res. 1992;52(3):643–5.
32. Doucet-O'Hare TT, Rodic N, Sharma R, Darbari I, Abril G, Choi JA, Young Ahn J, Cheng Y, Anders RA, Burns KH, et al. LINE-1 expression and retrotransposition in Barrett's esophagus and esophageal carcinoma. Proc Natl Acad Sci U S A. 2015;112(35):E4894–4900.
33. Doucet-O'Hare TT, Sharma R, Rodic N, Anders RA, Burns KH, Kazazian Jr HH. Somatically Acquired LINE-1 Insertions in Normal Esophagus Undergo Clonal Expansion in Esophageal Squamous Cell Carcinoma. Hum Mutat. 2016;37(9):942–54.

5

A map of mobile DNA insertions in the NCI-60 human cancer cell panel

John G. Zampella[1], Nemanja Rodić[2], Wan Rou Yang[2], Cheng Ran Lisa Huang[3], Jane Welch[3], Veena P. Gnanakkan[3], Toby C. Cornish[2], Jef D. Boeke[3,4,6] and Kathleen H. Burns[2,3,4,5]*

Abstract

Background: The National Cancer Institute-60 (NCI-60) cell lines are among the most widely used models of human cancer. They provide a platform to integrate DNA sequence information, epigenetic data, RNA and protein expression, and pharmacologic susceptibilities in studies of cancer cell biology. Genome-wide studies of the complete panel have included exome sequencing, karyotyping, and copy number analyses but have not targeted repetitive sequences. Interspersed repeats derived from mobile DNAs are a significant source of heritable genetic variation, and insertions of active elements can occur somatically in malignancy.

Method: We used Transposon Insertion Profiling by microarray (TIP-chip) to map Long INterspersed Element-1 (LINE-1, L1) and *Alu* Short INterspersed Element (SINE) insertions in cancer genes in NCI-60 cells. We focused this discovery effort on annotated Cancer Gene Index loci.

Results: We catalogued a total of 749 and 2,100 loci corresponding to candidate LINE-1 and *Alu* insertion sites, respectively. As expected, these numbers encompass previously known insertions, polymorphisms shared in unrelated tumor cell lines, as well as unique, potentially tumor-specific insertions. We also conducted association analyses relating individual insertions to a variety of cellular phenotypes.

Conclusions: These data provide a resource for investigators with interests in specific cancer gene loci or mobile element insertion effects more broadly. Our data underscore that significant genetic variation in cancer genomes is owed to LINE-1 and *Alu* retrotransposons. Our findings also indicate that as large numbers of cancer genomes become available, it will be possible to associate individual transposable element insertion variants with molecular and phenotypic features of these malignancies.

Significance statement

Transposable elements are repetitive sequences that comprise much of our DNA. They create both inherited and somatically acquired structural variants. Here, we describe a first generation map of LINE-1 and *Alu* insertions in NCI-60 cancer cell lines. This provides a resource for discovering and testing functional consequences of these sequences.

* Correspondence: kburns@jhmi.edu
[2]Department of Pathology, Johns Hopkins University School of Medicine, 733 North Broadway, Miller Research Building Room 469, Baltimore, MD 21205, USA
[3]McKusick-Nathans Institute of Genetic Medicine, 733 North Broadway, Miller Research Building Room 469, Baltimore, MD 21205, USA
Full list of author information is available at the end of the article

Background

The National Cancer Institute-60 (NCI-60) cell panel was developed in the 1980s as a tool for pharmacologic screens and has become the most extensively studied collection of human cancers [1]. The panel comprises 59 cell lines encompassing nine tissue origins of malignancy, including blood, breast, colon, central nervous system, kidney, lung, ovary, prostate, and skin [2]. They have become a resource for high throughput characterizations and systems biology based approaches to cancer.

NCI-60 cell genomes have been described by targeted [3] and whole exome sequencing [4], karyotyping [5], and assays to detect copy number alteration [6], loss of heterozygosity [7], and DNA methylation [8]. Large scale mRNA [9] and microRNA [10] expression, protein abundance [11] and phosphorylation [12], and metabolomic [13] studies have also been conducted. Because assays are applied across the panel of cell lines in each case, datasets from orthogonal studies can be related to one another. For example, gene expression patterns have been found to be predictive of chemotherapeutic sensitivities [9].

Interspersed repeats have not been incorporated in these or many other genome-wide surveys. These repetitive sequences are dynamic constituents of human genomes and important sources of structural variation [14–20]. RNA transcribed from active elements can be reverse transcribed and integrated into the genome at new sites by proteins encoded by LINE-1 (Long INterspersed Element)-1 [21–23]. The result is that relatively recent insertions of LINE-1 (L1Hs) and *Alu* SINEs (*Alu*Ya5, *Alu*Ya8, *Alu*Yb8, *Alu*Yb9) are sources of genetic polymorphisms where both the pre-insertion allele and the insertion allele coexist in human populations. Moreover, LINE-1 sequences are hypomethylated [24–28] and express protein in a wide variety of human cancers [29],

and somatic LINE-1 integrations have been reported in tumor genomes [15, 30–36].

It is well established that inherited and acquired mobile DNA insertions can affect gene expression; there is inherent potential for insertions to have effects on tumor biology. However, the large majority occur in intronic or intergenic regions. Strong biases in the distribution of insertion sites or recurrent 'hotspots' for insertions arising during tumor development are frequently not obvious, leading to the presumption that most are non-functional 'passenger mutations' [34, 36].

This is *not* such a tumor-normal comparison study, but rather, one aimed to identify potential functions of mobile DNAs in human cancer cells. Towards this end, we mapped LINE-1 and *Alu* insertions in the NCI-60 tumor cell panel. We used a method for interspersed repeat mapping, Transposon Insertion Profiling by microarray (TIP-chip), to identify insertion sites. We also use

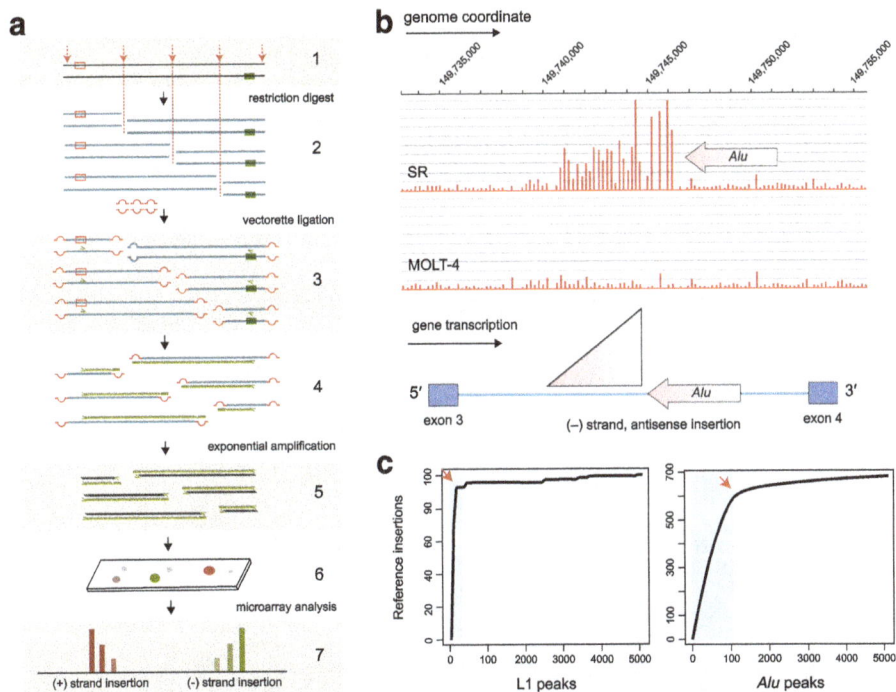

Fig. 1 Mapping transposable element (TE) insertion sites. **a.** A schematic illustrating the sequential steps of Transposon Insertion Profiling by microarray (TIP-chip). (1) An interval of double stranded genomic DNA with two TE insertions (boxes) oriented on opposing strands is shown; (2) the DNA is digested in parallel restriction enzyme reactions and ligated to vectorette oligonucleotides; (3) oligonucleotides complementary to the TE insertions prime first strand synthesis; (4) the elongating strands form reverse complements of the vectorette sequence; (5) there is exponential amplification of insertion site fragments; (6) these amplicons are labeled and hybridized to genomic tiling microarrays; and (7) 'peaks' of fluorescence intensity across several probes corresponding to contiguous genomic positions indicate a TE insertion. **b.** An example of a polymorphic *Alu* peak in two leukemia cell lines (SR and MOLT-4) in the third intron of the *TCOF1* (Treacher Collins-Franceschetti syndrome 1) gene on chromosome 5. The upper panels show TIP-chip data for the insertion, which is present in the SR line and not the MOLT-4 cells. The *Alu* insertion is a minus (-) strand insertion to the right of the probe with the greatest intensity; an arrow is drawn to indicate its position and orientation, but the arrow is not drawn to scale. *Alu* insertions approximate 300 bp, and the width of the peak in this case is 5 kb. **c.** Peaks were recognized using a sliding window algorithm which identified adjacent probes above a threshold fluorescence intensity value. The threshold value was progressively lowered to identify peaks in a rank order. The graphs show the number of reference insertions identified verses peak rank for a representative LINE-1 and *Alu* TIP-chip. The cut-off for defining a candidate insertion was established using the inflection points (red arrows) of these plots

previous characterizations of the cell panel to associate specific insertions with cellular phenotypes.

Results

Transposon insertion profiling by microarray

To map mobile DNA insertions, we used a method we have termed transposon insertion profiling by microarray (TIP-chip), which uses vectorette PCR to amplify unknown sequence adjacent to a known primer-binding site (Fig. 1a). We surveyed three major currently active mobile DNAs in humans (L1Hs, *Alu*Ya5/8; and *Alu*Yb8/

9) as previously described [14]. To focus on the potential functional impact of these sequences on cancer cell phenotypes, PCR amplicons were labeled and analyzed using a genomic tiling microarray designed to encompass 6,484 known Cancer Gene Index loci (+/- 10 kb) (Biomax™ Informatics), about 17 % of the genome. Peaks of signal intensity correspond to TE insertions (Fig. 1a, b); known LINE-1 and *Alu* elements incorporated in the reference genome assembly (hereafter, 'reference insertions') were used as a quality control metric and to set cut-offs for recognized peaks (Fig. 1c).

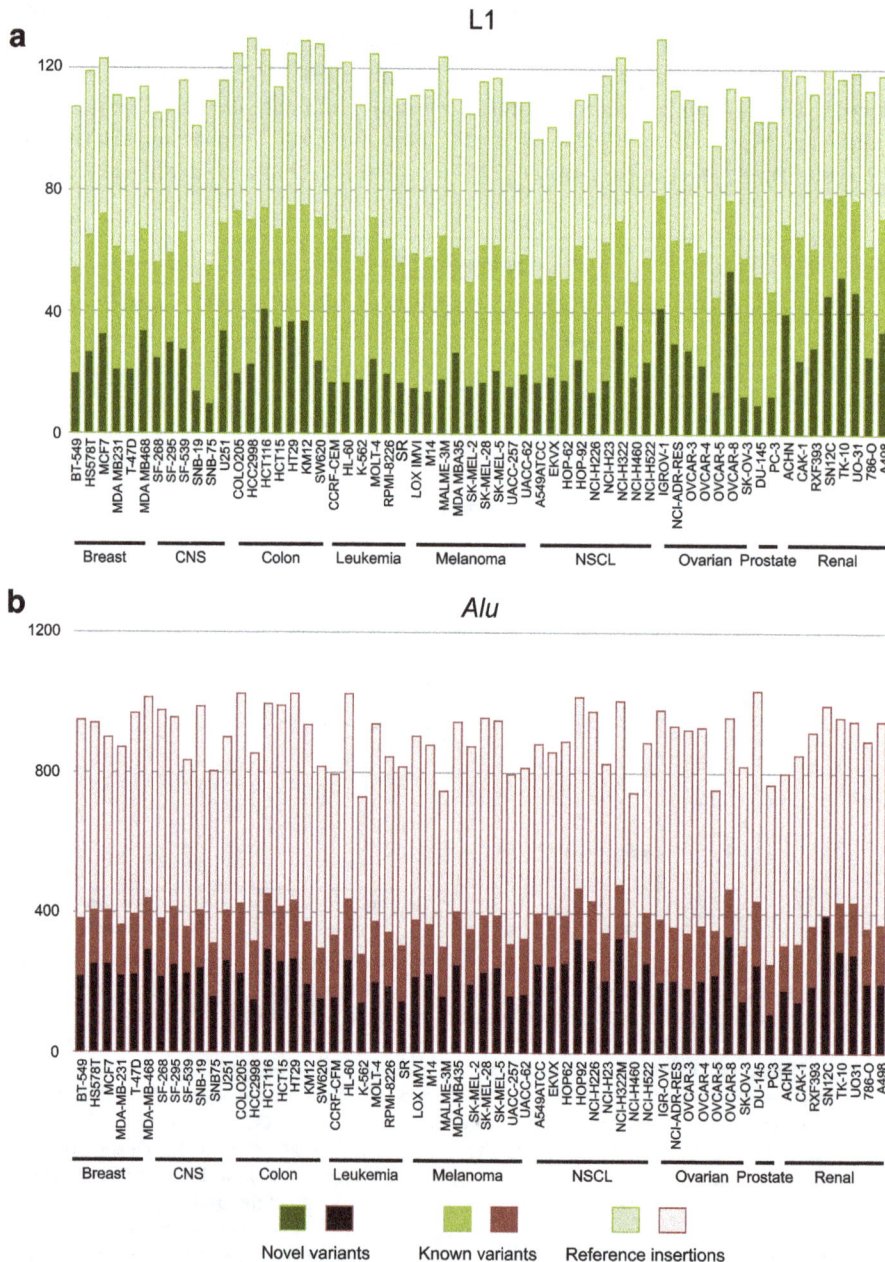

Fig. 2 Total TE insertions. The stacked bar plots show the relative numbers of novel variants, known variants, and reference insertions per cell line for LINE-1 (green, *upper panel*) and *Alu* (red, *lower panel*). The total number of insertions detected per cell line is similar across the tumor panel

A total of 749 and 2,100 peaks corresponding to candidate LINE-1 and *Alu* insertion sites respectively were recognized across the NCI-60 cell panel. These locations were cross-referenced to previously described insertions to define three categories: (*i.*) reference insertions, which include invariant insertions and insertion polymorphisms incorporated in the reference genome assembly; (*ii.*) inherited variants either previously described (known polymorphic) or newly discovered, but occurring in multiple, unrelated cell lines (novel polymorphic); and (*iii.*) novel, 'singleton' insertions seen uniquely in one cell line (Fig. 2a, b). The last category includes both insertions that were constitutive (germline) in the patient from whom the cell line was derived as well as somatic insertions acquired during tumor development or the propagation of these cell lines. A greater proportion of LINE-1 insertions were singletons (68 %) compared with *Alu* insertions (21 %). Density plots for both LINE-1 and *Alu* show most peaks fall into this last category, particularly for L1Hs, although a biphasic distribution was seen (Fig. 3a, b).

Our array encompassed 130 known reference LINE-1 and 1278 *Alu* insertions. A total of 112 LINE-1 and 1,160 *Alu* insertions detected were present in the reference genome assembly. A total of 697 LINE-1 and 1,147 *Alu* insertions were singleton or polymorphic (known and novel) segregating in human populations (Fig. 2a, b). Insertions incorporated in the reference genome that are known to be polymorphic are counted in both groups. A summary of insertion positions by tumor type and cell line can be found in Additional file 1: Table S1, Additional file 2: Table S2.

We found that each cell line had a unique transposable element (TE) insertion profile (Fig. 3a). After correcting for batch effects, a principal component analyses (PCA) did not show clustering by tumor type. As expected, however, pairs of cell lines derived from the same individual grouped together, and these pairs showed a high concordance of top-ranking peaks as compared to unrelated cell lines. We compared TE insertion profiles to described cytogenetic abnormalities. In some instances, insertions were informative of deletions; for example, a reference LINE-1 in the retinoblastoma 1 (*RB1*) locus was only absent in the MB468 breast cancer cell line, consistent with the homozygous deletion of *RB1* reported for this cell line [37].

Insertions in genes involved in oncogenesis

In TIP-chip, probe spacing does not resolve insertions to the precise base, and insertion strandedness was not predicted for all peak intervals in this study. Despite these limitations, we identified peak intervals that partially or entirely overlapped exon intervals for further inspection. Partial overlaps were almost entirely attributable to insertions near an exon. We identified 9 insertions within exons, and all were located within gene 3' untranslated regions (3' UTRs); none affected protein open reading frames.

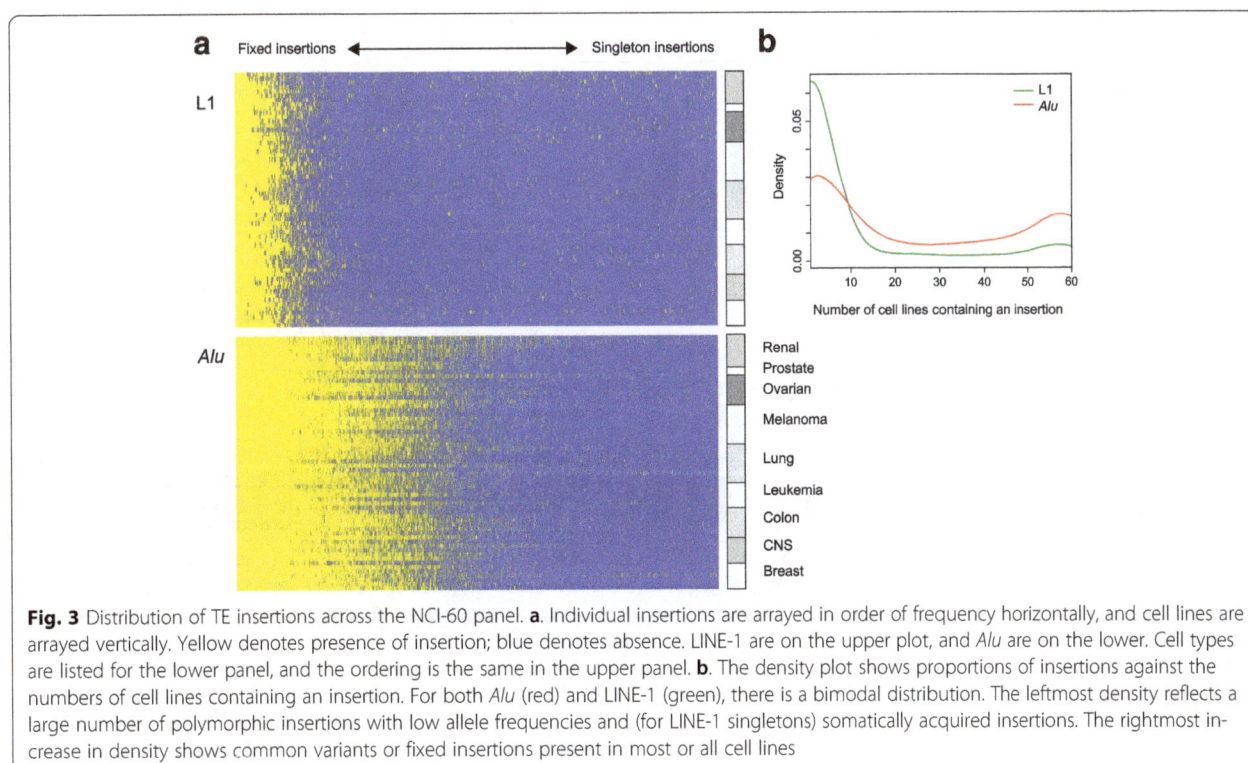

Fig. 3 Distribution of TE insertions across the NCI-60 panel. **a**. Individual insertions are arrayed in order of frequency horizontally, and cell lines are arrayed vertically. Yellow denotes presence of insertion; blue denotes absence. LINE-1 are on the upper plot, and *Alu* are on the lower. Cell types are listed for the lower panel, and the ordering is the same in the upper panel. **b**. The density plot shows proportions of insertions against the numbers of cell lines containing an insertion. For both *Alu* (red) and LINE-1 (green), there is a bimodal distribution. The leftmost density reflects a large number of polymorphic insertions with low allele frequencies and (for LINE-1 singletons) somatically acquired insertions. The rightmost increase in density shows common variants or fixed insertions present in most or all cell lines

To begin to approach potential functional conse-quences of intronic insertions, we analyzed insertion sites in sets of genes with described roles in cancer. We considered collections of genes with TE insertions while grouping together malignant cell lines by tissue of origin. Interestingly, in breast cancer cell lines, we observed a significant enrichment of singleton and polymorphic LINE-1 and *Alu* insertions in "STOP genes", defined in shRNA screens as suppressors of human mammary epi-thelial cell proliferation [38] ($p = 1.23 \times 10^{-9}$) (Fig. 4a). This result persisted when LINE-1 and *Alu* insertions were analyzed independently; LINE-1 singleton inser-tions but not *Alu* singleton insertions were also enriched in this gene set (Fig. 4b). Analysis of expression of these "STOP" genes shows that a preponderance of these genes are down-regulated; this result persists in those genes containing a TE insertion. The findings suggest that collectively, insertions may act to compromise ex-pression of these genes.

Consistent with this model, ovarian cancer cell lines showed a preponderance of insertions in genes that are down regulated in ovarian cancers as compared to nor-mal tissue. A random set of genes from the array is shown as a histogram for comparison (Fig. 4d). This pat-tern was absent in other tumor types.

We saw an enrichment of singleton and polymorphic TEs in genes recurrently mutated in experimental cancer models and in human tumors. For the former, we con-sidered common insertion sites (CIS) defined as gene loci recurrently interrupted by insertional mutagens in forward cancer gene screens in mice [39, 40] ($p = 1.46 \times 10^{-4}$). The latter was assessed using genes fre-quently mutated in human cancers taken from the Cata-logue Of Somatic Mutations In Cancer (COSMIC) database [41] ($p = 7.74 \times 10^{-10}$) (Fig. 4c). We also com-pared our insertion profiles to sites of reported somatic TE insertions in human cancers. We analyzed novel (singleton and polymorphic) insertions and discovered that we had overlaps in 22 of the 64 genes noted by Lee et al. [32] and 23 of 76 from Solimini et al. [38](Fig. 4c). We anticipate the possibility that common insertion site loci will be identified as more insertion site mapping studies are conducted in human tumors.

Functional associations of individual insertions

An advantage of working with the NCI-60 cell lines is that these are well studied. To integrate our insertion site maps with other findings in these cells, we performed COM-PARE analyses [42]. COMPARE is a pattern matching method developed specifically for NCI-60 cell lines that provides a *p*-value for each association (S5–25). Direct, local roles for TEs (in *cis*) were not observed for the ma-jority of correlations. However, COMPARE did reveal three insertions associated with DNA hypermethylation

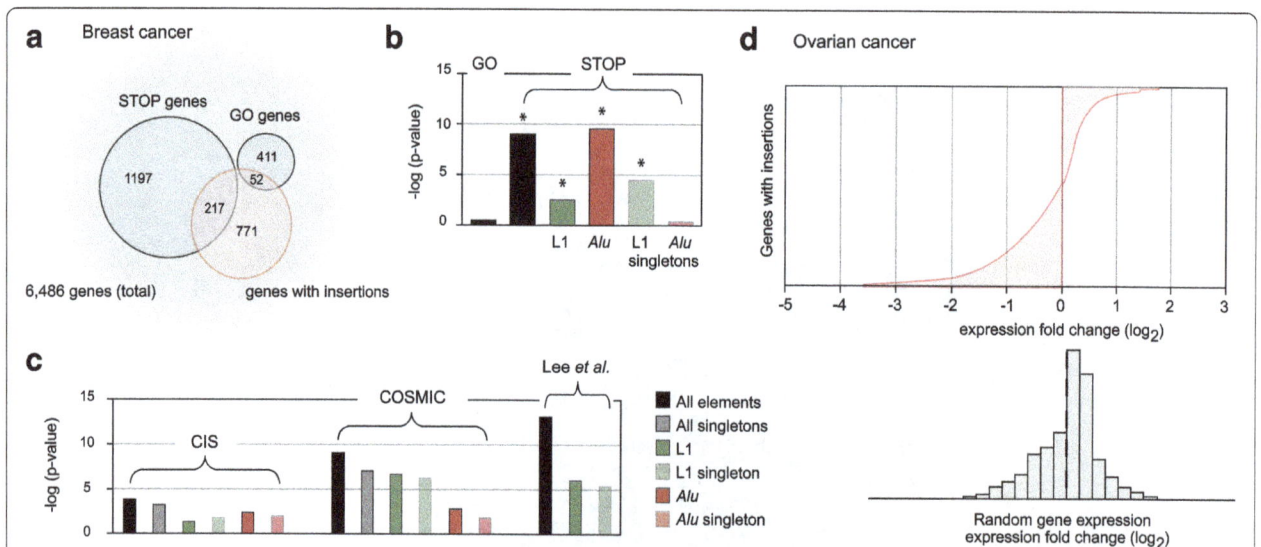

Fig. 4 TE enrichment analyses. **a**. STOP and GO genes have been implicated in breast cancer as genes that appear to inhibit and promote tumor development, respectively. Using a hypergeometric distribution to assess enrichment, we found that TE insertions are enriched in STOP genes on the array ($p = 1.23 \times 10^{-9}$) but not in GO genes ($p = 0.33$). **b**. The bar graph shows enrichment by type of TE plotted as the negative log of the p-value. No GO gene enrichment is seen. STOP gene enrichment is seen considering all LINE-1 ($p = 3.11 \times 10^{-3}$); all *Alu* ($p = 2.27 \times 10^{-10}$); as well as LINE-1 singletons ($p = 4.16 \times 10^{-5}$). **c**. Insertions were also enriched in common insertion sites (CIS) ($p = 1.46 \times 10^{-4}$); COSMIC commonly mutated cancer genes ($p = 7.74 \times 10^{-10}$); and genes reported to acquire somatic LINE-1 insertions in cancer by Lee et al. ($p = 5.34 \times 10^{-14}$). **d**. Genes with TE insertions in ovarian cancer cell lines are more likely than other genes to be downregulated in ovarian cancer samples as compared to normal tissue controls. Randomly selected genes are shown for comparison (*bottom panel*)

within 30 kb of the insertion site. For example, a polymorphic *Alu* insertion in the *SS18L1* (Synovial sarcoma translocation gene on chromosome 18-like 1) gene locus oriented anti-sense to the transcription of the gene, is associated with increased methylation of nearby CpG sites at the same gene locus ($p = 6.67 \times 10^{-6}$) (Fig. 5a).

Manhattan plots illustrate highly significant correlations found in *trans* (Fig. 5a–c). A subset of insertions

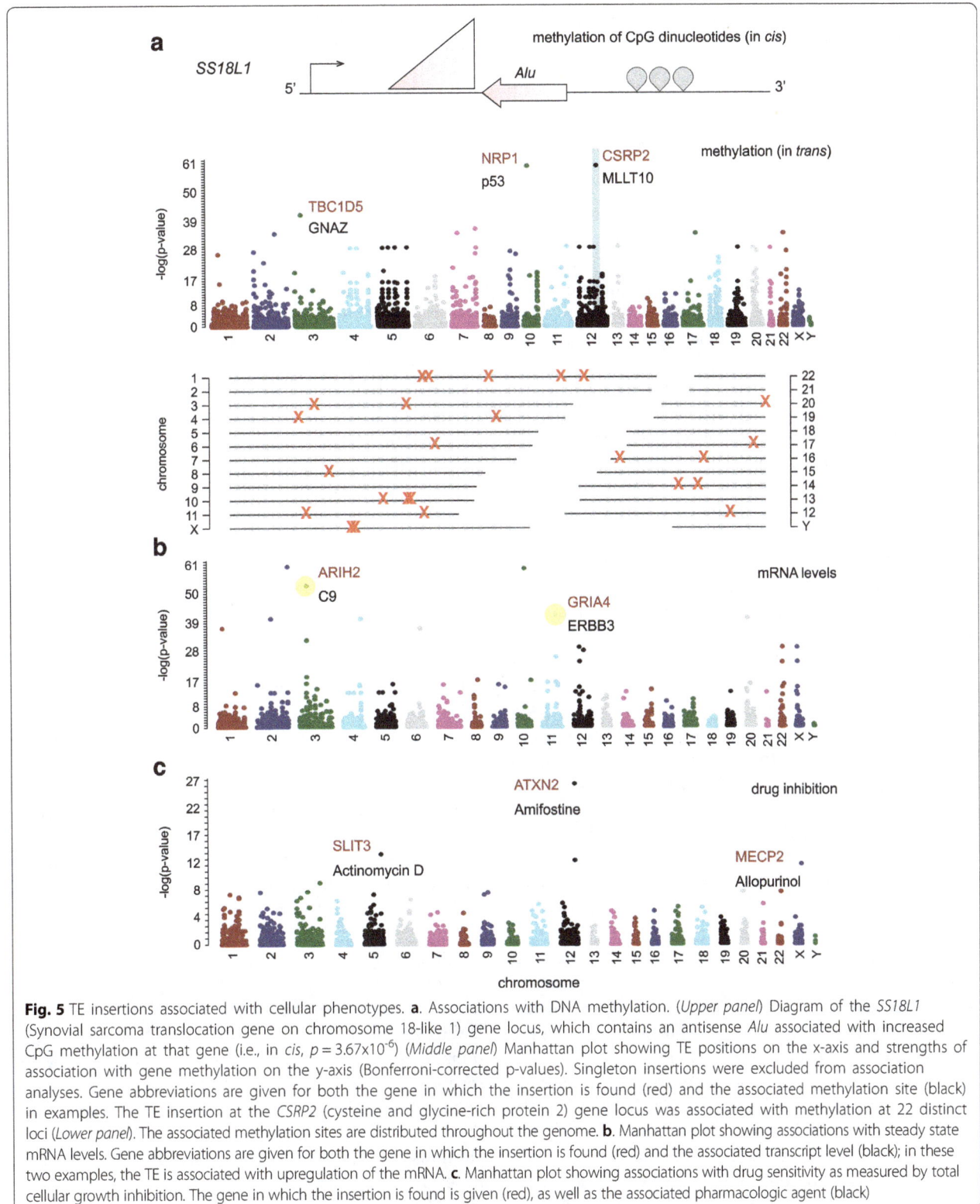

Fig. 5 TE insertions associated with cellular phenotypes. **a**. Associations with DNA methylation. (*Upper panel*) Diagram of the *SS18L1* (Synovial sarcoma translocation gene on chromosome 18-like 1) gene locus, which contains an antisense *Alu* associated with increased CpG methylation at that gene (i.e., in *cis*, $p = 3.67 \times 10^{-6}$) (*Middle panel*) Manhattan plot showing TE positions on the x-axis and strengths of association with gene methylation on the y-axis (Bonferroni-corrected p-values). Singleton insertions were excluded from association analyses. Gene abbreviations are given for both the gene in which the insertion is found (red) and the associated methylation site (black) in examples. The TE insertion at the *CSRP2* (cysteine and glycine-rich protein 2) gene locus was associated with methylation at 22 distinct loci (*Lower panel*). The associated methylation sites are distributed throughout the genome. **b**. Manhattan plot showing associations with steady state mRNA levels. Gene abbreviations are given for both the gene in which the insertion is found (red) and the associated transcript level (black); in these two examples, the TE is associated with upregulation of the mRNA. **c**. Manhattan plot showing associations with drug sensitivity as measured by total cellular growth inhibition. The gene in which the insertion is found is given (red), as well as the associated pharmacologic agent (black)

had multiple associations (vertical series of dots corresponding to one TE location), suggesting the possibility of pleomorphic effects of an insertion haplotype.

In addition, we encountered examples of single 'driver' mutations and cellular phenotypes that could be associated with multiple TE insertions. Five insertions correlated with a mutation in the *ERBB2* gene (v-erb-b2 erythroblastic leukemia viral oncogene homolog 2, the HER2/neu locus), and more than 10 insertions were associated with thymidylate synthase activity (p values $< 10^{-20}$). To probe relationships between multiple *trans* associated factors related to a single TE insertion, we performed pathway analyses on sets of genes, each encompassing the TE insertion locus and all RNAs and proteins with associated expression patterns. This yielded more than 250 curated pathways with enrichment p-values less than 10^{-4}, supporting the concept that these are biologically relevant as opposed to spurious associations. All COMPARE results are provided in the (Additional file 3: Table S3).

Discussion

Our genomes are filled with highly repetitive DNA sequences derived from TEs. Tailored methods for their detection, including TIP-chip [14], targeted insertion site sequencing [15, 17, 18, 31, 36, 43], and algorithms for finding variants in whole genome sequencing [20, 34, 44] are revealing this previously masked dimension of genomic data. Collectively, these studies confirm that TEs are rich sources of genetic diversity in human populations, and provide evidence that they are somatically unstable in a variety of tumor types. Of the two most active germline elements, LINE-1 and *Alu*, (which is mobilized in *trans* by LINE-1-encoded proteins), LINE-1 has been more well documented to be active in cancer. *Alu* insertions account for more inherited polymorphisms. For both types of TEs, the vast majority of catalogued insertions are intronic and intergenic without clear function.

To begin a systematic survey for functionally consequential LINE-1 and *Alu* integrations in human neoplasias, we mapped these variants in the NCI-60 cell panel. NCI-60 is a unique resource for this, encompassing a variety of cancer cell lines that have the advantages of being well studied and readily available. We mapped LINE-1 and *Alu* insertion positions using a microarray-based approach over a large census of cancer genes. Even as TIP-chip is replaced by sequencing, we expect these data will provide a useful reference.

TIP-chip across the NCI-60 panel revealed numerous novel candidate TEs, totaling about 500 L1Hs and 1000 *AluYa/Yb* insertions distributed across the 60 cell lines. These include insertions that are unique to a cell line ('singleton') and novel polymorphic insertions (found in unrelated cell lines). Although 'singletons' may be

enriched for tumor-specific, somatic insertion events, matched non-neoplastic cells for the corresponding patient cases are not available, and therefore we cannot definitively differentiate somatic from inherited variants. Similarly, these cell lines have undergone numerous passages since their creation, and somatic insertion events occurring in culture cannot be clearly recognized. We note a greater proportion of LINE-1 singletons (68 % of LINE-1 loci) than *Alu* singletons (21 % of *Alu* loci), consistent with ongoing LINE-1 retrotransposition in vivo or in vitro.

We approached the question of TE function by two avenues. We first tested for biases in the distribution of insertions with respect to known gene sets. We found a preferential accumulation of TE insertions in retained copies of 'STOP genes' in breast cancer cell lines; these gene loci function as inhibitors of mammary epithelial cell proliferation. Experimental models suggest that it is advantageous for tumor growth to compromise the function of these genes [38], and we speculate that TE insertions are enriched at these loci because they have a role in this process. These 'STOP genes' are downregulated in the breast cancer cell lines, as is the subset of 'STOP genes' containing TE insertions. We also found preferential TE accumulation in genes downregulated in ovarian cancers compared with normal ovarian tissue, which would be consistent with this model. Finally, genes with functional roles in cancer were also more commonly seen as insertion sites than expected. These included genes 'hit' recurrently by insertional mutagenesis in forward genetics screens in mice, the so-called common insertion sites (CIS), and in genes commonly mutated in human cancers (COSMIC catalog) [41].

We note that the exonizations of intronic LINE-1 [45] and *Alu* sequences [46] are being increasingly recognized using RNA-seq, and that many of the resulting transcripts have an altered protein coding capacity. It may be possible to identify aberrant mRNA species corresponding to these insertion loci and thus invoke a molecular mechanism to underlie this type of functional effect.

Our second approach relied on association studies. We used existing data in COMPARE analyses to test for relationships between TE insertion alleles and cellular phenotypes. In the case of DNA methylation only, *cis* effects could be seen relating individual TEs with local DNA hypermethylation. We identified three *Alu* integrations associated with DNA hypermethylation at the insertion site (+/- 30 kb). The most notable is a polymorphic *Alu* insertion in the first intron of the *SS18L1* (synovial sarcoma translocation gene on chromosome 18-like 1) gene locus associated with CpG hypermethylation at the same locus ($p = 3.67 \times 10^{-6}$). *SS18* and *SS18L1* encode transcriptional regulators and are breakpoints in chromosomal translocations in synovial sarcoma [47]. These translocations are not seen in the NCI-60 panel tumors, and whether the epigenetic signature associated with the *Alu* insertion impacts expression of this gene is unknown. So, while it is not clear at this point that *SS18L1* methylation is germane to the development of these malignancies, our ability to relate genotype and epigenetics at these sites demonstrate the value of this approach.

The large majority of statistically significant associations between insertions and cellular phenotypes appeared to involve indirect or *trans* effects that are difficult to test further. Pathway analyses suggest that many are not random, but reflect recognized, related gene sets. It may be that the indirect effects can be dissected for some insertion alleles; particularly promising may be those at loci of transcriptional regulators with definable target genes [29].

Conclusions

In summary, we profiled LINE-1 and *Alu* insertion sites in a panel of widely used cancer cell lines, the NCI-60. We expect maps such as these will be a useful resource for experimentalists with interests in how transposable element insertions interact with genes. Our analyses show that insertion sites can be integrated with other data to develop testable hypotheses about the function of mobile DNAs in cancer.

Methods

NCI-60 cell lines

The National Cancer Institute-60 (NCI-60) human cancer cells are a group of 60 cell lines representing nine different types of neoplasias(breast cancer, colon cancer, CNS tumor, leukemia, lung cancer, melanoma, ovarian cancer, prostate cancer, and renal cell carcinoma) composed of 54 individual cancer cases and three pairs of cell lines (ADR and OVCAR-8; MB-435 and M14; and SNB19 and U251) with each pair originating from the same patient [48, 49]. The NCI-60 panel has been extensively characterized in a breadth of molecular and pharmacologic assay [50]. Genomic DNA was obtained directly from the NCI.

Microarray design

A genomic tiling micorarrray was designed to cover the NCI Cancer Gene Index (disease list). A total of 6,484 RefSeq gene identifiers were extracted from the. XML file and converted to genomic coordinates corresponding to each transcript unit +/- 10 kb hg19 reference genome assembly (February 2009, GRCh37). UCSC Table Browser intervals were merged using GALAXY [51], and probes were chosen for the NimbleGen HD (2.1 M feature) array platform by the manufacturer (Roche NimbleGen, Madison, WI).

Transposon insertion profiling by microarray (TIP-chip)

Five micrograms of genomic DNA of each cell line was digested overnight in parallel reactions using four restriction enzymes (*Ase*I, *Bsp*HI, *Hind*III, and *Xba*I). Sticky ends were ligated to annealed, partially complementary vectorette oligonucleotide adapters. Each template was aliquoted into 3 separate vectorette PCR reactions for L1Hs, *Alu*Ya5/8, and *Alu*Yb8/9 mobile DNA families. These were then labeled with Cy3-dUTP for LINE-1 and Cy5-dUTP for *Alu* and hybridized to Nimblegen genome tiling arrays according to the manufacturer's instructions. Reference insertions are those incorporated in the Feb. 2009 assembly of the human genome (hg19, GRCh37 Genome Reference Consortium Human Reference 37, GCA_000001405.1).

Peak recognition

Each scanned array yielded a raw .tff file, which was processed using Nimblescan v2.5 (Roche Nimblegen, Madison, WI) to give genomic coordinates and probe intensities (.gff files). A PERL script removed probes overlapping repeats to reduce noise (RepeatMasking). Nimblescan called peaks using a sliding window threshold. Peaks were ranked by the threshold of the log2 transformed ratio of red (Alu) and green (L1) channels or the reciprocal (settings: percent (p) start = 90, p step = 1, #steps = 76, width of sliding window = 1500 bp, min probes > 4, all probes > 2). The top 5,000 L1 and Alu peaks were kept for evaluation.

Peak cut-off

Among these peaks, recovery of those corresponding to mobile DNA insertions in hg19 (reference insertions) was used as a proxy of assay performance. Reference insertion count was plotted against peaks recognized (Fig. 1c). A cut-off was imposed on the peak threshold value (p >70 for L1 and p > 60 for Alu) to include peaks up to the approximate inflection point of this curve in subsequent analyses. These threshold values were altered for outlier cell lines to reflect the curve inflection point. MYSQL was used to annotate peaks with respect to genes and known mobile DNA insertions (L1Hs, AluY, AluYa5, AluYa8, AluYb8, and AluYb9 using 1–2 kb margins). Lists of known insertions were obtained from previously published databases [14, 19, 52, 53].

Clustering and insertion profiles

Principle component analysis (PCA) (R-package) was used to remove batch effect. All insertions were sorted by density across the cell lines and plotted as a matrix.

Cell lines lacking high-frequency insertions were assessed for karyotype abnormalities manually.

COMPARE analysis

Reference and non-reference insertions were analyzed using a COMPARE analysis [42] associating each with the CellMiner database of NCI-60 cell profiling studies. These have included DNA mutations and methylation; RNA and miRNA expression; protein expression, enzymatic activity; and drug inhibition studies. Associations for those insertions found in one cell line (singleton) were considered only for *cis* effects and were discarded from other associations due to their high false-positive rates. *P*-values for other insertions were corrected using Bonferroni multiple test correction and plotted using the start position of peak intervals to generate Manhattan plots (adaptation of Genetics Analysis Package, R-package).

Pathway analysis

Gene loci containing candidate non-reference (polymorphic and singleton) LINE-1 and *Alu* insertions and associated gene names from RNA and protein COMPARE analysis were uploaded in batch to the MSigDb 'Investigate Gene Sets' from the Broad Institute Gene Set Enrichment Analysis web interface [54] (using the C2 curated gene sets). Pathways were selected if the insertion locus was part of the pathway and the p-value of the pathway was less than 10^{-4}. Interactome plots were used to visualize relationships between genes in pathways using Search Tool for the Retrieval of Interacting Genes/Protein (STRING) 9.0 [55]. Plots were adapted to show the gene locus containing the insertion (yellow) and the direction of related correlations (red for positive correlations with the insertion; purple for negative correlations).

Preferential integration sites

To investigate preferential transposable element insertion in genes implicated in oncogenesis and mouse common insertion sites, we used a hypergeometric distribution test (pHypr R-package) which controlled for genes tiled on the array. Results were plotted using the $-\log(p\text{-value})$.

Tumor-normal gene expression studies

Tumor vs normal gene expression for genes containing candidate non-reference TE insertions was assessed for each tumor type using large tumor/normal gene expression databases. Tumor gene to normal gene expression ratios were obtained using NCBI GEO2R [56]. GEO2R was used to log2 transform expression data if datasets were not in log2 formats. Value distribution of all databases was assessed for median-centering prior to evaluation. Expression values for all insertion-containing genes was plotted

as a horizontal bar plot. A random sample of 1000 genes from the array were evaluated in the same manner to serve as a control set. A histogram of random gene expression values was plotted. Databases (Breast = GSE5764, Ovarian = GSE26712, omitted samples with "no evidence of disease", Colon = GSE6988, omitted non-primary tumors, Melanoma = GSE7553, CNS = GSE4290, non-tumor used as "normal" and non-glioblastomas omitted, Prostate = GSE3325, Renal = GSE11151, non-conventional tumors omitted, NSCL = GSE19188).

STOP gene expression in breast cancer cell lines

Expression of STOP genes containing candidate non-reference TE insertions was assessed using log2 transformed Agilent mRNA expression data [57] obtained from the CellMiner for the Breast cancer cell lines. The expression was averaged across all cell lines, sorted, and plotted as a horizontal bar plot. STOP genes tiled on the array, but without a TE insertion was plotted as well. Tumor-Normal expression for STOP genes was performed according to the methods used above in Tumor-Normal gene expression studies.

Additional files

Additional file 1 A map of LINE-1 (L1) insertion site positions in the NCI-60 cell panel. Genomic coordinates of TIP-chip peaks are provided. Reference insertions are indicated in column D (hg19), and known polymorphic variants are indicated by a 'Y' in column E (Y/N, yes/no). For each cell line in columns G-BN, a '1' indicates that the insertion is present, while '0' indicates that the insertion is not found. (XLSX 85 kb)

Additional file 2 A map of *Alu* insertion site positions in the NCI-60 cell panel. Genomic coordinates of TIP-chip peaks are provided. Reference insertions are indicated in column D (hg19), and known polymorphic variants are indicated by a 'Y' in column E (Y/N, yes/no). For each cell line in columns G-BN, a '1' indicates that the insertion is present, while '0' indicates that the insertion is not found. (XLSX 380 kb)

Additional file 3 COMPARE analysis associating insertions with other cell characteristics. Different tabs are used for different datasets. Activity, enzyme activity measures; Decreased / Increased methylation, DNA methylation measures; Metabolome, metabolic intermediates; Drug Effect GI50, concentration for 50 % growth inhibition; Drug Effect TGI, concentration for total growth inhibition; miRNA, microRNA expression levels; RNA, mRNA expression levels; Mutations, somatically-acquired DNA mutations; Protein, protein expression. (XLSX 4049 kb)

Abbreviations

LINE-1: Long INterspersed Element-1; NCI: National Cancer Institute; SINE: Short INterspersed Element; TIP-chip: Transposon insertion profiling by microarray

Acknowledgements

We thank Peilin Shen, Jared Steranka, Youngran Park, Xuan Pham, and Ashley Castillo for technical assistance and thoughtful discussion of the project; Eitan Halper-Stromberg for computational assistance; and Beatriz Villaba-Martín for Fig. 1a. Susan L. Holbeck at the National Cancer Institute performed the COMPARE analysis, and Joel Bader provided approaches for pathway analyses.

Funding

This work was supported by R01CA163705, R01GM103999, and a Career Award for Medical Scientists from the Burroughs Welcome Foundation (to

KHB); and P50GM107632 (to JDB). JGZ was supported by the Howard Hughes Medical Institute (HHMI) Medical Research Fellows Program.

Authors' contributions

JGZ carried out the TIP-chip experiments; NR, WRY, CRLH, JW, VPG, and TCC optimized the protocols and/or conducted data analysis; JGZ, JDB, and KHB designed the study and secured funding for the project; JGZ and KHB drafted the manuscript. All authors contributed to the manuscript review.

Competing interests

The authors declare that they have no competing interests.

Author details

¹Department of Dermatology, Johns Hopkins University School of Medicine, 733 North Broadway, Miller Research Building Room 469, Baltimore, MD 21205, USA. ²Department of Pathology, Johns Hopkins University School of Medicine, 733 North Broadway, Miller Research Building Room 469, Baltimore, MD 21205, USA. ³McKusick-Nathans Institute of Genetic Medicine, 733 North Broadway, Miller Research Building Room 469, Baltimore, MD 21205, USA. ⁴High Throughput (HiT) Biology Center, 733 North Broadway, Miller Research Building Room 469, Baltimore, MD 21205, USA. ⁵The Sidney Kimmel Comprehensive Cancer Center, Johns Hopkins University School of Medicine, 733 North Broadway, Miller Research Building Room 469, Baltimore, MD 21205, USA. ⁶Present address: Institute for Systems Genetics, NYU Langone University School of Medicine, New York, NY 10016, USA.

References

1. Shoemaker RH. The NCI60 human tumour cell line anticancer drug screen. Nat Rev Cancer. 2006;6:813–23.
2. Stinson SF, Alley MC, Kopp WC, Fiebig HH, Mullendore LA, Pittman AF, Kenney S, Keller J, Boyd MR. Morphological and immunocytochemical characteristics of human tumor cell lines for use in a disease-oriented anticancer drug screen. Anticancer Res. 1992;12:1035–53.
3. Ikediobi ON, Davies H, Bignell G, Edkins S, Stevens C, O'Meara S, Santarius T, Avis T, Barthorpe S, Brackenbury L, et al. Mutation analysis of 24 known cancer genes in the NCI-60 cell line set. Mol Cancer Ther. 2006;5:2606–12.
4. Abaan OD, Polley EC, Davis SR, Zhu YJ, Bilke S, Walker RL, Pineda M, Gindin Y, Jiang Y, Reinhold WC, et al. The exomes of the NCI-60 panel: a genomic resource for cancer biology and systems pharmacology. Cancer Res. 2013; 73:4372–82.
5. Roschke AV, Tonon G, Gehlhaus KS, McTyre N, Bussey KJ, Lababidi S, Scudiero DA, Weinstein JN, Kirsch IR. Karyotypic complexity of the NCI-60 drug-screening panel. Cancer Res. 2003;63:8634–47.
6. Varma S, Pommier Y, Sunshine M, Weinstein JN, Reinhold WC. High resolution copy number variation data in the NCI-60 cancer cell lines from whole genome microarrays accessible through Cell Miner. PLoS ONE. 2014; 9:e92047.
7. Ruan X, Kocher JP, Pommier Y, Liu H, Reinhold WC. Mass homozygotes accumulation in the NCI-60 cancer cell lines as compared to HapMap Trios, and relation to fragile site location. PLoS ONE. 2012;7:e31628.
8. Shen L, Kondo Y, Ahmed S, Boumber Y, Konishi K, Guo Y, Chen X, Vilaythong JN, Issa JP. Drug sensitivity prediction by CpG island methylation profile in the NCI-60 cancer cell line panel. Cancer Res. 2007;67:11335–43.
9. Staunton JE, Slonim DK, Coller HA, Tamayo P, Angelo MJ, Park J, Scherf U, Lee JK, Reinhold WO, Weinstein JN, et al. Chemosensitivity prediction by transcriptional profiling. Proc Natl Acad Sci U S A. 2001;98:10787–92.
10. Patnaik SK, Dahlgaard J, Mazin W, Kannisto E, Jensen T, Knudsen S, Yendamuri S. Expression of microRNAs in the NCI-60 cancer cell-lines. PLoS ONE. 2012;7:e49918.
11. Nishizuka S, Charboneau L, Young L, Major S, Reinhold WC, Waltham M, Kouros-Mehr H, Bussey KJ, Lee JK, Espina V, et al. Proteomic profiling of the NCI-60 cancer cell lines using new high-density reverse-phase lysate microarrays. Proc Natl Acad Sci U S A. 2003;100:14229–34.
12. Federici G, Gao X, Slawek J, Arodz T, Shitaye A, Wulfkuhle JD, De Maria R, Liotta LA, Petricoin 3rd EF. Systems analysis of the NCI-60 cancer cell lines by alignment of protein pathway activation modules with "-OMIC" data fields and therapeutic response signatures. Mol Cancer Res. 2013;11:676–85.
13. Jain M, Nilsson R, Sharma S, Madhusudhan N, Kitami T, Souza AL, Kafri R, Kirschner MW, Clish CB, Mootha VK. Metabolite profiling identifies a key role for glycine in rapid cancer cell proliferation. Science. 2012;336:1040–4.
14. Huang CR, Schneider AM, Lu Y, Niranjan T, Shen P, Robinson MA, Steranka JP, Valle D, Civin CI, Wang T, et al. Mobile interspersed repeats are major structural variants in the human genome. Cell. 2010;141:1171–82.
15. Iskow RC, McCabe MT, Mills RE, Torene S, Pittard WS, Neuwald AF, Van Meir EG, Vertino PM, Devine SE. Natural mutagenesis of human genomes by endogenous retrotransposons. Cell. 2010;141:1253–61.
16. Beck CR, Collier P, Macfarlane C, Malig M, Kidd JM, Eichler EE, Badge RM, Moran JV. LINE-1 retrotransposition activity in human genomes. Cell. 2010; 141:1159–70.
17. Ewing AD, Kazazian Jr HH. High-throughput sequencing reveals extensive variation in human-specific L1 content in individual human genomes. Genome Res. 2010;20:1262–70.
18. Witherspoon DJ, Xing J, Zhang Y, Watkins WS, Batzer MA, Jorde LB. Mobile element scanning (ME-Scan) by targeted high-throughput sequencing. BMC Genomics. 2010;11:410.
19. Hormozdiari F, Alkan C, Ventura M, Hajirasouliha I, Malig M, Hach F, Yorukoglu D, Dao P, Bakhshi M, Sahinalp SC, Eichler EE. Alu repeat discovery and characterization within human genomes. Genome Res. 2011;21:840–9.
20. Stewart C, Kural D, Stromberg MP, Walker JA, Konkel MK, Stutz AM, Urban AE, Grubert F, Lam HY, Lee WP, et al. A comprehensive map of mobile element insertion polymorphisms in humans. PLoS Genet. 2011;7:e1002236.
21. Mathias SL, Scott AF, Kazazian Jr HH, Boeke JD, Gabriel A. Reverse transcriptase encoded by a human transposable element. Science. 1991;254: 1808–10.
22. Feng Q, Moran JV, Kazazian Jr HH, Boeke JD. Human L1 retrotransposon encodes a conserved endonuclease required for retrotransposition. Cell. 1996;87:905–16.
23. Dewannieux M, Esnault C, Heidmann T. LINE-mediated retrotransposition of marked Alu sequences. Nat Genet. 2003;35:41–8.
24. Alves G, Tatro A, Fanning T. Differential methylation of human LINE-1 retrotransposons in malignant cells. Gene. 1996;176:39–44.
25. Jurgens B, Schmitz-Drager BJ, Schulz WA. Hypomethylation of L1 LINE sequences prevailing in human urothelial carcinoma. Cancer Res. 1996;56: 5698–703.
26. Lin CH, Hsieh SY, Sheen IS, Lee WC, Chen TC, Shyu WC, Liaw YF. Genome-wide hypomethylation in hepatocellular carcinogenesis. Cancer Res. 2001; 61:4238–43.
27. Chalitchagorn K, Shuangshoti S, Hourpai N, Kongruttanachok N, Tangkijvanich P, Thong-ngam D, Voravud N, Sriuranpong V, Mutirangura A. Distinctive pattern of LINE-1 methylation level in normal tissues and the association with carcinogenesis. Oncogene. 2004;23:8841–6.
28. Estecio MR, Gharibyan V, Shen L, Ibrahim AE, Doshi K, He R, Jelinek J, Yang AS, Yan PS, Huang TH, et al. LINE-1 hypomethylation in cancer is highly variable and inversely correlated with microsatellite instability. PLoS ONE. 2007;2:e399.
29. Rodic N, Sharma R, Sharma R, Zampella J, Dai L, Taylor MS, Hruban RH, Iacobuzio-Donahue CA, Maitra A, Torbenson MS, et al. Long interspersed element-1 protein expression is a hallmark of many human cancers. Am J Pathol. 2014;184:1280–6.
30. Miki Y, Nishisho I, Horii A, Miyoshi Y, Utsunomiya J, Kinzler KW, Vogelstein B, Nakamura Y. Disruption of the APC gene by a retrotransposal insertion of L1 sequence in a colon cancer. Cancer Res. 1992;52:643–5.
31. Solyom S, Ewing AD, Rahrmann EP, Doucet T, Nelson HH, Burns MB, Harris RS, Sigmon DF, Casella A, Erlanger B, et al. Extensive somatic L1 retrotransposition in colorectal tumors. Genome Res. 2012;22:2328–38.
32. Lee E, Iskow R, Yang L, Gokcumen O, Haseley P, Luquette 3rd LJ, Lohr JG, Harris CC, Ding L, Wilson RK, et al. Landscape of somatic retrotransposition in human cancers. Science. 2012;337:967–71.
33. Shukla R, Upton KR, Munoz-Lopez M, Gerhardt DJ, Fisher ME, Nguyen T, Brennan PM, Baillie JK, Collino A, Ghisletti S, et al. Endogenous retrotransposition activates oncogenic pathways in hepatocellular carcinoma. Cell. 2013;153:101–11.
34. Tubio JM, Li Y, Ju YS, Martincorena I, Cooke SL, Tojo M, Gundem G, Pipinikas CP, Zamora J, Raine K, et al. Mobile DNA in cancer. Extensive transduction of nonrepetitive DNA mediated by L1 retrotransposition in cancer genomes. Science. 2014;345:1251343.
35. Doucet-O'Hare TT, Rodic N, Sharma R, Darbari I, Abril G, Choi JA, Young Ahn

J, Cheng Y, Anders RA, Burns KH, et al. LINE-1 expression and retrotransposition in Barrett's esophagus and esophageal carcinoma. Proc Natl Acad Sci U S A. 2015;112:E4894–900.

36. Rodic N, Steranka JP, Makohon-Moore A, Moyer A, Shen P, Sharma R, Kohutek ZA, Huang CR, Ahn D, Mita P, et al. Retrotransposon insertions in the clonal evolution of pancreatic ductal adenocarcinoma. Nat Med. 2015; 21:1060–4.

37. T'Ang A, Varley JM, Chakraborty S, Murphree AL, Fung YK. Structural rearrangement of the retinoblastoma gene in human breast carcinoma. Science. 1988;242:263–6.

38. Solimini NL, Xu Q, Mermel CH, Liang AC, Schlabach MR, Luo J, Burrows AE, Anselmo AN, Bredemeyer AL, Li MZ, et al. Recurrent hemizygous deletions in cancers may optimize proliferative potential. Science. 2012;337:104–9.

39. Akagi K, Suzuki T, Stephens RM, Jenkins NA, Copeland NG. RTCGD: retroviral tagged cancer gene database. Nucleic Acids Res. 2004;32:D523–7.

40. de Ridder J, Uren A, Kool J, Reinders M, Wessels L. Detecting statistically significant common insertion sites in retroviral insertional mutagenesis screens. PLoS Comput Biol. 2006;2:e166.

41. Bamford S, Dawson E, Forbes S, Clements J, Pettett R, Dogan A, Flanagan A, Teague J, Futreal PA, Stratton MR, Wooster R. The COSMIC (Catalogue of Somatic Mutations in Cancer) database and website. Br J Cancer. 2004;91: 355–8.

42. Paull KD, Shoemaker RH, Hodes L, Monks A, Scudiero DA, Rubinstein L, Plowman J, Boyd MR. Display and analysis of patterns of differential activity of drugs against human tumor cell lines: development of mean graph and COMPARE algorithm. J Natl Cancer Inst. 1989;81:1088–92.

43. Baillie JK, Barnett MW, Upton KR, Gerhardt DJ, Richmond TA, De Sapio F, Brennan PM, Rizzu P, Smith S, Fell M, et al. Somatic retrotransposition alters the genetic landscape of the human brain. Nature. 2011;479:534–7.

44. Burns KH, Boeke JD. Human transposon tectonics. Cell. 2012;149:740–52.

45. Denli AM, Narvaiza I, Kerman BE, Pena M, Benner C, Marchetto MC, Diedrich JK, Aslanian A, Ma J, Moresco JJ, et al. Primate-specific ORF0 contributes to retrotransposon-mediated diversity. Cell. 2015;163:583–93.

46. Schwartz S, Gal-Mark N, Kfir N, Oren R, Kim E, Ast G. Alu exonization events reveal features required for precise recognition of exons by the splicing machinery. PLoS Comput Biol. 2009;5:e1000300.

47. Storlazzi CT, Mertens F, Mandahl N, Gisselsson D, Isaksson M, Gustafson P, Domanski HA, Panagopoulos I. A novel fusion gene, SS18L1/SSX1, in synovial sarcoma. Genes Chromosomes Cancer. 2003;37:195–200.

48. Ellison G, Klinowska T, Westwood RF, Docter E, French T, Fox JC. Further evidence to support the melanocytic origin of MDA-MB-435. Mol Pathol. 2002;55:294–9.

49. Garraway LA, Widlund HR, Rubin MA, Getz G, Berger AJ, Ramaswamy S, Beroukhim R, Milner DA, Granter SR, Du J, et al. Integrative genomic analyses identify MITF as a lineage survival oncogene amplified in malignant melanoma. Nature. 2005;436:117–22.

50. Ross DT, Scherf U, Eisen MB, Perou CM, Rees C, Spellman P, Iyer V, Jeffrey SS, Van de Rijn M, Waltham M, et al. Systematic variation in gene expression patterns in human cancer cell lines. Nat Genet. 2000;24:227–35.

51. Afgan E, Baker D, van den Beek M, Blankenberg D, Bouvier D, Cech M, Chilton J, Clements D, Coraor N, Eberhard C, et al. The Galaxy platform for accessible, reproducible and collaborative biomedical analyses: 2016 update. Nucleic Acids Res. 2016;44(W1):W3–W10.

52. Mir AA, Philippe C, Cristofari G. euL1db: the European database of L1HS retrotransposon insertions in humans. Nucleic Acids Res. 2015;43(Database issue):D43–7.

53. Wang J, Song L, Grover D, Azrak S, Batzer MA, Liang P. dbRIP: a highly integrated database of retrotransposon insertion polymorphisms in humans. Hum Mutat. 2006;27:323–9.

54. Subramanian A, Tamayo P, Mootha VK, Mukherjee S, Ebert BL, Gillette MA, Paulovich A, Pomeroy SL, Golub TR, Lander ES, Mesirov JP. Gene set enrichment analysis: a knowledge-based approach for interpreting genome-wide expression profiles. Proc Natl Acad Sci U S A. 2005;102:15545–50.

55. Szklarczyk D, Franceschini A, Kuhn M, Simonovic M, Roth A, Minguez P, Doerks T, Stark M, Muller J, Bork P, et al. The STRING database in 2011: functional interaction networks of proteins, globally integrated and scored. Nucleic Acids Res. 2011;39:D561–8.

56. Edgar R, Domrachev M, Lash AE. Gene Expression Omnibus: NCBI gene expression and hybridization array data repository. Nucleic Acids Res. 2002; 30:207–10.

57. Liu H, D'Andrade P, Fulmer-Smentek S, Lorenzi P, Kohn KW, Weinstein JN, Pommier Y, Reinhold WC. mRNA and microRNA expression profiles of the NCI-60 integrated with drug activities. Mol Cancer Ther. 2010;9:1080–91.

6

Deciphering fact from artifact when using reporter assays to investigate the roles of host factors on L1 retrotransposition

Pamela R. Cook[1]* 🆔 and G. Travis Tabor[2]

Abstract

Background: The Long INterspersed Element-1 (L1, LINE-1) is the only autonomous mobile DNA element in humans and has generated as much as half of the genome. Due to increasing clinical interest in the roles of L1 in cancer, embryogenesis and neuronal development, it has become a priority to understand L1-host interactions and identify host factors required for its activity. Apropos to this, we recently reported that L1 retrotransposition in HeLa cells requires phosphorylation of the L1 protein ORF1p at motifs targeted by host cell proline-directed protein kinases (PDPKs), which include the family of mitogen-activated protein kinases (MAPKs). Using two engineered L1 reporter assays, we continued our investigation into the roles of MAPKs in L1 activity.

Results: We found that the MAPK p38δ phosphorylated ORF1p on three of its four PDPK motifs required for L1 activity. In addition, we found that a constitutively active p38δ mutant appeared to promote L1 retrotransposition in HeLa cells. However, despite the consistency of these findings with our earlier work, we identified some technical concerns regarding the experimental methodology. Specifically, we found that exogenous expression of p38δ appeared to affect at least one heterologous promoter in an engineered L1 reporter, as well as generate opposing effects on two different reporters. We also show that two commercially available non-targeting control (NTC) siRNAs elicit drastically different effects on the apparent retrotransposition reported by both L1 assays, which raises concerns about the use of NTCs as normalizing controls.

Conclusions: Engineered L1 reporter assays have been invaluable for determining the functions and critical residues of L1 open reading frames, as well as elucidating many aspects of L1 replication. However, our results suggest that caution is required when interpreting data obtained from L1 reporters used in conjunction with exogenous gene expression or siRNA.

Keywords: L1, LINE-1, Reporter, Host factor, p38, HSV-TK, SV40, Promoter, Renilla

Background

The only active, autonomous mobile DNA element in humans is the Long INterspersed Element-1 (LINE-1, L1) retrotransposon, which is responsible for generating almost half of the human genome via insertion of its own DNA and that of non-autonomous Short-INterspersed repeat Elements (SINES) [1]. These insertions, combined with 3′ transductions, nonallelic homologous recombination and mobilization of cellular mRNAs, have had a

defining impact on genomic architecture, and the consequences on gene regulation and human development are largely unknown [2–5]. L1 activity is restricted to certain cell types (reviewed in [6]), and retrotransposition is thought to occur mainly in embryonic cells [7, 8], pluripotent stem cells [9, 10], adult neuronal development [11–15], and cancer [16–19]. Clinical interest in L1 has increased due to its mutagenic and disease-causing potential [11, 20–23], as well as its association with cancer [16–19]. In addition, a growing number of studies suggest that transposable elements can be co-opted to serve fundamental physiological functions [24–30]. Recent work has thus been aimed at identifying cellular host factors required for L1

* Correspondence: pamela.cook@nih.gov
[1]Laboratory of Cell and Molecular Biology, National Institute of Diabetes and Digestive and Kidney Diseases, National Institutes of Health, 8 Center Drive, Bethesda, MD 20892, USA
Full list of author information is available at the end of the article

expression, repression and reactivation. With respect to this, our laboratory recently demonstrated that host proline-directed protein kinase (s) (PDPKs) phosphorylate the L1 protein ORF1p on multiple PDPK motifs required for L1 retrotransposition [31].

PDPK target motifs consist of serines or threonines with a proline in the +1 position (S/T-P motifs), which in ORF1p are: S18/P19; S27/P28; T203/P204; and T213/P214. The PDPK family includes mitogen-activated protein kinases (MAPKs), cyclin dependent kinases (CDKs) and glycogen synthase kinase 3 (GSK3). Prior to our finding that the phosphorylation of ORF1p by PDPKs is necessary for L1 activity, several studies reported associations between L1 and the PDPK p38 [32–34], a MAPK that exists in four different isoforms, α, β, γ and δ [35]. Moreover, the expression of one isoform, p38δ, can be induced in primary cell cultures via exogenous expression of ORF1p [34].

Given these associations between L1 and the PDPK p38, as well as our previous findings that host PDPKs are required for L1 retrotransposition, we decided to investigate the role of each p38 isoform on ORF1p phosphorylation and L1 activity. Although our studies are ongoing, we believe that dissemination of our present findings and their associated experimental pitfalls will be useful to the L1 research community. We report here that: 1) different populations of HeLa cells can result in different experimental outcomes; 2) two presumably complementary L1 retrotransposition reporter assays produced conflicting results when coupled with exogenously expressed p38δ; and 3) two different non-targeting control (NTC) small interfering RNA (siRNA) sequences differentially affected measured L1 activity.

Results

MAPK p38δ phosphorylates ORF1p on S/T-P motifs

We first determined whether activated wild type p38δ (WT, Invitrogen) could phosphorylate ORF1p on its S/T-P motifs, which are required for robust L1 activity [31]. In vitro radioactive kinase assays revealed that p38δ-WT exclusively phosphorylated bacterially purified ORF1p on these residues, as an ORF1p carrying mutations at all four motifs, S18A/S27A/T203G/T213G (AAGG), was not phosphorylated (Fig. 1a top). We next tested the ability of p38δ-WT to phosphorylate the ORF1p mutants S18A/S27A (AA) and T203G/T213G (GG), and found that the majority of phosphorylation occurred on the GG mutant, which retained both serine motifs (Fig. 1a top).

In order to compare the degree of phosphorylation at each motif, we constructed a series of mutants, each bearing only one intact S/T-P motif: SAGG (S27A/T203G/T213G); ASGG (S18A/T203G/T213G); AATG (S18A/S27A/T213G); and AAGT (S18A/S27A/T203G).

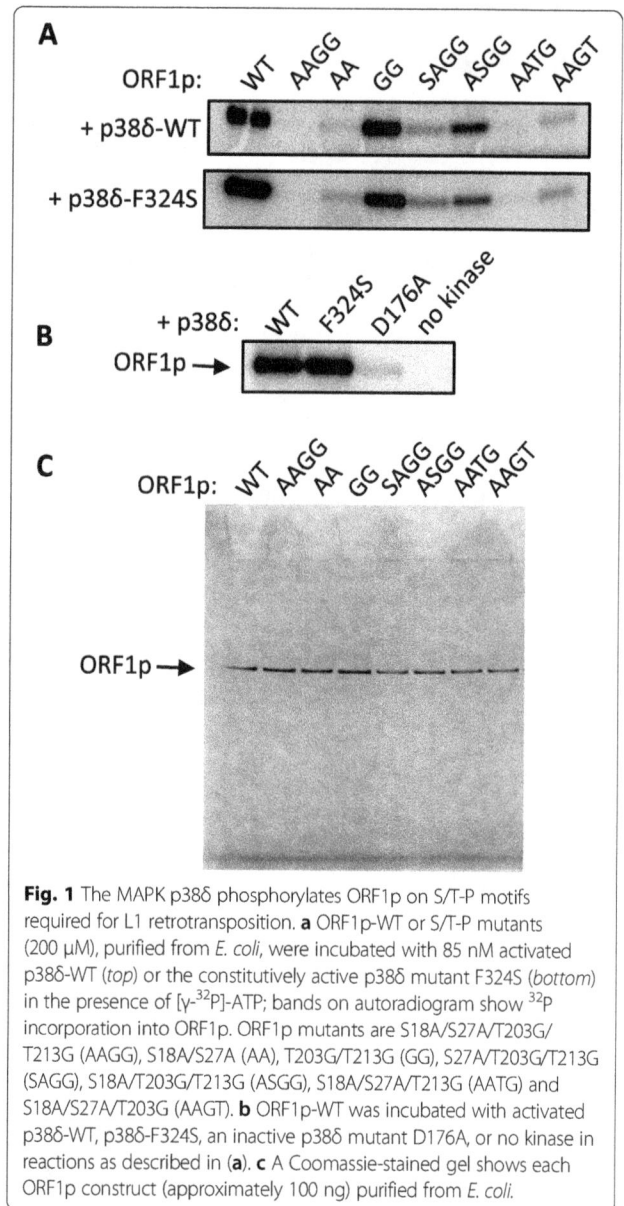

Fig. 1 The MAPK p38δ phosphorylates ORF1p on S/T-P motifs required for L1 retrotransposition. **a** ORF1p-WT or S/T-P mutants (200 μM), purified from *E. coli*, were incubated with 85 nM activated p38δ-WT (*top*) or the constitutively active p38δ mutant F324S (*bottom*) in the presence of [γ-^{32}P]-ATP; bands on autoradiogram show ^{32}P incorporation into ORF1p. ORF1p mutants are S18A/S27A/T203G/T213G (AAGG), S18A/S27A (AA), T203G/T213G (GG), S27A/T203G/T213G (SAGG), S18A/T203G/T213G (ASGG), S18A/S27A/T213G (AATG) and S18A/S27A/T203G (AAGT). **b** ORF1p-WT was incubated with activated p38δ-WT, p38δ-F324S, an inactive p38δ mutant D176A, or no kinase in reactions as described in (**a**). **c** A Coomassie-stained gel shows each ORF1p construct (approximately 100 ng) purified from *E. coli*.

S27 (ASGG) was phosphorylated by p38δ-WT to the greatest extent (Fig. 1a top). T213 (AAGT) was phosphorylated to approximately the same degree as S18 (SAGG), but p38δ-WT showed almost no activity on T203 (AATG). Of note, results from the kinase prediction program NetPhosK 1.0 [36] indicated that unspecified p38 isoforms were expected to target ORF1p at S18, T203 and T213, but not S27.

Constitutively active p38δ-F324S retains ORF1p substrate specificity

Various p38δ mutants that retain some degree of constitutive activity independent of phosphorylation by their activating upstream kinases in the MAPK pathway have been described [37]. In those studies, the constitutively active mutant p38δ-F324S retained the substrate specificity of activated p38δ-WT for glutathione S-transferase activating transcription factor 2 (GST-ATF2) in vitro when p38δ-F324S was purified from bacteria or immunoprecipitated from HEK293 cell lysates. We found that bacterially purified p38δ-F324S also exhibited wild type substrate specificity for ORF1p's S/T-P motifs (Fig. 1a bottom). In addition, we tested the mutant p38δ-D176A, which was reported to have no activity on GST-ATF2 when purified from bacteria but greater activity than p38δ-WT when immunoprecipitated from HEK293 cells [37]. Bacterially purified p38δ-D176A barely phosphorylated ORF1p in vitro compared to p38δ-WT or p38δ-F324S (Fig. 1b). Figure 1c shows each ORF1p construct, purified from *E. coli*, used for the in vitro kinase assays.

L1 reporter assays

Given our findings that p38δ specifically phosphorylated ORF1p S/T-P motifs, we proceeded to determine the effect of p38δ on L1 retrotransposition. To assess this, we used two previously characterized L1 reporter assays. The original L1 retrotransposition reporter, JM101 (a kind gift from Dr. John Moran), relies on the splicing of an artificial intron from an L1-borne neomycin-resistant gene and its L1-mediated conversion into genomic DNA to produce cell foci resistant to the neomycin analog G418 [38]. Specifically, the reporter contains a full-length L1 element driven by the cytomegalovirus (CMV) promoter and an *mneo* cassette that encodes the neomycin-resistant gene (*neo*), driven by a Simian virus 40 (SV40) promoter located within the 3′ untranslated region (UTR) of L1 (Fig. 2 top). The *neo* gene product, also known as aminoglycoside 3′-phosphotransferase-II (APH (3′)-II), phosphorylates and thereby inactivates G418. Selection with G418 is begun approximately three days following transfection of the reporter plasmid into retrotransposition-competent cells and is continued for 10–12 days. The arrangement of the *neo* gene in JM101 ensures that only cells that have undergone retrotransposition by the L1 reporter element will express APH (3′)-II. The coding sequence for *neo* and its promoter are located on the antisense strand in the 3′ untranslated region of L1. Within this sequence is the engineered artificial intron, but it can only be spliced from the L1 sense RNA driven by the L1 promoter due to the orientation of the splice donor (SD) and splice acceptor (SA) sequences. Once spliced, the L1 RNA is retrotransposed into cDNA and inserted into the genome. After synthesis of the complementary DNA strand, which contains the spliced *neo* template, the transcript for APH (3′)-II can be initiated from the antisense promoter.

The more recently developed single-vector dual luciferase L1 reporters (kind gifts from Dr. Wenfeng An) are based on the same principle as the *mneo* reporter, but

instead of *neo* they contain the gene for Firefly luciferase (Fluc). Fluc is also driven by an SV40 promoter and interrupted by an intron to monitor retrotransposition (Fig. 2, lower schematics) [39]. In addition, this reporter contains an internal control gene expressing Renilla luciferase (Rluc) driven by a Herpes simplex virus thymidine kinase (HSV-TK) promoter. Constitutively active Rluc expression is intended as a normalizing control for variations in cell plating, transfection efficiency and survival. Four days following transfection, cells are lysed and retrotransposition is reported as a function of Rluc-normalized Fluc luminescence. The three single-vector luciferase reporters used in this study were: pYX017, which contains an L1 element driven by a hybrid CAG promoter consisting of the CMV enhancer fused with a modified chicken beta-actin promoter and a splice element from the rabbit beta-globin gene [40]; pYX014, which contains only the native L1 promoter in the 5′UTR; and pYX015, a negative control, which is identical to pYX014 except that it carries two missense mutations in ORF1p and is thus retrotransposition-incompetent [38, 39].

Constitutively active p38δ increases G418-resistant colonies

Consistent with our in vitro results and our previous findings that the phosphorylation of ORF1p S/T-P motifs is required for robust L1 activity, we found that exogenous expression of the constitutively active p38δ-F324S (FS) appeared to increase L1 retrotransposition in the G418-based assay relative to an empty vector control (EV), while p38δ-D176A (DA), which failed to appreciably phosphorylate ORF1p in vitro, inhibited L1 (Fig. 3a top). Surprisingly, p38δ-WT (WT) also repressed formation of G418-resistant colonies (Fig. 3a top left). These effects did not appear to be a result of altered cell viability, as only p38δ-D176A somewhat affected cell growth (Fig. 3a bottom left). To determine whether the observed decrease in colony density resulting from p38δ-WT overexpression might be due to effects of the expression vector on cotransfection efficiencies, we cotransfected an expression plasmid for the enhanced green fluorescent protein (EGFP, a kind gift from Dr. Birong Shen) with either the pcDNA empty vector, p38δ-WT or p38δ-F324S. Neither p38δ-WT nor p38δ-F324S appreciably altered EGFP fluorescence compared to the empty vector (Fig. 3a right).

The inhibition of L1 by p38δ-WT may be explained by the fact that p38, like other MAPKs, relies on a complex network of docking interactions with many proteins, including substrates, upstream activating MAPK kinases, phosphatases and scaffolding and regulatory factors. These interactions collectively synchronize the activation and localization of p38 via feedback loops and crosstalk with other pathways [[41] and references therein]. Thus,

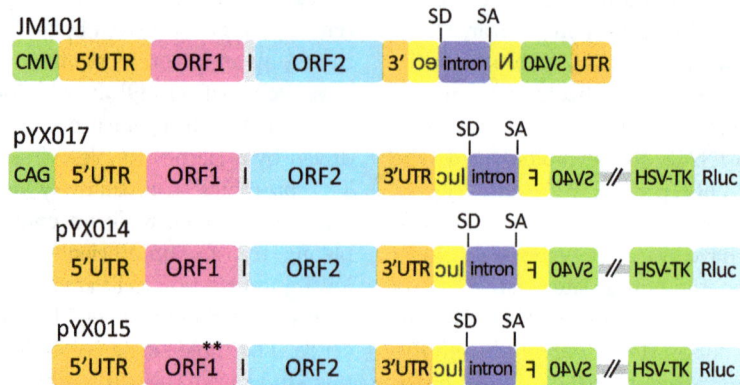

Fig. 2 Schematic of L1 reporter plasmids. All reporters contain a full-length L1 element with 5′ and 3′ UTRs (*orange*), ORF1 (*pink*), intergenic region (*gray*), ORF2 (*blue*) and a retrotransposition reporter (*yellow*) interrupted by an artificial intron (*purple*) with splice donor (SD) and acceptor (SA) sites. In JM101, L1 is driven by the CMV promoter (*green*), and in pYX017 by the hybrid CAG promoter (*green*). pYX014 contains only the native L1 promoter in the 5′UTR, and pYX015 is identical to pYX014 except for two missense mutations (R261A/R262A) [38] in ORF1p, rendering pYX015 incompetent for retrotransposition. The reporter in JM101 is an *mneo* cassette driven by the SV40 promoter (*green*) located within the 3′ UTR. The pYX017, pYX014 and pYX015 constructs contain a Firefly luciferase reporter (Fluc), also driven by SV40 (*green*), as well as a gene for Renilla luciferase (Rluc; aqua) driven by the HSV-TK promoter (*green*)

a pool of excess, unactivated p38δ-WT could perturb this regulatory system, or may simply compete with the population of endogenous activated p38, resulting in inhibition of L1. Consistent with this possibility are several studies that showed expression of a nonfunctional p38 has a dominant negative effect on endogenous p38 activity [42–46]. In addition, during some of our own preliminary experiments, we found on rare occasion that exogenous p38δ-WT slightly increased rather than decreased the number of G418-resistant colonies (unpublished data), further suggesting that the effect of exogenous p38δ-WT could depend on cellular conditions that affect the p38 pathway. For example, confluent stock cultures, as opposed to proliferating cultures, have been found to activate endogenous p38α, with effects lasting up to 48 h after re-plating [47]. However, our investigation of this and several other routine tissue culture variables, including the amount of time cells were exposed to trypsin during sub-culturing, the presence or absence of antibiotics in culture media, lot-to-lot variations in fetal bovine serum (FBS), passage number or overall time in culture, revealed no correlation with the effect of exogenous p38δ on L1 activity (unpublished data). A previous report indicated that individual HeLa clones can exhibit varying degrees of retrotransposition activity and that certain clones may grow to dominate a mixed culture over time [48]. This phenomenon may also bear on how exogenous host factors impact L1 activity.

Effects of MAPK p38δ-WT differ depending on the L1 reporter assay used

As part of our efforts to understand the effects of p38δ-WT on L1, we used the single-vector dual luciferase assay in parallel with the G418-based assay (i.e. cells were plated from a common suspension and transfected simultaneously using the same reagents). Data from dual luciferase assays are typically normalized to Rluc expression and reported as a ratio of Fluc/Rluc luminescence. Using this method in an experiment done in parallel with the G418-based assay in Fig. 3a, we found that p38δ-WT, p38δ-F324S, and, surprisingly, p38δ-D176A, increased L1 retrotransposition by 5, 7.7 and 7 fold, respectively (Fig. 3b). However, the Fluc/Rluc luminescence ratio is valid only if the expression of Rluc is independent of the experimental treatment. It is obvious from the individual luminescence data for Fluc and Rluc shown in Figs. 3c and d that p38δ expression dramatically affected Rluc luminescence. Such a decrease in Rluc in the absence of a corresponding decrease in cell survival or transfection efficiency would thus artificially inflate the Fluc/Rluc ratio. As shown previously, cell growth was not detectably affected by p38δ-WT or p38δ-F324S, and we detected no differences in cell densities in any wells during the course of the luciferase assay. Moreover, we found no effect from p38δ-WT or p38δ-F324S in the previous cotransfection efficiency control experiment using EGFP. Combined, these data strongly suggest that Rluc, driven by the HSV-TK promoter, is an inadequate normalizing control for these experiments.

Rluc expression notwithstanding, Fluc, like APH (3′)-II, reports on raw retrotransposition events and would thus be expected to produce results paralleling those of the G418 assay under similar experimental conditions. If we then consider only Fluc luminescence, the effects of p38δ-F324S and p38δ-D176 roughly coincide in direction if not degree with those observed in the G418 assay. However, p38δ-WT appears to affect the two reporters differently, inhibiting G418-resistant colony formation but slightly increasing Fluc luminescence (Figs. 3a, c and

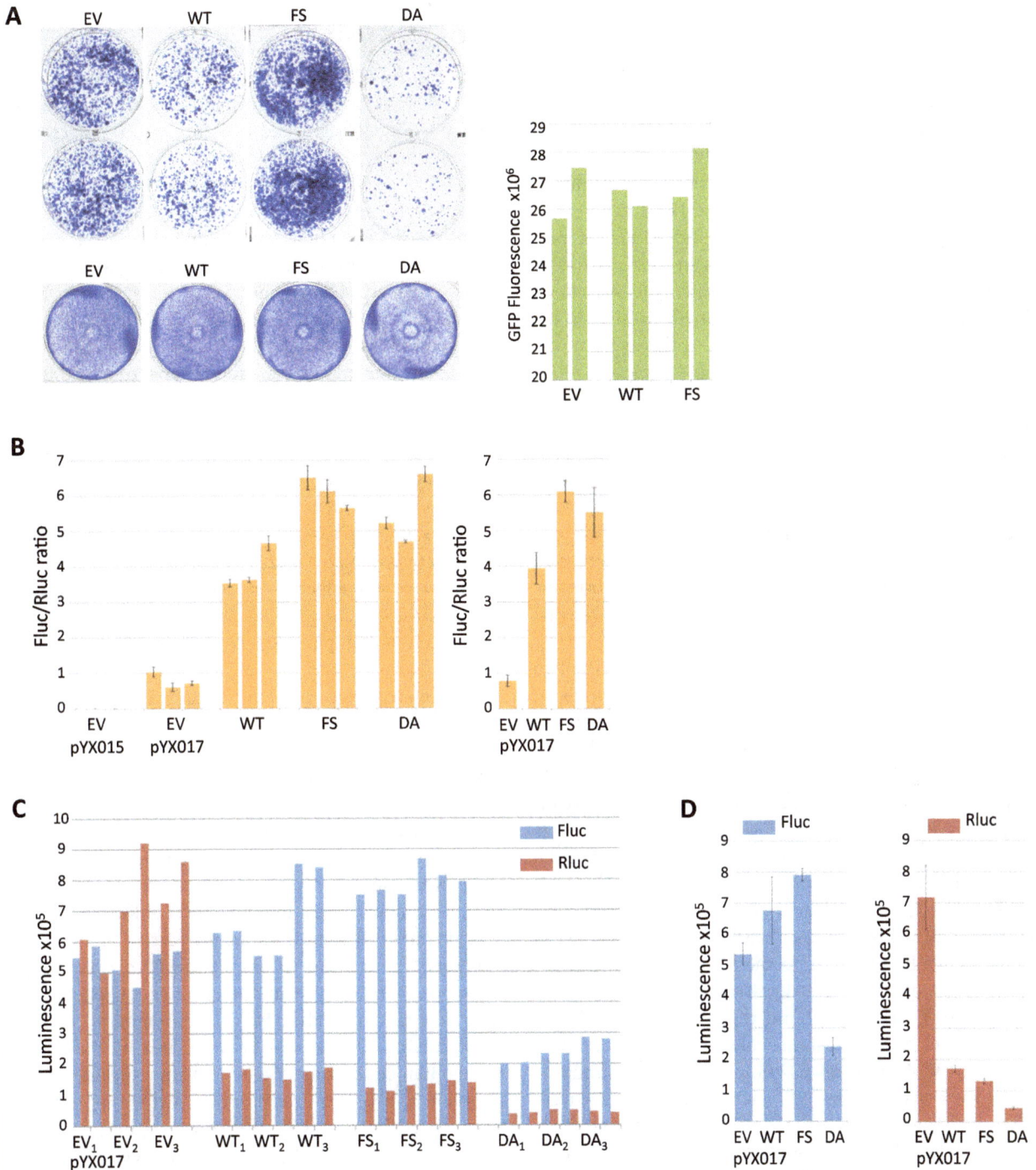

Fig. 3 Effects of p38δ on two different L1 reporter assays. **a** *Top rows* show duplicate wells of Giemsa-stained G418-resistant colonies resulting from transfection of the L1 reporter JM101 in the presence of pcDNA mammalian expression vectors for: empty vector (EV), p38δ-WT (WT), p38δ-F324S (FS) or p38δ-D176A (DA). Bottom row shows the effect of each pcDNA expression vector on cell growth. The right panel indicates fluorescence intensities obtained from cotransfection of EGFP with each indicated p38δ construct or empty vector; results from duplicate wells are shown. **b** Relative Fluc/Rluc luminescence ratios obtained from lysates of HeLa cells transfected with the L1 reporter plasmid pYX015 or pYX017 in the presence of indicated pcDNA mammalian expression vectors. Three biological replicates are shown for each experimental condition; error bars represent the SEM from two technical replicates (defined as two distinct samples taken from each biological sample). The graph at right shows the average of three biological replicates shown separately in the left panel; error bars indicate the SEM, *n* = 3 biological replicates. **c** Individual luminescence values are shown for Fluc (*blue*) and Rluc (*red*) used to calculate the Fluc/Rluc ratios from pYX017 in (**b**); technical replicates are side-by-side; biological replicates are indicated in subscript. **d** Mean Fluc and Rluc luminescence values were derived by first averaging the technical replicates for each biological sample (*n* = 2), and then averaging the resulting values of each biological replicate; error bars represent the SEM of biological replicates, *n* = 3

d left). As with the G418 assay, our preliminary experiments using the dual luciferase assay sometimes showed an outlier effect of p38δ-WT, but in this case the outlier was repression of Fluc (unpublished data). Although sub clonal HeLa populations may have been a contributing factor in those experiments, which utilized different stocks of cells, it would not explain differential effects of p38δ-WT on two reporters in experiments performed in parallel using a common suspension of HeLa cells.

Two questions thus arose: 1) why did p38δ-WT predominantly decrease colony numbers in the G418 assay but increase Fluc luminescence, while the effects of p38δ-F324S and p38δ-D176A remained consistent between the two reporters, and 2) what is the cause of decreased Rluc expression in the presence of p38δ?

With respect to the first question, it may be significant that variations were most evident in response to p38δ-WT since it, unlike F324S, would be dependent on a network of cellular factors for activation. This possibility notwithstanding, if the inhibitory effects of p38δ-WT in the G418-based assay arose from competition with endogenous p38δ, one would expect equivalent competition, not activation, with the pYX017 reporter. Since this was not what we observed, we then considered variables in the assays themselves that might explain the differential effects of p38δ-WT.

The first and most obvious difference between the two reporters is that L1 is driven by a CMV promoter in JM101 but a CAG promoter in pYX017, though the CAG promoter contains a CMV enhancer element (Fig. 2). CMV promoters can be affected by some p38 isoforms [49–53], but we did not observe a significant effect of p38δ-WT or p38δ-F324S on EGFP, which is also driven by a CMV promoter. To address whether the increase in Fluc luminescence stemmed from effects of p38δ on the CAG promoter, we used the pYX014 construct, which is identical to pYX017 except that it relies on the native L1 promoter in the 5′ UTR for L1 expression instead of CAG (Fig. 2). Using JM101 in parallel with pYX014, we again found that p38δ-WT inhibited formation of G418-resistant colonies (Fig. 4a), while both p38δ-WT and p38δ-F324S increased Fluc luminescence from pYX014 by 1.5 and 2.2 fold, respectively (Figs. 4b left and c), compared to 1.3 and 1.5 fold from pYX017 (Fig. 3d left). Since p38δ-WT increased Fluc in both pYX014 and pYX017, the effect of p38δ-WT appears to be independent of the CAG promoter in pYX017. We eliminated p38δ-D176A from this and further experiments given its effect on cell growth (Fig. 3) as well as the report that, despite its inactivity in vitro, it can be activated in HEK293 cells [37], making its effects on L1 uninterpretable, particularly given the inhibitory effect of p38δ-WT on G418-resistant colony formation.

Regarding the effect of p38δ on Rluc luminescence, we considered three possible explanations: 1) cell death; 2)

transcription or translation interference from pcDNA-p38δ; or 3) inhibition of the Rluc HSV-TK promoter.

As stated earlier, we found no evidence of cell death, despite a 76–94% decrease in Rluc luminescence using pYX017 (Figs. 3c and d right) and similar decreases with pYX014 (Fig. 4b right and c). Moreover, the decrease in Rluc luminescence from the retrotransposition defective pYX015 (Fig. 4c) ruled out the possibility that rampant L1 activity severely compromised the cells, an event the G418-based assay could have potentially missed.

The second option was that decreased Rluc luminescence resulted from generalized transcription and/or translation interference from the cotransfected plasmids. Competition for cellular factors can be relevant at multiple points, including promoter binding, transcription initiation, elongation or translation [54–57]. For example, the different levels of Rluc luminescence from pYX017 (Fig. 3) compared with pYX014 (Fig. 4) might suggest that the highly active heterologous CAG promoter in pYX017 competed with factors required by the HSV-TK promoter driving Rluc in pYX017. Also, the empty vector control lacked an optimized Kozak sequence, which may have rendered it less effective at competing for translational machinery than the p38δ constructs. To determine if the kinase-containing plasmids competed with pYX017 for factors necessary for Rluc expression, we cotransfected the L1 reporter with plasmids encoding constitutively active MAPK-kinases (MAPKKs) MKK3b-S288E/T222E (M3) or MKK6-S207E/T211E (M6), which are specific upstream activators of p38 isoforms [58–60]. Unlike p38δ, each MKK upregulated Rluc (Fig. 5a right and b). As expected, each MKK also increased Fluc (Fig. 5a left), presumably via activation of an endogenous p38. Neither of the MKKs had any effect on cell growth (Fig. 5c). These results strongly suggest that inhibition of Rluc by p38δ is a specific rather than indiscriminate effect.

The ability of p38δ to inhibit the Rluc HSV-TK promoter was not empirically determined by us, but multiple reports show that HSV-TK promoters, including those driving Renilla, can be perturbed by multiple experimental conditions [61–64]. These include the expression of the Sp1 transcription factor [64], which is upregulated by p38 [65]. We consider the potential effects of p38 on the HSV-TK and SV40 heterologous promoters, as well as other elements of the L1 reporters, in greater detail in the discussion.

Two non targeting control siRNAs differentially affect reported L1 activity

While investigating the effect of p38δ on L1 retrotransposition, we performed siRNA experiments using a SMARTpool mixture against p38δ (Dharmacon, M-003591-02-0005) and the NTC siRNA #3 (Dharmacon). Although the siRNA against p38δ dramatically reduced

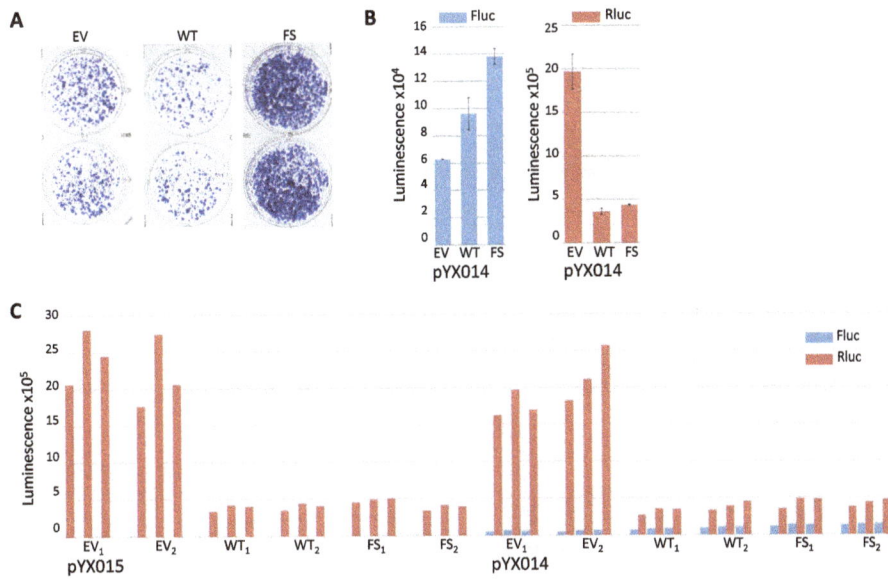

Fig. 4 p38δ increases Fluc independent of a heterologous promoter. **a** Duplicate wells containing G418-resistant colonies resulting from transfection of HeLa cells with the L1 reporter JM101 in the presence of pcDNA mammalian expression vectors for: empty vector (EV), p38δ-WT (WT) or p38δ-F3324S (FS). **b** Mean Fluc (*left*) and Rluc (*right*) luminescence values obtained from lysates of HeLa cells transfected with the L1 reporter plasmid pYX014 in the presence of indicated pcDNA mammalian expression vectors. Averages were derived from raw data shown in (**c**) by first averaging technical replicates for each biological sample ($n = 3$), and averaging biological replicates; error bars represent SEM of biological samples, $n = 2$. **c** Individual luminescence values are shown for Fluc (*blue*) and Rluc (*red*) used to calculate averages in (**b**); technical replicates are side-by-side; biological replicates are indicated with subscripts

the number of G418-resistant colonies relative to NTC #3, RT-PCR showed no significant knockdown of the p38δ transcript (data not shown). Interestingly, however, NTC #3 considerably increased colony density relative to the mock control (Fig. 6a left). EGFP fluorescence from cells pretreated with siRNA prior to transfection suggested that the siRNA had little impact on transfection efficiency (Fig. 6a right). Given these unexpected results, we tested an additional control siRNA, NTC #5, also from Dharmacon. In marked contrast to NTC #3, NTC #5 dramatically reduced G418-resistant colonies relative to the mock control (Fig. 6b top). Neither NTC dramatically affected cell growth, though NTC #3 had a slight inhibitory effect (Fig. 6b bottom). It is notable that unlike p38δ-WT, the NTC siRNAs exerted their respective effects similarly on both Fluc luminescence and G418-resistant colony formation (Fig. 6b top, c left and d). However, L1 activity as reported by the Fluc/Rluc ratio appears to be decreased by NTC #3 rather than increased (Fig. 6c). We did not further investigate potential causes for these results. Information on Dharmacon's website states that each NTC is reported to contain a minimum of 4 mismatches to all human, mouse and rat genes and to have minimal effects in genome-wide targeting via microarray analyses. We did not test Dharmacon's NTC #1, as it was reported to increase cell growth (personal communication, Dharmacon), nor NTC #2 or #4 due to their targeting of Firefly luciferase (Dharmacon website).

Discussion

Engineered L1 reporter assays have tremendously advanced the field of L1 research, allowing investigators to examine key details of the retrotransposition process [66]. Through mutational analyses, critical amino acids in ORF1p and ORF2p have been identified, leading to a greater understanding of the form and functions of these proteins and their roles in L1 retrotransposition. Investigations of L1 insertion sites, 5′ truncations, untranslated regions, native L1 promoters and the poly (A) tail have all been made possible by these assays, as have numerous comparative evolutionary studies of extinct L1 fossils in the human and mouse genomes. Our own work on the role of ORF1p phosphorylation would not have been possible without these reporters.

Importantly, we have not observed variation in relative differences between an L1-WT control and any L1 mutant in our history of working with L1 reporter plasmids. In other words, any mutant L1 construct we have made consistently exhibits the same degree of change in G418-resistant colonies relative to a WT control within a given experiment, independent of differences in cell populations. Thus, the L1 reporters are particularly reliable for investigating *cis* aspects of L1—the purpose for which the original reporter was designed. However, the results presented here strongly suggest that data derived from L1 reporters when used in conjunction with exogenous gene expression or siRNA to investigate the roles of host factors may be challenging to interpret. Although we have not exhaustively investigated possible

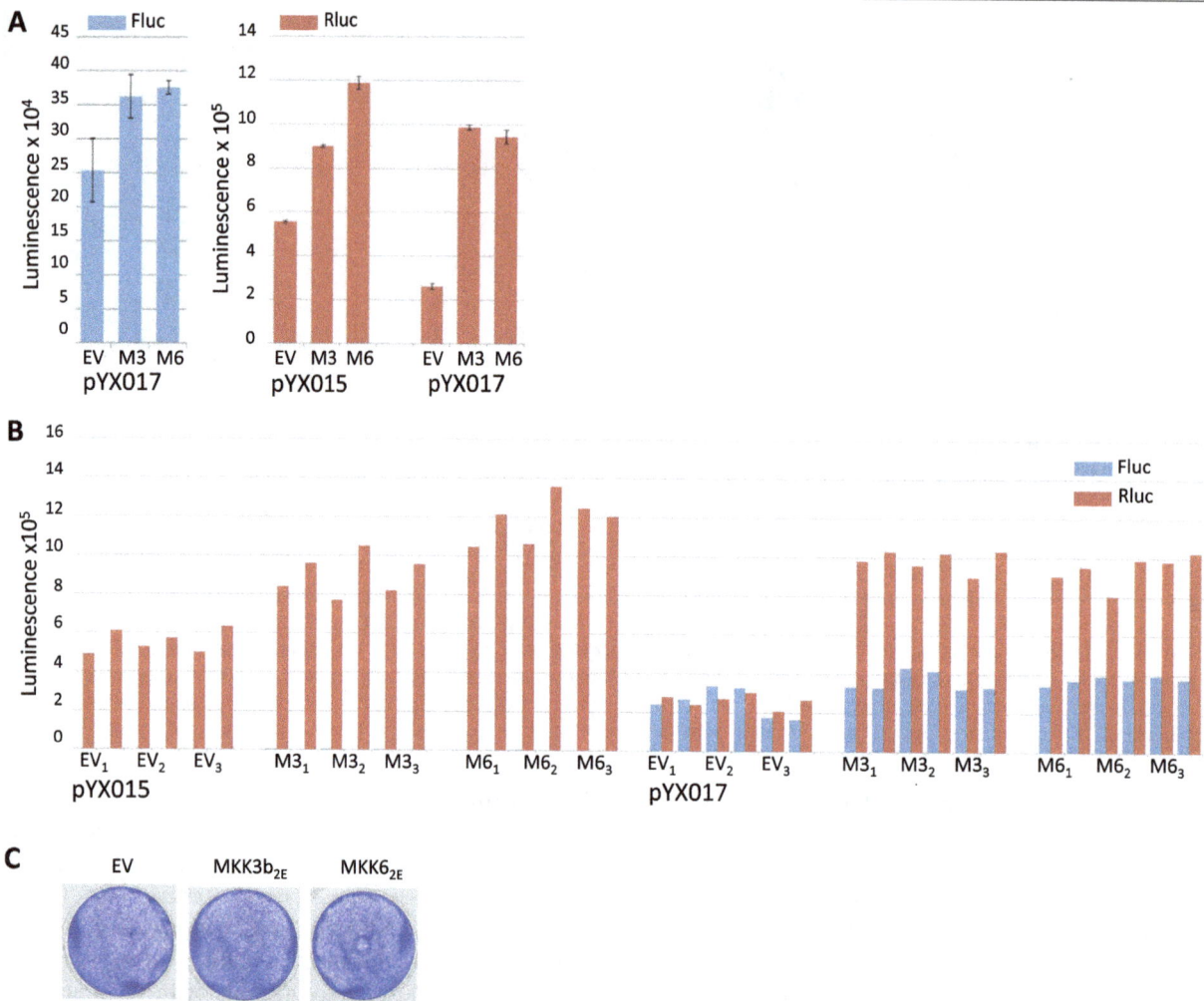

Fig. 5 MKK3b$_{2E}$ and pcDNA-MKK6$_{2E}$ increase Rluc luminescence. **a** Mean Fluc (*left*) and Rluc (*right*) luminescence values obtained from lysates of HeLa cells transfected with the L1 reporter plasmid pYX015 or pYX017 in the presence of pcDNA-MKK3b$_{2E}$ (M3) or pcDNA-MKK6$_{2E}$ (M6). Averages were derived from data shown in (**b**) by first averaging technical replicates for each biological sample ($n = 2$), then using this value to average biological replicates; error bars represent SEM of biological samples, $n = 3$. **b** Individual luminescence values are shown for Fluc (*blue*) and Rluc (*red*) obtained from lysates transfected with pYX015 or pYX017 and the indicated pcDNA expression vectors; technical replicates are side-by-side; biological replicates are indicated with subscripts. **c** Wells show effects on cell growth in response to expression of pcDNA-MKK3b$_{2E}$ (M3) or pcDNA-MKK6$_{2E}$ (M6)

factors that would account for our results, we feel these data are nonetheless informative and potentially timesaving for other researchers intending to use these approaches to investigate interactions between L1 and its host.

Our efforts to determine the effect of p38δ on L1 retrotransposition using engineered L1 reporters underscore the complexities inherent in such endeavors. The p38 signaling pathway itself is extremely complex, with different isoforms having unique, overlapping or competing functions depending on the cell type, or even within the same cell under different conditions [35, 67–69]. This complexity is compounded by the possibility that different p38 isoforms may have competing specificities and

functional outcomes on ORF1p and other substrates relevant to L1 activity, as well as on heterologous promoters in L1 reporters.

A case in point is the repression of Rluc by exogenous p38δ. Previous reports show that p38 can activate late HSV promoters [70] as well as the transcription factor Sp1 [65], which both binds [71] and activates the HSV-TK promoter [61, 64]. These studies would suggest that if p38δ had an effect on HSV-TK, it would be activation, not repression. However, this assumption would be an oversimplification given the complexity of p38 signaling and reports that p38 isoforms can compete with one other with opposing effects [68]. An alternative possibility is that over-expression of exogenous p38δ perturbed constitutive

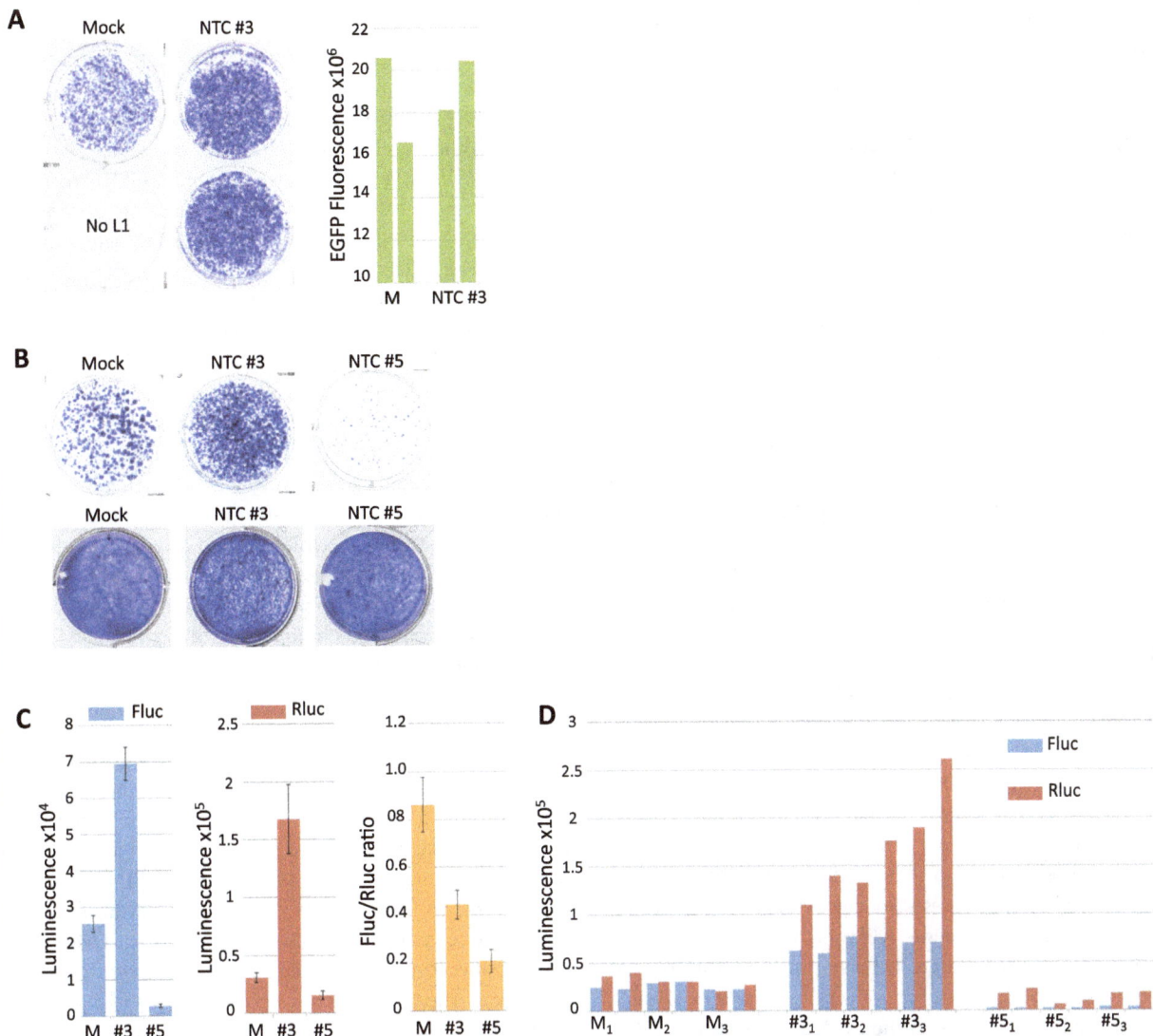

Fig. 6 NTC siRNAs have differential effects on L1 reporter assays. **a** Wells show G418-resistant colonies resulting from transfection of the L1 reporter JM101 in the presence of no siRNA (mock, with transfection reagent only) or 10 nM NTC #3 siRNA. Graph at right shows EGFP fluorescence from cells pretreated with 10 nM NTC #3 siRNA or mock (M); results from duplicate wells are shown. **b** *Top row* shows G418-resistant colonies resulting from the transfection of the L1 reporter JM101 in the presence or absence of 25 nM of indicated siRNA; bottom row shows effect of 25 nM of indicated siRNA on cell growth. **c** Mean Fluc (*left*) and Rluc (*second from right*) luminescence values obtained from lysates of HeLa cells transfected with the L1 reporter pYX017 in the presence of no siRNA (M) or 25 nM NTC #3 or NTC #5; averages were derived from data shown in (**d**) by first averaging technical replicates for each biological sample (*n* = 2), then using this value to average biological replicates; error bars represent SEM of biological samples, *n* = 3; average Fluc/Rluc ratios (*third from right*) are also shown. **d** Individual luminescence values are shown for Fluc (*blue*) and Rluc (*red*) obtained from lysates of HeLa cells transfected with pYX017 and the indicated siRNA; technical replicates are side-by-side; biological replicates are indicated with subscripts

activation of the HSV-TK promoter by interfering with a different endogenous p38 isoform. This possibility is supported by two observations. First, each p38δ construct repressed Rluc luminescence despite the fact that each has widely varying catalytic activities in vitro as well as different effects on L1 activation. Equivalent effects from each p38δ construct would be expected only if the effect were mediated by something other than their catalytic kinase activity; e.g., competition for docking interactions with limiting regulatory factors required by other p38 isoforms. Second, since MKK3b-2E and MKK6-2E selectively activate only p38 isoforms [72], their activation of Rluc strongly suggests that HSV-TK is indeed activated by an endogenous p38 isoform, but not p38δ. Combined, these data indicate that the ability of both active and inactive exogenous p38δ to repress the HSV-TK promoter derives from competition for host regulatory proteins by another, endogenous p38 isoform.

While most of our focus here has been on possible sources of artifact arising from the single vector dual luciferase assay, p38δ-WT and p38δ-F324S similarly activated Fluc in those assays; it was only in the G418-based assay where contradictory results between p38δ-WT and F324S were observed, with significant inhibition of apparent retrotransposition in response to p38δ-WT but strong activation by F324S. Since p38δ-WT gave conflicting results in these assays, it may be worth discussing potentially relevant variations between the assays.

One notable difference is the lack of the Epstein-Barr nuclear antigen 1 (EBNA1) gene and the Epstein-Barr virus (EBV) origin of replication on the single vector dual luciferase reporters, which were not required due to the shorter experimental time relative to the G418-based assay [39]. EBNA1, however, contains multiple phosphorylation sites required for the maintenance of plasmids and transcriptional activation [73, 74]. Specifically, the EBNA1 nuclear localization sequence contains two S/T-P motifs, whose phosphorylation is required for nuclear import [73–75]. Although at least one of these motifs is thought to be targeted by CDKs [75], it is possible that phosphorylation of one or both S/T-P motifs is perturbed by exogenous p38δ-WT expression via competition for regulatory factors.

Another difference between the two assays is their respective reporter genes. The G418-based assay relies on expression of APH (3′)-II to monitor L1 retrotransposition. However, in addition to inactivating aminoglycosides via phosphorylation, two APH isoforms have also been found to phosphorylate proteins. Although it is not known whether the neomycin resistance gene APH (3′)-II or the hygromycin resistant gene APH (4)-I, also present on JM101, can similarly target cellular proteins, caution has been urged in their use as selectable markers if such activity might interfere with the experimental design [76].

A source of potential artifact for both assays is the SV40 promoter, which drives the *neo* and Fluc reporter cassettes. As noted earlier, p38 is known to activate the transcription factor Sp1, which in addition to binding the HSV-TK promoter also binds and activates the SV40 early promoter [77]. Moreover, the SV40 promoter contains binding sites for AP-1 transcription factors [78, 79], which are activated by the isoform p38β but can be inhibited by p38γ or p38δ [68]. Thus, perturbed expression, in either direction, of an already spliced and integrated Fluc gene could falsely report on retrotransposition events. It is unclear, however, whether an increase above a given baseline expression of APH (3′)-II would alter colony viability or growth during G418 selection. Also of note, a recent study of the effects of heavy metals on L1 found that cobalt increased the activity of the SV40 promoter in HeLa cells but decreased its activity in human fibroblasts and the human neuroblastoma cell line BE (2)-M17 [80], indicating that heterologous promoters can be differentially affected

by the same variables in different cell lines. This raises the possibility that different clonal populations of the same cell type might also respond differentially to exogenous factors.

Regarding potential effects arising from the CMV promoter, although p38δ did not appear to affect expression of the CMV-driven EGFP, we imaged EGFP expressing cells 24 h post transfection for the purpose of monitoring transfection efficiencies, whereas G418 selection was begun three days post transfection. Thus, though EGFP appeared to report equivalent transfection efficiencies, it may not have accurately reflected cumulative effects of p38δ on a CMV promoter after 72 h. With respect to transfection efficiency controls, the potential for exogenous factors to impact these reporters remains an issue, as was demonstrated by the effects of p38δ on Rluc luminescence, which is the transfection efficiency reporter for the luciferase assay, versus no effect on from p38δ on EGFP fluorescence, which is also a common reporter for transfection efficiencies in a variety of assays.

The use of siRNA to probe the functions of cellular genes is a common technique, but the potential for off-target effects is a major drawback. This is typically accounted for by using NTC siRNA, with the assumption that NTC and target siRNAs produce equivalent off-target effects. While this may be true for some experimental systems, the dramatically different effects of NTC #3 and NTC #5 on L1 reporter output suggest a potential problem when these methods are used together. First, interpretations regarding the effect of a targeting siRNA based on comparison to a given NTC would be skewed if the siRNAs produced dissimilar off-target artifacts. This is true even if one confirms knockdown of the target gene. For example, if the target siRNA knocks down a gene of interest (GOI) by 50% and decreases L1 retrotransposition by 50%, one might conclude that knocking down the GOI decreases L1 activity if control siRNA #3 was the non-targeting control. In contrast, if one happened to use control siRNA #5, the conclusion would have been the opposite; i.e. that knockdown increased L1 activity.

In addition, it is possible that targeting siRNAs could induce the same types of artifacts we observed with the NTC siRNAs. For example, despite a hypothetical parallel 50% knockdown of the GOI and L1 activity, the decrease in L1 activity may have been due solely to off-target effects unrelated to gene knockdown. Similarly, it may be possible that off-target effects that increase apparent L1 activity could mask a genuine inhibitory effect mediated by gene knockdown. Our data with NTC #3 and #5 show that it is unreliable to control for such off-target effects by using non-targeting control siRNAs alone, as their effects can vary dramatically and may not be equivalent to those induced by targeting siRNAs. The most well- established method for confirming that results from targeting siRNA are due to GOI knockdown is the cotransfection of siRNA-resistant rescue plasmids. However, the interpretation of these results may still be complex in certain situations, as evidenced by our finding that p38δ-WT can both repress and activate L1 activity in different assays and cellular contexts.

In addition, our finding that non-targeting control siRNAs may affect L1 retrotransposition may have relevance not only for interpreting L1 assays but also for the development of therapeutic siRNA, a treatment option currently being optimized for numerous conditions including cancer [81–83]. As L1 is thought to have deleterious effects, caution is warranted in the design and testing of candidate molecules intended for clinical use.

Effects on heterologous promoters can be monitored in order to select one unaffected by experimental conditions. However, as some L1 reporters have up to three such promoters and may also be susceptible to artifacts arising from EBNA1 and the EBV origin of replication, this approach could be costly in terms of labor and resources and is therefore impractical for high throughput screening utilizing multiple experimental conditions. However, assuming suitable promoters could be identified for each experimental condition, a combination of native and constitutive L1 promoters with corresponding assays to monitor cell growth may be employed to successfully identify effects on L1 activity.

Several recently developed methods may offer some alternatives [84, 85]. The L1 element amplification protocol (LEAP assay) allows investigation of in vitro ORF2p enzymatic activity from L1 RNP particles purified from cells expressing engineered L1 reporters [86, 87]. The addition of purified host factors to these reactions would allow investigation of direct effects on ORF2p reverse transcriptase activity while avoiding some of the issues described herein. Next-generation sequencing methods [85, 88] including retrotransposon capture sequencing (RC-seq) [89, 90], as well as novel approaches for validation such as droplet digital PCR [91], offer the possibility of examining endogenous L1 elements in their native chromatin environment. These technical advances should facilitate investigation of the host factors that delimit L1 tissue specificity and various aspects of retrotransposition.

Conclusions

Our results indicate that the use of exogenous gene expression or siRNA with engineered L1 reporter assays may introduce confounding variables. Thus, investigation of the roles of host factors in L1 retrotransposition when using these techniques will require extra efforts to ensure that observed results are not artifacts.

Methods
Plasmid construction

Bacterial expression vectors for ORF1p (pET32aΔN-ORF1-6xHis) were made as follows. First, an existing ORF1 vector [92] with the backbone of pET32a was altered to remove the following: the pET32a N-terminal TRX and 6xHis tags, an engineered TEV sequence that had previously destroyed the multiple cloning region, a truncated ORF1 mutant and remnant sequence 3′ to ORF1 that was retained from prior subcloning. A remaining 3′ EcoRI site and the C-terminal 6xHis tag were left intact, and BamH1 site was inserted 5′ of the EcoRI site. These changes were made using the Quik-Change II kit (Agilent) with the forward deletion primer 5′TTAACTTTAAGAAGGAGATATACATGGATCCAAT CCCGGGACGCGTG and reverse deletion primer 5′ CACGCGTCCCGGGATTGGATCCATGTATATCCT TCTTAAAGTTAA. The resulting clone was designated pET32aΔN. Full-length ORF1 PCR-generated amplicons were created from the previously described pORF1-Flag mammalian expression vector [31] using a high-fidelity DNA polymerase with the forward primer 5′ CGCGGATCCATGGGGAAAAAACAGAACAG containing a 5′ BamH1 site, and reverse primer 5′ GCCGGAATTCGCCGCCGCCCATTTTGGCATGATTT TGC, which introduced a spacer of three glycines between the end of ORF1 and the 3′ EcoRI sequence (the Flag sequence was not retained). The ORF1p amplicon was inserted into pET32aΔN via the BamH1 and EcoRI sites. The BamH1 site was subsequently deleted to move the ATG start site of ORF1 to an optimal distance from the ribosomal binding site in pET32aΔN and destroy an alternate out-of-frame ATG start site that encompassed the 5′G of the BamH1 site. These changes were made using the QuikChange II kit (Agilent) with the forward primer 5′ GAAATAATTTTGTTTAACTTTAAGAAGGAGATATA-CATGGGGAAAAAACAGAACAG and the reverse primer 5′CTGTTCTGTTTTTTCCCCATATGTATATCTCC TTCTTAAAGTTAAACAAAATTATTTC. In an attempt to reduce translation initiation at internal non-canonical Shine-Dalgarno sequences in ORF1, we also created silent mutations at D123 and N126, changing the existing codons to GAC and AAC, respectively. ORF1p S/T-P motif mutations were created using sequential site-directed mutagenesis with the QuikChange II kit (Agilent).

Bacterial expression plasmids for p38δ-F324S and D176A (pRSET-A-6xHis-p38δ-StrepII) were made by first generating a p38δ-WT amplicon via PCR using a high-fidelity polymerase and the forward primer 5′ CGCGGATCCGCAATGAGCCTCATCCGGAAAAAGG GCTTCTACAAGCAGG and reverse primer 5′GCCG GAATTCTCACTTCTCGAACTGGGGGTGGCTCCAT GCGCCCAGCTTCATGCCACTCCG on the Addgene template plasmid # 20523 (pWZL Neo Myr Flag

MAPK13, a gift from William Hahn & Jean Zhao [93]). The amplicon containing a 5′ BamHI and Kozak sequence and a 3′ Gly/Ala spacer upstream of a StrepII tag, stop codon and EcoRI site was then inserted into pRSET-A (ThermoFisher) via the BamHI and 3′ EcoRI sites in the multiple cloning region. Point mutations were created via site-directed mutagenesis with the QuikChange II kit (Agilent).

The mammalian expression vector for p38δ-WT (pcDNA-Zeo (3.1+)-p38δ-StrepII) was made by PCR amplification of the Addgene plasmid # 20523 [93] using the same forward and reverse primers noted above for making pRSET-A-6xHis-p38δ-StrepII, followed by insertion into the multiple cloning region of pcDNA 3.1/Zeo (+) (ThermoFisher). Point mutations to make F324S and D176A were created via site-directed mutagenesis with the QuikChange II kit (Agilent).

Mammalian expression vectors for MKK3b$_{2E}$ (pcDNA3 Flag MKK3b (Glu) [58]; Addgene plasmid # 50449) and MKK6$_{2E}$ (pcDNA3-Flag MKK6 (Glu) [60]; Addgene plasmid # 13518) were both gifts from Roger Davis.

All cloned inserts were verified with DNA sequencing. DNA intended for cell culture transfections was purified using the endotoxin-free NucleoBond Xtra Midi plasmid DNA purification kit (Macherey-Nagel).

Protein expression

ORF1p proteins were expressed in Rosetta (DE3) cells (Novagen) transformed with pET32aΔN-ORF1-His. Overnight starter cultures of 15–25 ml LB medium with 100 μg/ml ampicillin and 34 μg/ml chloramphenicol were grown at 37 °C on a rotary shaker at 250 rpm. The following day, cultures were expanded 20 to 50 fold with LB medium containing the indicated antibiotics and grown at 37 °C on a rotary shaker at 250 rpm to an OD$_{600}$ of approximately 0.6. Cultures were then induced with 1 mM isopropyl-beta-D-thiogalactopyranoside (IPTG), grown for an additional 4–6 h, pelleted via centrifugation and frozen at -80 °C. At the time of purification, cells were thawed and resuspended in 5 ml per gram pellet of a buffer containing 100 mM Tris–HCl (pH 8.0), 100 mM NaCl, and 1 mg/ml lysozyme and incubated on ice for 30 min. Following lysozyme digest, lysates were supplemented with 400 mM NaCl (for final concentration of 500 mM), 2 mM dithiothreitol (DTT) and 15 mM imidazole. The lysates were pulled through a 19–21gauge syringe approximately 12 times and centrifuged at 10,000 × g at 4 °C for 20 min. Cleared lysates were applied to Ni-NTA superflow resin (Qiagen) previously equilibrated with lysis buffer (post lysozyme concentrations), rotated for 1 h at 4 °C, washed 4 times with 20 mM Tris–HCl (pH 7.4), 500 mM NaCl, and 25 mM imidazole, then eluted 4 times with 20 mM Tris–HCl (pH 7.4), 500 mM NaCl, 250 mM imidazole, 10% glycerol and 2 mM DTT at a ratio of 1 μl elution

buffer per 1 ml of original culture volume. Proteins were dialyzed overnight against 50 mM Tris–HCl (pH 80), 350 mM NaCl, 15 mM KCl, 5 mM MgCl$_2$, 20% glycerol, 2 mM DTT, and 1 mM phenylmethylsulfonyl fluoride (PMSF).

p38δ-F324S and p38δ-D176A proteins were expressed in Rosetta2 (DE3) cells (Novagen) transformed with pRSET-A-His-p38δ-StrepII and processed as described above for ORF1p except 150 mM NaCl was used in the dialysis buffers. Note: we found that omission of DTT in the elution and/or dialysis steps of p38δ purification resulted in an inactive protein, consistent with a previous report [94].

All proteins were quantified via denaturing gel electrophoresis with a standard curve of bovine serum albumin followed by staining with Coomassie G-250 PageBlue (ThermoFisher) and analysis with ImageJ [95].

Kinase assays

In vitro kinase reactions contained 85 nM p38δ or p38δ dialysis buffer and 200 μM ORF1p in 50 mM Tris–HCl (pH 7.4), 10 mM MgCl$_2$, 0.1 mM EGTA, 150 mM NaCl, 2 mM DTT, and 2 mM ATP spiked with approximated 0.5×10^6 c.p.m./nmol [γ-^{32}P]-ATP (PerkinElmer). Reactions were incubated at 37 °C for 15 min and stopped with the addition of loading buffer supplemented with EDTA to a final concentration of 50 mM. Samples were heated to 98 °C for 10 min then separated via denaturing gel electrophoresis. Gels were dried and exposed using Phosphorimaging.

Cell culture

HeLa-JVM cells (a kind gift from Dr. John Moran) were cultured in Dulbecco's Modified Eagle Media (DMEM) with high glucose and pyruvate (Gibco, ThermoFisher) supplemented with 10% Fetal Bovine Serum (Gibco, ThermoFisher, certified heat inactivated, US origin) and 100 Units/ml penicillin and 100 μg/ml streptomycin from a combined formulation (Gibco, ThermoFisher). The cells were maintained at 37 °C in a standard incubator and passaged using 0.05% Trypsin-EDTA (Gibco, ThermoFisher).

L1 reporter assays

Culture plates were seeded with HeLa-JVM cells in antibiotic-free DMEM with 10% FBS at a density to achieve approximately 50% confluency in 24 h, at which time cells were transfected using a ratio of 3 μl Fugene6 (Promega) per 1 μg DNA. For the G418-based assay, cells were seeded in 6-well plates and transfected with 500 ng JM101 and 500 ng pcDNA per well, allowed to grow for 72 h, then selected with media containing 400 μg/ml G418 sulfate (Geneticin, Gibco, Thermo-Fisher) for 10–12 days. Cells were then washed with phosphate buffered saline (PBS) and fixed with 2% formaldehyde and 0.2% glutaraldehyde in PBS for at least 30 min at room temperature. Cells were then washed twice with PBS, stained with KaryoMAX Giemsa (Gibco, ThermoFisher) for 1 h at room temperature, rinsed briefly twice with 50% ethanol and then water. For luciferase assays, cells were seeded in 24-well plates and transfected with 200 ng of reporter and 200 ng pcDNA-p38δ per well or 25 ng pcDNA-MKK3b$_{2E}$ or pcDNA-MKK6$_{2E}$. Lysates were harvested 4 days post transfection and processed in 96-well plates with the Dual-Luciferase Reporter Assay System (Promega) according to manufacturer's protocol.

Transfection efficiency assays

HeLa-JVM cells were plated in 8-well glass bottom μ-Slides (ibidi GmbH, Martinsried, Germany) in antibiotic-free DMEM with 10% FBS at a density to achieve approximately 60% confluency per well in 24 h. Wells with siRNA were reverse transfected as described in the following section. After a 24-h incubation, cells were transfected as described above with a pcDNA-EGFP expression plasmid (for siRNA wells) or cotransfected with pcDNA-EGFP and each pcDNA-p38δ expression plasmid. The ratio of DNA to surface area was identical to that used in the 6-well plates. After 24 h, cells were rinsed twice with PBS, then DMEM sans phenol red plus 10% FBS was added to each well. Cells were visualized with a Keyence BioRevo BZ-II 9000 digital microscope fitted with a Nikon PlanApo 4×/0.20 objective lens and 49002 ET-EGFP filter set from Chroma (Bellows Falls, VT). Tiled images covering approximately 70% of each well were stitched with Keyence BZ-II Analyzer software, and total fluorescence in each stitched image was quantified in Fiji software using the Integrated Density function.

siRNA knockdown

HeLa-JVM cells were plated in antibiotic-free DMEM with 10% FBS at a density to achieve approximately 60% confluency in 24 h and reverse transfected per manufactures' protocol using Lipofectamine RNAiMAX (ThermoFisher) at a ratio of 1 μl RNAiMAX per 8 pmols siRNA. All siRNAs were purchased from Dharmacon: NTC #3, NTC #5 and SMARTpool siRNA against p38δ (Dharmacon, M-003591-02-0005). Following reverse transfection, cells were incubated for 24 h, then siRNA-containing media was removed and replaced with fresh antibiotic-free plating media with 10% FBS at the time of transfection with L1 reporters as described above.

Abbreviations

APH(3')-II: Aminoglycoside 3'-phosphotransferase-II; CDKs: Cyclin dependent kinases; CMV: Cytomegalovirus; DTT: Dithiothreitol; EBNA1: Epstein-Barr nuclear antigen 1; EBV: Epstein-Barr virus; EGFP: Enhanced green fluorescent protein; FBS: Fetal bovine serum; Fluc: Firefly luciferase; GSK3: Glycogen synthase kinase 3; GST-ATF2: glutathione S-transferase activating transcription factor 2; HSV-TK: Herpes simplex virus thymidine kinase; IPTG: Isopropyl-beta-D-thiogalactopyranoside; L1, LINE-1: Long INterspersed Element-1; LEAP assay: L1 element amplification protocol; MAPKKs: MAPK-kinases; MAPKs: Mitogen-activated protein kinases; NTC: Non-targeting control; PDPKs: Proline-directed protein kinases; PMSF: Phenylmethylsulfonyl fluoride; RC-seq: Retrotransposon capture sequencing; Rluc: Renilla luciferase; S/T-P: Serine/threonine-proline; SA: Splice acceptor; SD: Splice donor; SINES: Short-INterspersed repeat Elements; siRNA: Small interfering RNA; SV40: Simian virus 40; UTR: Untranslated region

Acknowledgements

We would like to thank Dr. Charles E. Jones, Dr. Sandy Martin and Dr. Anthony V. Furano for their critical reading of the manuscript and useful discussion. We also thank Dr. John Moran for his gifts of JM101 and HeLa-JVM cells, Dr. Wenfeng An for his gifts of L1 dual luciferase reporters, and Dr. Birong Shen for her gift of the EGFP plasmid. We also thank Jeff M. Reece from the NIDDK light microscopy core for his invaluable help analyzing the EGFP data.

Funding

This work was supported by the laboratory of Anthony V. Furano at the National Institute of Diabetes and Digestive and Kidney Diseases, National Institutes of Health; Grant 1-ZIA-DK057601-19.

Authors' contributions

Both authors contributed equally to the experimental work. PRC drafted the manuscript, and GTT assisted in revisions and editing.

Competing interests

The authors declare they have no competing interests.

Author details

[1]Laboratory of Cell and Molecular Biology, National Institute of Diabetes and Digestive and Kidney Diseases, National Institutes of Health, 8 Center Drive, Bethesda, MD 20892, USA. [2]National Institute of Child Health and Human Development, National Institutes of Health, 35 Convent Drive, Bethesda, MD 20892, USA.

References

1. Treangen TJ, Salzberg SL. Repetitive DNA and next-generation sequencing: computational challenges and solutions. Nat Rev Genet. 2012;13(1):36–46.
2. Kazazian Jr HH, Goodier JL. LINE drive. retrotransposition and genome instability. Cell. 2002;110(3):277–80.
3. Babatz TD, Burns KH. Functional impact of the human mobilome. Curr Opin Genet Dev. 2013;23(3):264–70.
4. Faulkner GJ. Retrotransposons: mobile and mutagenic from conception to death. FEBS Lett. 2011;585(11):1589–94.
5. Boissinot S, Davis J, Entezam A, Petrov D, Furano AV. Fitness cost of LINE-1 (L1) activity in humans. Proc Natl Acad Sci U S A. 2006;103(25):9590–4.
6. Rosser JM, An W. L1 expression and regulation in humans and rodents. Front Biosci. 2012;4:2203–25.
7. Kano H, Godoy I, Courtney C, Vetter MR, Gerton GL, Ostertag EM, Kazazian Jr HH. L1 retrotransposition occurs mainly in embryogenesis and creates somatic mosaicism. Genes Dev. 2009;23(11):1303–12.
8. van den Hurk JA, Meij IC, Seleme MC, Kano H, Nikopoulos K, Hoefsloot LH, Sistermans EA, de Wijs IJ, Mukhopadhyay A, Plomp AS, et al. L1 retrotransposition can occur early in human embryonic development. Hum Mol Genet. 2007;16(13):1587–92.
9. Arokium H, Kamata M, Kim S, Kim N, Liang M, Presson AP, Chen IS. Deep sequencing reveals low incidence of endogenous LINE-1 retrotransposition in human induced pluripotent stem cells. PLoS One. 2014;9(10):e108682.
10. Wissing S, Munoz-Lopez M, Macia A, Yang Z, Montano M, Collins W, Garcia-Perez JL, Moran JV, Greene WC. Reprogramming somatic cells into iPS cells

activates LINE-1 retroelement mobility. Hum Mol Genet. 2012;21(1):208–18.

11. Reilly MT, Faulkner GJ, Dubnau J, Ponomarev I, Gage FH. The role of transposable elements in health and diseases of the central nervous system. J Neurosci. 2013;33(45):17577–86.

12. Kurnosov AA, Ustyugova SV, Nazarov VI, Minervina AA, Komkov AY, Shugay M, Pogorelyy MV, Khodosevich KV, Mamedov IZ, Lebedev YB. The evidence for increased L1 activity in the site of human adult brain neurogenesis. PLoS One. 2015;10(2):e0117854.

13. Thomas CA, Paquola AC, Muotri AR. LINE-1 retrotransposition in the nervous system. Annu Rev Cell Dev Biol. 2012;28:555–73.

14. Richardson SR, Morell S, Faulkner GJ. L1 retrotransposons and somatic mosaicism in the brain. Annu Rev Genet. 2014;48:1–27.

15. Upton KR, Gerhardt DJ, Jesuadian JS, Richardson SR, Sanchez-Luque FJ, Bodea GO, Ewing AD, Salvador-Palomeque C, van der Knaap MS, Brennan PM, et al. Ubiquitous L1 mosaicism in hippocampal neurons. Cell. 2015; 161(2):228–39.

16. Carreira PE, Richardson SR, Faulkner GJ. L1 retrotransposons, cancer stem cells and oncogenesis. The FEBS J. 2014;281(1):63–73.

17. Rodic N, Burns KH. Long interspersed element-1 (LINE-1): passenger or driver in human neoplasms? PLoS Genet. 2013;9(3):e1003402.

18. Helman E, Lawrence MS, Stewart C, Sougnez C, Getz G, Meyerson M. Somatic retrotransposition in human cancer revealed by whole-genome and exome sequencing. Genome Res. 2014;24(7):1053–63.

19. Tubio JM, Li Y, Ju YS, Martincorena I, Cooke SL, Tojo M, Gundem G, Pipinikas CP, Zamora J, Raine K. Mobile DNA in cancer. Extensive transduction of nonrepetitive DNA mediated by L1 retrotransposition in cancer genomes. Science. 2014;345(6196):1251343.

20. Beck CR, Garcia-Perez JL, Badge RM, Moran JV. LINE-1 elements in structural variation and disease. Annu Rev Genomics Hum Genet. 2011;12:187–215.

21. Hancks DC, Kazazian Jr HH. Active human retrotransposons: variation and disease. Curr Opin Genet Dev. 2012;22(3):191–203.

22. Kaer K, Speek M. Retroelements in human disease. Gene. 2013;518(2):231–41.

23. Belancio VP, Hedges DJ, Deininger P. Mammalian non-LTR retrotransposons: for better or worse, in sickness and in health. Genome Res. 2008;18(3):343–58.

24. Gifford WD, Pfaff SL, Macfarlan TS. Transposable elements as genetic regulatory substrates in early development. Trends Cell Biol. 2013;23(5):218–26.

25. Emera D, Wagner GP. Transposable element recruitments in the mammalian placenta: impacts and mechanisms. Brief Funct Genomics. 2012;11(4):267–76.

26. Macia A, Blanco-Jimenez E, Garcia-Perez JL. Retrotransposons in pluripotent cells: Impact and new roles in cellular plasticity. Biochim Biophys Acta. 2015; 1849(4):417–26.

27. Roberts JT, Cardin SE, Borchert GM. Burgeoning evidence indicates that microRNAs were initially formed from transposable element sequences. Mob Genet Elements. 2014;4:e29255.

28. Mita P, Boeke JD. How retrotransposons shape genome regulation. Curr Opin Genet Dev. 2016;37:90–100.

29. McLaughlin Jr RN, Young JM, Yang L, Neme R, Wichman HA, Malik HS. Positive selection and multiple losses of the LINE-1-derived L1TD1 gene in mammals suggest a dual role in genome defense and pluripotency. PLoS Genet. 2014;10(9):e1004531.

30. Chuong EB, Elde NC, Feschotte C. Regulatory evolution of innate immunity through co-option of endogenous retroviruses. Science. 2016;351(6277):1083–7.

31. Cook PR, Jones CE, Furano AV. Phosphorylation of ORF1p is required for L1 retrotransposition. Proc Natl Acad Sci U S A. 2015;112(14):4298–303.

32. Ishizaka Y, Okudaira N, Tamura M, Iijima K, Shimura M, Goto M, Okamura T. Modes of retrotransposition of long interspersed element-1 by environmental factors. Front Microbiol. 2012;3:191.

33. Okudaira N, Iijima K, Koyama T, Minemoto Y, Kano S, Mimori A, Ishizaka Y. Induction of long interspersed nucleotide element-1 (L1) retrotransposition by 6-formylindolo [3,2-b] carbazole (FICZ), a tryptophan photoproduct. Proc Natl Acad Sci U S A. 2010;107(43):18487–92.

34. Kuchen S, Seemayer CA, Rethage J, von Knoch R, Kuenzler P, Beat AM, Gay RE, Gay S, Neidhart M. The L1 retroelement-related p40 protein induces p38delta MAP kinase. Autoimmunity. 2004;37(1):57–65.

35. Zarubin T, Han J. Activation and signaling of the p38 MAP kinase pathway. Res. 2005;15(1):11–8.

36. Blom N, Sicheritz-Ponten T, Gupta R, Gammeltoft S, Brunak S. Prediction of post-translational glycosylation and phosphorylation of proteins from the amino acid sequence. Proteomics. 2004;4(6):1633–49.

37. Avitzour M, Diskin R, Raboy B, Askari N, Engelberg D, Livnah O. Intrinsically active variants of all human p38 isoforms. FEBS J. 2007;274(4):963–75.

38. Moran JV, Holmes SE, Naas TP, DeBerardinis RJ, Boeke JD, Kazazian Jr HH. High frequency retrotransposition in cultured mammalian cells. Cell. 1996;87(5):917–27.

39. Xie Y, Rosser JM, Thompson TL, Boeke JD, An W. Characterization of L1 retrotransposition with high-throughput dual-luciferase assays. Nucleic Acids Res. 2011;39(3):e16.

40. Niwa H, Yamamura K, Miyazaki J. Efficient selection for high-expression transfectants with a novel eukaryotic vector. Gene. 1991;108(2):193–9.

41. Mayor Jr F, Jurado-Pueyo M, Campos PM, Murga C. Interfering with MAP kinase docking interactions: implications and perspective for the p38 route. Cell Cycle. 2007;6(5):528–33.

42. Hsieh YH, Wu TT, Huang CY, Hsieh YS, Hwang JM, Liu JY. p38 mitogen-activated protein kinase pathway is involved in protein kinase Calpha-regulated invasion in human hepatocellular carcinoma cells. Cancer Res. 2007;67(9):4320–7.

43. Seternes OM, Johansen B, Hegge B, Johannessen M, Keyse SM, Moens U. Both binding and activation of p38 mitogen-activated protein kinase (MAPK) play essential roles in regulation of the nucleocytoplasmic distribution of MAPK-activated protein kinase 5 by cellular stress. Mol Cell Biol. 2002;22(20):6931–45.

44. Somwar R, Koterski S, Sweeney G, Sciotti R, Djuric S, Berg C, Trevillyan J, Scherer PE, Rondinone CM, Klip A. A dominant-negative p38 MAPK mutant and novel selective inhibitors of p38 MAPK reduce insulin-stimulated glucose uptake in 3 T3-L1 adipocytes without affecting GLUT4 translocation. J Biol Chem. 2002;277(52):50386–95.

45. Thandavarayan RA, Watanabe K, Ma M, Gurusamy N, Veeraveedu PT, Konishi T, Zhang S, Muslin AJ, Kodama M, Aizawa Y. Dominant-negative p38alpha mitogen-activated protein kinase prevents cardiac apoptosis and remodeling after streptozotocin-induced diabetes mellitus. Am J Physiol Heart Circ Physiol. 2009; 297(3):H911–919.

46. Twait E, Williard DE, Samuel I. Dominant negative p38 mitogen-activated protein kinase expression inhibits NF-kappaB activation in AR42J cells. Pancreatology. 2010;10(2-3):119–28.

47. Faust D, Dolado I, Cuadrado A, Oesch F, Weiss C, Nebreda AR, Dietrich C. p38alpha MAPK is required for contact inhibition. Oncogene. 2005;24(53):7941–5.

48. Streva VA, Faber ZJ, Deininger PL. LINE-1 and Alu retrotransposition exhibit clonal variation. Mob DNA. 2013;4(1):16.

49. Bruening W, Giasson B, Mushynski W, Durham HD. Activation of stress-activated MAP protein kinases up-regulates expression of transgenes driven by the cytomegalovirus immediate/early promoter. Nucleic Acids Res. 1998; 26(2):486–9.

50. Chen J, Stinski MF. Role of regulatory elements and the MAPK/ERK or p38 MAPK pathways for activation of human cytomegalovirus gene expression. J Virol. 2002;76(10):4873–85.

51. Radhakrishnan P, Basma H, Klinkebiel D, Christman J, Cheng PW. Cell type-specific activation of the cytomegalovirus promoter by dimethylsulfoxide and 5-aza-2'-deoxycytidine. Int J Biochem Cell Biol. 2008;40(9):1944–55.

52. Svensson RU, Barnes JM, Rokhlin OW, Cohen MB, Henry MD. Chemotherapeutic agents up-regulate the cytomegalovirus promoter: implications for bioluminescence imaging of tumor response to therapy. Cancer Res. 2007;67(21):10445–54.

53. Xiao J, Deng J, Lv L, Kang Q, Ma P, Yan F, Song X, Gao B, Zhang Y, Xu J. Hydrogen Peroxide Induce Human Cytomegalovirus Replication through the Activation of p38-MAPK Signaling Pathway. Viruses. 2015;7(6):2816–33.

54. Curtin JA, Dane AP, Swanson A, Alexander IE, Ginn SL. Bidirectional promoter interference between two widely used internal heterologous promoters in a late-generation lentiviral construct. Gene Ther. 2008;15(5):384–90.

55. Huliak I, Sike A, Zencir S, Boros IM. The objectivity of reporters: interference between physically unlinked promoters affects reporter gene expression in transient transfection experiments. DNA Cell Biol. 2012;31(11):1580–4.

56. Rosen H, Di Segni G, Kaempfer R. Translational control by messenger RNA competition for eukaryotic initiation factor 2. J Biol Chem. 1982;257(2):946–52.

57. Shearwin KE, Callen BP, Egan JB. Transcriptional interference–a crash course. Trends Genet. 2005;21(6):339–45.

58. Enslen H, Brancho DM, Davis RJ. Molecular determinants that mediate selective activation of p38 MAP kinase isoforms. EMBO J. 2000;19(6):1301–11.

59. Enslen H, Raingeaud J, Davis RJ. Selective activation of p38 mitogen-activated protein kinase isoforms by the MAP kinase kinases MKK3 and MKK6. J Biol Chem. 1998;273(3):1741–8.

60. Raingeaud J, Whitmarsh AJ, Barrett T, Derijard B, Davis RJ. MKK3- and MKK6-regulated gene expression is mediated by the p38 mitogen-activated protein kinase signal transduction pathway. Mol Cell Biol. 1996;16(3):1247–55.

61. Shifera AS, Hardin JA. Factors modulating expression of Renilla luciferase from control plasmids used in luciferase reporter gene assays. Anal

Biochem. 2010;396(2):167–72.

62. Theile D, Spalwisz A, Weiss J. Watch out for reporter gene assays with Renilla luciferase and paclitaxel. Anal Biochem. 2013;437(2):109–10.

63. Ho CK, Strauss 3rd JF. Activation of the control reporter plasmids pRL-TK and pRL-SV40 by multiple GATA transcription factors can lead to aberrant normalization of transfection efficiency. BMC Biotechnol. 2004;4:10.

64. Osborne SA, Tonissen KF. pRL-TK induction can cause misinterpretation of gene promoter activity. Biotechniques. 2002;33(6):1240–2.

65. D'Addario M, Arora PD, McCulloch CA. Role of p38 in stress activation of Sp1. Gene. 2006;379:51–61.

66. Rangwala SH, Kazazian Jr HH. The L1 retrotransposition assay: a retrospective and toolkit. Methods. 2009;49(3):219–26.

67. Keshet Y, Seger R. The MAP kinase signaling cascades: a system of hundreds of components regulates a diverse array of physiological functions. Methods Mol Biol. 2010;661:3–38.

68. Pramanik R, Qi X, Borowicz S, Choubey D, Schultz RM, Han J, Chen G. p38 isoforms have opposite effects on AP-1-dependent transcription through regulation of c-Jun. The determinant roles of the isoforms in the p38 MAPK signal specificity. J Biol Chem. 2003;278(7):4831–9.

69. Risco A, Cuenda A. New Insights into the p38gamma and p38delta MAPK Pathways. J Signal Transduct. 2012;2012:520289.

70. Mezhir JJ, Advani SJ, Smith KD, Darga TE, Poon AP, Schmidt H, Posner MC, Roizman B, Weichselbaum RR. Ionizing radiation activates late herpes simplex virus 1 promoters via the p38 pathway in tumors treated with oncolytic viruses. Cancer Res. 2005;65(20):9479–84.

71. Jones KA, Yamamoto KR, Tjian R. Two distinct transcription factors bind to the HSV thymidine kinase promoter in vitro. Cell. 1985;42(2):559–72.

72. Remy G, Risco AM, Inesta-Vaquera FA, Gonzalez-Teran B, Sabio G, Davis RJ, Cuenda A. Differential activation of p38MAPK isoforms by MKK6 and MKK3. Cell Signal. 2010;22(4):660–7.

73. Duellman SJ, Thompson KL, Coon JJ, Burgess RR. Phosphorylation sites of Epstein-Barr virus EBNA1 regulate its function. J Gen Virol. 2009;90(Pt 9):2251–9.

74. Kitamura R, Sekimoto T, Ito S, Harada S, Yamagata H, Masai H, Yoneda Y, Yanagi K. Nuclear import of Epstein-Barr virus nuclear antigen 1 mediated by NPI-1 (Importin alpha5) is up- and down-regulated by phosphorylation of the nuclear localization signal for which Lys379 and Arg380 are essential. J Virol. 2006;80(4):1979–91.

75. Kang MS, Lee EK, Soni V, Lewis TA, Koehler AN, Srinivasan V, Kieff E. Roscovitine inhibits EBNA1 serine 393 phosphorylation, nuclear localization, transcription, and episome maintenance. J Virol. 2011;85(6):2859–68.

76. Daigle DM, McKay GA, Wright GD. Inhibition of aminoglycoside antibiotic resistance enzymes by protein kinase inhibitors. J Biol Chem. 1997;272(40):24755–8.

77. Dynan WS, Tjian R. The promoter-specific transcription factor Sp1 binds to upstream sequences in the SV40 early promoter. Cell. 1983;35(1):79–87.

78. Lee W, Haslinger A, Karin M, Tjian R. Activation of transcription by two factors that bind promoter and enhancer sequences of the human metallothionein gene and SV40. Nature. 1987;325(6102):368–72.

79. Lee W, Mitchell P, Tjian R. Purified transcription factor AP-1 interacts with TPA-inducible enhancer elements. Cell. 1987;49(6):741–52.

80. Habibi L, Shokrgozar MA, Tabrizi M, Modarressi MH, Akrami SM. Mercury specifically induces LINE-1 activity in a human neuroblastoma cell line. Mutat Res Genet Toxicol Environ Mutagen. 2014;759:9–20.

81. Young SW, Stenzel M, Jia-Lin Y. Nanoparticle-siRNA: A potential cancer therapy? Crit Rev Oncol Hematol. 2016;98:159–69.

82. Haussecker D. Current issues of RNAi therapeutics delivery and development. J Control Release. 2014;195:49–54.

83. Ren YJ, Zhang Y. An update on RNA interference-mediated gene silencing in cancer therapy. Expert Opin Biol Ther. 2014;14(11):1581–92.

84. Garcia-Perez JL, editor. Transposons and Retrotransposons: Methods and Protocols. New York: Springer; 2016.

85. Xing J, Witherspoon DJ, Jorde LB. Mobile element biology: new possibilities with high-throughput sequencing. Trends Genet. 2013;29(5):280–9.

86. Kopera HC, Flasch DA, Nakamura M, Miyoshi T, Doucet AJ, Moran JV. LEAP: L1 Element Amplification Protocol. Methods Mol Biol. 2016;1400:339–55.

87. Kulpa DA, Moran JV. Cis-preferential LINE-1 reverse transcriptase activity in ribonucleoprotein particles. Nat Struct Mol Biol. 2006;13(7):655–60.

88. Streva VA, Jordan VE, Linker S, Hedges DJ, Batzer MA, Deininger PL. Sequencing, identification and mapping of primed L1 elements (SIMPLE) reveals significant variation in full length L1 elements between individuals. BMC Genomics. 2015;16:220.

89. Baillie JK, Barnett MW, Upton KR, Gerhardt DJ, Richmond TA, De Sapio F, Brennan PM, Rizzu P, Smith S, Fell M, et al. Somatic retrotransposition alters the genetic landscape of the human brain. Nature. 2011;479(7374):534–7.

90. Klawitter S, Fuchs NV, Upton KR, Munoz-Lopez M, Shukla R, Wang J, Garcia-Canadas M, Lopez-Ruiz C, Gerhardt DJ, Sebe A, et al. Reprogramming triggers endogenous L1 and Alu retrotransposition in human induced pluripotent stem cells. Nat comm. 2016;7:10286.

91. White TB, McCoy AM, Streva VA, Fenrich J, Deininger PL. A droplet digital PCR detection method for rare L1 insertions in tumors. Mob DNA. 2014;5(1):30.

92. Callahan KE, Hickman AB, Jones CE, Ghirlando R, Furano AV. Polymerization and nucleic acid-binding properties of human L1 ORF1 protein. Nucleic Acids Res. 2012;40(2):813–27.

93. Boehm JS, Zhao JJ, Yao J, Kim SY, Firestein R, Dunn IF, Sjostrom SK, Garraway LA, Weremowicz S, Richardson AL, et al. Integrative genomic approaches identify IKBKE as a breast cancer oncogene. Cell. 2007;129(6):1065–79.

94. Templeton DJ, Aye MS, Rady J, Xu F, Cross JV. Purification of reversibly oxidized proteins (PROP) reveals a redox switch controlling p38 MAP kinase activity. PLoS One. 2010;5(11):e15012.

95. Schneider CA, Rasband WS, Eliceiri KW. NIH Image to ImageJ: 25 years of image analysis. Nat Methods. 2012;9(7):671–5.

7

The intron-enriched HERV-K(HML-10) family suppresses apoptosis, an indicator of malignant transformation

Felix Broecker[1,2,3]*, Roger Horton[1], Jochen Heinrich[2], Alexandra Franz[1,4], Michal-Ruth Schweiger[1,5], Hans Lehrach[1,6] and Karin Moelling[1,2]

Abstract

Background: Human endogenous retroviruses (HERVs) constitute 8% of the human genome and contribute substantially to the transcriptome. HERVs have been shown to generate RNAs that modulate host gene expression. However, experimental evidence for an impact of these regulatory transcripts on the cellular phenotype has been lacking.

Results: We characterized the previously little described HERV-K(HML-10) endogenous retrovirus family on a genome-wide scale. HML-10 invaded the ancestral genome of Old World monkeys about 35 Million years ago and is enriched within introns of human genes when compared to other HERV families. We show that long terminal repeats (LTRs) of HML-10 exhibit variable promoter activity in human cancer cell lines. One identified HML-10 LTR-primed RNA was in opposite orientation to the pro-apoptotic Death-associated protein 3 (*DAP3*). In HeLa cells, experimental inactivation of HML-10 LTR-primed transcripts induced *DAP3* expression levels, which led to apoptosis.

Conclusions: Its enrichment within introns suggests that HML-10 may have been evolutionary co-opted for gene regulation more than other HERV families. We demonstrated such a regulatory activity for an HML-10 RNA that suppressed DAP3-mediated apoptosis in HeLa cells. Since HML-10 RNA appears to be upregulated in various tumor cell lines and primary tumor samples, it may contribute to evasion of apoptosis in malignant cells. However, the overall weak expression of HML-10 transcripts described here raises the question whether our result described for HeLa represent a rare event in cancer. A possible function in other cells or tissues requires further investigation.

Keywords: Apoptosis, Cancer, DAP3, Death-associated protein 3, Endogenous retrovirus, Gene regulation, Genome evolution, HERV, HERV-K(HML-10)

Background

About half of the human genome is composed of transposable elements (TEs) [1], and recent evidence suggests even a fraction of up to two thirds [2]. The most abundant TEs in the human genome are retroelements (REs) that amplify via a 'copy-and-paste' mechanism involving reverse transcription of an RNA intermediate [1, 3].

One class of REs, HERVs, comprises remnants of ancient retroviral germ line cell infections that became evolutionary fixed in the genome. About 450,000 HERV elements constitute 8% of the human genome and are classified into about 30 families [1, 4]. HERVs are structurally similar to proviruses of present-day retroviruses where the *gag*, *pol* and *env* genes are flanked by two long terminal repeats (LTRs) that act as promoters [4]. HERVs and other REs have been shown to influence gene regulation by providing regulatory elements such as enhancers, promoters, splice- and polyadenylation sites, for various host genes [3]. REs of all classes often contain functional promoters and consequently

* Correspondence: felixbroecker@gmx.net
[1]Max Planck Institute for molecular Genetics, Ihnestr. 63-73, 14195 Berlin, Germany
[2]Institute of Medical Microbiology, University of Zurich, Gloriastr. 32, 8006 Zurich, Switzerland
Full list of author information is available at the end of the article

The intron-enriched HERV-K(HML-10) family suppresses apoptosis, an indicator of malignant...

73

contribute to a large fraction of the human transcriptome [5]. Numerous REs are located within introns of host genes and might be involved in antisense gene regulation in *cis* [1]. The potential significance of RE-mediated *cis*-antisense gene regulation is suggested by the genome-wide presence of about 48,000 transcription start sites (TSSs) within HERVs and other REs that are in reverse orientation to overlapping host genes [6].

Promoter activity, a prerequisite for REs to exert antisense-mediated gene regulation, has been shown for representative LTRs of HERV-E [7], HERV-W [8], HERV-H [9–12], HERV-L [9], HERV-I [13] and HERV-K(HML-2), HML standing for human mouse mammary tumor virus-like [14–17]. The latter HERV family, HML-2 in the following, is the phylogenetically most recent and most active one in the human genome [3, 4], with about 50% of LTRs being transcriptionally active [15]. Antisense gene regulation in *cis* has been shown for HML-2 LTRs located within introns of the *SLC4A8* (a sodium bicarbonate co-transporter) and *IFT172* (intraflagellar transport protein 172) genes [14]. In addition, the *PLA2G4A* gene that encodes a phospholipase with a possible implication in tumorigenesis is negatively regulated by a HERV-E LTR-primed transcript [7]. These three individual cases are presently the only experimentally verified examples of the influence of LTR-primed transcripts on gene regulation.

A HERV family phylogenetically related to HML-2 is HERV-K(HML-10), HML-10 in the following [4]. The prototypical HML-10 provirus located within an intron of the long variant of the Complement Component 4 (*C4*) gene has been shown to possess promoter activity in its 3′LTR [18, 19]. Since this provirus remains the only one studied in detail to date, we here characterized the HML-10 family in more detail. We found that HML-10 invaded the ancestral genome of the Old World monkey (OWM) lineage about 35 Mya. A survey of the human genome revealed that HML-10 sequences were significantly enriched within host gene introns, indicating their evolutionary recruitment for gene regulatory functions. Three intron-located HML-10 proviruses exerted LTR promoter activity in the human HEK293T and HepG2 tumor cell lines in vitro. Transcriptional orientation and strength varied substantially between the cell lines and promoter activity was suppressed by interferon-gamma (IFNγ). One of the proviral LTRs showed transcriptional activity in opposite orientation to the encompassing pro-apoptotic *DAP3* gene that encodes a signaling protein of the Death Receptor (DR) pathway [20, 21]. We provide evidence that HML-10 LTR-primed transcripts negatively regulate *DAP3* expression in HeLa cells, as their inactivation by antisense oligonucleotides (ASOs) led to a 10-fold increase in *DAP3* mRNA levels and efficiently promoted apoptosis. Our findings support the functional relevance of LTR-primed *cis*-regulatory transcripts for human gene regulation and the cellular phenotype and function.

Results

HML-10 elements are 35 million years old and enriched within human genes

To identify potential priming of *cis*-acting regulatory transcripts by HERVs, we mined the GRCh38/hg38 human genome assembly [1] for sequences of the previously little described HML-10 family. The prototype member of HML-10 is an intron-located provirus in the long form of the *C4* gene that exhibits LTR promoter activity in vitro [18, 19]. Expression of this provirus has been detected via microarray before, for instance, in brain, breast, kidney and skin tissue, blood cells as well as various human cancer cell lines [22–27].

The provirus inside of the *C4* gene is currently the only HML-10 sequence described in the literature [18, 19]. With a size of about 6400 basepairs (bp) it contains the retroviral *gag*, *pol* and *env* genes, an A/T-rich stretch of unknown function between *pol* and *env* and two flanking LTRs [18] (Fig. 1a). Most HERV elements found in the human genome today have undergone homologous recombination between their two proviral LTRs, leaving behind solitary LTRs [1, 3, 4] that in this case have a size of about 550 bp. We identified seventy HML-10 elements within the human genome (Table 1). Of these, seven are proviruses with the structure 5′LTR-*gag*-*pol*-A/T-rich-*env*-3′LTR (with element no. 58 lacking the 5′LTR) and 63 are solitary LTRs. Some of the elements are truncated at either end or harbor other REs, mostly Alus. HERV sequences can be amplified by chromosomal duplication events following integration [4]. To reveal whether the identified HML-10 elements represent independent integration events, we determined their target site duplications (TSDs). The TSDs were expected to differ between independently acquired HML-10 elements. It has been shown previously that the provirus in the *C4* gene (element no. 22) created a 6 bp TSD [18]. Confirming these findings, we could identify TSDs of 5 or 6 bp for most (59 of 70) HML-10 elements (Table 1). All identified TSDs had a unique sequence, whereby the two copies of element no. 22 showed an identical 6 bp TSD with the expected sequence [18]. Alignment of the flanking regions of each HML-10 element (±1000 bp) revealed no sequence homology except for the two proviruses of element no. 22 as well as between elements nos. 27 and 45 (Additional file 1: Figure S1). Thus, one of the latter two has arisen through chromosomal duplication and the other 69 HML-10 elements listed in Table 1 are likely the result of independent retroviral integration events.

To reveal the evolutionary history of HML-10, we first searched for HML-10 sequences in genomes of different mammalian species. HML-10 was identified in all investigated genomes of the OWM lineage, but was absent in the genomes of New World monkeys (NWMs) and the more distantly related species mouse lemur, bushbaby and mouse

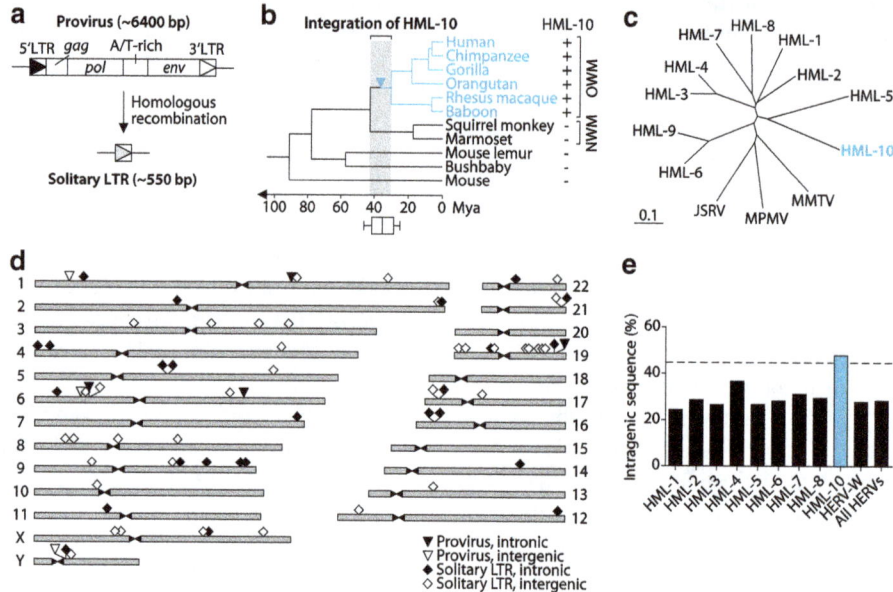

Fig. 1 Characteristics of the HML-10 endogenous retrovirus family. **a** Structure of HML-10 proviruses [18]. **b** Estimation of the evolutionary age of HML-10 with divergence times as reported before [78]. The *box-and-whiskers* plot shows age estimation by sequence comparison of LTRs from six complete proviruses (elements nos. 1, 3, 20, 22, 25 and 68 in Table 1) in the human genome [28]. The *arrowhead* indicates the integration events in the OWM lineage. **c** Neighbor-joining tree of Pol protein sequences of different endogenous and exogenous betaretroviruses [18, 72]. The *horizontal bar* represents 0.1 substitutions per amino acid position. **d** Chromosomal distribution of HML-10 elements in the human genome. Details can be found in Table 1. **e** Comparison of genomic fractions of intragenic elements (located within the boundaries of RefSeq [33] genes) between HML-10 and other HERV families in the human genome. All observed distributions differed significantly from the expected value for random integration that is shown as *dotted horizontal line*, with *P*-values ≤ 0.01 inferred by chi-square tests

(Fig. 1b). The OWM genomes contained between 80 and 96 HML-10 sequences (Additional file 2: Table S1). Of note, about 600 sequences annotated as HML-10 by RepeatMasker were found in the investigated NWM genomes that however shared little sequence homology with the ones found in OWMs. Thus, the annotated HML-10 elements in OWM and NWM genomes likely represent two distinct HERV families.

The evolutionary age of HML-10 was estimated by calculating the nucleotide sequence divergence between both LTRs of each of the six complete proviruses (Table 1), applying a mutation rate of 2.28 substitutions per site and year $\times 10^{-9}$ as described [28]. This analysis yielded an evolutionary age of 35.3 ± 7.8 million years (mean $\pm$ SD, see box-and-whiskers plot in Fig. 1b). Phylogenetic neighbor-joining analysis of 68 complete human HML-10 LTRs, including both LTRs of each of six complete proviruses, revealed a near-monophyletic tree (Fig. 2), indicating a single integration period. Therefore, the infectious progenitor of HML-10 likely invaded the ancestral genomes of OWM during a brief period around 35 Mya (Fig. 1b). The same age has been attributed before to other endogenous human betaretrovirus families, including HML-2 [4], HML-3 [29], HML-4 [30] and HML-6 [31]. In contrast, the HML-5 infectious progenitor was

active about 55 Mya [32] and HML-2 has remained active after the divergence of humans and chimpanzees about six Mya [4]. Neighbor-joining analysis of *pol* sequences of various endogenous and exogenous betaretroviruses showed that HML-10 is closely related to HML-1 through HML-9 HERVs and the extant exogenous retroviruses JSRV (Jaagsiekte sheep retrovirus), MPMV (Mason-Pfizer monkey virus) and MMTV (mouse mammary tumor virus) (Fig. 1c).

HML-10 elements are non-randomly distributed among human chromosomes (Fig. 1d). Most notably, the relatively small chromosome 19 harbored the highest number of elements (11 of 70). This was a first indication that HML-10 sequences were preferentially located near host genes, since chromosome 19 is the most gene-dense one [1]. Of the 70 HML-10 elements, 29 (41.4%) were found within introns of human genes (as annotated by RefSeq [33]), and 16 of the remaining 41 intergenic elements were located in proximity (±10,000 bp) to at least one RefSeq gene (Table 1). The relatively frequent location of HML-10 in the vicinity of host genes is a feature that is not shared with other HERV families studied in this regard. Namely, only 28% of all HERV-W elements are located within introns of genes [8] and HML-2 was found to be enriched outside genes, although *de novo* infection and integration of a resurrected HML-2 retrovirus favored actively transcribed regions [34],

Table 1 HML-10 elements in the human genome

No.	Chr.band	Coordinates (strand)	Structure	TSD	Integrated REs	Position	RefSeq genes (rel. orientation)	RefSeq genes ± 10,000 bp
1	1p36.13	chr1:19926886-19932710 (-)	Provirus	WWAAAT	-	Intergenic	-	PLA2G2E
2	1p35.3	chr1:29156730-29157306 (-)	LTR	GTTAC	-	Intronic	SFRS4 (s)	-
3	1q22	chr1:155691832-155699521 (-)	Provirus	ATTAAG	AluSp; MER11B	Intronic	DAP3 (as)	YY1AP1
4	1q23.1	chr1:158534566-158535143 (+)	LTR	GTCCAA	-	Intergenic	-	-
5	1q32.3	chr1:213045116-213046160 (-)	LTR[a]	TAGTGG	-	Intergenic	-	RPS6KC1
6	2p11.2	chr2:86245199-86245755 (+)	LTR	TATAC	-	Intronic	REEP1 (as)	-
7	2q37.3	chr2:238659222-238659778 (+)	LTR	not found	-	Intergenic	-	-
8	2q37.3	chr2:240556198-240556747 (+)	LTR	not found	-	Intronic	ANKMY1 (as)	DUSP28
9	3p14.2	chr3:58724279-58724839 (+)	LTR	CAGCAG	-	Intergenic	-	-
10	3q12.2	chr3:101089113-101089658 (+)	LTR	AGGCAC	-	Intergenic	-	-
11	3q21.3	chr3:128828801-128835173 (+)	LTR	TGCAT	AluY; LTR7B; HERVH	Intergenic	-	-
12	3q24	chr3:146283898-146284161 (-)	LTR[b]	not found	-	Intergenic	-	-
13	4p16.3	chr4:1709255-1710065 (+)	LTR	ATGGGG	AluY	Intronic	SLBP (as)	TMEM129
14	4p16.1	chr4:8441728-8442282 (-)	LTR	YYTTTA	-	Intronic	TRMT (as)	ACOX3
15	4q31.21	chr4:143777776-143778330 (+)	LTR	TCARCC	-	Intergenic	-	-
16	5q14.1	chr5:78245022-78245575 (-)	LTR	TCYWCA	-	Intronic	AP3B1 (s)	-
17	5q14.1	chr5:78248004-78248093 (-)	LTR[b]	not found	-	Intronic	AP3B1 (s)	-
18	5q31.3	chr5:142088759-142089306 (-)	LTR	TTGGTG	-	Intergenic	–	-
19	6p24.1	chr6:12989575-12990131 (-)	LTR	GAAAAC	-	Intronic	PHACTR1 (as)	-
20	6p22.1	chr6:27187520-27196279 (+)	Provirus	AAGATM	3× AluY; AluYc; LTR13A	Intergenic	-	-
21	6p22.1	chr6:28607908-28608446 (+)	LTR	CATGTT	-	Intergenic	-	-
22	6p21.33	chr6:31984691-31991052 (-) chr6:32017429-32023790 (-)	Provirus	TGTCTG TGTCTG	-	Intronic	C4A/C4B (as)	CYP21A1P; STK19; TNXB
23	6p21.32	chr6:32512829-32513385 (+)	LTR	GGGGRG	-	Intergenic	-	HLA-DRB5
24	6q21	chr6:114011436-114018365 (+)	LTR	CCCTAT	LTR7B; HERVH	Intergenic	-	-
25	6q22.31	chr6:122504844-122512093 (-)	Provirus	GGACAT	3× AluY	Intronic	PKIB (as)	-
26	7q36.2	chr7:154936774-154937317 (-)	LTR	ACTCCA	-	Intronic	PAXIP1 (as)	PAXIP1-AS2
27	8p22	chr8:17915846-17916036 (-)	LTR[b]	CCCMTA	-	Intergenic	-	PCM1
28	8p21.3	chr8:22985089-22985644 (-)	LTR	CCTCYY	-	Intergenic	-	RHOBTB2
29	8q11.1	chr8:46533684-46534254 (+)	LTR	CATTTC	-	Intergenic	-	-
30	8q21.13	chr8:82206225-82206776 (+)	LTR	CASCCK	-	Intergenic	-	-
31	9p13.3	chr9:34539821-34540365 (-)	LTR	GGCATG	-	Intergenic	-	-
32	9q22.1	chr9:87984145-87984721 (-)	LTR	TATGGC	-	Intergenic	-	CDK20
33	9q22.31	chr9:92523403-92523958 (-)	LTR	WATTGT	-	Intronic	CENPP (as); ECM2 (s)	-
34	9q31.3	chr9:109072071-109072632 (+)	LTR	CMAAAG	-	Intronic	TMEM245 (as)	-
35	9q34.11	chr9:129101280-129101834 (+)	LTR	GGGGAA	-	Intronic	CRAT (as)	PPP2R4
36	9q34.13	chr9:132289026-132289571 (+)	LTR	CTCTYA	-	Intronic	SETX (as)	-
37	10p11.21	chr10:37685164-37685717 (+)	LTR	GAATC	-	Intergenic	-	-
38	11p11.2	chr11:43747422-43747974 (+)	LTR	GTTCTG	-	Intronic	HSD17B12 (s)	-
39	12p13.1	chr12:12810288-12810845 (-)	LTR	ATCTA	-	Intergenic	-	DDX47
40	12q24.33	chr12:132949015-132949570 (+)	LTR	GTATC	-	Intronic	ZNF605 (as)	-
41	13q13.3	chr13:37090225-37090778 (+)	LTR	CCTGTT	-	Intergenic	-	-

Table 1 HML-10 elements in the human genome *(Continued)*

42	14q31.1	chr14:79706443-79707024 (+)	LTR	TTGGTC	-	Intronic	NRXN3 (s)	-
43	16p13.13	chr16:10785748-10785943 (+)	LTR[b]	not found	-	Intronic	FAM18A (as)	-
44	16p13.13	chr16:10788495-10789043 (-)	LTR	GAGAYC	-	Intronic	FAM18A (s)	-
45	17p13.1	chr17:8056835-8057031 (+)	LTR[b]	not found	-	Intergenic	-	ALOX15B
46	17p13.1	chr17:8082291-8082848 (+)	LTR	CCAGG	-	Intronic	ALOX12B (as)	MIR4314
47	17q11.2	chr17:32913177-32913734 (+)	LTR	GGTATR	-	Intergenic	-	-
48	19p13.3	chr19:2863773-2863865 (+)	LTR[b]	not found	-	Intergenic	-	ZNF555; ZNF556
49	19p13.2	chr19:7100429-7100974 (-)	LTR	GTCTC	-	Intergenic	-	-
50	19p12	chr19:21987441-21988020 (+)	LTR	ATAAYA	-	Intronic	ZNF208 (as)	-
51	19p12	chr19:23876577-23877146 (-)	LTR	CTCCCC	-	Intergenic	-	-
52	19q13.11	chr19:34595165-34595718 (-)	LTR	TGTAGG	-	Intergenic	-	SCGBL
53	19q13.12	chr19:36636542-36637090 (-)	LTR	not found	-	Intergenic	-	ZNF382; ZNF461
54	19q13.2	chr19:42667555-42668111 (-)	LTR	GTGTG	-	Intergenic	-	-
55	19q13.31	chr19:44283216-44283766 (-)	LTR	GTAAG	-	Intergenic	-	ZNF233; ZNF235
56	19q13.32	chr19:46030891-46031467 (+)	LTR	CAAGGT	-	Intergenic	-	IGFL4; PGLYRP1
57	19q13.41	chr19:51900477-51900716 (-)	LTR[b]	not found	-	Intronic	ZNF649 (s); ZNF619 (as)	ZNF649-AS1
58	19q13.41	chr19:52460866-52466497 (-)	Provirus[b]	AAAAC	-	Intronic	ZNF578 (as)	ZNF534
59	21q22.3	chr21:43698519-43699073 (+)	LTR	TTTAG	-	Intergenic	-	RRP1B
60	21q22.3	chr21:43738447-43739417 (+)	LTR	CTAAT	AluSx; AluY	Intronic	PDXK (as)	-
61	22q11.21	chr22:20818927-20819480 (-)	LTR	TAAGA	-	Intronic	PI4KA (s)	-
62	22q13.31	chr22:45188740-45189285 (+)	LTR	TGCAAC	-	Intergenic	-	NUP50; KIAA0930
63	Xp11.23	chrX:48047190-48047228 (-)	LTR[b]	not found	-	Intergenic	-	ZNF630-AS1
64	Xp11.22	chrX:51636506-51637062 (+)	LTR	GCTCTA	-	Intergenic	-	-
65	Xq22.2	chrX:103469871-103470407 (-)	LTR	GRGGAG	-	Intergenic	–	–
66	Xq22.3	chrX:107232277-107232369 (-)	LTR[b]	not found	-	Intronic	PIH1D3 (as)	-
67	Xq27.1	chrX:139517757-139518302 (+)	LTR	CTTAAG	-	Intergenic	-	-
68	Yq11.221	chrY:12993871-13001093 (-)	Provirus	TGSATT	AluY; LTR2B	Intergenic	-	-
69	Yq11.221	chrY:13333756-13334302 (+)	LTR	CRYAGC	-	Intronic	UTY (as)	-
70	Yq11.221	chrY:16492757-16493302 (+)	LTR	TCCAAR	-	Intergenic	-	-

Chromosome bands and coordinates are according to the hg38 assembly [1] and RepeatMasker annotation [70]
As, antisense orientation; s, sense orientation; TSD, target site duplication; [a]denotes a tandemly repeated LTR; [b]denotes truncated elements.
Ambiguous nucleotides are indicated as follows: K, G or T; M, A or C; S, C or G; R, A or G; W, A or T; Y, C or T

a common feature of present-day retroviruses as well [35]. Based on the published literature about HERV-W and HML-2, we compared the integration preferences of these two HERV families with HML-10 as well as all other HML families, except for HML-9 that was not annotated by RepeatMasker, at the genome-wide level. HML-10 sequences were found with higher frequency within the boundaries of RefSeq genes (47.94%) than expected by random distribution (44.79%), whereby sequences of HML-2 (28.76%), HERV-W (27.95%) and of all annotated HERV elements combined (28.46%) were less abundant within genes (Fig. 1e). The intragenic sequence fractions of the other HML families were below the expected value for random integration and ranged between 24.35% (HML-1) and 36.75% (HML-4). Thus, the frequent location of HML-10

within host genes appears to be a unique feature of this family and suggests an important and conserved function for gene regulation. The intronic HML-10 elements showed a distinct bias for integration in reverse orientation relative to the respective encompassing gene, with 20 being in reverse (antisense) and 7 in parallel (sense) orientation (Table 1). Two elements were reverse to one gene and parallel to another overlapping one. The integration bias of HML-10 indicates that the reverse orientation was evolutionarily favored, which is in line with previous findings of other HERV families [36–38]. One explanation for this observation is that parallel intronic proviruses are more likely to disrupt the encompassing gene due to the presence of transcription termination sites in the LTRs, which leads to negative selection of such integration events [39].

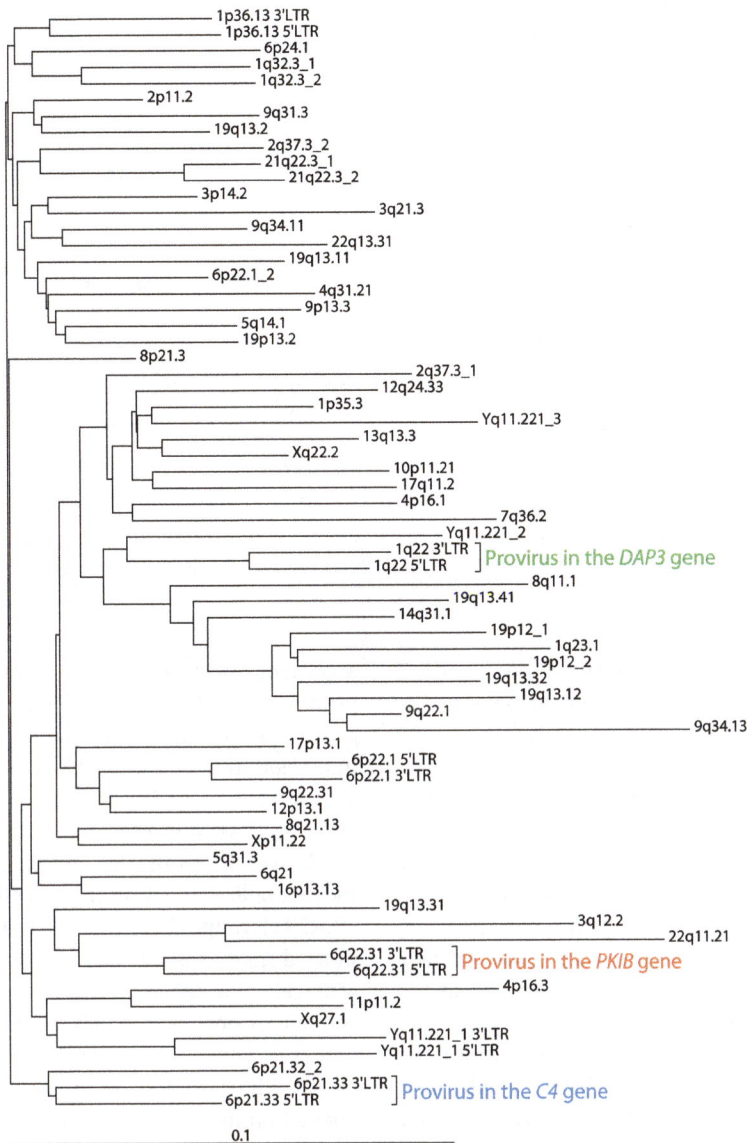

Fig. 2 Neighbor-joining tree of 68 complete HML-10 LTRs in the human genome. HML-10 LTR sequences (see Table 1) were retrieved from the human genome GRCh38/hg38 assembly [1] according to RepeatMasker [70] annotation. The horizontal bar represents 0.1 substitutions per nucleotide position

Contrarily, reverse oriented proviruses may even be beneficial by protecting from newly infecting retroviruses by antisense RNA mechanisms [19] and by contributing regulatory elements such as LTR promoters that can modulate gene expression in *cis*, as shown before [7, 14].

HML-10 exerts differential LTR promoter activity in tumor cell lines
To further investigate the potential of HML-10 in generating *cis*-regulatory transcripts, we determined LTR promoter activities of three complete proviruses located in reverse orientation within introns of host genes (Fig. 3). These were elements nos. 3, 22 and 25, within the

DAP3, *C4* and *PKIB* (Protein kinase inhibitor beta) genes, respectively (Table 1). The *PKIB* gene harbors numerous other intronic HERV sequences not belonging to the HML-10 family that together with other REs constitute over 50% of its genomic sequence. Three additional HML-10 proviruses are located outside of genes, elements nos. 1, 20 and 68, and one found in an intron of the zinc finger protein gene *ZNF578*, no. 58, lacks the 5′ LTR. We focused on the three complete and intronic proviruses, referred to as HML-10(DAP3), HML-10(C4) and HML-10(PKIB), that comprise six LTRs for promoter analysis, since these could potentially generate *cis*-regulatory transcripts. We preferred proviruses over

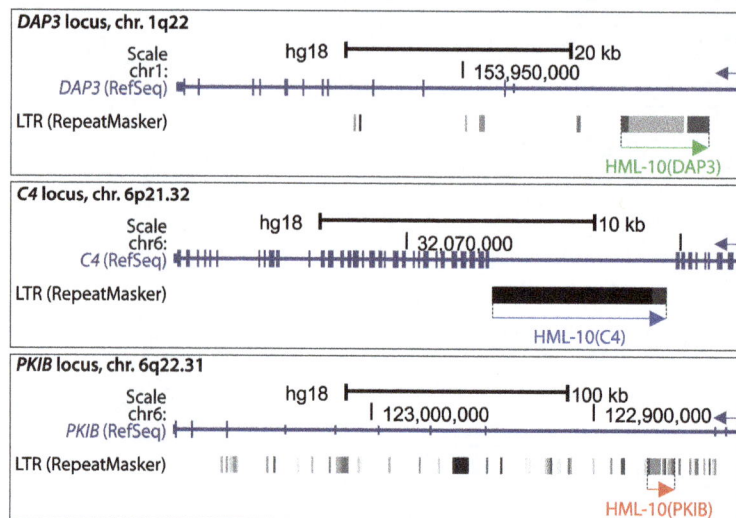

Fig. 3 Genomic organization of the HML-10(DAP3), HML-10(C4) and HML-10(PKIB) proviruses (from *top* to *bottom*). The grey rectangles in the LTR (RepeatMasker) track shows all annotated HERV elements including the indicated HML-10 proviruses. Images were retrieved and modified from the UCSC Genome Browser [68]

solitary LTRs since proviral LTRs of the related HML-2 and HERV-W families have been shown to be stronger promoters than the respective solitary LTRs [8, 15]. We also found that the two LTRs of each HML-10 provirus clustered in the neighbor-joining tree (Fig. 2). Thus, despite their high sequence similarities these LTRs have resisted homologous recombination, suggesting their functional importance. HML-10 provirus RNA has been detected in various human tissues and cell lines by microarray analyses [22–27, 40–43] that, however, lack the information whether transcription is initiated in the 5′LTR or upstream of the provirus.

To assess their promoter activities, we cloned the LTRs of HML-10(DAP3), HML-10(C4) and HML-10(PKIB) into the promoter-free pGL3-Enhancer luciferase reporter vector, as described [19] (Fig. 4a). As HERV LTRs can be bidirectional promoters [5, 17, 44], we also included the retroviral antisense orientation for each of the six LTRs. LTR promoter activity of HML-10(C4) has been demonstrated before with reporter assays in the human hepatocellular carcinoma cell line HepG2 and in COS7 monkey kidney cells [19]. Additionally, HML-10 *pol* transcripts have been identified in human hepatocellular carcinoma cells and in human embryonic kidney HEK293 cells by microarray analysis [26] (Table 2). Based on these findings, we transfected our pGL3-Enhancer constructs into HepG2 and HEK293T cells (HEK293 expressing SV40 virus T-antigen) to measure their promoter activities (Fig. 4b). The pGL3-Control vector bearing the SV40 promoter served as positive control and empty, promoter-free pGL3-Enhancer as negative control. HML-10(C4) showed significant transcriptional activity exclusively in the 3′LTR in HepG2 in both retroviral sense and antisense orientations. This is in

accordance with a previous study that has demonstrated promoter activity in the 3′LTR, but not in the 5′LTR of this provirus in the same cell line [19]. In HEK293T, we found transcription from the 5′LTR in retroviral sense orientation and from the 3′LTR in retroviral antisense orientation. HML-10(DAP3) exerted bidirectional promoter activity in its 5′LTR in both cell lines, whereas HML-10(PKIB) showed bidirectional promoter activity in its 3′LTR, but in HEK293T only. Therefore, all three investigated proviruses showed transcriptional activity in at least one of their LTRs, with cell type-specific strength and orientation (Fig. 4b). While LTR promoter activity in retroviral antisense orientation was unlikely to primarily affect gene regulation, all three HERVs exerted promoter activity in retroviral sense orientation in one of their LTRs, which is antisense relative to the respective encompassing gene. Thus, the proviruses have the potential for antisense-mediated regulation of the encompassing *DAP3*, *C4* and *PKIB* genes in *cis* in a cell type-specific manner.

Promoter activity of the HML-10(C4) 3′LTR has been reported previously to be suppressed by IFNγ in HepG2 [19], which we reproduced (Fig. 4c). Likewise, the 5′LTR promoter of HML-10(DAP3) in retroviral sense orientation (antisense relative to the *DAP3* gene) was dose-dependently suppressed by IFNγ. We speculate this to be mediated by an IFNγ activated site (GAS) matching the consensus motif 5′-TTNCNNNAA-3′, a putative binding site for STAT1 homodimers that form during IFNγ signaling [45]. This motif is present in all analyzed LTRs (Fig. 4d) as well as the SV40 promoter (data not shown). The latter is known to be inhibited by IFNγ [46] and served as positive control for IFNγ-mediated suppression (Fig. 4c). In contrast, the herpes simplex virus

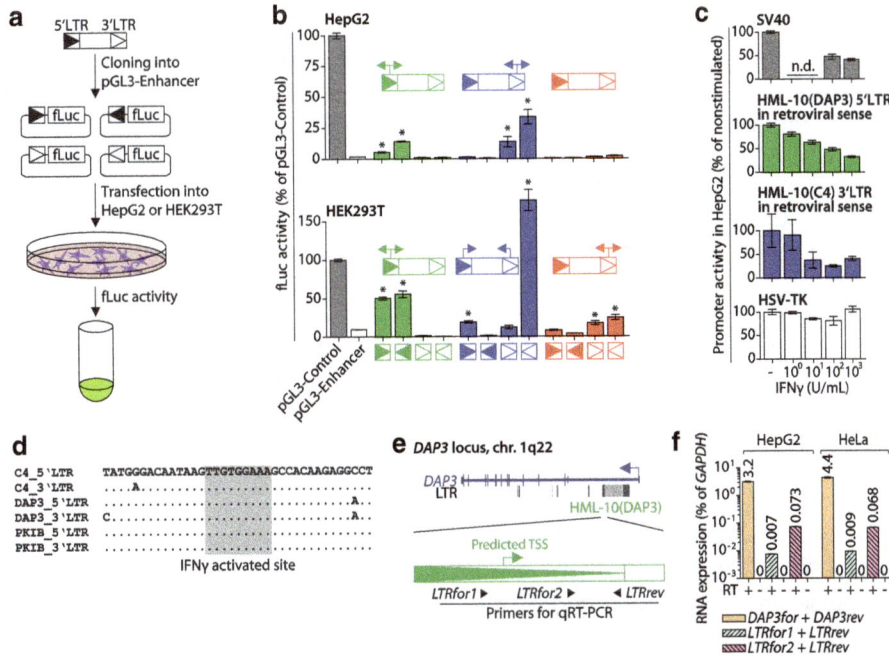

Fig. 4 Promoter activities of HML-10 LTRs. **a** LTRs of the HML-10(DAP3), HML-10(C4) and HML-10(PKIB) proviruses were cloned in both orientations into the promoter-free pGL3-Enhancer vector and transfected into HepG2 or HEK293T cells. Firefly luciferase (fLuc) activities were determined 24 h after transfection **b** Promoter activities expressed as fLuc activity normalized to renilla luciferase (rLuc) activity of the co-transfected pGL4.74 vector in the indicated cell lines. The pGL3-Control vector bearing the SV40 promoter (grey bars) served as positive and empty pGL3-Enhancer (white bars) as negative control. Promoter activities were normalized to pGL3-Control set to 100%. The bars show mean ± SEM of three independent experiments in duplicates. *P-value ≤ 0.05, Student's t-Test compared to pGL3-Enhancer. **c** For HepG2 cells the effect of IFNγ stimulation on two selected LTRs as well as the SV40 and HSV-TK promoters is shown. LTR and SV40 activity is expressed as fLuc normalized to rLuc signals, HSV-TK activity is expressed as rLuc activity only. The bars show mean ± SEM of at least three independent experiments and were normalized to unstimulated (-) cells set to 100%. n.d., not determined. **d** Identification of a conserved IFNγ activated site (GAS) of the consensus sequence 5′-TTNCNNNAA-3′ [45]. **e** Locations of primers used to detect transcripts originating from the 5′LTR of HML-10(DAP3). The predicted TSS was identified as described in the text and Additional file 1: Figure S1. **f** Detection of DAP3 mRNA and HML-10(DAP3) transcripts in HepG2 and HeLa cells by qRT-PCR. cDNA samples prepared without reverse transcriptase (RT) for the indicated primer pairs, but with RT for GAPDH, served as controls. Values are normalized to GAPDH mRNA levels. Bars show mean ± SD of two measurements. In most cases, the SD is too small to be visible

Table 2 Detection of HML-10(DAP3) pol transcripts by previously reported microarray studies [79]

Tissue/cell type	HML-10(DAP3) RNA expressed	HML-10(DAP3) RNA not expressed	HML-10(DAP3) RNA variably expressed
Normal tissue and non-tumor cell lines (n = 23)	Cervix [22], epidermal keratinocytes (HaCaT) [26], thyroid [22], umbilical vein endothelial cells (HUVEC) [40], uterus [22] (5/23)	Blood [22, 23], breast [25], colon [22], heart [22], liver [22], lung [22], mamma [22, 23], neural stem cells (HNSC.100) [40], ovary [22], placenta [22], prostate [22], rectum [22], skeletal muscle [22], stomach [22], testes [22] (15/23)	Brain [22, 24], kidney [22, 27], skin [22, 41] (3/23)
Primary tumors and tumor cell lines (n = 13)	CAKI (renal carcinoma) [26], GliNS1 (neural tumor stem line) [40], HEK293 (embryonic kidney) [26], HeLa (cervix adenocarcinoma) [26], HuH-7 (hepatocellular carcinoma) [26], MIA PaCa-2 (pancreatic carcinoma) [26], SK-N-MC (neuroblastoma) [42, 43], SK-N-SH (neuroblastoma) [42], T47D (ductal breast epithelial carcinoma) [26], U-251MG (glioblastoma) [42] (10/13)	Breast cancer [25], U-138MG (glioblastoma) [42] (2/13)	Renal cell carcinoma [27] (1/13)

Numbers in parentheses indicate how many of the analyzed tissue/cell types show the respective expression status of HML-10(DAP3) RNA thymidine kinase (HSV-TK) promoter that was used to normalize promoter activities was unaffected by IFNγ (Fig. 4c). The GAS motif is highly conserved among proviral HML-10 LTRs in the human genome (Fig. 4d) and the solitary LTRs (data not shown), which supports its functional relevance. Hence, IFNγ-mediated promoter suppression is likely a general feature of HML-10 LTRs, in line with the known antiviral activity of interferons [19]. This is of particular interest for possible HML-10-mediated negative regulation of the encompassing genes, since mRNA expression of C4 and DAP3 is known to be induced by IFNγ [20, 47] and DAP3 is implicated in IFNγ-dependent apoptosis [20].

Based on our promoter activity studies, HML-10(DAP3) was the most interesting candidate for further investigation, since its 5′LTR is the only one investigated that promoted transcription in the retroviral sense orientation, which is antisense to DAP3, in both cell lines (Fig. 4b). The involvement of an HML-10(DAP3)-primed transcript in regulating the encompassing gene is suggested by the fact that DAP3 expression is induced [20], whereas the LTR promoter is suppressed by IFNγ (Fig. 4c). In addition, HML-10(DAP3) RNA has been detected previously in various human cancer-derived cell lines but not in most healthy tissues (Table 2). This indicates a possible role in the regulation of DAP3 gene expression in cancer cells and some distinct tissues, including cervix, thyroid and uterus as well as epidermal keratinocytes and umbilical vein endothelial cells. Our promoter activity studies indicated that transcription of HML-10(DAP3) originated from the 5′LTR (Fig. 4b). For further proof, we determined the most likely TSS within this promoter. Since LTR-dependent transcription relies on host RNA polymerase (RNA pol) II [5, 48] we sought to identify the two integral core elements of this promoter, Initiator (Inr) elements and TATA boxes [49]. The TSS within LTRs of the related HML-2 family has been identified previously within an Inr element with a TATA box about 10 bp upstream of the Inr [50]. We identified a similar configuration a single time in the HML-10(DAP3) 5′LTR in retroviral sense orientation, an Inr element 11 bp downstream of a TATA box (Additional file 3: Figure S2). This Inr sequence contained the most likely TSS. We also identified a downstream promoter element (DPE) matching the consensus 5′-RGWYVT-3′ sequence [49], a putative binding site for the transcription factor TFIID of the RNA pol II core promoter, at nucleotide position +19 relative to the putative TSS. To get experimental proof that HML-10(DAP3) transcription is initiated within this putative TSS, we performed quantitative real-time PCR (qRT-PCR) measurements in HepG2 cells with a reverse primer located downstream of the TSS (LTRrev) and two different forward primers, one located upstream (LTRfor1) and one downstream (LTRfor2) of the TSS (Fig. 4e and Additional file 3: Figure S2). If transcription was initiated from the TSS, we would expect higher expression measured using LTRfor2 + LTRrev than with LTRfor1 + LTRrev primer combinations. This was indeed the case, whereby weak signals seen with LTRfor1 + LTRrev likely resulted from amplification of the intron of the DAP3 pre-mRNA (Fig. 4f). To avoid false signals from genomic DNA for these lowly abundant transcripts, we subjected the RNA preparations to DNase treatment prior to reverse transcription and included control samples without reverse transcriptase that did not result in detectable amplification. We thus verified expression of the HML-10(DAP3) RNA that is present at about 40-fold lower levels than the DAP3 mRNA, and provide further evidence that it originates from the 5′LTR around the predicted TSS. These findings confirmed the weak but significant transcription of this LTR in the promoter activity studies in retroviral sense orientation in the same cell line (Fig. 4b). Our findings are in agreement with reported microarray data that demonstrated expression of the retroviral transcript in various cell lines, which extends into the pol gene of the HML-10(DAP3) provirus (Table 2). However, although the primer combinations were designed to only amplify the HML-10(DAP3) sequence, as judged by in silico PCR analysis, we cannot completely rule out that transcripts of other potentially active HML-10 elements were co-amplified.

The pro-apoptotic effect of DAP3 has been described previously in HeLa cells [20] in which HML-10(DAP3) RNA has been identified by microarray analysis (Table 2). Accordingly, we detected HML-10-primed transcripts by qRT-PCR in HeLa where it was present at comparable levels as in HepG2 (Fig. 4f). We therefore selected HeLa cells to determine the functional relevance of HML-10(DAP3) RNA on the expression of DAP3.

Inactivation of HML-10(DAP3) RNA induces DAP3 expression and apoptosis in HeLa cells

Having confirmed the presence of HML-10(DAP3) RNA in HeLa cells and its likely origin within the proviral 5′ LTR, we sought to determine its function within the cell. We expected the retroviral RNA to suppress DAP3 gene expression in cis similar to the previously described LTR-primed regulatory transcripts [7, 14]. To determine its potential regulatory function we aimed at inactivating the retroviral RNA by means of sequence-specific ASOs. We opted for ASOs rather than siRNAs that are both known to be active in the nucleus [51, 52], the common site of action of LTR-primed transcripts [5], as siRNAs may directly influence DAP3 expression levels through the passenger strand that would be antisense to the DAP3 pre-mRNA. ASO-mediated inactivation of the HML-10(DAP3) RNA was expected to activate DAP3 gene expression.

We designed four ASOs downstream of the putative TSS, ASOs 1–4, to counteract the retroviral RNA (Fig. 5a). At 24 h after transfecting the ASOs at 25 or 50 nM into HeLa cells, we determined HML-10(DAP3) and *DAP3* expression at the RNA level by qRT-PCR. Transfecting the ASOs caused an increase in *DAP3* mRNA levels, as expected, but not a decrease in HML-10(DAP3) RNA (Fig. 5b). These observations likely indicate that the ASOs blocked association of *DAP3* pre-mRNA with the retroviral RNA, but did not significantly mediate cleavage of the latter. Although RNase H1/H2-dependent hybrid-specific RNA degradation has been reported to be induced by ASOs [51, 53], cleavage efficiency is largely sequence-dependent and the HML-10(DAP3) RNA might resist degradation. For these reasons, measuring *DAP3* mRNA levels was the only feasible way to assess the impact of inactivating the retroviral RNA. Transfecting ASOs 1–4 resulted in increased *DAP3* mRNA levels with varying efficiencies (Fig. 5b). When used at 25 nM, ASOs 1–4 increased *DAP3* mRNA levels about 5-fold as compared to non-transfected control cells. The most efficient ASO 2

exerted a dose-dependent increase of the *DAP3* mRNA up to 10-fold at 50 nM. Both control ASOs, one with a random sequence (Mock) and one immediately upstream of the 5′LTR, did not significantly change *DAP3* expression levels, demonstrating a sequence-dependent effect and that the HML-10(DAP3) RNA originates from the 5′ LTR. Although ASOs 1–4 were designed to only map to the *DAP3* locus, we consider the possibility that HML-10 RNA species transcribed at other loci that may act *in trans* on *DAP3* expression may have been inactivated by these ASOs as well. Overall, the use of ASOs to counteract HML-10-primed transcripts confirmed their negative impact on *DAP3* mRNA expression levels.

DAP3 is an adapter protein that links the intracellular portion of DRs to the Fas-Associated Death Domain (FADD) in the DR pathway of extrinsic apoptosis [21]. Consequently, we expected the HML-10(DAP3) RNA to suppress apoptosis via this pathway. Overexpressing *DAP3* has been shown to induce apoptosis in HeLa cells [20]. We wondered whether upregulation of *DAP3* by the most effective ASO 2 at 50 nM (Fig. 5b) was

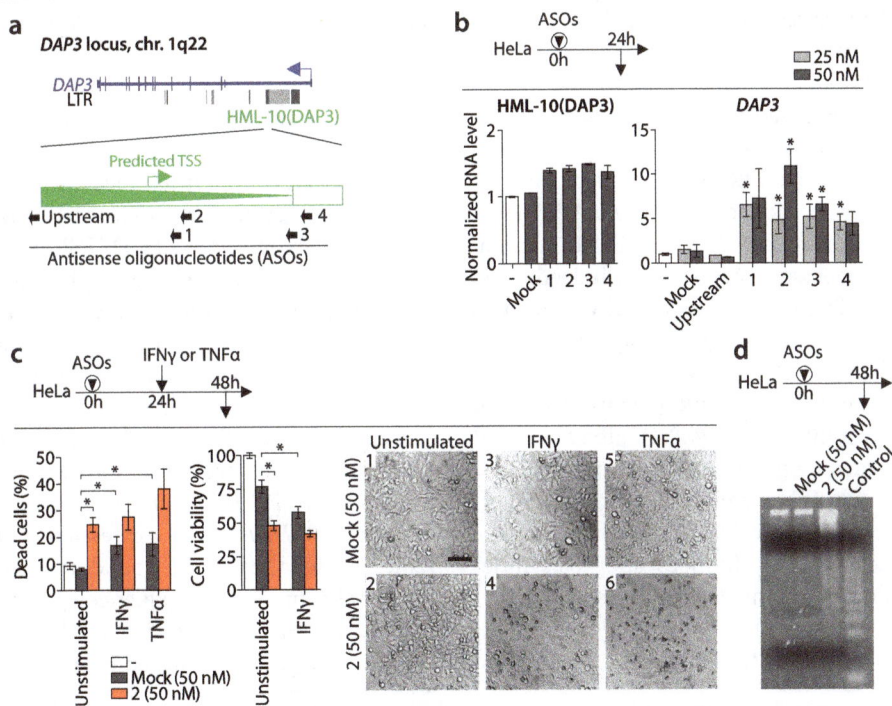

Fig. 5 Inactivating the HML-10(DAP3) RNA induces *DAP3* expression and apoptosis in HeLa cells. **a** Target regions of sequence-specific ASOs are indicated. ASOs 1-4 are in antisense orientation to the retroviral transcript and in sense orientation to the *DAP3* transcript. The ASO designated as Upstream served as control. **b** Cells were transfected with 25 or 50 nM of the indicated ASOs. At 24 h after transfection, expression levels of HML-10(DAP3) (*left*) and *DAP3* mRNA (*right*) were determined by qRT-PCR. *Bars* show mean ± SEM of three independent experiments. RNA levels were normalized to *GAPDH* and levels of non-transfected cells were set to 1. *P-value ≤ 0.05, Student's t-Test against Mock. **c** Cells were transfected with the indicated ASOs at 50 nM, after 24 h stimulated with 1000 U/mL IFNγ or 100 ng/mL TNFα, or left unstimulated. After additional 24 h, *Trypan Blue* exclusion as indicator of dead cells (*left*), MTS cell viability assays (*center*) or light microscopic analysis (*right*) was performed. *Bars* show mean ± SEM of three independent experiments in duplicates. *P-value ≤ 0.05, Student's t-Test. The *scale bar* in light microscopy panel 1 is 100 μm. **d** Cells were transfected with the indicated ASOs at 50 nM. At 48 h after transfection, genomic DNA of these cells was prepared with the Apoptotic DNA Ladder Kit (Roche). The control DNA is from apoptotic U937 cells provided with the kit

sufficient to cause apoptosis. To this end, we compared the effect of ASO 2 with known apoptosis-inducing stimuli, tumor necrosis factor-alpha (TNFα) and IFNγ, on HeLa cells. Both cytokines significantly induced cell death associated with diminished cell viability, as well as cell rounding characteristic of apoptosis (Fig. 5c). Likewise, HeLa cells transfected with ASO 2 showed similar signs of apoptosis but not those transfected with Mock ASO. The fraction of dead cells was significantly higher for ASO 2-transfected cells when compared to Mock-transfected cells (24.8% vs. 8.0%, $P = 10^{-4}$), and cell viability was lower (47.9% vs. 76.8% relative to non-transfected cells, $P = 10^{-4}$). In addition, transfection of ASO 2, but not of Mock, induced features of apoptosis, such as detachment from the tissue culture dish, rounding and shrinking (Fig. 5c, light microscopy panels 1 and 2). This was supported by another test for apoptosis, genomic DNA fragmentation, that occurred upon transfection of ASO 2 (Fig. 5d). These findings provided evidence that ASO 2-mediated induction of *DAP3* mRNA led to increased expression of DAP3 protein that is required for apoptosis and DNA fragmentation. Thus, inactivating HML-10(DAP3) RNA increased *DAP3* expression sufficiently to induce apoptosis, which demonstrates the functional relevance of this retroviral transcript.

In parallel, we assessed whether inactivating HML-10(DAP3) RNA also increased the susceptibility to apoptosis by TNFα. We expected this since TNFα stimulates extrinsic apoptosis via the DR pathway that involves DAP3 [21]. Thus, inactivation of HML-10(DAP3) RNA with resulting *DAP3* overexpression and TNFα stimulation may promote apoptosis synergistically. Indeed, ASO 2-transfected HeLa cells that were additionally stimulated with TNFα exhibited increased signs of apoptosis compared to unstimulated ASO 2-transfected cells (Fig. 5c, light microscopy panels 2 and 6), and contained a larger fraction of dead cells (38.1% vs. 24.8%), albeit without statistical significance (Fig. 5c). Stimulating ASO 2-transfected cells with IFNγ had less pronounced effects on the fraction of dead cells and viability (Fig. 5c), which may be because IFNγ induces apoptosis independent of DR signaling. Conclusively, we showed that *DAP3* expression is negatively regulated by the HML-10(DAP3) RNA to an extent that apoptosis is inhibited in HeLa cells.

Discussion

Here we have characterized the previously little described HML-10 family of endogenous retroviruses in the human genome and studied its potential in regulating host gene expression. We found that the infectious progenitor of HML-10 invaded the ancestral genome of OWMs about 35 Mya (Fig. 1b). With 70 identified elements, HML-10 is a relatively small HERV family when compared, for

instance, to the intensely investigated HML-2 that constitutes about 2500 sequences in the human genome [4]. It is known that HERVs, after *de novo* integration, may increase in number due to chromosomal duplication events [4]. However, sequence comparisons of the TSDs (Table 1) and the flanking regions (Additional file 1: Figure S1) indicated that only one of the 70 identified HML-10 elements is the result of a chromosomal duplication, whereas the other 69 elements most likely arose by independent retroviral integrations. We found an unusually high abundance of HML-10 within introns of host genes when compared to other HERV sequences including those of phylogenetically related HML families (Fig. 1e), indicating that this family in particular has been evolutionary co-opted for gene regulatory functions. Since LTR promoter activity of the provirus in the *C4* gene has been demonstrated previously [19], we hypothesized HML-10 to express LTR-primed regulatory transcripts in *cis* similar to recently reported HML-2 [14] and HERV-E [7] LTRs.

To assess their potential in expressing such regulatory RNAs, LTRs of three selected, intron-located HML-10 proviruses were subjected to promoter activity studies in HepG2 and HEK293T cells (Fig. 4b). Interestingly, both strength and orientation of LTR transcription differed substantially between the cell lines. Based on the promoter activity studies, all three investigated HML-10 proviruses had the potential to negatively regulate their encompassing genes by priming antisense RNAs. The HML-10(DAP3) provirus located in the *DAP3* gene showed LTR promoter activity in retroviral sense orientation (antisense relative to *DAP3*) in both cell lines and was therefore selected for further analysis (Fig. 4b). DAP3 is a signaling protein involved in the DR pathway of extrinsic apoptosis that induces apoptosis when overexpressed [20, 21]. Promoter activity of the HML-10(DAP3) 5′LTR in retroviral sense orientation (antisense relative to the *DAP3* gene) was suppressed by IFNγ, as reported previously for the HML-10 provirus in the *C4* gene [19] (Fig. 4c). This might, at least partially, explain how *DAP3* gene expression is induced by IFNγ [20]. In HeLa cells, we found that counteracting the retroviral transcript by sequence-specific ASOs led to an increase of *DAP3* expression levels sufficient to induce apoptosis (Fig. 5b,c). Two control ASOs, one targeting a region upstream of and one with a randomized sequence, did neither induce *DAP3* mRNA expression nor apoptosis, verifying that the ASO transfection procedure itself did not exert any non-specific effects on these two read-outs. Thus, the HML-10(DAP3) RNA suppressed apoptosis in HeLa. HML-10 transcripts originating from other loci may have been inactivated by the ASOs as well and consequently might also contribute to the reduction of *DAP3* expression *in trans*. ASO-mediated inactivation confirmed that the HML-10-primed transcripts, despite being about 60-fold weaker expressed than the DAP3 mRNA in this cell line (Fig. 4f), had a substantial impact on *DAP3* expression levels. Indeed, regulatory non-coding RNAs are often weakly expressed [54] and capable of substantially down-regulating gene expression even if 10-100 fold less abundant than their respective mRNA [55]. Among the mechanisms that have been proposed for this kind of gene regulation is the induction of repressive epigenetic modifications that lead to heterochromatin formation, or transcriptional collision of opposing RNA polymerases [54]. It has been shown previously that preventing association between weakly expressed regulatory RNAs and their corresponding mRNA (as opposed to degradation of the regulatory RNA) is sufficient to substantially induce mRNA expression [55], which might explain why we did not observe ASO-mediated degradation of HML-10 RNA but nevertheless an increase in *DAP3* mRNA expression levels (Fig. 5b).

Promoter activity studies (Fig. 4b), qRT-PCR experiments (Fig. 4f), and the fact that the ASO immediately upstream of the LTR did not affect *DAP3* expression levels (Fig. 5b) provided evidence that the retroviral RNA originates from the proviral 5′LTR. We determined the most likely TSS within this LTR by *in silico* sequence analysis (Additional file 3: Figure S2). Attempts to experimentally verify this TSS by 5′RACE-PCR as described previously [14] were not successful, as orientation-specific cDNA synthesis did not yield sufficient starting material for subsequent PCR reactions (see Methods section for details). Insufficient orientation-specific cDNA synthesis may have been due to the low abundance of the HML-10(DAP3) RNA as seen by qRT-PCR (Fig. 4f) and is a known issue with rare transcripts [56]. Thus, the actual TSS of the retroviral transcript may differ from the predicted one but our experiments provide evidence that it is located between the target regions of ASO upstream and ASO 1 (Fig. 5a). Our findings indirectly confirmed the expression of HML-10(DAP3) RNA in HeLa cells, which was supported by reported microarray experiments (Table 2). Further direct proof could be obtained by sequencing cDNA clones and identifying genomic markers that are unique to the HML-10(DAP3) copy, such as the *AluSp* or *MER11B* repeats that are integrated into this provirus (Table 1).

Suppression of apoptosis, as mediated by the HML-10(DAP3) RNA in HeLa cells, is a general hallmark of cancer cells [57]. Thus, retroviral transcripts may contribute to the malignant cellular phenotype of this cell line by counteracting *DAP3* expression, and thereby suppressing apoptosis. Aberrant expression levels of *DAP3* have been suggested to play a role in some cases of malignant disease [58–63]. The data shown here indicate that most of the HML-10 LTRs are even weaker expressed than the one analyzed. We hypothesize that the LTRs, which are normally strong promoters in infectious retroviruses, have been silenced by mutation during evolution. Thus, they likely play a limited role in cancer promotion.

The data presented in Fig. 4b-c suggests that the expression of LTR-primed transcripts varies substantially in intensity and direction depending on the cell type as well as the action of cytokines. Moreover, despite its weak expression at about 60-fold lower levels than *DAP3* mRNA (Fig. 4f) the HML-10-primed RNA had a strong impact on *DAP3* gene regulation (Fig. 5b). Consistent contributions of this and other HERV-primed RNAs to various tissues or tumors may therefore be hard to identify. However, the presence of HML-10(DAP3) RNA in many tumor cell lines and the absence in most healthy tissues (Table 2) suggest that its upregulation may be a relevant feature in some human cancer diseases. This is in line with the observation that transcriptional activation of HERVs and other REs by epigenetic DNA demethylation is a frequent characteristic of malignant cells [64–66].

Conclusions

This work provides experimental support for recent evidence that HERVs and other REs play a role in gene regulation and cellular processes relevant to mammalian tumor cell formation. In the case presented here, transcripts of the previously little described HML-10 family suppressed the pro-apoptotic *DAP3* gene and consequently, apoptosis in HeLa cells. Therefore, we could verify a direct link between HERV expression and cellular phenotype in this cell line. A potential role of these LTRs in promoting a malignant phenotype possibly by inducing resistance to apoptosis as described here in other cell lines or tissues requires further investigation.

Methods

Identification of HML-10 elements in the human genome

The Table Browser function [67] of the UCSC Genome Browser [68] was used to identify HML-10 elements in the human genome. We queried the Repbase sequence of HML-10 LTRs, *LTR14* [69], in the RepeatMasker track [70] of the GRCh38/hg38 human genome assembly [1]. This search yielded 86 hits. Manual inspection of these hits revealed the 70 unique HML-10 elements listed in Table 1.

Estimation of the evolutionary age of HML-10 proviruses

For each of the six complete HML-10 proviruses (elements Nos. 1, 3, 20, 22, 25 and 68 in Table 1), both LTR sequences (5′ and 3′LTRs) were aligned with Clustal X 2.0 [71]. The evolutionary age of each provirus was calculated from the number of mutations between both

LTRs by applying an estimated nucleotide substitution rate of 2.28 per site and year × 10^{-9} as described [28].

Construction of phylogenetic neighbor-joining trees

The *pol* sequences of HML-10 and other betaretroviruses were retrieved from published literature [18, 72]. The fasta protein sequences can be found in Additional file 4. Sequences were aligned with Clustal X 2.0 [71] using standard parameters of the *Multiple Alignment Mode*. The neighbor-joining tree was visualized with TreeView 1.6.6 [73]. The phylogenetic tree of HML-10 LTR nucleotide sequences and those of the flanking sequences shown in Additional file 1: Figure S1 were constructed similarly. All nucleotide sequences were retrieved from the UCSC Genome Browser [68] and the current release of the human genome, GRCh38/hg38 [1].

Identification of target site duplications

The sequences immediately up- and downstream of RepeatMasker-annotated HML-10 elements (Table 1) were searched for homologous sequences in the retroviral sense orientation. Homologous sequences of at least 5 bp were defined as TSDs, allowing for one (5 bp TSDs) or two (6 bp TSDs) nucleotide mismatches.

Location of HERV sequences relative to human genes

The fractions of intragenic HERV sequences were determined with the UCSC Table Browser [67] using the GRCh38/hg38 human genome assembly [1]. HERV elements were identified as described below in this paragraph within the RepeatMasker track [70]. The output of these searches was used to generate custom tracks covering the sequences of the respective HERV families. Using the *intersection* function, the overlap of HERV sequences with a custom track representing full-length RefSeq genes was determined, yielding the following values (displayed as: HERV family, Repbase annotation, sequence covered, sequence intersected with RefSeq genes): HML-1, *LTR14A / LTR14B / LTR14C*, 274,910 bp, 66,940 bp (24.35%); HML-2, *LTR5A / LTR5B*, 595,281 bp, 171,219 bp (28.76%); HML-3, *MER9B / MER9a1 / MER9a2 / MER9a3*, 568,179 bp, 151,429 bp (26.65%); HML-4, *LTR13 / LTR13A*, 545,702 bp, 200,556 bp (26.75%); HML-5, *LTR22 / LTR22A / LTR22B / LTR22B1 / LTR22B2 / LTR22C / LTR22C0 / LTR22C2 / LTR22E*, 396,533 bp, 105,855 bp (26.70%); HML-6, *LTR3 / LTR3A / LTR3B*, 130,701 bp, 37,058 bp (28.35%); HML-7, *MER11D*, 194,536 bp, 60,756 bp (31.23%); HML-8, *MER11A / MER11B / MER11C*, 2,222,448 bp, 656,281 bp (29.53%); HML-10, *LTR14*, 40,556 bp, 19,443 bp (47.94%); HERV-W, *LTR17*, 482,257 bp, 134,803 bp (27.95%). All RepeatMasker-annotated HERV elements covered 266,970,452 bp, of which 75,967,800 bp (28.46%) intersected with RefSeq genes. The fraction of the total genome (3,088,269,808 bp) accounted for RefSeq genes was 1,320,982,363 bp (44.97%).

Cell lines and culture conditions

HeLa (ATCC CCL-2), HepG2 (ATCC HB-8065) and HEK293T cell lines were cultivated in complete growth medium; Dulbecco's Modified Eagle's Medium (DMEM) (Invitrogen, Carlsbad, CA, USA) supplemented with 10% heat-inactivated fetal calf serum (Invitrogen) and 100 U/mL penicillin, 100 µg/mL streptomycin and 0.25 µg/mL amphotericin (Antibiotic-Antimycotic by Invitrogen). Cells were incubated at 37 °C with 5% CO_2. Subcultivation ratios ranged between 1:2 and 1:10.

Primers

All primers were synthesized by Microsynth AG, Balgach, Switzerland. Primer sequences are listed in Additional file 5. Primer sequences were designed such that they only amplified the desired regions, as verified by the *in silico* PCR analysis tool of UCSC on https://genome.ucsc.edu/cgi-bin/hgPcr/.

Construction of pGL3-Enhancer luciferase reporter vectors

LTRs of HML-10(C4), HML-10(DAP3) and HML-10(PKIB) were amplified by standard PCR from genomic DNA of the QBL cell line (No. 4070713) obtained from the Health Protection Agency Culture Collections (ECACC, Salisbury, UK), using primer pairs with a HindIII or XhoI cleavage site at their 5′ ends. HML-10(C4) primer pairs: 5′LTR(s), *C4_5LTRforHindIII + C4_5LTRrevXhoI*; 5′LTR(as), *C4_5LTRforXhoI + C4_5LTRrevHindIII*; 3′LTR(s), *C4_3LTRforHindIII + C4_3LTRrevXhoI*; 3′LTR(as), *C4_3LTRforXhoI + C4_3LTRrevHindIII*. HML-10(DAP3) primer pairs: 5′LTR(s), *DAP3_5LTRforHindIII + DAP3_5LTRrevXhoI*; 5′LTR(as), *DAP3_5LTRforXhoI + DAP3_5LTRrevHindIII*; 3′LTR(s), *DAP3_3LTRforHindIII + DAP3_3LTRrevXhoI*; 3′LTR(as), *DAP3_3LTRforXhoI + DAP3_3LTRrevHindIII*. HML-10(PKIB) primer pairs: 5′LTR(s), *PKIB_5LTRforHindIII + PKIB_5LTRrevXhoI*; 5′LTR(as), *PKIB_5LTRforXhoI + PKIB_5LTRrevHindIII*; 3′LTR(s), *PKIB_3LTRforHindIII + PKIB_3LTRrevXhoI*; 3′LTR(as), *PKIB_3LTRforXhoI + PKIB_3LTRrevHindIII*). Cycling conditions were 10 min. 95 °C; (30 s. 95 °C, 30 s. 60 °C, 30 s. 72 °C) × 40; 7 min. 72 °C. LTRs were cloned into the pGL3-Enhancer vector (Promega, Madison, WI, USA), containing the fLuc gene as reporter, after digestion with HindIII and XhoI (New England Biolabs, Ipswich, MA, USA) and phosphatase treatment. Vectors were ligated with T4 DNA Ligase (New England Biolabs). All vector constructs were heat-shock-transformed into competent *E. coli* JM109 (Promega). Positive colonies were detected by ampicillin resistance on selective agar plates. Selected clones were grown in ampicillin-containing LB medium and plasmid DNA was isolated with the QIAamp Plasmid DNA Mini Kit (Qiagen, Hilden, Germany).

Plasmid DNAs were screened for correct inserts by restriction enzyme digestions using appropriate enzyme combinations and subsequent agarose gel electrophoresis as well as by capillary sequencing (Microsynth, Balgach, Switzerland).

Determination of LTR promoter activities

Freshly passaged HepG2 or HEK293T cells were seeded into 24-well tissue culture plates (4×10^4 cells per well in complete growth medium) and cultivated overnight to ~80% confluence. Cells were transfected with 50 ng/well of pGL3-Enhancer constructs, empty pGL3-Enhancer, or pGL3-Control, 4 ng/well of pGL4.74 vector for normalization (Promega) and 346 ng/well of unrelated carrier DNA using DreamFect Gold transfection reagent (OZ Biosciences, Marseille, France) following the manufacturer's recommendations. Vector pGL4.74 contains the renilla luciferase (rLuc) gene under control of the herpes simplex virus thymidine kinase (HSV-TK) promoter. Medium was replaced with fresh prewarmed complete growth medium 6 h post-transfection. At 24 h post-transfection, medium was aspirated, cells were rinsed with prewarmed PBS, lysed, and fLuc and rLuc activities in each sample were determined with the Dual-Glo Luciferase Assay System (Promega) in a Sirius Luminometer (Berthold Detection Systems, Pforzheim, Germany). fLuc activities were normalized to rLuc activities for each sample. To assess the effect of IFNγ stimulation on promoter activities, selected pGL3-Enhancer constructs were transfected into HepG2 cells as above and were stimulated with different amounts of recombinant human IFNγ (PeproTech, Rocky Hill, NJ, USA) by addition to the growth medium immediately after medium change 6 h post-transfection. fLuc activities were determined 30 h post-transfection.

Inactivation of HML-10(DAP3) RNA with ASOs

The ASOs were 25-mer DNA molecules with phosphorothioate bonds at the flanking three nucleotides on both sides to confer exonuclease resistance. ASOs for inactivating the HML-10(DAP3) RNA were designed to be complementary to regions within the 5′LTR or the proviral body downstream of the predicted TSS. We used only sequences that uniquely mapped to their respective target region and nowhere else in the human genome. A Mock ASO with a randomized sequence and one complementary to a region shortly upstream of the 5′LTR were used as negative controls. ASOs were purchased from Microsynth. Their sequences are listed in Additional file 6.

qRT-PCR

Freshly passaged HepG2 or HeLa cells were seeded in 96-well plates (10^4 cells per well in complete growth medium) and grown overnight to ~80% confluence. Cells were transfected with 25 or 50 nM of the indicated

ASOs using DreamFect Gold transfection reagent (OZ Biosciences) according to the manufacturer's recommendations. Medium was replaced with fresh prewarmed complete growth medium 6 h post-transfection. At 24 h post-transfection, total RNA was extracted using the QIAamp RNA Blood Mini Kit (Qiagen), including an on-column DNA digestion step with the RNase-free DNase Set (Qiagen). First strand cDNA was synthesized using the High Capacity cDNA Reverse Transcription Kit (Applied Biosystems, Foster City, CA, USA) with random hexamer primers. qRT-PCR was performed using the TaqMan Universal PCR Master Mix (Applied Biosystems) with the addition of 1:10000 (v/v) SYBR Green I (Sigma-Aldrich, St. Louis, MO, USA) and primers specific for *DAP3* (*DAP3for* + *DAP3rev*) or *GAPDH* (*GAPDHfor* + *GAPDHrev*) mRNAs, HML-10(DAP3) RNA (*LTRfor2* + *LTRrev*) or *LTRfor1* + *LTRrev* as control reaction. Cycling conditions were 2 min. 50 °C; 10 min. 95 °C; (15 s. 95 °C, 1 min. 58 °C) × 65. Specificity of the PCR reactions was assessed by checking for correct amplicon lengths and amplification artifacts by agarose gel electrophoresis. All shown RNA levels were calculated by relative quantification (double delta Ct method) using *GAPDH* as reference, with primer efficiencies calculated from serial dilutions of HepG2 cDNA samples. Control samples without addition of reverse transcriptase gave no amplification signals.

Strand-specific cDNA synthesis

A number of 10^6 freshly passaged HepG2 or HeLa cells were seeded into wells of 6-well plates and grown overnight to ~80% confluence. Total RNA was extracted using the QIAamp RNA Blood Mini Kit (Qiagen). First strand cDNA was synthesized using either the Reverse Transcriptase of the High Capacity cDNA Reverse Transcription Kit (Applied Biosystems), or the Thermoscript Reverse Transcriptase (Invitrogen) with primers specific for the HML-10(DAP3) transcript (Additional file 5). Different incubation times and temperatures (ranging from 25 to 60 °C) were evaluated. To assess reverse transcription efficiencies, qRT-PCR was done using TaqMan Universal PCR Master Mix (Applied Biosystems) with the addition of 1:10000 (v/v) SYBR Green I (Sigma-Aldrich) and primers *LTRfor2* + *LTRrev*. Cycling conditions were 2 min. 50 °C; 10 min. 95 °C; (15 s. 95 °C, 1 min. 58 °C) × 65. No specific amplification was detected, while positive controls with cDNA prepared with random hexamer primers and with genomic human DNA yielded HML-10(DAP3)-specific amplicons.

Trypan Blue exclusion and cell viability (MTS) assays

Freshly passaged HeLa cells were seeded in 48-well plates (2×10^4 cells per well in complete growth medium) and cultivated overnight to ~70% confluence. Cells were transfected with 50 nM of the indicated ASOs using the DreamFect Gold transfection reagent (OZ Biosciences) according to the manufacturer's recommendations. Medium was replaced with fresh prewarmed complete growth medium 6 h post-transfection. At 24 h post-transfection, cells were stimulated with 1000 U/mL recombinant human IFNγ (PeproTech) or 100 ng/mL recombinant human TNFα (Biomol, Hamburg, Germany) for 24 h by addition to the growth medium, or left unstimulated. For Trypan Blue exclusion assays, cells were harvested 48 h post-transfection, resuspended in 50 μL PBS, mixed 1:1 (v/v) with 50 μL 0.4% (v/v) Trypan Blue Stain (Invitrogen) and incubated for 1 min. Total cell number and number of stained cells of each sample were counted in a hemocytometer. About 100–200 total cells per sample were counted. To obtain the fraction of dead cells, the number of stained cells was divided by the total cell number. For cell viability (MTS) assays, one tenth of the growth medium volume of MTS reagent (CellTiter 96 AQueous One Solution Cell Proliferation Assay by Promega) was added to each well 48 h post-transfection. Cells were incubated for approximately 1 h before the absorbance at 495 nm of the supernatants was measured with a NanoDrop ND-1000 Spectrophotometer (Thermo Scientific, Waltham, MA, USA). Fresh growth medium with the addition of one tenth of MTS reagent was used as blank.

Detection of apoptosis by DNA laddering

Freshly passaged HeLa cells were seeded into 6-well plates (10^6 cells per well in complete growth medium) and cultivated overnight to ~70% confluence. Cells were transfected with 50 nM of the indicated ASOs using the DreamFect Gold transfection reagent (OZ Biosciences) according to the manufacturer's recommendations. Medium was replaced with fresh prewarmed complete growth medium 6 h post-transfection. At 48 h post-transfection, cells were lysed and DNA was prepared with the Apoptotic DNA Ladder Kit (Roche, Mannheim, Germany) according to the manufacturer's recommendations. Samples were analyzed using a 1% agarose TAE gel and DNA was visualized with ethidium bromide.

Additional files

Additional file 1: Figure S1. (a) Neighbor-joining trees of the flanking 1000 bp upstream (left tree) and downstream (right tree) relative to retroviral orientation. Two pairs of flanking sequences that cluster are highlighted by colored boxes, whereby elements nos. 3 and 43 only clustered in their upstream regions. Element numbers according to Table 1 are shown in parentheses. The horizontal bars represent 0.1 substitutions per nucleotide position. (b) Comparison of the genomic loci representing HML-10 elements nos. 3 and 43 (see blue boxes in panel (a)) that integrated into a L1MB7 repeat or into a MER4A1 repeat, respectively. Both HML-10 elements therefore likely represent independent integration events. (c) Comparison of the genomic loci representing HML-10 elements nos. 27 and 45 (red boxes in panel (a)) that are both flanked by LTR5_Hs and HERVK-int repeats in the same configuration. The overall sequence identity of the

depicted regions is >88%. Therefore, one of these HML-10 was likely the result of a chromosomal duplication event that copied the other one. Images in panels (b) and (c) were obtained from the UCSC Genome Browser [68]. (PDF 1217 kb)

Additional file 2: Table S1. Presence or absence of HML-10 in different mammalian genomes. The number of LTR14 hits that represent HML-10 elements in the indicated genomes were assessed with the UCSC Table Browser [67] by querying LTR14 in the respective RepeatMasker tracks [69, 70]. BLAT [74] searches within the UCSC Genome Browser [68] using the consensus sequence of LTR14 obtained from the DFAM database (www.dfam.org) [75] were performed verify the presence of HML-10. (PDF 262 kb)

Additional file 3: Figure S2. Identification of a putative TSS in the HML-10(DAP3) 5'LTR. The sequence shown is chr1:153935293–153936036 of the human genome hg18 assembly [1]. The HML-10(DAP3) 5'LTR (according to RepeatMasker annotation [70]) is highlighted in bold letters. Six Inr sequences highlighted blue were identified by sequence homology searches of the consensus YYANWYY sequence [76]. Five TATA boxes highlighted red were identified with the TFBind program [77] on http://tfbind.hgc.jp using a similarity threshold of 0.8. Inr2 and TATA4 are overlapping. Only one Inr element (Inr1) is located in close proximity downstream of a TATA box (TATA3). The putative TSS within Inr1 is underlined. A downstream promoter element (DPE) highlighted violet matching the consensus RGWYVT sequence [49] is located 19 bp downstream of the putative TSS. Primer locations for LTRfor1, LTRfor2 and LTRrev (see Additional file 4) are indicated by arrows. An IFNγ activated sequence (GAS) is highlighted green. (PDF 364 kb)

Additional file 4: Protein fasta sequences of Pol proteins of different endogenous and exogenous betaretroviruses. (TXT 11 kb)

Additional file 5: Primer sequences. (XLS 42 kb)

Additional file 6: Sequences of ASOs. Asterisks denote phosphorothioate bonds. (XLS 39 kb)

Abbreviations

5'RACE-PCR: 5' rapid amplification of cDNA ends-PCR; ASO: Antisense oligonucleotide; BLAT: BLAST-like Alignment Tool; DAP3: Death-associated protein 3; DMEM: Dulbecco's Modified Eagle's Medium; DPE: Downstream promoter element; DR: Death Receptor; fLuc: Firefly luciferase; GAS: IFNγ activated sequence; HERV: Human endogenous retrovirus; HML: Human MMTV-like; HSV-TK: Herpes simplex virus thymidine kinase; HUVEC: Human umbilical vein endothelial cells; IFNγ: Interferon-gamma; Inr: Initiator element; JSRV: Jaagsiekte sheep retrovirus; LTR: long terminal repeat; MMTV: Mouse mammary tumor virus; MPMV: Mason-Pfizer monkey virus; Mya: Million years ago; NWM: New World monkey; OWM: Old World monkey; qRT-PCR: Quantitative real-time PCR; RE: Retroelement; rLuc: Renilla luciferase; RT: Reverse transcriptase; TE: Transposable element; TNFα: Tumor necrosis factor-alpha; TSD: Target site duplication; TSS: Transcription start site

Acknowledgements

We thank Dr. Anja Twiehaus for support in some of the experiments. Prof. Peter H. Seeberger of the Max Planck Institute of Colloids and Interfaces, Potsdam, Germany, is thanked for his generous support of FB.

Funding

This work was supported by the Max Planck Society and a grant from the Volkswagenstiftung (Lichtenberg Program) to MRS. KM provided some financial support.

Author's contributions

KM initiated the project. KM, JH and FB designed the experiments. FB performed the experiments and prepared the data with assistance from KM, JH, RH, AF and MRS. HL enabled us to perform these studies. FB and KM wrote the manuscript. All authors read and approved the final version of the manuscript.

Competing interests

The authors declare that they have no competing interests.

Author details

[1]Max Planck Institute for molecular Genetics, Ihnestr. 63-73, 14195 Berlin, Germany. [2]Institute of Medical Microbiology, University of Zurich, Gloriastr. 32, 8006 Zurich, Switzerland. [3]Current affiliation: Max Planck Institute of Colloids and Interfaces, Am Mühlenberg 1, 14424 Potsdam, Germany. [4]Current affiliation: University of Zurich, Institute of Molecular Life Sciences, Winterthurerstr. 190, 8057 Zurich, Switzerland. [5]Current affiliation: Functional Epigenomics, CCG, Cologne University Hospital, University of Cologne, Weyertal 115b, 50931 Cologne, Germany. [6]Dahlem Centre for Genome Research and Medical Systems Biology, Fabeckstr. 60-62, 14195 Berlin, Germany.

References

1. Lander ES, Linton LM, Birren B, Nusbaum C, Zody MC, Baldwin J, et al. Initial sequencing and analysis of the human genome. Nature. 2001;409:860–921.
2. de Koning AP, Gu W, Castoe TA, Batzer MA, Pollock DD. Repetitive elements may comprise over two-thirds of the human genome. PLoS Genet. 2011;7:e1002384.
3. Gogvadze E, Buzdin A. Retroelements and their impact on genome evolution and functioning. Cell Mol Life Sci. 2009;66:3727–42.
4. Bannert N, Kurth R. The evolutionary dynamics of human endogenous retroviral families. Annu Rev Genomics Hum Genet. 2006;7:149–73.
5. Faulkner GJ, Kimura Y, Daub CO, Wani S, Plessy C, Irvine KM, et al. The regulated retrotransposon transcriptome of mammalian cells. Nat Genet. 2009;41:563–71.
6. Conley AB, Miller WJ, Jordan IK. Human cis natural antisense transcripts initiated by transposable elements. Trends Genet. 2008;24:53–6.
7. Gosenca D, Gabriel U, Steidler A, Mayer J, Diem O, Erben P, et al. HERV-E-mediated modulation of PLA2G4A transcription in urothelial carcinoma. PLoS One. 2012;7:e49341.
8. Li F, Nelläker C, Yolken RH, Karlsson H. A systematic evaluation of expression of HERV-W elements; influence of genomic context, viral structure and orientation. BMC Genomics. 2011;12:22.
9. Schön U, Diem O, Leitner L, Günzburg WH, Mager DL, Salmons B, et al. Human endogenous retroviral long terminal repeat sequences as cell type-specific promoters in retroviral vectors. J Virol. 2009;83:12643–50.
10. Huh JW, Kim DS, Kang DW, Ha HS, Ahn K, Noh YN, et al. Transcriptional regulation of GSDML gene by antisense-oriented HERV-H LTR element. Arch Virol. 2008;153:1201–5.
11. Sin HS, Huh JW, Kim DS, Kang DW, Min DS, Kim TH, et al. Transcriptional control of the HERV-H LTR element of the GSDML gene in human tissues and cancer cells. Arch Virol. 2006;151:1985–94.
12. Sjøttem E, Anderssen S, Johansen T. The promoter activity of long terminal repeats of the HERV-H family of human retrovirus-like elements is critically dependent on Sp1 family proteins interacting with a GC/GT box located immediately 3' to the TATA box. J Virol. 1996;70:188–98.
13. Chang NT, Yang WK, Huang HC, Yeh KW, Wu CW. The transcriptional activity of HERV-I LTR is negatively regulated by its cis-elements and wild type p53 tumor suppressor protein. J Biomed Sci. 2007;14:211–22.
14. Gogvadze E, Stukacheva E, Buzdin A, Sverdlov E. Human-specific modulation of transcriptional activity provided by endogenous retroviral insertions. J Virol. 2009;83:6098–105.
15. Buzdin A, Kovalskaya-Alexandrova E, Gogvadze E, Sverdlov E. At least 50% of human-specific HERV-K (HML-2) long terminal repeats serve in vivo as active promoters for host nonrepetitive DNA transcription. J Virol. 2006;80:10752–62.
16. Ruda VM, Akopov SB, Trubetskoy DO, Manuylov NL, Vetchinova AS, Zavalova LL, et al. Tissue specificity of enhancer and promoter activities of a HERV-K(HML-2) LTR. Virus Res. 2004;104:11–6.
17. Domansky AN, Kopantzev EP, Snezhkov EV, Lebedev YB, Leib-Mosch C, Sverdlov ED. Solitary HERV-K LTRs possess bi-directional promoter activity and contain a negative regulatory element in the U5 region. FEBS Lett. 2000;42:191–5.
18. Dangel AW, Mendoza AR, Baker BJ, Daniel CM, Carroll MC, Wu LC, et al. The dichotomous size variation of human complement C4 genes is mediated by a novel family of endogenous retroviruses, which also establishes species-specific genomic patterns among Old World primates. Immunogenetics. 1994;40:425–36.
19. Mack M, Bender K, Schneider PM. Detection of retroviral antisense transcripts and promoter activity of the HERV-K(C4) insertion in the MHC class III region. Immunogenetics. 2004;56:321–32.
20. Kissil JL, Deiss LP, Bayewitch M, Raveh T, Khaspekov G, Kimchi A. Isolation of DAP3, a novel mediator of interferon-gamma-induced cell death. J Biol Chem. 1995;270:27932–6.
21. Miyazaki T, Reed JC. A GTP-binding adapter protein couples TRAIL receptors to apoptosis-inducing proteins. Nat Immunol. 2001;2:493–500.
22. Seifarth W, Frank O, Zeilfelder U, Spiess B, Greenwood AD, Hehlmann R, et al. Comprehensive analysis of human endogenous retrovirus transcriptional activity in human tissues with a retrovirus-specific microarray. J Virol. 2005;79:341–52.
23. Seifarth W, Spiess B, Zeilfelder U, Speth C, Hehlmann R, Leib-Mösch C. Assessment of retroviral activity using a universal retrovirus chip. J Virol Methods. 2003;112:79–91.
24. Frank O, Giehl M, Zheng C, Hehlmann R, Leib-Mösch C, Seifarth W. Human endogenous retrovirus expression profiles in samples from brains of patients with schizophrenia and bipolar disorders. J Virol. 2005;79:10890–901.
25. Frank O, Verbeke C, Schwarz N, Mayer J, Fabarius A, Hehlmann R, et al. Variable transcriptional activity of endogenous retroviruses in human breast cancer. J Virol. 2008;82:1808–18.
26. Schön U, Seifarth W, Baust C, Hohenadl C, Erfle V, Leib-Mösch C. Cell type-specific expression and promoter activity of human endogenous retroviral long terminal repeats. Virology. 2001;279:280–91.
27. Haupt S, Tisdale M, Vincendeau M, Clements MA, Gauthier DT, Lance R, et al. Human endogenous retrovirus transcription profiles of the kidney and kidney-derived cell lines. J Gen Virol. 2011;92:2356–66.
28. Johnson WE, Coffin JM. Constructing primate phylogenies from ancient retrovirus sequences. Proc Natl Acad Sci U S A. 1999;96:10254–60.
29. Mayer J, Meese EU. The human endogenous retrovirus family HERV-K(HML-3). Genomics. 2002;80:331–43.
30. Seifarth W, Baust C, Murr A, Skladny H, Krieg-Schneider F, Blusch J, et al. Proviral structure, chromosomal location, and expression of HERV-K-T47D, a novel human endogenous retrovirus derived from T47D particles. J Virol. 1998;72:8384–91.
31. Medstrand P, Mager DL, Yin H, Dietrich U, Blomberg J. Structure and genomic organization of a novel human endogenous retrovirus family: HERV-K (HML-6). J Gen Virol. 1997;78:1731–44.
32. Lavie L, Medstrand P, Schempp W, Meese E, Mayer J. Human endogenous retrovirus family HERV-K(HML-5): status, evolution, and reconstruction of an ancient betaretrovirus in the human genome. J Virol. 2004;78:8788–98.
33. Pruitt KD, Tatusova T, Maglott DR. NCBI Reference Sequence (RefSeq): a curated non-redundant sequence database of genomes, transcripts and proteins. Nucleic Acids Res. 2005;33(Database issue):D501–4.
34. Brady T, Lee YN, Ronen K, Malani N, Berry CC, Bieniasz PD, et al. Integration target site selection by a resurrected human endogenous retrovirus. Genes Dev. 2009;23:633–43.
35. Bushman FD. Targeting survival. Integration site selection by retroviruses and LTR-retrotransposons. Cell. 2003;115:135–8.
36. Smit AF. Interspersed repeats and other momentos of transposable elements in mammalian genomes. Curr Opin Genet Dev. 1999;9:657–63.
37. Medstrand P, van de Lagemaat LN, Mager DL. Retroelement distrubutions in the human genome: Variations associate with age and proximity to genes. Genome Res. 2002;12:1483–95.
38. Villesen P, Aagaard L, Wiuf C, Pedersen FS. Identification of endogenous retroviral reading frames in the human genome. Retrovirology. 2004;1:32.
39. Conley AB, Jordan IK. Cell type-specific termination of transcription by transposable element sequences. Mob DNA. 2012;3:15.
40. Assinger A, Yaiw KC, Göttesdorfer I, Leib-Mösch C, Söderberg-Nauclér C. Human cytomegalovirus (HCMV) induces human endogenous retrovirus (HERV) transcription. Retrovirology. 2013;10:132.
41. Maliniemi P, Vincendeau M, Mayer J, Frank O, Hahtola S, Karenko L, et al. Expression of human endogenous retrovirus-w including syncytin-1 in cutaneous T-cell lymphoma. PLoS One. 2013;8:e76281.
42. Diem O, Schäffner M, Seifarth W, Leib-Mösch C. Influence of antipsychotic drugs on human endogenous retrovirus (HERV) transcription in brain cells. PLoS One. 2012;7:e30054.
43. Frank O, Jones-Brando L, Leib-Mosch C, Yolken R, Seifarth W. Altered transcriptional activity of human endogenous retroviruses in neuroepithelial cells after infection with Toxoplasma gondii. J Infect Dis. 2006;194:1447–9.
44. Dunn CA, Romanish MT, Gutierrez LE, van de Lagemaat LN, Mager DL.

Transcription of two human genes from a bidirectional endogenous retrovirus promoter. Gene. 2006;366:335–42.

45. Boehm U, Klamp T, Groot M, Howard JC. Cellular responses to interferon-gamma. Annu Rev Immunol. 1997;15:749–95.

46. Harms JS, Splitter GA. Interferon-gamma inhibits transgene expression driven by SV40 or CMV promoters but augments expression driven by the mammalian MHC I promoter. Hum Gene Ther. 1995;6:1291–7.

47. Kulics J, Colten HR, Perlmutter DH. Counterregulatory effects of interferon-gamma and endotoxin on expression of the human C4 genes. J Clin Invest. 1990;85:943–9.

48. Köhrer K, Grummt I, Horak I. Functional RNA polymerase II promoters in solitary retroviral long terminal repeats (LTR-IS elements). Nucleic Acids Res. 1985;13:2631–45.

49. Butler JE, Kadonaga JT. The RNA polymerase II core promoter: a key component in the regulation of gene expression. Genes Dev. 2002;16:2583–92.

50. Kowalskaya E, Buzdin A, Gogvadze E, Vinogradova T, Sverdlov E. Functional human endogenous retroviral LTR transcription start sites are located between the R und U5 regions. Virology. 2006;346:373–8.

51. Zong X, Huang L, Tripathi V, Peralta R, Freier SM, Guo S, et al. Knockdown of nuclear-retained long noncoding RNAs using modified DNA antisense oligonucleotides. Methods Mol Biol. 2015;1262:321–31.

52. Gagnon KT, Li L, Chu Y, Janowski BA, Corey DR. RNAi factors are present and active in human cell nuclei. Cell Rep. 2014;6:211–21.

53. Kwok T, Heinrich J, Jung-Shiu J, Meier MG, Mathus S, Moelling K. Reduction of gene expression by a hairpin-loop structured oligodeoxynucleotide: alternative to siRNA and antisense. Biochim Biophys Acta. 2009;1790:1170–8.

54. Faghihi MA, Wahlestedt C. Regulatory roles of natural antisense transcripts. Nat Rev Mol Cell Biol. 2009;10:637–43.

55. Modarresi F, Faghihi MA, Lopez-Toledano MA, Fatemi RP, Magistri M, Brothers SP, van der Brug MP, Wahlestedt C. Inhibition of natural antisense transcripts in vivo results in gene-specific transcriptional upregulation. Nat Biotechnol. 2012;30:453–9.

56. Brink AA, Oudejans JJ, Jiwa M, Walboomers JM, Meijer CJ, van den Brule AJ. Multiprimed cDNA synthesis followed by PCR is the most suitable method for Epstein-Barr virus transcript analysis in small lymphoma biopsies. Mol Cell Probes. 1997;11:39–47.

57. Hanahan D, Weinberg RA. Hallmarks of cancer: the next generation. Cell. 2011;144:646–74.

58. Jia Y, Ye L, Ji K, Zhang L, Hargest R, Ji J, et al. Death associated protein-3, DAP-3, correlates with preoperative chemotherapy effectiveness and prognosis of gastric cancer patients following perioperative chemotherapy and radical gastrectomy. Br J Cancer. 2014;110:421–9.

59. Wazir U, Jiang WG, Sharma AK, Mokbel K. The mRNA expression of DAP3 in breast cancer: correlation with clinicopathological parameters. Anticancer Res. 2012;32:671–4.

60. Sasaki H, Ide N, Yukiue H, Kobayashi Y, Fukai I, Yamakawa Y, et al. Arg and DAP3 expression was correlated with human thymoma stage. Clin Exp Metastasis. 2004;21:507–13.

61. Jacques C, Fontaine JF, Franc B, Mirebeau-Prunier D, Triau S, Savagner F, et al. Death-associated protein 3 is overexpressed in human thyroid oncocytic tumours. Br J Cancer. 2009;101:132–8.

62. Davidsson J, Andersson A, Paulsson K, Heidenblad M, Isaksson M, Borg A, et al. Tiling resolution array comparative genomic hybridization, expression and methylation analyses of dup(1q) in Burkitt lymphomas and pediatric high hyperdiploid acute lymphoblastic leukemias reveal clustered near-centromeric breakpoints and overexpression of genes in 1q22-32.3. Hum Mol Genet. 2007;16:2215–25.

63. Mariani L, Beaudry C, McDonough WS, Hoelzinger DB, Kaczmarek E, Ponce F, et al. Death-associated protein 3 (Dap-3) is overexpressed in invasive glioblastoma cells in vivo and in glioma cell lines with induced motility phenotype in vitro. Clin Cancer Res. 2001;7:2480–9.

64. Wilson AS, Power BE, Molloy PL. DNA hypomethylation and human diseases. Biochim Biophys Acta. 2007;1775:138–62.

65. Yoder JA, Walsh CP, Bestor TH. Cytosine methylation and the ecology of intragenomic parasites. Trends Genet. 1997;8:335–40.

66. Wilkins AS. The enemy within: an epigenetic role of retrotransposon in cancer initiation. Bioessays. 2010;32:856–65.

67. Karolchik D, Hinrichs AS, Furey TS, Roskin KM, Sugnet CW, Haussler D, et al. The UCSC Table Browser data retrieval tool. Nucleic Acids Res. 2004;32:D493–6.

68. Kent WJ, Sugnet CW, Furey TS, Roskin KM, Pringle TH, Zahler AM, et al. The human genome browser at UCSC. Genome Res. 2002;12:996–1006.

69. Jurka J, Kapitonov VV, Pavlicek A, Klonowski P, Kohany O, Walichiewicz J. Repbase Update, a database of eukaryotic repetitive elements. Cytogenet Genome Res. 2005;110:462–7.

70. Smit AFA, Hublez R, Green P. RepeatMasker Open-3.0. 1996-2010 <www.repeatmasker.org>.

71. Larkin MA, Blackshields G, Brown NP, Chenna R, McGettigan PA, McWilliam H, et al. Clustal W and Clustal X version 2.0. Bioinformatics. 2007;23:2947–8.

72. Jern P, Sperber GO, Blomberg J. Use of endogenous retroviral sequences (ERVs) and structural markers for retroviral phylogenetic inference and taxonomy. Retrovirology. 2005;2:50.

73. TreeView 1.6.6 Software. http://taxonomy.zoology.gla.ac.uk/rod/treeview.html. Accessed Oct 2015.

74. Kent WJ. BLAT-the BLAST-like alignment tool. Genome Res. 2002;12:656–64.

75. Hubley R, Finn RD, Clements J, Eddy SR, Jones TA, Bao W, Smit AF, Wheeler TJ. The Dfam database of repetitive DNA families. Nucleic Acids Res. 2016; 44(D1):D81–9.

76. Xi H, Yu Y, Fu Y, Foley J, Halees A, Weng Z. Analysis of overrepresented motifs in human core promoters reveals dual regulatory roles of YY1. Genome Res. 2007;17:798–806.

77. Tsunoda T, Takagi T. Estimating transcription factor bindability on DNA. Bioinformatics. 1999;15:622–30.

78. Steiper ME, Young ME. Primate molecular divergence dates. Mol Phylogenet Evol. 2006;41:384–94.

79. Seifarth W, Krause U, Hohenadl C, Baust C, Hehlmann R, Leib-Mösch C. Rapid identification of all known retroviral reverse transcriptase sequences with a novel versatile detection assay. AIDS. 2000;16:721–9.

8

Insertion and deletion polymorphisms of the ancient *AluS* family in the human genome

Maria S. Kryatova[1,2†], Jared P. Steranka[1,2†], Kathleen H. Burns[1,2*] and Lindsay M. Payer[1*]

Abstract

Background: Polymorphic *Alu* elements account for 17% of structural variants in the human genome. The majority of these belong to the youngest *AluY* subfamilies, and most structural variant discovery efforts have focused on identifying *Alu* polymorphisms from these currently retrotranspositionally active subfamilies. In this report we analyze polymorphisms from the evolutionarily older *AluS* subfamily, whose peak activity was tens of millions of years ago. We annotate the *AluS* polymorphisms, assess their likely mechanism of origin, and evaluate their contribution to structural variation in the human genome.

Results: Of 52 previously reported polymorphic *AluS* elements ascertained for this study, 48 were confirmed to belong to the *AluS* subfamily using high stringency subfamily classification criteria. Of these, the majority (77%, 37/48) appear to be deletion polymorphisms. Two polymorphic *AluS* elements (4%) have features of non-classical *Alu* insertions and one polymorphic *AluS* element (2%) likely inserted by a mechanism involving internal priming. Seven *AluS* polymorphisms (15%) appear to have arisen by the classical target-primed reverse transcription (TPRT) retrotransposition mechanism. These seven TPRT products are 3′ intact with 3′ poly-A tails, and are flanked by target site duplications; L1 ORF2p endonuclease cleavage sites were also observed, providing additional evidence that these are L1 ORF2p endonuclease-mediated TPRT insertions. Further sequence analysis showed strong conservation of both the RNA polymerase III promoter and SRP9/14 binding sites, important for mediating transcription and interaction with retrotransposition machinery, respectively. This conservation of functional features implies that some of these are fairly recent insertions since they have not diverged significantly from their respective retrotranspositionally competent source elements.

Conclusions: Of the polymorphic *AluS* elements evaluated in this report, 15% (7/48) have features consistent with TPRT-mediated insertion, thus suggesting that some *AluS* elements have been more active recently than previously thought, or that fixation of *AluS* insertion alleles remains incomplete. These data expand the potential significance of polymorphic *AluS* elements in contributing to structural variation in the human genome. Future discovery efforts focusing on polymorphic *AluS* elements are likely to identify more such polymorphisms, and approaches tailored to identify deletion alleles may be warranted.

Keywords: Retrotransposon, Mobile element, SINE, *Alu*, *AluS*, Polymorphism, Structural variation, Mobilome

* Correspondence: kburns@jhmi.edu; lhorvat1@jhmi.edu
†Equal contributors
[1]Department of Pathology, Johns Hopkins University School of Medicine, Miller Research Building (MRB) Room 447, 733 North Broadway, Baltimore, MD 21205, USA
Full list of author information is available at the end of the article

Background

While we have long appreciated differences between individual genomes, it is only recently that robust sequencing efforts have allowed us to begin to build a comprehensive catalog of human structural variants [1, 2]. Mobile element insertions are an important source of structural variation in the human genome, with *Alu* elements specifically accounting for 17% of structural variants [2, 3]. *Alu* elements are non-autonomous retrotransposons, relying on the protein machinery of Long INterspersed Element-1 (LINE-1, L1) for their propagation [4]. Classically, new *Alu* insertions occur by target-primed reverse transcription (TPRT). This mechanism of insertion requires the L1 encoded protein ORF2p, which contains an endonuclease domain and reverse transcriptase domain [4–6]. L1 ORF2p endonuclease has a preference to cleave the negative strand at 5′ TTTT/AA 3′ sites, but is capable of targeting a range of sequences [7–10]. The T-rich sequence on the cleaved negative strand then primes with the poly-A tail of the *Alu* transcript, allowing reverse transcriptase to synthesize a copy of the *Alu* [3]; premature termination of reverse transcription results in the integration of a 5′ truncated element. Because the positive strand is nicked downstream of the initial cleavage site, the newly integrated *Alu* element is flanked by direct repeats, resulting from a duplication of the sequence at the insertion site when the staggered break is repaired [3]. Thus, an *Alu* insertion having arisen by TPRT exhibits the following defining features [11]: (1) an intact 3′ end, (2) a 3′ poly-A tail, and (3) flanking target site duplications (TSDs).

Only a small subset of the 1.1 million *Alu* insertions in the human genome are capable of retrotransposition, and recent retrotransposition events have created thousands of polymorphic insertions [1, 3, 11–14]. Polymorphic *Alu* elements almost exclusively belong to the youngest *AluY* subfamilies [2, 3, 7, 11, 14, 15]. While there have been reports of polymorphic elements from the evolutionarily older *AluS* subfamily in humans [2, 13, 15, 16], polymorphic *AluS* insertions are generally not considered to be an important contributor to structural variation and most structural variant discovery efforts have not specifically focused on identifying these elements. In this report we present examples of polymorphic *AluS* elements, provide annotations of the sequences, and consider the mechanisms that likely created the polymorphisms. Thus, our work expands the potential significance of *AluS* elements in contributing to structural variation in the human genome and emphasizes the importance of identifying additional *AluS* polymorphisms.

Results

Identification of polymorphic *AluS* elements in the human genome

How retrotransposon variants in the human genome affect gene expression or phenotype remains poorly elucidated. To better understand the functional effects of these elements, we focus on polymorphic elements near loci associated with disease risk and pathogenesis [17]. We compiled a catalog of previously reported polymorphic *Alu* elements (see Methods) and from this list selected 112 *Alu* variants that map near genome-wide association study (GWAS) signals to Sanger sequence and fully annotate [17]. As expected, most (96%), are from the youngest *Alu* subfamilies, 46% *AluYa5* and 23% *AluYb8*, the most recently retrotranspositionally active subfamilies whose members account for the overwhelming majority of previously reported polymorphic *Alu* insertions [3]. Intriguingly though, 4% ($n = 4$) belong to the evolutionarily older *AluS* subfamily, which was most active 35–60 million years ago [18, 19] and is considered to have limited in vivo retrotransposition capability in humans in the modern era [7]. These results suggest that polymorphic *AluS* elements may contribute to structural variation in the human genome more than previously thought.

Structural variants involving *Alu* elements may either be deletion or insertion polymorphisms. Since the *AluS* subfamily is considered to have been largely inactive for tens of millions of years [18, 19], we expected that some portion of *AluS* polymorphisms would reflect deletion polymorphisms, arising when *AluS* elements are (imperfectly) excised by an interstitial deletion. On the other hand, *Alu* insertion polymorphisms classically arise by TPRT. Ongoing retrotransposition has resulted in thousands of *Alu* insertion polymorphisms in the human genome, mostly confined to the *AluY* subfamilies [3]. Remarkably, we found that two of the four polymorphic *AluS* elements near GWAS signals described above are full-length elements that have all three defining features of a TPRT-mediated insertion [11]: (1) an intact 3′ end, (2) a 3′ poly-A tail, and (3) flanking TSDs. Therefore, these *AluS* polymorphisms appear to have arisen by TPRT.

To further expand the list of polymorphic *AluS* elements, we considered data from the most comprehensive effort to characterize structural variation in humans – the 1000 Genomes Project. In the most recent analysis there were 49 polymorphic *AluS* elements reported [2]. This list includes one of the *AluS* elements discussed above, thus bringing the total to 52 polymorphic *AluS* elements to characterize in more detail.

Confirming *AluS* subfamily classification

We first set out to confirm the *AluS* subfamily assignment of these 52 polymorphic elements using high stringency criteria (see Methods). Subfamily classification was performed using multiple established methods whenever possible.

For the 49 *AluS* polymorphisms that are annotated in the reference genome (hg19), we compared the subfamily calls made by RepeatMasker [20], the RepeatMasker track of the UCSC Genome Browser (hg19), and the 1000 Genomes Project [2], (Additional file 1: Table S1). In 22 cases there was complete agreement among these sources. In 22 cases there was minor disagreement, limited to a discrepancy in classification among *AluS* subfamilies. In the remaining five cases there was a more substantial disagreement regarding subfamily classification. In the first of the five cases, there was a discrepancy between subfamily classification by the 1000 Genomes Project as an *AluSz* element and subfamily classification by both RepeatMasker and the RepeatMasker track of the UCSC Genome Browser (hg19) as an *AluJb* element; this element was ultimately classified as an *AluJ* element and excluded from later analysis. In the other four cases there was disagreement regarding the classification of the element in question as *AluS* versus *AluY*. To resolve this issue, we identified five diagnostic nucleotides that definitively distinguish between *AluS* and *AluY* consensus sequences [21] when considering the six *AluS* subfamilies and the six most common *AluY* subfamilies (Fig. 1a), [3]. The sequences of the four polymorphic *Alu* elements were manually evaluated with respect to these positions. Three elements were severely 5′ truncated making subfamily classification difficult. In particular, two elements were so short that they did not contain the necessary diagnostic positions to determine *AluY* versus *AluS* assignment, thus explaining the disagreement among the methods described above. Therefore, while these elements do have characteristics of classical TPRT insertions and may belong to the *AluS* subfamily, they were not included in our remaining analysis as they could not be confirmed to be *AluS* elements. The other element was less severely truncated than the two described above. It contained one diagnostic position distinguishing between *AluS* and *AluY* consensus sequences, which matched that of an *AluY* element, and was therefore also excluded from further analysis. The final element in question (located at 11q14.1) was full-length, which allowed for the evaluation of all five diagnostic positions between *AluS* and *AluY* consensus sequences illustrated in Fig. 1a. The polymorphic sequence matched the *AluS* consensus sequence at the two diagnostic positions in the left monomer, the *AluY* consensus sequence at two of the three diagnostic positions in the right monomer, and neither consensus sequence at the remaining nucleotide (Fig. 1b, Additional file 2: Figure S1). While such a chimeric element may have arisen by recombination between adjacent *AluS* and *AluY* elements [3], the fact that this polymorphic element is flanked by identical TSDs makes this possibility unlikely. Given the full-length

nature of this *Alu* polymorphism, we considered seven additional positions that largely, although not definitively, distinguish *AluY* and *AluS* elements (Additional file 2: Figure S1). When considering all twelve positions, this element is consistent with only an *AluS* subfamily consensus sequence at six positions (highlighted in green) and consistent with only an *AluY* subfamily consensus sequence at three positions (highlighted in red). At one position (highlighted in gray) this element is consistent with both *AluS* and *AluY* subfamily consensus sequences and at two positions (highlighted in yellow) it is consistent with neither *AluS* nor *AluY* subfamily consensus sequences; evaluation at these positions was, thus, uninformative. Due to predominating *AluS* features, the polymorphic *Alu* element at 11q14.1 was ultimately classified as an *AluS* element and included in subsequent analysis.

Confirmation of subfamily classification of the three *AluS* polymorphisms not annotated in the reference genome was handled slightly differently. For two such *AluS* polymorphisms no subfamily assignment was made in the original report of the polymorphism; the element was only classified as belonging to the *Alu* family [14, 22] and subfamily assignment (to an *AluS* subfamily in both cases) was thus solely made using RepeatMasker [20]. For the third element, there was agreement between the original report [13] and RepeatMasker [20] with respect to subfamily assignment to the *AluSg* subfamily (Additional file 1: Table S1).

Further analysis thus focused on the 48 *Alu* polymorphisms confirmed to be *AluS* elements using these high stringency criteria.

AluS deletion polymorphism candidates and Non-classical Alu insertion candidates

The overwhelming majority of previously reported polymorphic *AluS* elements in humans were classified as deletion polymorphisms [2, 13]. However, based on our analysis of the *AluS* variants mapping near GWAS signals, it appears that some extant *AluS* polymorphisms have features of a TPRT-mediated *Alu* insertion event. Therefore, we set out to categorize the 48 polymorphic *AluS* elements as insertion or deletion polymorphisms.

Deletion polymorphisms may arise when fixed *Alu* elements are deleted through recombination, thus becoming polymorphic in the population. Such *Alu* deletions are often imprecise. While the pre-deletion allele contains the *Alu* element, the post-deletion allele lacks the *Alu* element, either in part or in its entirety, along with adjacent genomic sequence as well in some cases, depending on the recombination or end-joining event (Fig. 2). Thus, we defined deletion polymorphism candidates as polymorphisms that are not limited to the *Alu* element (i.e., due to the inclusion of adjacent genomic sequence) or that contain only a portion of the *Alu*

a

Alu Left monomer   A-rich   Right monomer   A$_n$

	AluS	AluY
	C	T
	T	A/C
	C	G
	T	G
	G	C

Position (AluS/AluY) [95/93] [100/98] [197/196] [200/199] [219/218]

b

```
                                                           95  100
3p21.31  -GGCTGGGCACGGTGGCTCACGCCTGTAATCCCAGCACTTTGGGCGGCTGAGGTGGGTGGATCACTTGAGGT
4p15.1   -GGCCGGGAGCAGTGGCTCATGCCTGTAATCCCAGCACTTTGGGAGGCCAAGGCGGGCAGATCAC--GAGGT
5q23.1   TGGCCGAGCACAGTGGCTCACGCCTGTAATCCCAGCACTTTGGGAGGCTGAGGTAGGCAGATCACTTGAGGT
15q15.3  --GCTGGGCGCAATGGCTCACGCCTGTAATCCCAGCACTTTGGGAGGCCAAGGCAGGCAGATCACT--AGGT
16q22.1  TGGCCGGGCGTAGTGGCTCACACCTGTAATCCCAGCGCTTTGGGAGGCCGAGGTGGGCGGATCATCTAAGGT
20p12.2  -GGCCGGGCGCGGTGGCTCACGCCTGTAATCTCAGCACTTTGGGAGGCTGAGGTGGGCGGATCACCTGAGGT
11q14.1  -GGTCGGGCGTGGTGGCTCACGCCTGTAATCCCAGCACTTTGGGAGGCCGAGGCAGGCGGATCACCTGAGGT

                                                           95  100
3p21.31  CAGGAGTTCAAGACCAGCCTGGGCAACATGATGAAACCCTGTCTCTACTAAAAATACAAAAAAATTAGCCAGG
4p15.1   CAGAAGTTTGAGACCAGCCTGGCCAA--TAGTGAAATCC-ATCTGTACTAAAAATAC-AAAAATTAGCCAGG
5q23.1   TGGGAGTTCAAGACCAGCCTGGCCAACATGGTGAAACCCCGTCGCTACTGAAAAAAAAAAAAAATTAGCCGGG
15q15.3  CAGGAGTTTAAGAACAGCCTGGCCAACGTGGTGAAACCCCGTCTNTACTAAAAATAC-AAAAATTAGCTGGG
16q22.1  TGGGAGTTTGAGACCAGCCTGACCAGCATGGTGAAACCCTGTCTCTACTAAAAATAC-AAAAATTAGCTGGG
20p12.2  CAGGAGTTCGAGACCAGCCTGACCAACATAGTGAAACCCCGTCTCTACT-AAAATAC-AAAAATTAGCCGGG
11q14.1  CAGGAGTTCGAGA-CAGCCTGACCAAAATGGTGAAACCCCGTCTCTACTAAAAATACAAAAAATTAGCCGGG

                                                               197 200
3p21.31  CGTGGTGGCACATGTCTGTAATTCCAGCTACTCAAGAGGCTGAGGCA-GGAGAATCGCTTGAACCTGGGAGG
4p15.1   CATGGTGGTGGGTGCCTGTAATCTCAGCTACTCAGGAGGCTGAGGCA-GGAGAATTACTTGAACCCTGGAGG
5q23.1   CATGGTGGCGGGTGCCTGTAATCTCAACTTCTCAGGAGGCTGAGGCA-AGAGAATCGCTTGAACCTGGGAGG
15q15.3  CGTCCTGGCGTGCACATGGCTGTATTCCCAGCTACTTGGGAGGCTGAGGCA-GCAGAATCGCTTGAACCCAGGAGG
16q22.1  TGTGGTGGCGCATGCCTGTAATCCCAACTACTCGGGAGGCTGAGGCA-GGAGAATCGCTAGAACCCAGGAGG
20p12.2  CGCAGTGGCGGGCGCCTGTAATCTCAGCTACTTGGGAGCCTGAGGCA-GGAGAATCGCTTGAACCCAGGAGG
11q14.1  CGTGGTGGTG-GCGCCTGTAGTCCCAGCTACTCGGGAGGCTGAGGCA-GGAGAATGGCATGAACCCAGGAGG

         219
3p21.31  TAGAGGTTGCAGTGAGCCAAGATCACACCACTTTAC--TCCAGCCTGTGCAATAAAGCGAAACTCTATCTCA
4p15.1   TGGAGGCTGCAGTGGGCCGAGAGTGCACCATTGCAC--TCTAGCCCAGGCGACAGTGCGAGACTCTGTCTCA
5q23.1   CGGAGGTTGCAGTGAGCTGAGATCATACCACTGCAC--TCCTGCCTGGGCAACAAAGTGAGACTCCATCTCA
15q15.3  CGGAGACTGCAGTGAGCTGAGATTGTGCCACTGCATTGTCCAGCCTGGGTGACAGAGTGAGACTCAGTTTCA
16q22.1  CGGAGGTTGCAGTGAGCCCAGATCGTGCCATTGCAC--TCCAGCCTGGGC--TGGAGCAAAACTCCATCTCA
20p12.2  CGGAGGTTGCAGTGACCCAAGATCGCGCCATTGCAC--TCCAGCCTG-GCAACAGAGCGAGACT--GCCTCA
11q14.1  CGGAGCTTGCAGTGAGCAGAAATCGTGCCACTGCAC--TCCAGCCTGGGCGACAGAGCAAGACGCCGTCTCA
```

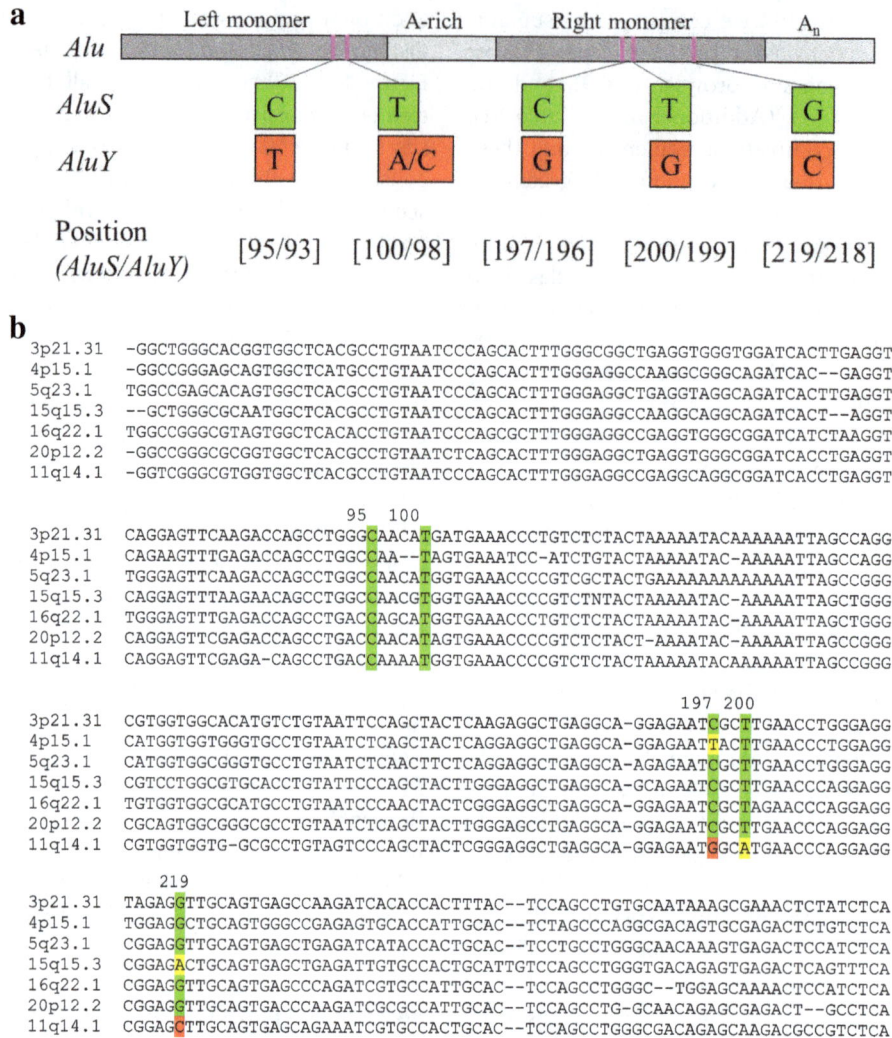

Fig. 1 Diagnostic nucleotides differentiate *AluS* and *AluY* elements. **a** Five diagnostic nucleotides that distinguish RepBase consensus sequences of all six *AluS* subfamilies (*AluSc, AluSg, AluSp, AluSq, AluSx, AluSz,*) from the most common *AluY* subfamilies (*AluY, AluYa5, AluYa8, AluYb8, AluYb9, AluYc1*) were identified [3]. **b** Five diagnostic positions indicated in part (a) in the context of *Alu* sequence confirms *AluS* subfamily classification. The seven *AluS* TPRT insertion candidates are shown. *AluS* specific nucleotides at the diagnostic positions are highlighted in *green*, *AluY* specific nucleotides at the diagnostic positions are highlighted in *red*, and nucleotides at the diagnostic positions that are neither *AluS* nor *AluY* specific are highlighted in *yellow*. Further analysis of the polymorphic *Alu* element at 11q14.1, which has features of both *AluS* and *AluY* elements, that led to its ultimate classification as an *AluS* element is shown in Additional file 2: Figure S1

element at that locus (i.e., only part of the *Alu* is variably present among individuals and the rest of the *Alu* is fixed). Of the 48 polymorphic *AluS* elements, 39 were initially identified to be deletion polymorphism candidates based on this definition (Fig. 2).

We considered the possibility that some of these 39 deletion polymorphism candidates we identified may instead reflect non-classical *Alu* insertions (NCAI). While we were confident that the 33 cases in which the polymorphism does not include the entire *Alu* element at that locus, so that part of the *Alu* is polymorphic and part is fixed (top five post-deletion allele categories in Fig. 2b), represent deletion polymorphisms, the six cases

in which the polymorphism includes the entire *Alu* element as well as flanking genomic sequence (bottom two post-deletion allele categories in Fig. 2b) were evaluated more closely to determine if these could represent NCAI. These six polymorphisms could potentially be NCAI because over half of previously reported NCAI had 2 bp to 2 kb of non-*Alu* sequence inserted along with the *Alu* fragment [23].

NCAI have several other characteristic features that we considered in evaluating these six polymorphisms. NCAI are typically 3′ truncated, thus lacking a poly-A tail, and also lack flanking TSDs [23]. Because they arise by an endonuclease-independent (ENi) mechanism, no

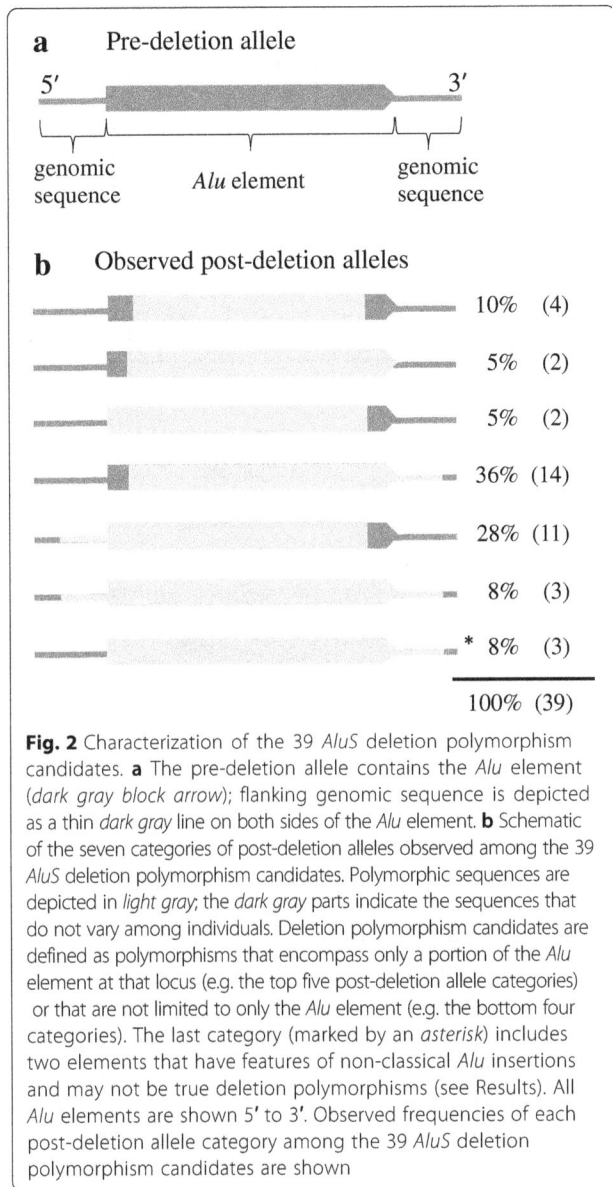

a   Pre-deletion allele

b   Observed post-deletion alleles

10% (4)

5% (2)

5% (2)

36% (14)

28% (11)

8% (3)

* 8% (3)

100% (39)

Fig. 2 Characterization of the 39 *AluS* deletion polymorphism candidates. **a** The pre-deletion allele contains the *Alu* element (*dark gray block arrow*); flanking genomic sequence is depicted as a thin *dark gray* line on both sides of the *Alu* element. **b** Schematic of the seven categories of post-deletion alleles observed among the 39 *AluS* deletion polymorphism candidates. Polymorphic sequences are depicted in *light gray*; the *dark gray* parts indicate the sequences that do not vary among individuals. Deletion polymorphism candidates are defined as polymorphisms that encompass only a portion of the *Alu* element at that locus (e.g. the top five post-deletion allele categories) or that are not limited to only the *Alu* element (e.g. the bottom four categories). The last category (marked by an *asterisk*) includes two elements that have features of non-classical *Alu* insertions and may not be true deletion polymorphisms (see Results). All *Alu* elements are shown 5′ to 3′. Observed frequencies of each post-deletion allele category among the 39 *AluS* deletion polymorphism candidates are shown

L1 ORF2p cleavage sites are observed at the insertion site [23]. Previous studies found that most NCAI are also associated with deletions at the insertion site ranging from 1 bp to ~7 kb [11, 23–25].

The three polymorphisms that include both 5′ and 3′ flanking genomic sequence in addition to the *AluS* element were all confirmed to be deletion polymorphisms. All three *AluS* elements included in these polymorphisms are full-length with 3′ poly-A tails, and are flanked by identical TSDs ranging from 11 to 17 bp; these features exclude the possibility that these are NCAI [23]. We next considered the three polymorphisms that include 3′ flanking genomic sequence in addition to the *AluS* element in its entirety. One of these was confirmed to be a deletion polymorphism, by virtue

of being a full-length *AluSz* element with a 3′ poly-A tail; flanking TSDs could not be identified due to the presence of another *Alu* insertion immediately 5′ of this element. The two remaining polymorphisms in this category appear to be NCAI (Fig. 3, Additional file 3: Table S2). Both of these elements are *Alu* fragments, truncated at both the 5′ and 3′ ends, one with 95 bp of homology to the left monomer and A-rich region of the *AluSc* consensus sequence and the other with 77 bp of homology to the right monomer of the *AluSq2* consensus sequence. Flanking TSDs are not observed in either case. No previously reported L1 ORF2p endonuclease cleavage sites [7] are present at either insertion site, consistent with ENi insertion [23]. Both polymorphisms include short stretches of non-*Alu* sequence at the 3′ end (11 bp with the *AluSc* fragment and 21 bp with the *AluSq2* fragment); the polymorphism that includes the *AluSq2* fragment is also associated with a 14 bp deletion at the insertion site. These features are thus consistent with those previously reported for NCAI [23]. Both of these NCAI candidates were PCR validated to be polymorphic in the population (Additional file 3: Table S2).

In summary, of the 48 polymorphic *AluS* elements, 77% (37/48) are deletion polymorphism candidates (Additional file 4: Table S3) and 4% (2/48) appear to be NCAI. The remaining nine insertion polymorphism candidates were next evaluated in more detail.

Polymorphic *AluS* insertion likely arising by internal priming

One *AluS* insertion polymorphism candidate, which was PCR validated to be polymorphic in the population (Additional file 3: Table S2), does not have all of the defining characteristics of retrotransposon insertions occurring by TPRT [11]. The polymorphic *AluSq2* element at 8p11.23 is full-length and flanked by TSDs but completely lacks a 3′ poly-A tail (Fig. 3, Additional file 3: Table S2). Thus, it likely arose through an insertion mechanism other than TPRT. Specifically, its features are characteristic of an element that inserted by internal priming [26]. While in classic TPRT, reverse transcription begins at the 3′ end of the poly-A tail [27], in this case reverse transcription likely began at the 5′ end of the poly-A tail, thus accounting for the insertion of a full-length element, only lacking the poly-A tail. While poly-A tail length tends to decrease over time after insertion toward a more stable equilibrium value, the poly-A tail is unlikely to be completely eliminated and no such cases have been reported even among older *Alu* subfamilies [27]. Thus, insertion of this tail-less polymorphic *AluSq2* element most likely occurred by internal priming. Absence of an L1 ORF2p endonuclease cleavage site at the insertion site of this element is consistent with the fact that the internal priming

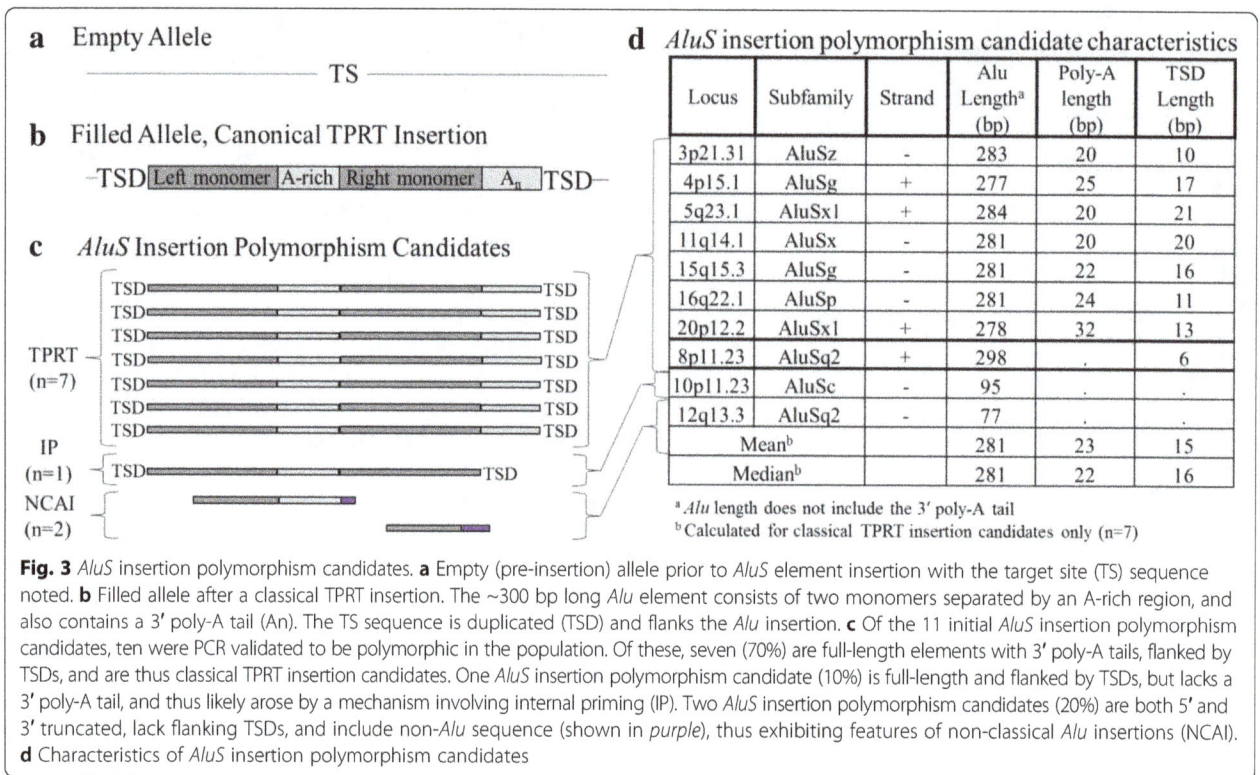

Fig. 3 *AluS* insertion polymorphism candidates. **a** Empty (pre-insertion) allele prior to *AluS* element insertion with the target site (TS) sequence noted. **b** Filled allele after a classical TPRT insertion. The ~300 bp long *Alu* element consists of two monomers separated by an A-rich region, and also contains a 3′ poly-A tail (An). The TS sequence is duplicated (TSD) and flanks the *Alu* insertion. **c** Of the 11 initial *AluS* insertion polymorphism candidates, ten were PCR validated to be polymorphic in the population. Of these, seven (70%) are full-length elements with 3′ poly-A tails, flanked by TSDs, and are thus classical TPRT insertion candidates. One *AluS* insertion polymorphism candidate (10%) is full-length and flanked by TSDs, but lacks a 3′ poly-A tail, and thus likely arose by a mechanism involving internal priming (IP). Two *AluS* insertion polymorphism candidates (20%) are both 5′ and 3′ truncated, lack flanking TSDs, and include non-*Alu* sequence (shown in *purple*), thus exhibiting features of non-classical *Alu* insertions (NCAI). **d** Characteristics of *AluS* insertion polymorphism candidates

mechanism of insertion does not always rely on L1 ORF2p endonuclease and may occur at the site of staggered double-strand breaks (DSBs), thus creating flanking TSDs [26].

Polymorphic *AluS* insertions likely arising by target-primed reverse transcription

Eight *AluS* polymorphisms have all the defining features of retrotransposon insertions that have arisen by TPRT [11]). However, we were only able to PCR validate seven of them to be polymorphic in the population (Additional file 3: Table S2). Therefore, further analysis only focused on the seven validated *AluS* TPRT insertion candidates (Figs. 1b and 3).

These seven *AluS* elements are full-length insertions (277–284 bp), with intact 3′ poly-A tails (20–32 bp), and are flanked by TSDs (10–21 bp) (Fig. 3, Additional file 3: Table S2). To evaluate further the possibility of TPRT-mediated insertion, we searched for the L1 ORF2p endonuclease cleavage site at the insertion site of each of the seven elements (Fig. 4, Additional file 3: Table S2). The endonuclease cleavage sites for the seven loci fall within the distribution previously reported by Konkel et al. [7] (Fig. 4, Additional file 3: Table S2). Thus, we see features consistent with TPRT-mediated insertion in seven polymorphic *AluS* elements of 48 total *AluS* polymorphisms evaluated in this report (15%).

Percent divergence from subfamily consensus sequence and estimated Age of TPRT insertion candidates

These seven *AluS* insertion candidates could reflect fairly recent TPRT-mediated insertions, or could be old insertions slow to reach fixation in human populations. To consider the age of these sequences relative to other *AluS* elements, we compared them to their respective subfamily consensus sequences. Relatively new insertions would not have had much time to accumulate random mutations (i.e., neutral substitutions) and would conserve many of the features of *Alu* elements required for retrotransposition.

We evaluated the degree of divergence of each *AluS* TPRT insertion candidate from its subfamily consensus sequence, and found that the divergence ranges from 5.2

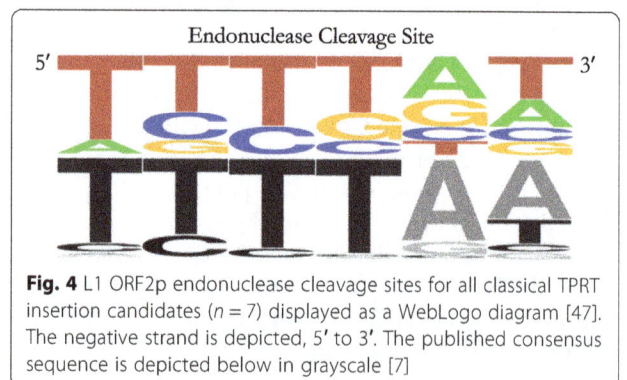

Fig. 4 L1 ORF2p endonuclease cleavage sites for all classical TPRT insertion candidates (*n* = 7) displayed as a WebLogo diagram [47]. The negative strand is depicted, 5′ to 3′. The published consensus sequence is depicted below in grayscale [7]

to 11.2%, with a mean of 8.9% and a median of 10.2% (Fig. 5a). We found that these seven TPRT insertion candidates are significantly less diverged from their respective subfamily consensus sequence than are all the *AluS* elements annotated in the reference genome (n = 686,955) from their respective subfamily consensus sequence (p = 0.0038, permutation test), (Fig. 5b-c). This supports the hypothesis that these insertions likely occurred more recently than at the peak *AluS* activity approximately 35–60 million years ago [18, 19].

We also estimated the age of the TPRT insertion candidates using previously reported substitution rates at CpG and non CpG sites. Since *Alu* elements are rich in CpG sites, which are known to have an appreciably higher mutation rate than non-CpG sites [28], CpG and non-CpG based age estimates are both valuable metrics in addition to percent divergence from the consensus sequence. Age estimates were calculated for each TPRT insertion candidate as previously described [28–32] (see Methods). CpG based age estimates range from 13.9 to 48.6 million years and non-CpG based age estimates range from 22.7 to 37.5 million years (Fig. 5a). Despite the range of estimated ages among the TPRT insertion candidates, overall they appear to have

inserted more recently than at the peak of *AluS* activity 35–60 million years ago [18, 19].

Conservation of functionally significant *Alu* sequence features in TPRT insertion candidates

To evaluate the degree of conservation of functionally significant *Alu* sequence features, we focused on three specific regions – the RNA polymerase III promoter A and B boxes within the left monomer [33], the SRP9/14 major and minor binding sites within the left and right monomers [16], and AC dinucleotides within the left and right monomers previously reported to be important for maintaining *Alu* RNA secondary structure [34]. Since these sequences are critical for successful retrotransposition, they would presumably be present and functional in the source element templating each TPRT insertion variant, and would be highly conserved in recent insertions.

The RNA polymerase III promoter, which is important for efficient transcription [33], is indeed well conserved (Fig. 6a, Additional file 5: Figure S2). Nine of the 11 nucleotides in the A box consensus sequence are fully conserved in all seven *AluS* elements, with infrequent departures from the consensus at the other two positions; five *AluS* elements have an A box that exactly

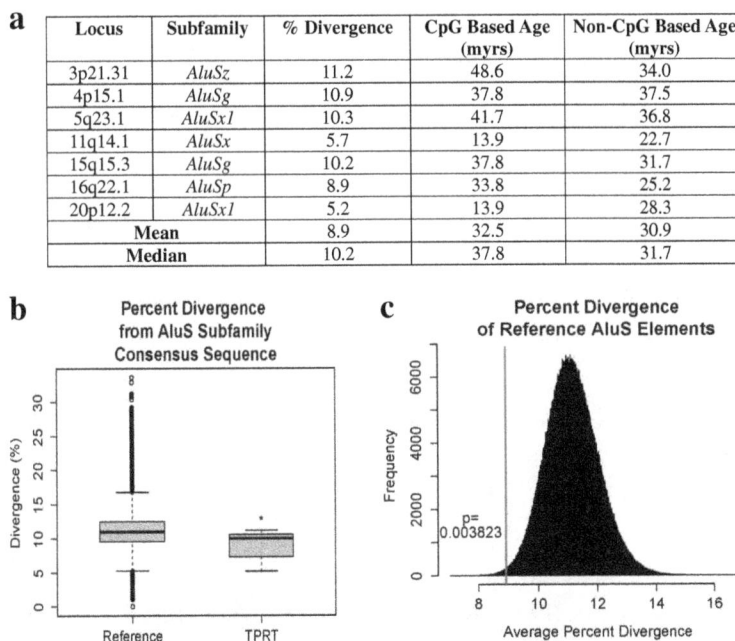

a

Locus	Subfamily	% Divergence	CpG Based Age (myrs)	Non-CpG Based Age (myrs)
3p21.31	*AluSz*	11.2	48.6	34.0
4p15.1	*AluSg*	10.9	37.8	37.5
5q23.1	*AluSx1*	10.3	41.7	36.8
11q14.1	*AluSx*	5.7	13.9	22.7
15q15.3	*AluSg*	10.2	37.8	31.7
16q22.1	*AluSp*	8.9	33.8	25.2
20p12.2	*AluSx1*	5.2	13.9	28.3
Mean		8.9	32.5	30.9
Median		10.2	37.8	31.7

Fig. 5 Estimated age and degree of divergence from subfamily consensus sequence of *AluS* TPRT insertion candidates. **a** For each of the *AluS* TPRT insertion candidates (n = 7) percent divergence from the respective *AluS* subfamily consensus sequence is shown along with estimated ages for the elements based on CpG and non-CpG substitution rates. **b** Boxplot of percent divergence from the respective *AluS* subfamily consensus sequence of all *AluS* elements annotated in the reference genome (n = 686,955) and the TPRT insertion candidates (n = 7). **c** The TPRT insertion candidates (n = 7) are significantly less diverged from their respective *AluS* subfamily consensus sequence than are all the *AluS* elements annotated in the reference genome (n = 686,955) from their respective subfamily consensus sequence (p = 0.0038, permutation test). The distribution of the mean percent divergence of 1×10^6 random samples of n = 7 drawn from the total 686,955 *AluS* elements annotated in the reference genome is shown. The mean percent divergence of the seven TPRT insertion candidates is shown as a *vertical red line*

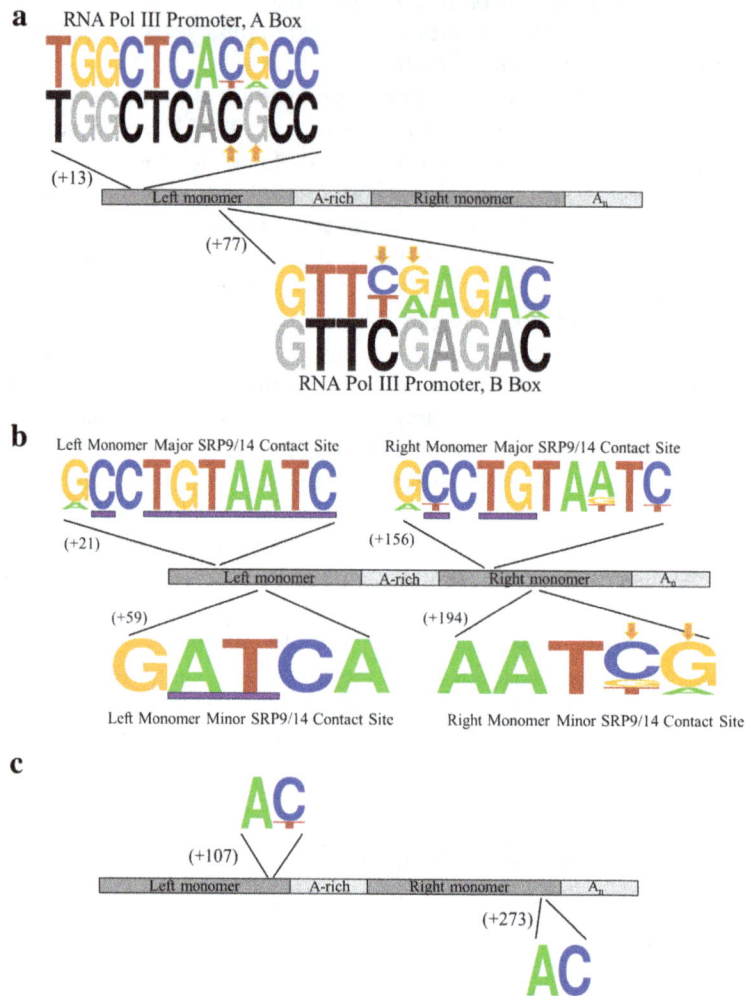

Fig. 6 Conservation of functionally significant *Alu* sequence features in the seven classical TPRT insertion candidates. **a** WebLogo diagram [47] of the RNA polymerase III promoter A and B boxes, with the published consensus sequences depicted below in grayscale [33]. CpG sites are indicated by an *orange* arrow. **b** WebLogo diagram [47] of the SRP9/14 binding sites. Previously reported most highly conserved sites within the SRP9/14 binding sites of elements capable of retrotransposition are underlined by a *purple* bar [16]. CpG sites are indicated by an *orange* arrow. **c** WebLogo diagram [47] of two AC dinucleotides in the *Alu* sequence reported to play a critical role in maintaining the closed loop conformation of *Alu* RNA that is important for interaction with SRP9/14 and efficient retrotransposition [34]

matches the consensus sequence, and the remaining two elements have one mismatch, each at a different position. Six of the nine nucleotides in the B box consensus sequence are fully conserved in all seven *AluS* elements, with some variation at the other three positions; two *AluS* elements have a B box that exactly matches the consensus sequence, four elements have one mismatch, and one element has three mismatches. Two *AluS* elements have both an A and a B box that exactly matches the consensus sequence. Across the A and B boxes, four of the five imperfectly conserved positions are at CpG sites, which are known to have a higher mutation rate due to spontaneous deamination of methylated cytosines at these positions [28].

The SRP9/14 binding sites, which, as predicted by the ribosome-binding model, are important for interaction with SRP9/14, and subsequently with the ribosome and L1 retrotransposition machinery [16, 35] are also highly conserved (Fig. 6b, Additional file 5: Figure S2). In the left monomer, all of the previously reported most highly conserved nucleotides in the major and minor binding sites of elements capable of retrotransposition were fully conserved in all seven *AluS* elements. In the right monomer, two of the three nucleotides previously reported to be most highly conserved in the major binding site of elements capable of retrotransposition were fully conserved in all seven *AluS* elements, and the third nucleotide was conserved in six of the seven *AluS* elements. In all, six of the seven *AluS* elements have full conservation of all the nucleotides previously reported to be most highly conserved within the SRP9/14 binding sites of elements capable of retrotransposition.

Finally, there was strong conservation of both the left and right monomer AC dinucleotides (Fig. 6c, Additional file 5: Figure S2), [34]. These positions have been reported to play a critical role in stabilizing the closed loop conformation of *Alu* RNA that is important for interaction with SRP9/14 via the SRP9/14 binding sites [34] thus lending further support to the ribosome-binding model [35]. The left monomer AC dinucleotide was largely conserved among the seven *AluS* TPRT insertion candidates; there was no variation at the A position, and variation at the C position in only one of the seven *AluS* TPRT insertion candidates. There was perfect conservation of the right monomer AC dinucleotide in all seven *AluS* elements.

Discussion

In this report, we annotate 52 previously reported polymorphic *AluS* elements, confirm that 48 of them do indeed belong to the *AluS* subfamily using high stringency criteria, and comment on their likely mechanism of origin. While most of these appear to be deletion polymorphisms consistent with previous reports [2, 13], we present evidence that seven of these polymorphic *AluS* elements have features consistent with insertion by TPRT. This implies that some *AluS* elements may have been more active recently than previously thought [7, 19] and expands the significance of *AluS* retrotransposition in contributing to structural variation in the human genome.

The overwhelming majority of *AluS* insertions in the human genome are fixed, consistent with the fact that the *AluS* subfamily was most active 35–60 million years ago [18, 19]. Based on this, most *AluS* elements that are polymorphic among humans are expected to be deletion polymorphisms. Deletion events are highly unlikely to include only the *Alu* element in its entirety; while a recombination event between the flanking TSDs would yield a precise deletion of the intervening *Alu* element, such events are extremely rare [36]. Thus, we defined a deletion polymorphism candidate as one that was not limited to only the *Alu* element or did not contain the *Alu* element at that locus in its entirety. As expected, the majority (77%, 37/48) of polymorphic *AluS* elements evaluated in this study are deletion polymorphism candidates.

The remaining 23% (11/48) of *AluS* polymorphisms characterized in this report are insertion polymorphism candidates, of which ten were PCR validated to be polymorphic in the population. Three mechanisms of *Alu* insertion have been previously reported – TPRT [4], internal priming (IP) [26], and NCAI/ENi [23]– and are all represented among the ten *AluS* insertion polymorphism candidates (Fig. 3). Two insertion polymorphisms have features of NCAI [23]. The *AluS* elements included in

these polymorphisms are both 5′ and 3′ truncated and lack flanking TSDs; no previously reported L1 ORF2p endonuclease cleavage sites could be identified at the insertion site. Both polymorphisms include non-*Alu* sequence at the 3′ end and one polymorphism is associated with a deletion at the insertion site. These are all characteristic features of ENi insertion, a mechanism in which *Alu* transcripts are utilized by the cell to repair DSBs, thus leading to insertion polymorphisms [23]. One *AluS* insertion polymorphism appears to have arisen by a mechanism involving IP [26]. While this element is full-length and flanked by TSDs, it lacks a 3′ poly-A tail and no L1 ORF2p endonuclease cleavage site could be identified at the insertion site. These features are consistent with insertion by IP, which may be an alternative mechanism to repair staggered DSBs [26].

The majority of *AluS* insertion polymorphism candidates (70%, 7/10) have all the features of classical TPRT insertions, which may suggest that *AluS* elements were more active recently than previously thought. Our observation is consistent with the "stealth driver" model, which posits that some subfamily members retain low levels of activity over tens of millions of years, long after the subfamily's peak activity [37]. The seven TPRT insertion candidates do not appear to all be products of a single persistent "stealth-driver" element, however. These elements belong to four different subfamilies (three *AluSx* elements, two *AluSg* elements, one *AluSp* element, and one *AluSz* element), which strongly implies that multiple source elements contributed to these insertions.

The ability of *AluS* elements to retrotranspose in humans is considered minimal due to the fact that few polymorphic insertions with features consistent with TPRT-mediated insertion have been identified [7]. However, a de novo *AluSq/Sp* insertion in an exon of *BRCA1* with features of TPRT-mediated insertion has been reported, thus suggesting that some *AluS* elements may still be retrotransposition competent and may, moreover, influence disease risk [38]. Furthermore, *AluS* consensus sequences and some genomic *AluS* elements have been shown to be active in in vitro retrotransposition assays [16]. Importantly, those elements diverging more than 10% from their consensus sequence are inactive, thus highlighting the importance of *Alu* sequence integrity for retrotransposition capability [16]. If the seven candidate TPRT insertions we evaluated did indeed occur fairly recently, then the sequences would not have had much time to accumulate random mutations and would not yet significantly diverge from their respective subfamily consensus sequences. This is indeed what we observe, and the seven TPRT insertion candidates, with a mean percent divergence of 8.9%, are significantly less diverged from their respective subfamily consensus sequence than are all the *AluS* elements annotated in the reference genome from their respective subfamily consensus sequence ($p = 0.0038$, permutation test).

Similarly, specific sequence features known to be functionally important for retrotransposition, namely the RNA polymerase III promoter, SRP9/14 binding sites, and AC dinucleotides involved in maintaining *Alu* RNA secondary structure, are highly conserved. The RNA polymerase III promoter is important for efficient transcription of the element [33], and the AC dinucleotides play a critical role in maintaining the closed loop conformation of *Alu* RNA that is important for interaction with SRP9/14 via the SRP9/14 binding sites, which thus allows the *Alu* transcript to associate with the ribosome and ultimately positions it in close proximity to the L1 encoded proteins required for retrotransposition [16, 34, 35, 39, 40]. The high overall level of conservation suggests these seven TPRT candidates have not diverged significantly from elements that would be capable of retrotransposition, consistent with a recent insertion templated by a TPRT competent element.

While polymorphic *AluS* elements have previously been reported, most discovery efforts aimed to identify only young *Alu* subfamilies and did not include *AluS* subfamily consensus sequences in their algorithms [2, 12, 41]. The *AluS* polymorphisms reported by the 1000 Genomes Project (phase 3) [2] were identified using a pipeline for mapping insertion/deletion polymorphisms rather than the Mobile Element Locator Tool; thus, their identification of polymorphic *AluS* elements was limited to those that are in the reference genome. Notably, a recent study by Hormozdiari et al. [13] that did intend to identify polymorphic *AluS* elements, reports that 9.4% of identified polymorphic *Alu* elements belong to the *AluS* subfamily. However, the authors inferred that the majority of these were likely deletion polymorphisms or insertions arising by endonuclease-independent mechanisms as opposed to novel TPRT events [13]. Here, we present evidence that while, in agreement with previous studies, the majority (77%, 37/48) of the polymorphic *AluS* elements evaluated in this report are likely deletion polymorphisms, 15% (7/48) have features of TPRT-mediated insertions. While these may represent insertions that occurred at the peak of *AluS* activity and have by some means been slow to progress to fixation in the population, the strong conservation of sequence features important for retrotransposition (e.g. RNA polymerase III promoter, SRP9/14 binding sites) implies that some of these may be fairly recent insertions. As most previous discovery efforts did not specifically target identification of *AluS* polymorphisms [2, 12, 41], the majority of the polymorphic *AluS* TPRT candidates described here were identified because there were no reads mapping to the

annotated *Alu* element in the reference genome. Therefore, there is a potential for more polymorphic *AluS* elements, especially less common variants not already annotated in the reference genome, to be identified with targeted discovery efforts.

Conclusions

In summary, we present evidence of polymorphic *AluS* elements in the human genome with features consistent with TPRT-mediated insertion events. These findings imply that multiple *AluS* subfamilies may have been more active recently than previously thought, consistent with the "stealth driver" model [37]. Our analysis also substantiates the concept that *AluS* element deletions are an important contributor to structural variation in humans. In fact, since most methodologies for finding insertion variants have focused on identifying polymorphic *AluY* elements, we expect there to be more yet uncharacterized polymorphic *AluS* insertions and deletions in the human genome. Our findings stress the importance of future structural variant discovery efforts to identify polymorphic *AluS* elements.

Methods

Cataloging polymorphic *Alu* elements

We compiled a list of previously reported polymorphic *Alu* elements [12–14, 22, 41–44], totaling 13,572 *Alu* variants across the human genome. To focus on *Alu* variants with potential functional consequences, we selected polymorphic *Alu* elements within linkage disequilibrium (LD) blocks ($r^2 \geq 0.8$) around GWAS trait-associated SNPs (TAS) with $p < 10^{-9}$ from the NHGRI-EBI GWAS catalog [45] and Sanger sequenced 112 of them. We used RepeatMasker [20] to make subfamily assignments [17].

We also considered the list of polymorphic *AluS* elements reported by the 1000 Genomes Project, phase 3 [2]. In this report [2], the *Alu* polymorphisms of interest were identified using deletion discovery algorithms, and then classified as *AluS* using the AluScan algorithm as part of the Mobile Element Locator Tool (MELT).

Classifying polymorphic *Alu* elements by subfamily

Subfamily assignment of the 52 *AluS* polymorphisms ascertained for this study was confirmed using high stringency criteria, using multiple established methods when possible. Subfamily assignments for all 52 elements were made using RepeatMasker [20].

For the 49 elements annotated in the reference genome, subfamily assignments from RepeatMasker [20], the UCSC Genome Browser RepeatMasker track (hg19), and the 1000 Genomes Project (obtained using the AluScan algorithm as part of the Mobile Element Locator Tool (MELT)) [2] were compared (Additional file 1: Table S1). When there was disagreement in classification among *AluS* subfamilies, the final assignment was made in accordance with RepeatMasker [20], both because it is a well trusted tool and for the sake of consistency, since it is the main tool available to classify the three non-reference genome polymorphic *AluS* elements presented in this report. When there was more substantial disagreement with respect to *AluS* versus *AluY* assignment, the elements were manually classified. RepBase consensus sequences [21] for the following subfamilies were obtained (*AluSc, AluSg, AluSp, AluSq, AluSx, AluSz, AluY, AluYa5, AluYa8, AluYb8, AluYb9, AluYc1*) and aligned using the MUSCLE multiple sequence alignment tool [46], (Additional file 2: Figure S1). Five diagnostic nucleotides that distinguish all six *AluS* subfamilies from the six *AluY* subfamilies included in the alignment were identified (Fig. 1a), [3]. The polymorphic *Alu* sequences were evaluated at these positions and the final assignment was made based on which consensus sequence the element more closely resembled.

For the three elements not annotated in the reference genome, the RepeatMasker [20] classification was compared to the subfamily classification indicated in the original report of the polymorphism, if available (Additional file 1: Table S1). When no subfamily classification was made in the original report, subfamily assignment was made solely using RepeatMasker.

Annotating reference genome polymorphic *Alu* elements

Forty-five of the 48 polymorphic *Alu* elements confirmed to belong to the *AluS* subfamily are in the reference genome (hg19). To confirm the extent of the polymorphism, we compared the annotated polymorphic sequence [2] to the reference genome (hg19) and the RepeatMasker track (UCSC Genome Browser, hg19). Deletion polymorphisms were initially identified as those not limited to only the *Alu* element (i.e., due to the inclusion of adjacent genomic sequence) or those that did not contain the *Alu* element at that locus in its entirety (i.e., part of the *Alu* is polymorphic among individuals and the rest of the *Alu* is present in everyone). Deletion polymorphism candidates that included the entire *Alu* element as well as additional 5′ and/or 3′ flanking genomic sequence were further investigated to determine whether they could represent NCAI; this was done by evaluating the *Alu* element included in the polymorphism for degree of truncation, presence of a 3′ poly-A tail, and flanking TSDs. Further analysis focused only on the insertion polymorphism candidates. When present, the 3′ poly-A tail was identified. Flanking genomic sequence was obtained from the reference genome and used to manually identify TSDs; TSD length was maximized at the expense of poly-A tail length when applicable.

Annotating non-reference genome polymorphic *Alu* elements

Three of the 48 polymorphic *Alu* elements confirmed to belong to the *AluS* subfamily are not in the reference genome (hg19). These loci were PCR amplified with primers flanking the insertion site from individuals from the Centre d'Etude du Polymorphisme Humain (CEPH) from Utah (CEU) HapMap Reference panel to obtain sequences for the filled and empty alleles with respect to each *Alu* element. These PCR products were then cloned and Sanger sequenced. The resulting sequences of the filled and empty alleles were aligned to each other and the reference genome (hg19) to determine the extent of the polymorphism, including the position of the breakpoints with respect to the *Alu* sequence. The identified polymorphic sequence was then analyzed by RepeatMasker [20] to determine which parts of the polymorphic sequence had *Alu* homology and make a subfamily assignment. When present, the 3′ poly-A tail and TSDs were manually identified and annotated as described in the section above.

PCR validation of insertion polymorphism candidates

Primers flanking the insertion site of the 11 insertion polymorphism candidate elements were designed, the regions were PCR amplified, and the PCR amplicons were resolved using gel electrophoresis. Polymorphic *AluS* elements were confirmed by amplification of two different sized alleles from the locus, that, when viewed on the agarose gel, differ in size corresponding to the size of the respective *AluS* element. Loci at which two alleles, the pre-insertion allele and *AluS*-containing allele, could be observed were validated polymorphisms. Additional file 3: Table S2 includes the DNA samples and PCR primer sequences used for validation of each locus.

Percent divergence from *AluS* subfamily consensus sequence

For the TPRT insertion candidates classified as *AluS* elements by RepeatMasker [20] ($n = 6$), the percent divergence from the respective subfamily consensus sequence was obtained from that analysis. For the *Alu* element at 11q14.1, the percent divergence from the RepBase *AluSx* consensus sequence [21] was determined using the MUSCLE multiple sequence alignment tool [46]. To extend analysis to all *AluS* elements the genome, we used the Table Browser function in the UCSC Genome Browser to obtain for all *AluS* elements annotated in the RepeatMasker track (hg19) the percent divergence from the respective *AluS* subfamily consensus sequence. A permutation test was performed using R version 3.2.3 to determine whether the mean percent divergence of the seven TPRT insertion candidates

$(\bar{x} = 8.914)$ was significantly lower than the mean percent divergence of all *AluS* elements in the reference genome. From the total 686,955 *AluS* elements annotated in the reference genome, 1×10^6 random samples of $n = 7$ were drawn with replacement, and the mean percent divergence of each sample was calculated to obtain a distribution of the means (X), (Fig. 5c). The *p*-value ($p = 0.003823$) was calculated as the fraction of random samples with means less than the observed percent divergence of the seven TPRT insertion candidates ($\Pr(X < \bar{x})$).

Alu element age estimates

Two estimates for the age of the *Alu* elements were calculated based on CpG and non-CpG substitution rates as previously reported [28–32]. Briefly, *AluS* TPRT insertion candidate sequences (without the poly-A tail) were aligned to the respective *AluS* subfamily RepBase consensus sequence [21] using the MUSCLE multiple sequence alignment tool [46]. The number of substitutions at CpG and non-CpG sites were counted; for the CpG sites, only C to T and G to A substitutions were counted. The substitution densities were then calculated by dividing the number of observed CpG (or non-CpG) substitutions by the total number of CpG (or non-CpG) sites in the consensus sequence. The age of these elements was then calculated using a neutral rate of evolution of $k = 1.5 \times 10^{-9}$ per nucleotide position per year for the non-CpG sites [31] and a six-fold higher rate of evolution ($k = 9 \times 10^{-9}$ per nucleotide position per year) for the CpG sites as determined by Xing et al. [28].

Additional files

Additional file 1: Table S1. Comparison of subfamily classification of *Alu* polymorphisms ($n = 52$) using multiple established methods. (XLSX 14 kb)

Additional file 2: Figure S1. Diagnostic nucleotides differentiate *AluS* and *AluY* subfamily consensus sequences. **a**. To identify diagnostic nucleotides differentiating *AluS* and *AluY* subfamily consensus sequences, RepBase consensus sequences of all six *AluS* subfamilies (*AluSc, AluSg, AluSp, AluSq, AluSx, AluSz,*) and the most common *AluY* subfamilies (*AluY, AluYa5, AluYa8, AluYb8, AluYb9, AluYc1*) were aligned. Five diagnostic nucleotides that distinguish all six *AluS* subfamilies from six *AluY* subfamilies included in the alignment were identified (highlighted in magenta). Seven additional positions that largely, but not definitively, distinguish between *AluS* and *AluY* elements are also illustrated (highlighted in cyan). **b**. Full-length polymorphic *Alu* element at 11q14.1 that has features of both *AluS* and *AluY* elements. Manual evaluation at the 12 diagnostic nucleotides that differentiate *AluS* and *AluY* elements led to its final classification as an *AluS* element due to predominating *AluS* features. This element is consistent with only an *AluS* subfamily consensus sequence at six positions (highlighted in green) and consistent with only an *AluY* subfamily consensus sequence at three positions (highlighted in red). At one position (highlighted in gray) this element is consistent with both *AluS* and *AluY* subfamily consensus sequences and at two positions (highlighted in yellow) it is consistent with neither *AluS* nor *AluY* subfamily consensus sequences; evaluation at these positions was, thus, uninformative. (PDF 72 kb)

Additional file 3: Table S2. Polymorphic *AluS* insertion candidates ($n = 11$). (XLSX 17 kb)

Additional file 4: Table S3. Sequences of deletion and *Alu*-containing alleles of *AluS* deletion polymorphisms ($n = 37$) [2]. (XLSX 16 kb)

Additional file 5: Figure S2. Functionally significant *Alu* sequence features annotated in TPRT insertion candidates ($n = 7$). (PDF 203 kb)

Abbreviations

CEPH: Centre d'Etude du Polymorphisme Humain; CEU: Centre d'Etude du Polymorphisme Humain (CEPH) from Utah; DSB: Double-strand break; ENi: Endonuclease-independent; GWAS: Genome-wide association study; IP: Internal priming; LD: Linkage disequilibrium; LINE-1, L1: Long INterspersed Element-1; MELT: Mobile Element Locator Tool; NCAI: Non-classical *Alu* insertion; NHGRI-EBI: National Human Genome Research Institute – European Bioinformatics Institute; ORF2p: Open reading frame 2 protein; SNP: Single nucleotide polymorphism; SRP: Signal recognition particle; TAS: Trait-associated SNP; TPRT: Target-primed reverse transcription; TSD: Target site duplication

Acknowledgements

We thank Daniel Ardeljan for helping with the statistical analysis and for critically reading the manuscript.

Funding

This work was supported by US National Institute of Health awards R01CA163705 and R01GM103999 (KHB), and the Howard Hughes Medical Institute (HHMI) Medical Research Fellows Program (MSK).

Authors' contributions

MSK annotated and analyzed sequences, and wrote the manuscript. JPS conducted molecular biology experiments and annotated sequences. KHB and LMP conceived of the study, oversaw the project, and critically revised the manuscript. All authors read and approved the final manuscript.

Competing interests

The authors declare that they have no competing interests.

Author details

[1]Department of Pathology, Johns Hopkins University School of Medicine, Miller Research Building (MRB) Room 447, 733 North Broadway, Baltimore, MD 21205, USA. [2]McKusick-Nathans Institute of Genetic Medicine, Johns Hopkins University School of Medicine, Miller Research Building (MRB) Room 447, 733 North Broadway, Baltimore, MD 21205, USA.

References

1. Mills RE, Walter K, Stewart C, Handsaker RE, Chen K, Alkan C, Abyzov A, Yoon SC, Ye K, Cheetham RK, et al. Mapping copy number variation by population-scale genome sequencing. Nature. 2011;470(7332):59–65.
2. Sudmant PH, Rausch T, Gardner EJ, Handsaker RE, Abyzov A, Huddleston J, Zhang Y, Ye K, Jun G, Fritz MHY, et al. An integrated map of structural variation in 2,504 human genomes. Nature. 2015;526(7571):75–+.
3. Batzer MA, Deininger PL. Alu repeats and human genomic diversity. Nat Rev Genet. 2002;3(5):370–9.
4. Dewannieux M, Esnault C, Heidmann T. LINE-mediated retrotransposition of marked Alu sequences. Nat Genet. 2003;35(1):41–8.
5. Feng Q, Moran JV, Kazazian Jr HH, Boeke JD. Human L1 retrotransposon encodes a conserved endonuclease required for retrotransposition. Cell. 1996;87(5):905–16.
6. Moran JV, Holmes SE, Naas TP, DeBerardinis RJ, Boeke JD, Kazazian Jr HH. High frequency retrotransposition in cultured mammalian cells. Cell. 1996;87(5):917–27.
7. Konkel MK, Walker JA, Hotard AB, Ranck MC, Fontenot CC, Storer J, Stewart C, Marth GT, Genomes C, Batzer MA. Sequence analysis and characterization of active human Alu subfamilies based on the 1000 genomes pilot project. Genome Biol Evol. 2015;7(9):2608–22.

8. Cost GJ, Boeke JD. Targeting of human retrotransposon integration is directed by the specificity of the L1 endonuclease for regions of unusual DNA structure. Biochemistry. 1998;37(51):18081–93.

9. Repanas K, Zingler N, Layer LE, Schumann GG, Perrakis A, Weichenrieder O. Determinants for DNA target structure selectivity of the human LINE-1 retrotransposon endonuclease. Nucleic Acids Res. 2007;35(14):4914–26.

10. Jurka J. Sequence patterns indicate an enzymatic involvement in integration of mammalian retroposons. Proc Natl Acad Sci U S A. 1997;94(5):1872–7.

11. Xing J, Zhang Y, Han K, Salem AH, Sen SK, Huff CD, Zhou Q, Kirkness EF, Levy S, Batzer MA, et al. Mobile elements create structural variation: analysis of a complete human genome. Genome Res. 2009;19(9):1516–26.

12. Witherspoon DJ, Xing J, Zhang Y, Watkins WS, Batzer MA, Jorde LB. Mobile element scanning (ME-Scan) by targeted high-throughput sequencing. BMC Genomics. 2010;11:410.

13. Hormozdiari F, Alkan C, Ventura M, Hajirasouliha I, Malig M, Hach F, Yorukoglu D, Dao P, Bakhshi M, Sahinalp SC, et al. Alu repeat discovery and characterization within human genomes. Genome Res. 2011;21(6):840–9.

14. Stewart C, Kural D, Stromberg MP, Walker JA, Konkel MK, Stutz AM, Urban AE, Grubert F, Lam HY, Lee WP, et al. A comprehensive map of mobile element insertion polymorphisms in humans. PLoS Genet. 2011;7(8):e1002236.

15. Bennett EA, Coleman LE, Tsui C, Pittard WS, Devine SE. Natural genetic variation caused by transposable elements in humans. Genetics. 2004;168(2):933–51.

16. Bennett EA, Keller H, Mills RE, Schmidt S, Moran JV, Weichenrieder O, Devine SE. Active Alu retrotransposons in the human genome. Genome Res. 2008;18(12):1875–83.

17. Payer LM, Steranka JS, Yang WR, Kryatova MS, Medabalimi S, Ardeljan D, Liu C, Boeke JD, Avramopoulos D, Burns KH. Structural variants caused by *Alu* insertions are associated with risks for many human diseases. Proc Natl Acad Sci USA. 2017. Pending.

18. Johanning K, Stevenson CA, Oyeniran OO, Gozal YM, Roy-Engel AM, Jurka J, Deininger PL. Potential for reposition by old Alu subfamilies. J Mol Evol. 2003;56(6):658–64.

19. Shen MR, Batzer MA, Deininger PL. Evolution of the master Alu gene(s). J Mol Evol. 1991;33(4):311–20.

20. Smit AFA, Hubley R, Green P. RepeatMasker Open-4.0.6. 2013-2016. http://www.repeatmasker.org.

21. Bao WD, Kojima KK, Kohany O. Repbase update, a database of repetitive elements in eukaryotic genomes. Mob DNA. 2015;6:6.

22. Lee E, Iskow R, Yang L, Gokcumen O, Haseley P, Luquette 3rd LJ, Lohr JG, Harris CC, Ding L, Wilson RK, et al. Landscape of somatic retrotransposition in human cancers. Science. 2012;337(6097):967–71.

23. Srikanta D, Sen SK, Huang CT, Conlin EM, Rhodes RM, Batzer MA. An alternative pathway for Alu retrotransposition suggests a role in DNA double-strand break repair. Genomics. 2009;93(3):205–12.

24. Salem AH, Kilroy GE, Watkins WS, Jorde LB, Batzer MA. Recently integrated Alu elements and human genomic diversity. Mol Biol Evol. 2003;20(8):1349–61.

25. Callinan PA, Wang J, Herke SW, Garber RK, Liang P, Batzer MA. Alu retrotransposition-mediated deletion. J Mol Biol. 2005;348(4):791–800.

26. Srikanta D, Sen SK, Conlin EM, Batzer MA. Internal priming: an opportunistic pathway for L1 and Alu retrotransposition in hominins. Gene. 2009;448(2):233–41.

27. Roy-Engel AM, Salem AH, Oyeniran OO, Deininger L, Hedges DJ, Kilroy GE, Batzer MA, Deininger PL. Active alu element "A-tails": Size does matter. Genome Res. 2002;12(9):1333–44.

28. Xing J, Hedges DJ, Han K, Wang H, Cordaux R, Batzer MA. Alu element mutation spectra: molecular clocks and the effect of DNA methylation. J Mol Biol. 2004;344(3):675–82.

29. Batzer MA, Kilroy GE, Richard PE, Shaikh TH, Desselle TD, Hoppens CL, Deininger PL. Structure and variability of recently inserted Alu family members. Nucleic Acids Res. 1990;18(23):6793–8.

30. Carroll ML, Roy-Engel AM, Nguyen SV, Salem AH, Vogel E, Vincent B, Myers J, Ahmad Z, Nguyen L, Sammarco M, et al. Large-scale analysis of the Alu Ya5 and Yb8 subfamilies and their contribution to human genomic diversity. J Mol Biol. 2001;311(1):17–40.

31. Labuda D, Striker G. Sequence conservation in Alu evolution. Nucleic Acids Res. 1989;17(7):2477–91.

32. Xing J, Salem AH, Hedges DJ, Kilroy GE, Watkins WS, Schienman JE, Stewart CB, Jurka J, Jorde LB, Batzer MA. Comprehensive analysis of two Alu Yd subfamilies. J Mol Evol. 2003;57 Suppl 1:S76–89.

33. Conti A, Carnevali D, Bollati V, Fustinoni S, Pellegrini M, Dieci G. Identification of RNA polymerase III-transcribed Alu loci by computational screening of RNA-Seq data. Nucleic Acids Res. 2015;43(2):817–35.

34. Ahl V, Keller H, Schmidt S, Weichenrieder O. Retrotransposition and crystal structure of an Alu RNP in the ribosome-stalling conformation. Mol Cell. 2015;60(5):715–27.

35. Boeke JD. LINEs and Alus–the polyA connection. Nat Genet. 1997;16(1):6–7.

36. van de Lagemaat LN, Gagnier L, Medstrand P, Mager DL. Genomic deletions and precise removal of transposable elements mediated by short identical DNA segments in primates. Genome Res. 2005;15(9):1243–9.

37. Han KD, Xing J, Wang H, Hedges DJ, Garber RK, Cordaux R, Batzer MA. Under the genomic radar: the stealth model of Alu amplification. Genome Res. 2005;15(5):655–64.

38. Teugels E, De Brakeleer S, Goelen G, Lissens W, Sermijn E, De Greve J. De novo Alu element insertions targeted to a sequence common to the BRCA1 and BRCA2 genes. Hum Mutat. 2005;26(3):284.

39. Dewannieux M, Heidmann T. Role of poly(A) tail length in Alu retrotransposition. Genomics. 2005;86(3):378–81.

40. Mills RE, Bennett EA, Iskow RC, Devine SE. Which transposable elements are active in the human genome? Trends Genet. 2007;23(4):183–91.

41. Witherspoon DJ, Zhang Y, Xing J, Watkins WS, Ha H, Batzer MA, Jorde LB. Mobile element scanning (ME-Scan) identifies thousands of novel Alu insertions in diverse human populations. Genome Res. 2013;23(7):1170–81.

42. Wang J, Song L, Grover D, Azrak S, Batzer MA, Liang P. dbRIP: a highly integrated database of retrotransposon insertion polymorphisms in humans. Hum Mutat. 2006;27(4):323–9.

43. Iskow RC, McCabe MT, Mills RE, Torene S, Pittard WS, Neuwald AF, Van Meir EG, Vertino PM, Devine SE. Natural mutagenesis of human genomes by endogenous retrotransposons. Cell. 2010;141(7):1253–61.

44. Shukla R, Upton KR, Munoz-Lopez M, Gerhardt DJ, Fisher ME, Nguyen T, Brennan PM, Baillie JK, Collino A, Ghisletti S, et al. Endogenous retrotransposition activates oncogenic pathways in hepatocellular carcinoma. Cell. 2013;153(1):101–11.

45. Welter D, MacArthur J, Morales J, Burdett T, Hall P, Junkins H, Klemm A, Flicek P, Manolio T, Hindorff L, et al. The NHGRI GWAS Catalog, a curated resource of SNP-trait associations. Nucleic Acids Res. 2014;42(Database issue):D1001–6.

46. Edgar RC. MUSCLE: multiple sequence alignment with high accuracy and high throughput. Nucleic Acids Res. 2004;32(5):1792–7.

47. Crooks GE, Hon G, Chandonia JM, Brenner SE. WebLogo: a sequence logo generator. Genome Res. 2004;14(6):1188–90.

9

Dynamic silencing of somatic L1 retrotransposon insertions reflects the developmental and cellular contexts of their genomic integration

Manoj Kannan[1,2,3], Jingfeng Li[2,4,5], Sarah E. Fritz[6,7], Kathryn E. Husarek[6,8], Jonathan C. Sanford[6,9], Teresa L. Sullivan[2], Pawan Kumar Tiwary[2,10], Wenfeng An[11,12], Jef D. Boeke[11,13] and David E. Symer[2,4,14,15*]

Abstract

Background: The ongoing mobilization of mammalian transposable elements (TEs) contributes to natural genetic variation. To survey the epigenetic control and expression of reporter genes inserted by L1 retrotransposition in diverse cellular and genomic contexts, we engineered highly sensitive, real-time L1 retrotransposon reporter constructs.

Results: Here we describe different patterns of expression and epigenetic controls of newly inserted sequences retrotransposed by L1 in various somatic cells and tissues including cultured human cancer cells, mouse embryonic stem cells, and tissues of pseudofounder transgenic mice and their progeny. In cancer cell lines, the newly inserted sequences typically underwent rapid transcriptional gene silencing, but they lacked cytosine methylation even after many cell divisions. L1 reporter expression was reversible and oscillated frequently. Silenced or variegated reporter expression was strongly and uniformly reactivated by treatment with inhibitors of histone deacetylation, revealing the mechanism for their silencing. By contrast, *de novo* integrants retrotransposed by L1 in pluripotent mouse embryonic stem (ES) cells underwent rapid silencing by dense cytosine methylation. Similarly, *de novo* cytosine methylation also was identified at new integrants when studied in several distinct somatic tissues of adult founder mice. Pre-existing L1 elements in cultured human cancer cells were stably silenced by dense cytosine methylation, whereas their transcription modestly increased when cytosine methylation was experimentally reduced in cells lacking DNA methyltransferases DNMT1 and DNMT3b. As a control, reporter genes mobilized by *piggyBac* (*PB*), a DNA transposon, revealed relatively stable and robust expression without apparent silencing in both cultured cancer cells and ES cells.

Conclusions: We hypothesize that the *de novo* methylation marks at newly inserted sequences retrotransposed by L1 in early pre-implantation development are maintained or re-established in adult somatic tissues. By contrast, histone deacetylation reversibly silences L1 reporter insertions that had mobilized at later timepoints in somatic development and differentiation, e.g., in cancer cell lines. We conclude that the cellular contexts of L1 retrotransposition can determine expression or silencing of newly integrated sequences. We propose a model whereby reporter expression from somatic TE insertions reflects the timing, molecular mechanism, epigenetic controls and the genomic, cellular and developmental contexts of their integration.

* Correspondence: david.symer@osumc.edu
[2]Laboratory of Immunobiology, Mouse Cancer Genetics Program and Basic Research Laboratory, Center for Cancer Research, National Cancer Institute, Frederick, MD 21702, USA
[4]Department of Cancer Biology and Genetics, The Ohio State University, Columbus, OH, USA
Full list of author information is available at the end of the article

Background

Approximately half of the human and mouse genomes is comprised of various classes of transposable elements (TEs). These TE insertions have mobilized by distinct mechanisms and accumulated over evolutionary time [1–4]. Until recently, such mobilization was thought to occur almost exclusively in germline cells or early in embryogenesis [5]. However, recent studies established that L1 retrotransposons, along with other classes of mobile genetic elements, also can move actively in somatic cells, i.e., in mouse, rat and human neural progenitor cells, in the developing brain, and in certain human cancers [6–11].

This ongoing movement of endogenous TEs including L1 retrotransposons can result in diverse genetic consequences. These include insertional and deletional (indel) gains and losses of genomic fragments, exon shuffling, insertional mutagenesis of genes, probably chromosomal translocations and inversions, and expression of retrotransposon-initiated fusion transcripts (RIFTs), among others [12–22]. Much of our existing knowledge about TE-related genetic disruption was derived from specific examples of *de novo* insertions causing diseases in mouse and man [23–25]. By contrast, the epigenetic marks established at newly mobilized TEs have not been well characterized.

Cytosine methylation is a key epigenetic regulatory mark localized predominantly within extant L1 retrotransposons and other TEs in mammalian genomes. It has been strongly associated with their transcriptional silencing and regulation, and may affect expression of adjacent genes [26, 27]. Cytosine methylation can be inherited either through mitotic or meiotic cell divisions, and in general are stably maintained. In normal somatic cells, L1 retrotransposons are heavily methylated at CpG dinucleotides, but in most cancers they become hypomethylated, potentially resulting in increased transcription and mobilization [9, 28–30].

A recent study of host epigenetic responses to L1 retrotransposition in various somatic cells including embryonal carcinoma (EC) cells showed that newly integrated L1 reporters were silenced by *de novo* transcriptional gene silencing (TGS) [31]. The epigenetic modifications at newly inserted L1 retrotransposons included histone deacetylation, but not *de novo* cytosine methylation. By contrast, more strongly repressive epigenetic marks including cytosine methylation have been identified at recently inserted L1 elements that were transmitted via meiotic cell division through the mouse germ line in a transgenic mouse model [32]. Similarly, reporter genes that were transduced by retrovirus mobilization or integrated randomly as a transgene typically were methylated rapidly after integration in mammalian cells [33, 34]. Such silencing has been associated with the source and sequence content of the reporter genes themselves.

In classic examples of variable epigenetic silencing at mammalian TEs, changes in epigenetic marks (e.g., methylcytosine density) at pre-existing, integrated endogenous retroviruses (ERVs) have resulted in highly variable expression of nearby genes, resulting in variable phenotypes in genetically identical siblings. Variable but heritable phenotypes in the classic pseudo-agouti mouse model illustrate this impact of epigenetic regulation of an existing TE on neighboring gene expression [35–37]. The term "variegation" describes this epigenetically mediated variability in phenotypes. Typically such phenotypic variation is due to the relatively unstable inheritance of epigenetic controls at a so-called "metastable epiallele", from a cell to its daughter cells [38]. A related and profoundly important question asks how widespread and functionally significant are the heritable impacts of TEs on gene expression and regulation [39].

To investigate how different cellular contexts and mechanisms of transposition may impact reporter expression and epigenetic silencing of newly mobilized TE insertions, we developed new and highly sensitive real-time reporters. We compared the reporter expression and silencing of newly integrated L1 insertions vs. of new *piggyBac* (*PB*) insertions. We corroborated recently reported results about cytosine methylation established at new L1 integrants transmitted through the mouse germ line [32]. We observed variable reporter expression and epigenetic controls established at newly integrated L1 elements that mobilized in different genomic, cellular and developmental contexts.

Results

A sensitive, real-time reporter for L1 retrotransposition reveals dynamic silencing of new genomic insertions

To define the genetic consequences of *de novo* L1 retrotransposition, several research groups have engineered L1 donor constructs to track mobilization [19, 40–42]. In each of these retrotransposition assays, the L1 donor was marked in its 3′ untranslated region (UTR) by a reporter gene disrupted by an artificial intron (AI). Identification of the spliced reporter gene in host genomic DNA indicated that the newly integrated element had undergone expression and splicing as an RNA intermediate, confirming that it was a bona fide product of L1 retrotransposition [40].

To assess expression of L1 reporters that are newly mobilized by retrotransposition, we constructed novel, real-time reporter assays. Their expression levels would not be influenced by positive or negative selective pressures imposed on the cells. We first chose *TEM1*, encoding a beta-lactamase, to generate an exquisitely sensitive reporter assay in living cells. Its expression levels can be quantified over a very large dynamic range extending over four orders of magnitude (Additional file 1: Figure S1: [43]). This greatly exceeds sensitivity of other real-time reporters used in L1 retrotransposition assays. We also chose green fluorescent protein (GFP) [41] as a second, convenient but less sensitive reporter for particular assays. Mimicking the design of other retrotransposition reporter constructs, we introduced the AI into donor cassettes to disrupt the *TEM1* or GFP open reading frames (ORFs), respectively. The AI would be spliced from L1 RNA transcripts at the time of retrotransposition, so a newly integrated reporter cDNA would lack the AI and therefore could be expressed as the intact gene [40].

We marked the full-length, retrotransposition competent, human L1.3 retrotransposon with a novel *TEM1*-AI reporter cassette. The resulting construct was subcloned into a stably maintained, episomal vector, i.e., pCEP4, resulting in pDES46 (Fig. 1a). Cultured human cancer cells, i.e., HCT116 and HeLa, were transfected. To assure stable maintenance of the donor plasmid episome, transfected cells were selected for Hygromycin resistance. In parallel we introduced a transient plasmid expressing GFP alone, to measure transfection efficiency.

To measure beta-lactamase expression levels as a surrogate for active retrotransposition, we incubated transfected cells with CCF2-AM fluorescent substrate. Two enzymatic activities are required in this real-time assay. First, constitutive cellular esterases cleave the –AM esters, resulting in trapping of the charged CCF substrate in the cytoplasm (Additional file 1: Figure S1A). Next, the reporter beta-lactamase cleave the beta-lactam ring in CCF, disrupting fluorescence resonance energy transfer (FRET) and changing the fluorescence emission wavelength from green to blue. After incubation with CCF2-AM, cells were examined by fluorescence microscopy [43]. Uncleaved CCF2 inside the cells would fluoresce green, implying low or absent beta-lactamase expression. In turn, this indicated either no retrotransposition, integration of only truncated TEM1 reporter gene, or silencing of the newly integrated reporter gene (Additional file 1: Figure S1). The presence of blue fluorescence indicated high levels of beta-lactamase expression and therefore robust L1 retrotransposition without silencing of the newly integrated reporter gene (Fig. 1b; Additional file 1: Figure S1).

We confirmed that L1 retrotransposition was the mechanism of reporter mobilization and integration in the transfected cells. We used PCR to amplify the integrated *TEM1* reporter gene across the spliced (excised) AI junction [13, 40], using primers DES657 and DES658 (Additional file 1: Table S1). To determine the detailed structures of several diverse integrants, we employed a PCR-based assay (Additional file 1: Figure S2 and Table S2) [44]. First, we restricted genomic DNA using common 4 bp-cutting restriction endonucleases. Appropriate adaptors were ligated onto compatible overhangs, and PCR primers annealing to the L1 and the adaptor respectively

were used [22, 44]. Insertion-host genomic junctions were recovered as PCR products and were assessed by Sanger sequencing, resulting in the recovery of 9 independent integrant sites. Although several integrants were not long enough to include any L1 sequence or even full-length reporter genes per se, they nevertheless were *bona fide* L1 retrotransposition events (Additional file 1: Figure S2 and Additional file 1: Table S2), as their structures included target site duplications (TSDs), a poly(A) tail, and occasional 5′ inversions [13, 45, 46]. At an L1 integrant on chromosome 2, an intact, spliced *TEM1* gene had been inserted (Additional file 1: Figure S2). Its spliced structure and its TSDs confirmed that it had been retrotransposed.

After selection on Hygromycin for 10 d, bulk populations of transfected cells were screened for beta-lactamase expression by treating them with CCF2-AM (Fig. 1b). Fluorescence microscopy revealed that the beta-lactamase expression was "variegated". Some individual cells fluoresced bright blue, indicating robust expression of beta-lactamase as the reporter for the occurrence of at least one retrotransposition event. Other immediately adjacent cells in the same colony appeared green, indicating that the integrated reporter gene was not expressed and therefore was potentially silenced. Many other cells displayed various intermediate shades of blue and green, implying partial silencing or less-than-maximal expression of the *TEM1* reporter (Fig. 1b).

To quantify *TEM1* reporter expression in individual cells incubated with CCF2-AM, we used flow cytometry [47]. Wide-ranging fluorescence emissions ranging from green to blue were visualized using fluorescence microscopy for individual cells in the bulk population of transfected cells (Figs. 1b and c), and quantified using flow cytometry. Blue/green ratios were calculated as a surrogate score for beta-lactamase activity [43]. These ratios ranged from <10 to well over 150, thereby demonstrating a large dynamic range over which beta-lactamase was differentially expressed in individual cells (Fig. 1c; Additional file 1: Figure S1). To standardize the blue and green signals in flow cytometry, all-green and all-blue cell populations also were assayed in each experiment (Additional file 1: Figure S3).

To test heritability of reporter expression or silencing through multiple rounds of mitotic cell division, we isolated individual cells from these mixed populations by limiting dilution, and then grew up subcloned progeny cells. The resulting cellular clones were stained with CCF2-AM and then visualized by fluorescence microscopy (Fig. 1d). Resulting subcloned daughter cells recurrently stained various shades of blue and green, documenting continued variability in reporter expression. Occasional colonies contained mostly blue cells, indicating high levels of reporter expression amongst most of the cloned daughter cells. However, even in such

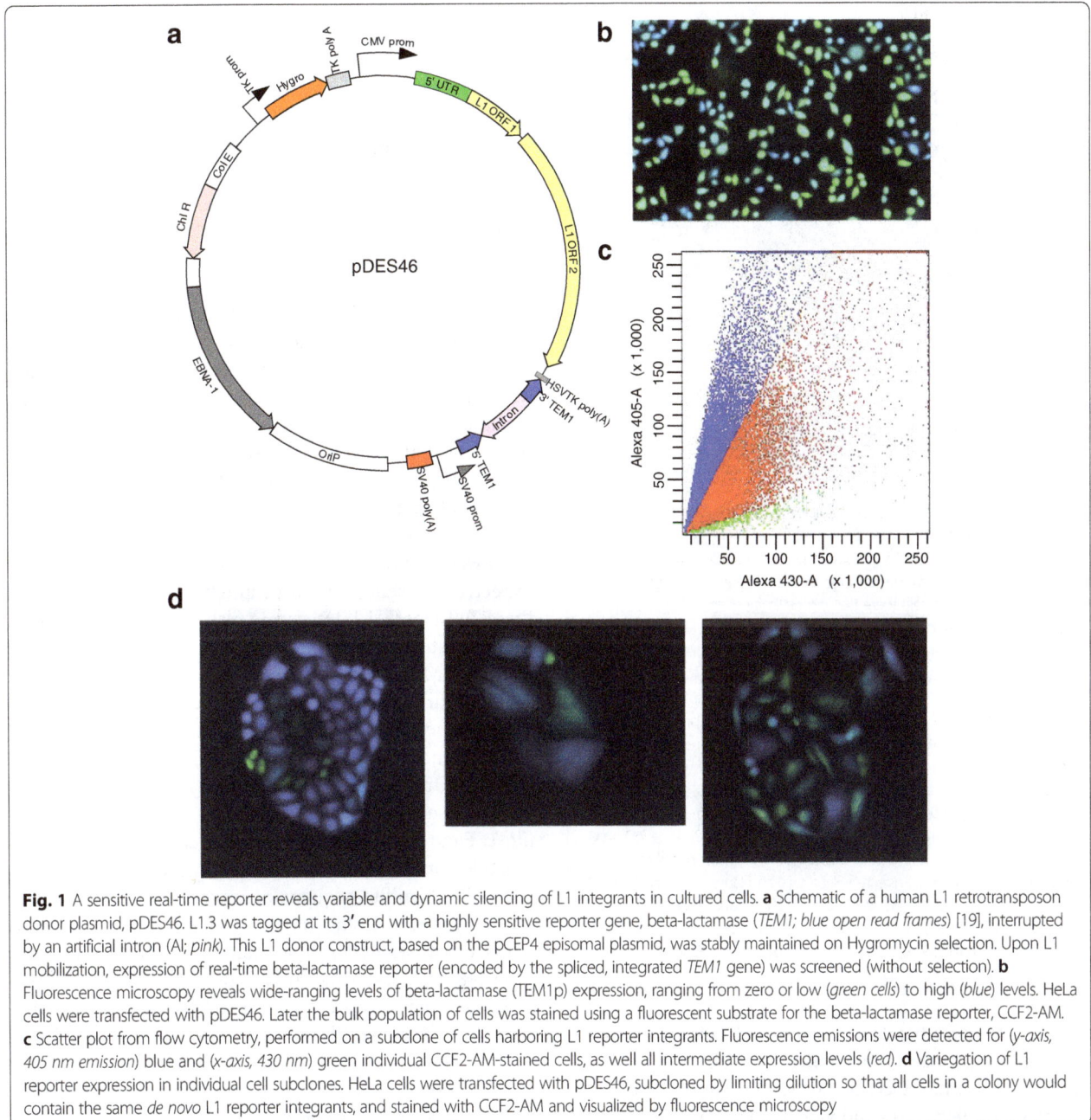

Fig. 1 A sensitive real-time reporter reveals variable and dynamic silencing of L1 integrants in cultured cells. **a** Schematic of a human L1 retrotransposon donor plasmid, pDES46. L1.3 was tagged at its 3′ end with a highly sensitive reporter gene, beta-lactamase (*TEM1; blue open read frames*) [19], interrupted by an artificial intron (AI; *pink*). This L1 donor construct, based on the pCEP4 episomal plasmid, was stably maintained on Hygromycin selection. Upon L1 mobilization, expression of real-time beta-lactamase reporter (encoded by the spliced, integrated *TEM1* gene) was screened (without selection). **b** Fluorescence microscopy reveals wide-ranging levels of beta-lactamase (TEM1p) expression, ranging from zero or low (*green cells*) to high (*blue*) levels. HeLa cells were transfected with pDES46. Later the bulk population of cells was stained using a fluorescent substrate for the beta-lactamase reporter, CCF2-AM. **c** Scatter plot from flow cytometry, performed on a subclone of cells harboring L1 reporter integrants. Fluorescence emissions were detected for (*y-axis, 405 nm emission*) blue and (*x-axis, 430 nm*) green individual CCF2-AM-stained cells, as well all intermediate expression levels (*red*). **d** Variegation of L1 reporter expression in individual cell subclones. HeLa cells were transfected with pDES46, subcloned by limiting dilution so that all cells in a colony would contain the same *de novo* L1 reporter integrants, and stained with CCF2-AM and visualized by fluorescence microscopy

predominantly blue subclones, occasional green cells arose, indicating the stochastic and dynamic establishment of reporter silencing. Upon a second round of cell subcloning by sequential limiting dilution, again we observed mitotically heritable patterns of reporter expression, revealing mostly stable (blue) or variable (mixed) expression of L1 reporter integrants. This result again suggested that while the states of reporter expression or silencing were mostly heritable, they also could oscillate (Fig. 1d). This variability in expression implied that the newly inserted L1 reporters are epigenetically regulated.

To assess whether the variable L1 reporter expression could be attributed to variable numbers and locations of newly integrated L1 reporters, we used Southern blotting to investigate eight related cell lines that had been subcloned from the same initial population of HeLa transfectants (Fig. 2). These clones and their subclones displayed different levels of L1 reporter expression, i.e., either mostly high (staining blue), or variegated (mixed, oscillating) cells including many that were green (i.e., low beta-lactamase expressing). A radiolabeled probe specific for the β-lactamase reporter was used to detect the reporter gene copies in the clones. Almost all subclones

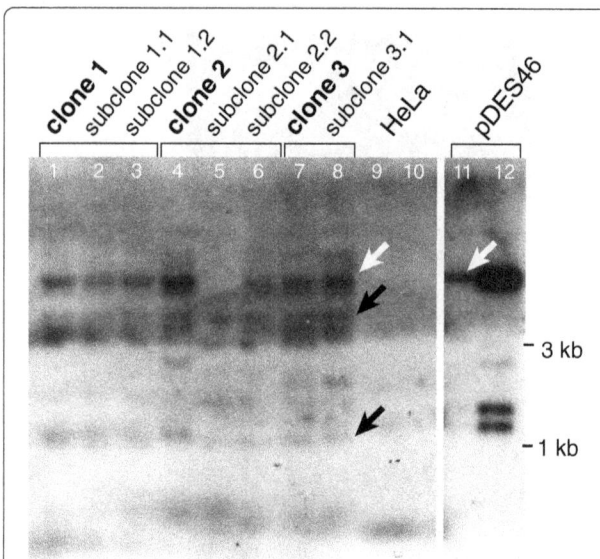

Fig. 2 Southern blot of subcloned cell lines expressing various levels of beta-lactamase reporter reveals their genetic similarity. HeLa cells were transfected with pDES46. To evaluate reporter expression, they were stained with CCF2-AM and examined by fluorescence microscopy. Individual, subcloned cells were derived by limiting dilution. For Southern blots, DNA was extracted, 10 mcg was restricted with EcoRI, electrophoresed on a gel, blotted, and probed for *TEM1* reporter. *Reporter expression phenotypes:* lanes 1–6, clones 1 and 2 and their derived subclones, variegating phenotype with mixed beta-lactamase (*TEM1*) expression; lanes 7–8, clone 3 and its derived subclone, predominantly blue cells with high levels of beta-lactamase (*TEM1*) expression. *Cell line names:* Clone 1: cell line 5B 0.3c/w C9 (parent of sub-clones corresponding to lanes 2 and 3); subclone 1.1: 5B #1G10; subclone 1.2: 5B #4D9; clone 2: 6I 0.3c/w C8 (parent of sub-clones corresponding to lanes 5 and 6); subclone 2.1: 6I #8G2; subclone 2.2: 6I #9 F8; clone 3: 1 1c/w B11 (parent of sub-clone in lane 8); subclone 3.1: 1 #5H5. *Control samples:* lanes 9–10, negative, untransfected HeLa cells; and lanes 11–12, pDES46 plasmid DNA (50 and 500 pg). *White arrows: TEM1* bands shared between cell clones and pDES46 L1 donor plasmid; *black arrows*, integrated *TEM1* bands present in all clones, but not in EcoRI- digested pDES46

shared 7 or more bands of various molecular weights, including a dominant one similar in size to the donor pDES46 episome. Occasional single bands also were detected in a few of the individual clones (Fig. 2), without apparent correlation with expression levels, suggesting gains or losses of additional retrotransposition events. Although the reporter expression levels diverged markedly between these cell clones, the pattern of *TEM1* bands was mostly similar amongst the cell clones and subclones. This suggested that the variable reporter expression was likely due to epigenetic variation between the clones, since their genetic variation appeared to be limited.

Lack of de novo cytosine methylation at new L1 integrants in cultured cancer cells

In previous studies of human L1 retrotransposition in cultured cancer cells, expression of the integrated *Neo^R*

reporter was enforced by positive selection [13, 40]. Selection with the drug G418 imposed a requirement for strong expression of the *Neo* resistance gene, since cells lacking it would be killed. Thus we reasoned that epigenetic marks observed at newly retrotransposed *Neo^R* reporters could have been in favor of active, euchromatic marks.

In addition to finding many truncated *de novo* L1 insertions, we previously mapped two full-length L1 insertions, on chrs. 2 and 14 of transfected HCT116 cells [13]. These newly inserted sequences that were retrotransposed by L1 each included 5′ transduction of adjacent 5′ CMV promoter fragments. Their identification provided us with a unique opportunity to study *de novo* cytosine methylation, established both at the inserted reporter and several kilobases upstream in the 5′ transduced sequences and in the proximal L1 5′ UTR. We measured DNA methylation using conventional bisulfite conversion followed by PCR amplification and sequencing. We found virtually no DNA methylation at the 5′ UTR of both insertions, as only 2.9 and 0.4% of all CpG dinucleotides at those locations were methylated, respectively (Fig. 3). In addition, as expected, the spliced reporter integrants at the 3′ ends of these full-length L1 integrants also were almost entirely unmethylated; only 0.4% of all their CpG dinucleotides were methylated (Fig. 3).

To confirm that the host cells still harbored effective maintenance methyltransferase activity, we measured cytosine methylation within the proximal portion of 5′ UTRs of pre-existing genomic L1Hs elements. They were densely methylated (~64% on average, Additional file 1: Figure S4), confirming that methylated CpG dinucleotides (meCpG) are maintained in the cultured cancer cells. The bisulfite sequencing assay may underestimate unmethylated cytosine content at CpG dinucleotides, as many cytosines undergo spontaneous deamination over time [48, 49], resulting in TpG dinucleotides. Such pre-existing mutations are indistinguishable from unmethylated CpG dinucleotides upon treatment with bisulfite.

Impact of genome-wide hypomethylation on expression of endogenous L1 elements

In mutant double knockout (DKO) HCT116 cells, which lack both the maintenance DNA methyltransferase gene *DNMT1* and the *de novo* methyltransferase *DNMT3b* [50], cytosine methylation at pre-existing L1 elements was markedly reduced. Only ~6.5% of all CpG dinucleotides remained methylated in DKO cells (Additional file 1: Figure S4), reflecting a ~90% reduction in L1 methylation.

Because cytosine methylation is a strong and stable mediator of transcriptional gene silencing (TGS) [27], we compared L1 transcript levels in DKO cells vs. their parental (wildtype) controls. We conducted expression profiling by performing long-read serial analysis of gene expression (LongSAGE) [51]. We corroborated the

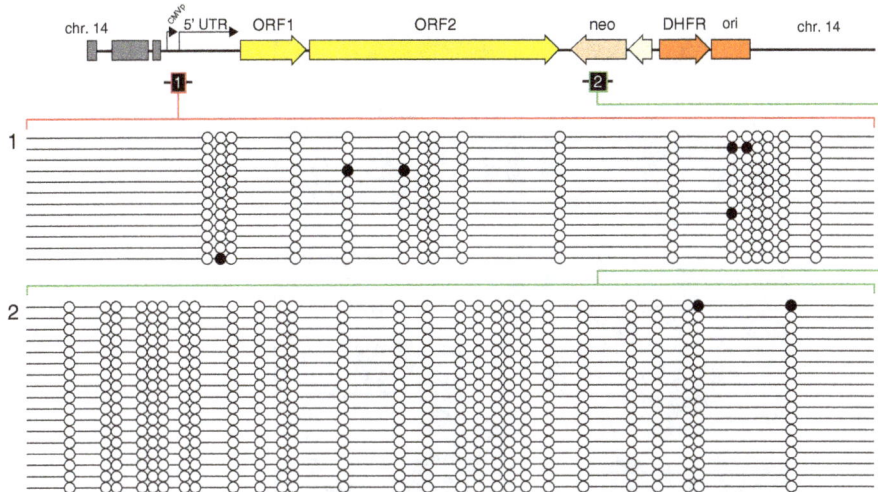

Fig. 3 Lack of cytosine methylation at full-length L1 insertions in cultured cancer cells. *Top*: Schematic of a full-length *de novo* human L1.3 insertion (in chr. 14 in derivative cell clone 7H2 after transfection of the HCT116 parental cell line, as previously mapped [13]). *Black bars*: PCR amplicons analyzed by bisulfite sequencing. Minimal *de novo* cytosine methylation was observed at the newly retrotransposed (1) 5' transduced sequence from the distal CMV promoter and proximal L1 5' UTR; and (2) in the spliced *Neo* reporter gene. PCR amplicons were generated using primers DES512 x DES530 (1); and DES515 x DES524 (2), respectively. Similar results were obtained at a second independent full-length L1 integrant on chromosome 3 in the 2A2 subclone of transfected HCT116 cells [13]

results using several other transcript profiling methods (manuscript in preparation). Transcription of pre-existing genomic L1 templates (measured at their 3′ ends) was only minimally detectable in HCT116 cells (Additional file 1: Table S3), as expected [52]. By contrast, L1 transcript levels were increased modestly (approximately 3-fold) in the DKO cells (Additional file 1: Table S3), consistent with derepression of the TGS regulating their expression.

DNA methylation does not silence newly integrated L1 reporters in cultured cancer cells

We conducted bisulfite sequencing to examine cytosine methylation levels at several independently retrotransposed, spliced *TEM1* reporter integrants. After their retrotransposition, inserted L1 reporter sequences were retained in the host cell genomes in the absence of imposed positive selection. As was the case with L1-Neo^R integrants (Fig. 3), both the integrated L1 reporter *TEM1* (including 18 CpG dinucleotides in an amplicon spanning the splice site in an artificial intron), and an L1 integrant including a portion of the SV40 promoter and *TEM1* reporter (including 20 CpGs) were almost completely unmethylated (Additional file 1: Figure S5). Therefore de novo cytosine methylation played no role in silencing or variegated expression of the TEM1 reporter in these cultured cancer cells (Fig. 1). These results corroborated what we observed after positive selection on Neo^R (also driven by the SV40 promoter as described in [40] and [13]), suggesting that regardless of imposed selection, only minimal methylation is established at new L1 integrants.

Several of the newly integrated sequences retrotransposed by L1 that were recovered from HeLa cells had inserted into repetitive elements pre-existing in the host genome (Additional file 1: Table S2). Nevertheless, bisulfite sequencing analysis of new reporter insertions in bulk showed that most were unmethylated (data not shown). This result suggests that the epigenetic controls established at *de novo* L1 insertions do not reflect spreading of the repressive marks already maintained at neighboring, extant repetitive elements.

Histone deacetylation is strongly associated with L1 reporter silencing in cultured cancer cells

The heritability and the variability in L1 reporter expression in cultured cancer cells suggested that new L1 insertions are epigenetically silenced. To evaluate the possibility that histone tail lysine acetylation could be involved in L1 reporter silencing, we investigated several subcloned cell lines harboring reporter integrants whose expression was variegated or mostly repressed. We treated the cells with various histone deacetylase (HDAC) inhibitors including 100 nM trichostatin A (TsA), 10 mM sodium butyrate, 1 μM scriptaid, 1 nM apicin, and 5 mM nicotinamide (Fig. 4 and Additional file 1: Figure S6). Each of these agents was added in standard growth medium to the cultured cells. Upon incubation for 24 h, expression of the silenced reporter gene was reactivated in virtually all cells. Treated cells showed consistently high levels of *TEM1* reporter expression, as demonstrated by their uniform blue fluorescence upon staining with CCF2-AM (Fig. 4 and Additional file 1: Figure S6). Thus a broad range of HDAC inhibitors from different

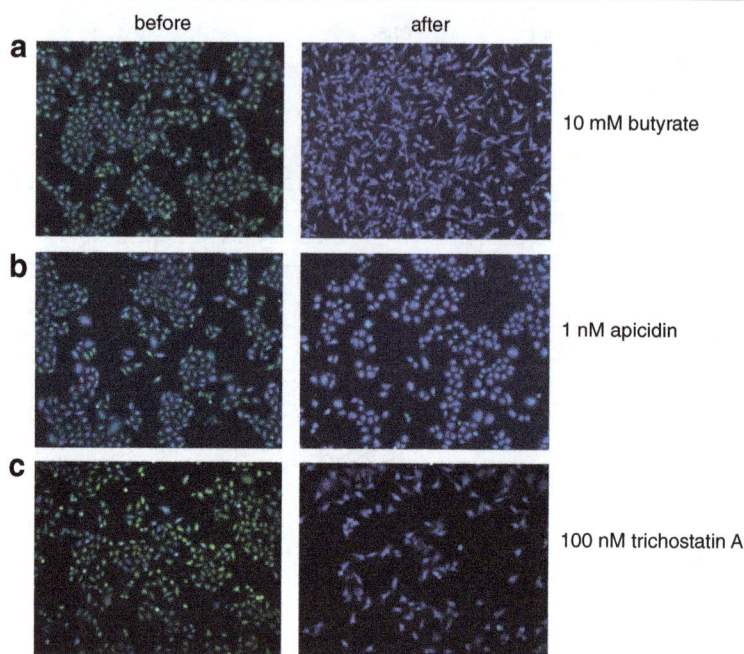

Fig. 4 Variable L1 reporter expression in cultured cancer cells is associated with changes in histone acetylation. Cultured human cervical cancer (HeLa) cells harboring *de novo* L1 reporter integrants were assayed for reporter beta-lactamase expression by incubating them with the fluorescent substrate, CCF2-AM. *Left*: Before and *right*: after incubation for 24 h with various histone deacetylase inhibitors including: **a** 10 mM butyrate; **b** 1 nM apicidin; and **c** 100 nM TsA. Similar responses also were observed upon treatment for 24 h with 1 uM scriptaid and with 5 mM nicotinamide respectively (Additional file 1: Figure S6)

mechanistic categories was active in de-repressing the silenced reporters.

Upon withdrawal of the HDAC inhibitor TsA from derepressed cells by washout of the drug, silencing of the *TEM1* reporter was gradually re-established over 1 – 3 days (Additional file 1: Figure S7). This resetting of L1 reporter silencing demonstrated that it can be dynamically re-established and is reversible. In addition, the state of reporter expression generally appears to be heritable (Fig. 1d). Thus we conclude that the establishment and maintenance of L1 retrotransposon silencing in cultured human cancer cells is consistent with a *de novo* epigenetic mechanism involving dynamic changes in histone lysine deacetylation, but not cytosine methylation.

New L1 integrants undergo rapid and dense DNA methylation in mouse ES cells

To study epigenetic controls at *de novo* L1 integrants in other cellular and developmental contexts, we induced new mobilization of a highly active synthetic L1 retrotransposon in mouse ES cells. The Bruce4 parental ES cell line [53] was transfected with linearized pJH435, resulting in an inducible, codon-optimized synthetic mouse L1 (smL1, ORFeus) donor present in the genome of the resulting Truck_305 cells [54]. We activated L1 retrotransposition by infecting the Truck_305 cells with an adenovirus expressing Cre recombinase, to remove

the floxed beta-geo gene from their donor construct. This resulted in the juxtaposition of ORFeus ORF1 and ORF2 directly downstream of the CAG promoter, thereby activating smL! transcription and potentially their mobilization [54].

After exposing the Truck_305 mES cells to Cre, individual colonies were picked without regard to GFPuv expression, to derive subclonal populations potentially harboring new smL1 insertions. Genomic DNA was isolated from several of these mES cell clones. We did not routinely screen for GFPuv expression using excitation light in the UV range, because a high fluorescence background was observed in cultured cells. In addition, the UV light required to excite GFPuv would be expected to damage the cells when examined, so they could not be cultured further. Instead, linear amplification mediated-PCR (LAM-PCR) was performed to recover any new L1 ORFeus integrants, regardless of their expression of GFPuv. They were sequenced and mapped, and custom bisulfite sequencing primers were designed. Genomic DNA was modified with sodium bisulfite, and then PCR amplification was performed using primers internal to the reporter, or alternatively to target individual integrants. Amplicons were cloned and sequenced. The results showed heavy methylation of the retrotransposed reporter gene, assayed either in bulk or from individual L1 integrants (Fig. 5).

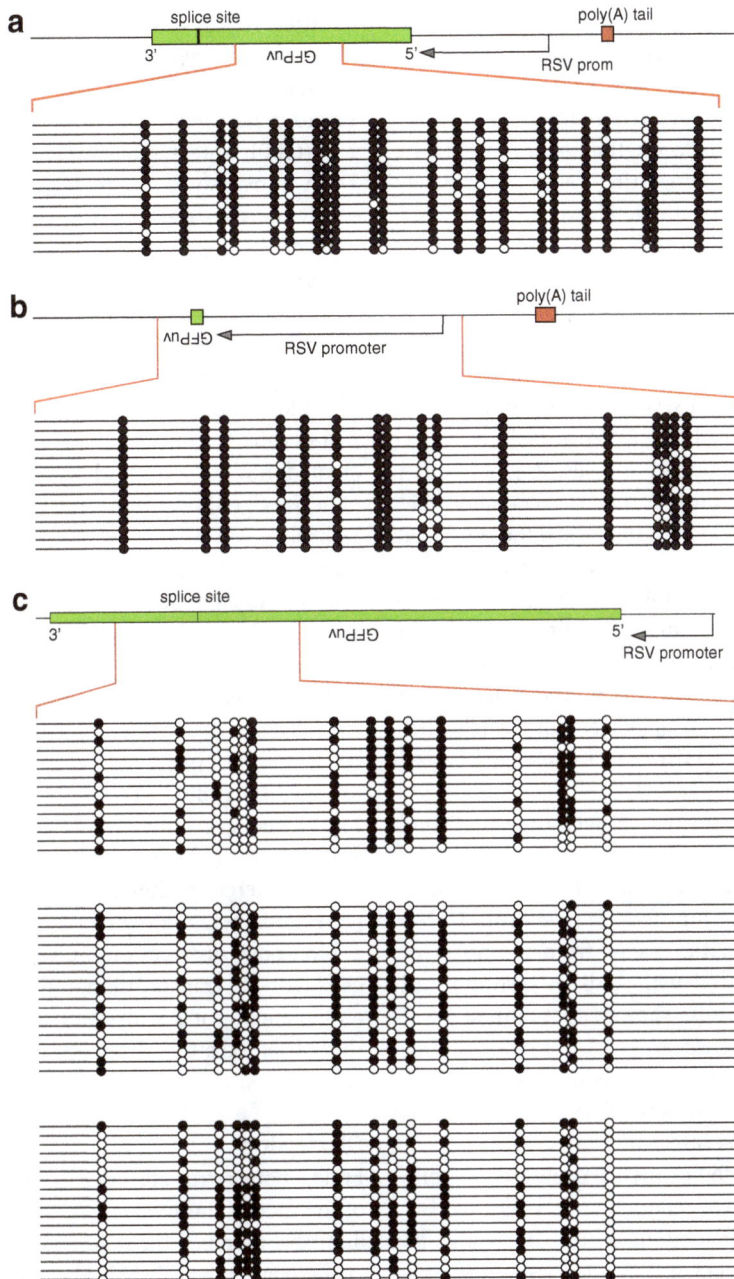

Fig. 5 New L1 integrants in mouse embryonic stem (mES) cells undergo dense cytosine methylation. Dense cytosine methylation at new L1 integrants in mouse ES cells was revealed by bisulfite sequencing. Initially, Bruce 4 cells were transfected with pJH435, encoding an inactivated L1 ORFeus donor element marked with GFPuv-AI reporter regulated by a respiratory syncytial virus (RSV) promoter. Upon activation of L1 donor expression by transient infection of the culture using adenoviral Cre recombinase, individual colonies were picked and cultured on feeder cells for > 2 months. Of these mES subclones, most harbored newly retrotransposed L1 reporter integrants, as shown by a PCR-based assay documenting spliced integrated copies of the reporter GFPuv-AI (not shown). Their cytosine methylation status was assessed either in bulk or at individual loci using bisulfite sequencing. **a** For mES subclone 1B6-A07, we used primers DES3301 x DES3314, which anneal within the gene encoding GFPuv. This PCR amplicon does not cross the AI splice site, so unspliced donor L1 sequences also can be amplified. **b** For mES subclone 1B6-A08, primers DES3298 x DES3299 were used. **c** For mES clones 1B06/B02, 1C6 and 2D4, primers DES3321 x DES3322 were used to assay 15 CpG dinucleotides in a 234 nt amplicon across the GFPuv-AI splice junction

To evaluate silencing of a second, independent reporter gene, we also transfected pJL5, a donor plasmid encoding L1 ORFeus marked by *TEM1*-AI in its 3′ UTR, into mES cells (not shown). Minimal expression of integrated L1 reporters was observed, consistent with dense *de novo* cytosine methylation resulting in strong silencing. An alternative explanation for low beta-lactamase expression was that L1 ORFeus did not retro-transpose efficiently in mES cells. However, this same donor element mobilized very actively in various cultured cancer cell lines [19, 55]. Upon limiting dilution, we observed rare mES subclones that exhibited stable, robust reporter expression, suggesting derepressed newly retrotransposed reporter insertions. However, their DNA methylation was not investigated further.

New L1 integrants undergo rapid, dense DNA methylation in various tissues in vivo

To study the epigenetic modifications established at new L1 integrants in vivo, we obtained several tissues from a transgenic mouse model, in which L1 ORFeus had retrotransposed initially in "pseudofounder" animals. Pseudofounder mice were defined as those mice in the first generation that harbored new, spliced L1 insertions but lacked the unspliced donor element. Their progeny also harbored some of the same, newly retrotransposed L1 ORFeus insertions as were present in the initial pseudofounders themselves, showing that these genomic L1 insertions could be transmitted through the germ line. The new L1 insertions likely had retrotransposed from the donor episome, immediately after its injection as a transgene and before its loss due to cell division during early embryogenesis [55].

We measured *de novo* cytosine methylation established at the newly mobilized L1 ORFeus integrants in the pseudofounders, as well as at some of the same integrants transmitted to offspring [55, 56]. Genomic DNAs isolated from various somatic tissues from members of three pedigrees were treated with sodium bisulfite and sequenced (Fig. 6). The results showed that almost all of the CpGs in independent *de novo* L1 integrants were methylated in two independent pseudofounder mice, F235 and F234 (Fig. 6), in a variety of somatic tissues including the tail and various internal organs.

We also used locus-specific primers to conduct bisulfite sequencing PCR at particular genomic targets (Fig. 6). Tail DNA samples from N2 generation mouse B386 and its progeny B864 and B867 (N3 generation) were used for this more focused study. As was the case with the bulk assays, almost all CpG dinucleotides at the specific genomic target sites were methylated (Fig. 6d). Dense cytosine methylation also was observed in other somatic tissues and at independent L1 insertions in mouse B389 and its offspring, as well as in other mice (Fig. 6; data not shown).

Taken together, these results demonstrated that: a) new L1 insertions occurring early in embryogenesis underwent dense, *de novo* methylation during development; b) methylation was maintained through differentiation into a range of tissues in the developing organism; and c) methylation at such new L1 insertions was maintained and/or re-established upon their transmission through the germline.

PiggyBac reporters are not variegated or silenced in cultured cancer cells and mouse ES cells

To compare reporter integrants mobilized by different mechanisms into distinct genomic targets, the same reporter genes, including *TEM1* and green fluorescent protein (GFP), were engineered to be mobilized as cargo by *PB* DNA transposons. A large majority of HeLa cells harboring newly transposed integrants displayed stable, robust expression of the reporters when mobilized by *PB* (Fig. 7). We observed a bimodal distribution of reporter expression, i.e., with individual cells displaying either robust expression or no expression (Fig. 7) and minimal or no variegation.

As a negative control, we transfected the reporter gene plasmid alone, without *PB* transposase. In resulting transfected cells, no integration events were detected, the transient donor plasmid harboring the unintegrated reporter gene was gradually lost, and no reporter gene expression was observed after several days in culture. In cells that had been transfected with *PB* transposase and the reporter donor, and then subcloned by limiting dilution, we occasionally observed a small fraction of the cells expressing stably diminished levels of the reporter protein.

In additional control experiments, we also launched *PB* transposons carrying comparable reporter genes in mES cells. As was observed in cancer cell lines, expression levels of integrated *PB* reporters in mES cells remained at high levels even after many days of culture (Fig. 7). No variegation or decreases in reporter expression were detected. Therefore, they underwent no or only minimal epigenetic silencing. Flow cytometry experiments confirmed a biomodal distribution of reporter expression in mES cells with *PB* transposition, indicating a lack of variegation and silencing of *PB* insertions (not shown), distinct from that observed after L1 retrotransposition.

Discussion

Endogenous retrotransposons comprise a substantial portion of the mouse and human genomes. Several distinct TE families have modified the mammalian genome profoundly over evolutionary time [57, 58]. The genetic and genomic changes caused by endogenous mobilization of human or mouse L1 retrotransposons have been well studied. By contrast, the epigenetic regulation of *de novo* L1 integrants has not been evaluated fully in the wide-ranging biological contexts in which retrotransposition can occur [39, 59].

Here we investigated the expression and epigenetic silencing of newly integrated L1 reporters in cultured human cancer cells, mouse ES cells, and in several tissues of pseudofounder transgenic mice and their progeny. The results revealed distinctive patterns of L1 reporter expression and associated epigenetic marks, which are associated with the genomic, cellular and developmental contexts of mobilization.

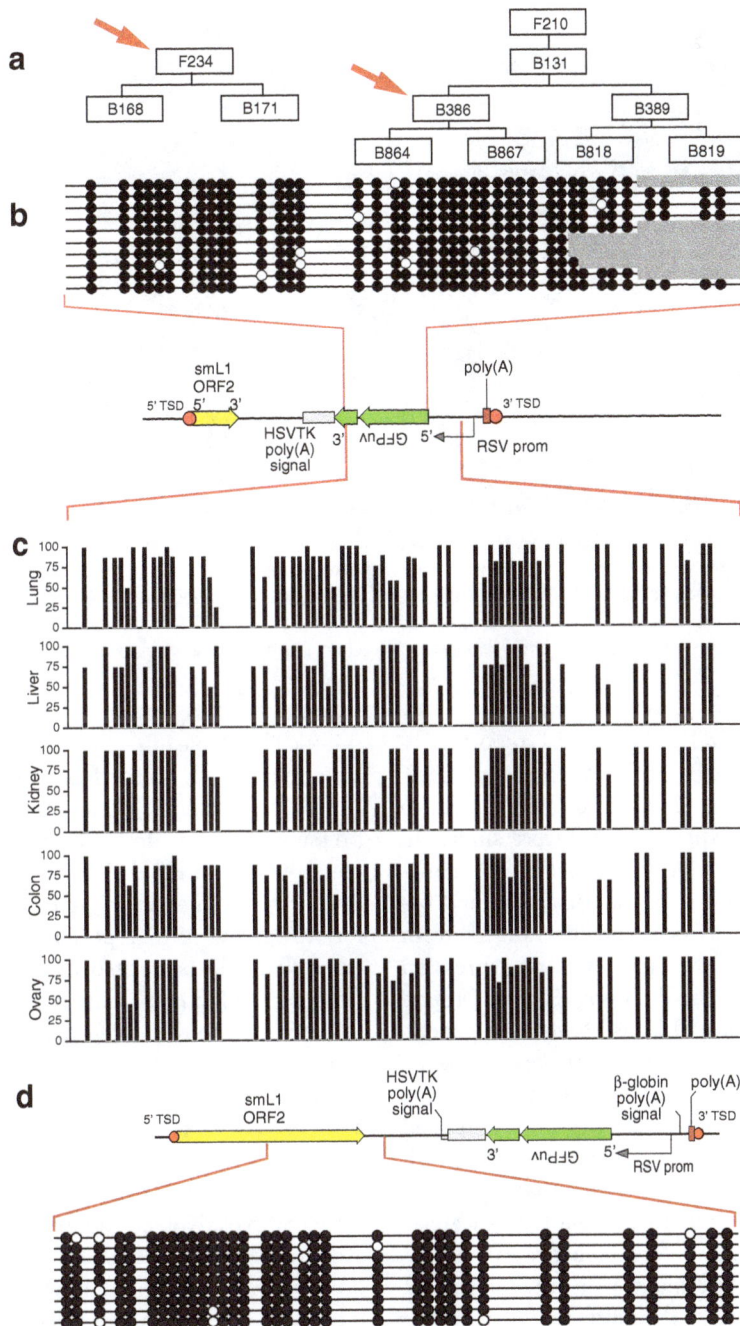

Fig. 6 *De novo* silencing of somatic L1 insertions by dense cytosine methylation during embryogenesis. We evaluated cytosine methylation at new L1 insertions genome-wide by performing bisulfite sequencing. **a** Pedigrees of (*left*) pseudofounder mice F235 and F234, and (*right*) offspring mouse B386 (*red arrows*). At least some of the *de novo* L1 integrants initially integrated in pseudofounder mice, despite the absence of the L1 donor, were transmitted to their offspring. **b** Dense *de novo* methylation at a new L1 insertion in pseudofounder mouse F235, revealed by bisulfite sequencing using primers DES2219 x DES2221 (Additional file 1: Table S1). Each row represents an individual sequence read. *Filled circles*, methylcytosines; *open circles*, unmethylated cytosines. *Gray shading, right*: sequence data not available. *Bottom*: Schematic of *de novo* L1 integrant. Cytosine methylation within the spliced GFPuv reporter gene was assayed (*red connecting lines*). Other integrants at independent genomic loci also may have been assayed by the same amplicons. **c** Dense de novo methylation in various somatic tissues (*left*) in pseudofounder mouse F234, including lung, liver, kidney, colon and ovary, using the PCR amplicon DES2219 x DES2221. Data are presented as cumulative percentages methylated (*y-axis*) for 61 CpG dinuceotides at the indicated positions in genomic template DNA (*x-axis*). **d** Dense cytosine methylation at a new L1 insertion (*schematic*, top), initially identified in founder mouse F210, which was transmitted through the germ line to its offspring. Tail tissue from progeny mouse B386 (*panel A, right, red arrow*) was assayed using primers DES2016 x DES2018

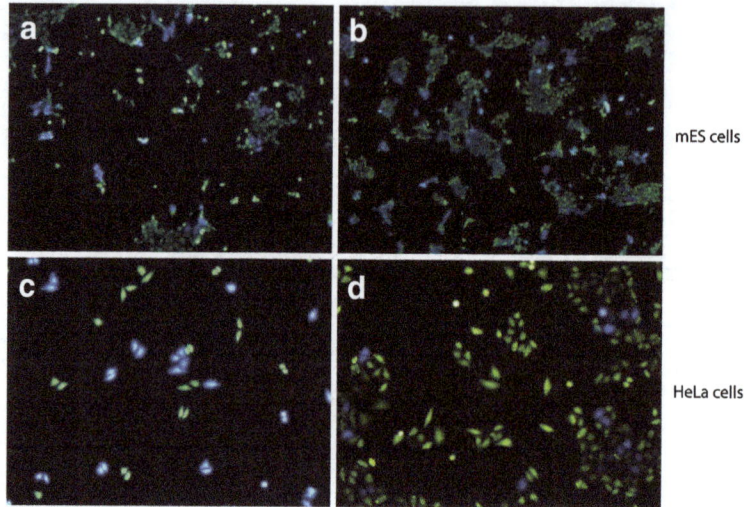

Fig. 7 Lack of variegation and silencing of reporters newly mobilized by *PB* transposons. **a, b** Stable expression of *TEM1* beta-lactamase reporter mobilized by *PB* DNA transposon in mouse ES (mES) cells. E14Tg2a.4 cells were transfected with *PB* vectors carrying the TEM-1 reporter as cargo. Upon staining with CCF2-AM substrate, resulting mES cells fluoresced either blue (stable expression of reporter) or green (no reporter), with minimal variegation observed. **c, d** Stable expression of *TEM1* beta-lactamase reporter mobilized by *PB* in HeLa cells. No variegation was observed. Treatment with HDAC inhibitors did not increase the percentage of cells fluorescing blue (not shown), indicating a lack of silencing of reporter insertions that had been mobilized by *PB*

Newly retrotransposed and integrated L1 reporters in cultured human cancer cells frequently were silenced. This silencing was heritable, as daughter cells tended to display levels of reporter expression that were similar to their parents after mitotic cell divisions (Fig. 1). However, L1 reporter expression also was variegated and oscillated dynamically, as it occasionally ranged from almost completely silenced to high levels of expression over just a single or a few cell divisions (Fig. 1). Despite virtually identical patterns of L1 insertoins at a genetic level, cell clones displayed marked differences in phenotypes, suggesting epigenetic regulation of reporter expression (Fig. 2). New L1 reporter integrants remained almost completely unmethylated, even after many cell divisions, regardless of their expression levels (Fig. 3). L1 reporters were silenced rapidly by histone tail deacetylation, as shown by strong, uniform reactivation of reporter expression upon treatment with any of several diverse HDAC inhibitors (Fig. 4, Additional file 1: Figure S6). Histone deacetylation-mediated L1 silencing was reestablished within 2–3 days in most cells upon removal of those inhibitors (Additional file 1: Figure S7).

We speculate that this variegated expression of reporter sequences retrotransposed by L1 in cultured cancer cells may reflect the timing of their integration, i.e., at later stages of somatic development (Fig. 8), when cellular *de novo* methyltransferases are expressed at low levels [60]. Retrotransposon silencing typically may be both incomplete and stochastic [59]. Thus the observed variegation in new reporter expression may be explained by this stochastic nature of L1 silencing. Although variegation typically is

mitotically stable, the inheritance of a heterochromatic state may rapidly switch from a repressed chromatin state to an open state [61]. The variegated pattern of cells expressing high and low L1 reporter level observed here could be due to rapid changes in chromatin structure induced by L1 integration [62]. An alternative explanation for variegation of L1 reporter expression in cultured cancer cells could be related to genome-wide hypomethylation observed in most human cancers.

In contrast to the variegated, HDAC-mediated silencing of newly retrotransposed L1 integrants in cultured cancer cells, new L1 insertions in mouse ES cells were silenced by dense *de novo* CpG methylation (Figs. 5 and 8). Totipotent mouse embryonic stem (mES) cells can be viewed as a surrogate for the undifferentiated cells present in early embryos. Notably, abundant *de novo* methyltransferases are expressed in both ES cells and early postimplantation embryos [63]. In addition, we observed that new L1 integrants, present in the differentiated somatic tissues of adult pseudofounder mice, also were stably silenced by dense cytosine methylation (Figs. 6 and 8). This dense cytosine methylation may represent faithful maintenance of the initial epigenetic marks established very early during embryogenesis when the insertions occurred (Fig. 8). Dense cytosine methylation in mES cells may mimic epigenetic controls established in early development in vivo (Figs. 5 and 6). One plausible explanation is that *de novo* DNA methyltransferases, which are highly expressed in early embryogenesis and in ES cells [60], could target the newly inserted L1 cDNA sequences, which initially would be unmethylated at time of integration. In developing

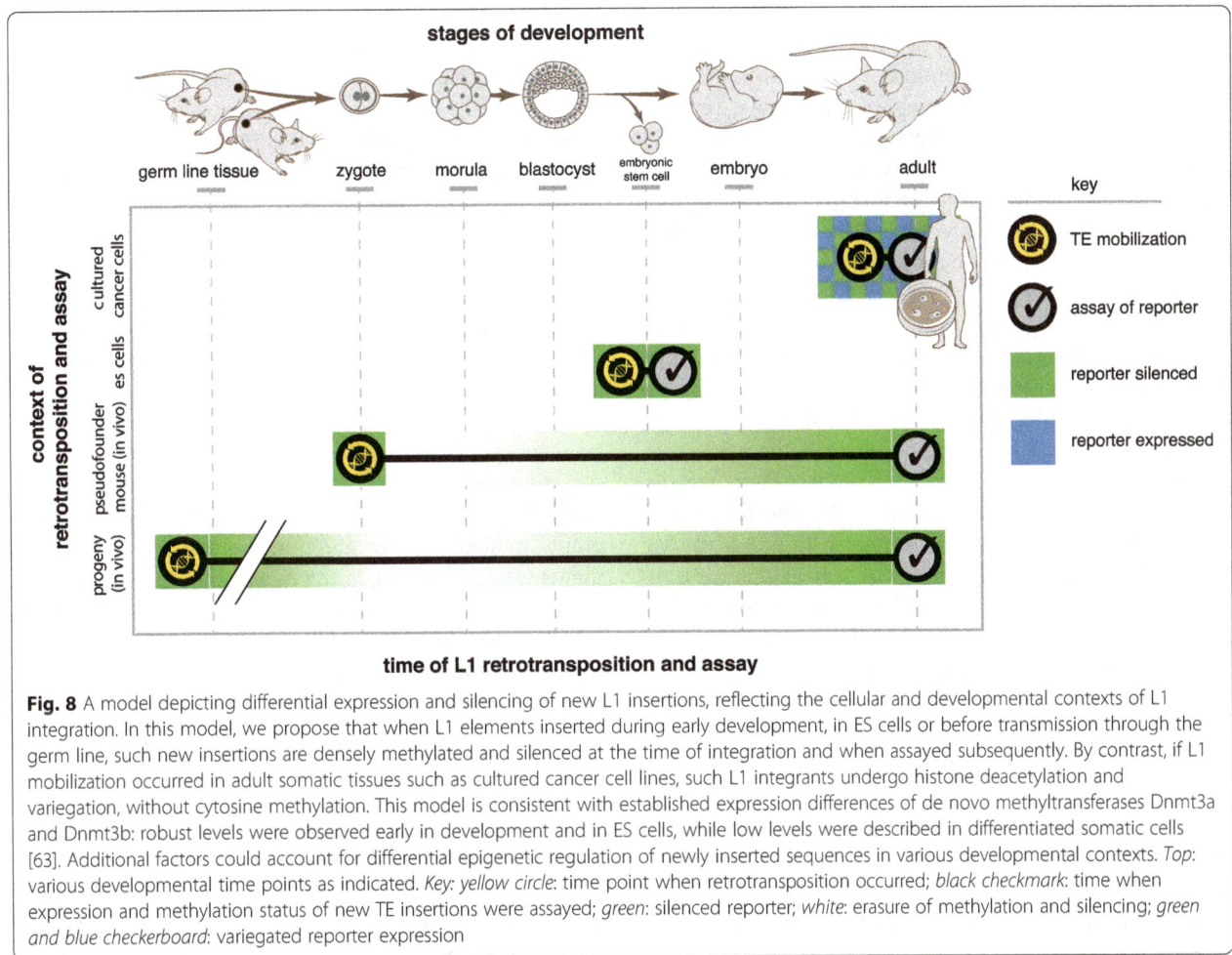

Fig. 8 A model depicting differential expression and silencing of new L1 insertions, reflecting the cellular and developmental contexts of L1 integration. In this model, we propose that when L1 elements inserted during early development, in ES cells or before transmission through the germ line, such new insertions are densely methylated and silenced at the time of integration and when assayed subsequently. By contrast, if L1 mobilization occurred in adult somatic tissues such as cultured cancer cell lines, such L1 integrants undergo histone deacetylation and variegation, without cytosine methylation. This model is consistent with established expression differences of de novo methyltransferases Dnmt3a and Dnmt3b: robust levels were observed early in development and in ES cells, while low levels were described in differentiated somatic cells [63]. Additional factors could account for differential epigenetic regulation of newly inserted sequences in various developmental contexts. *Top*: various developmental time points as indicated. *Key: yellow circle*: time point when retrotransposition occurred; *black checkmark*: time when expression and methylation status of new TE insertions were assayed; *green*: silenced reporter; *white*: erasure of methylation and silencing; *green and blue checkerboard*: variegated reporter expression

embryos, these enzymes normally re-establish DNA methyation after a wave of hypomethylation erases most methylcytosine marks. A recent study of extant L1 expression in human ES cells suggested that predominantly those elements localized in expressed genes were expressed, while others located outside of such genes were not [64]. However, this differential expression of L1 elements may reflect overlapping expression of flanking gene exons which would include intronic L1s, rather than the specific expression of intronic L1s per se. In addition, the activation of endogenous L1 expression, upon reprogramming somatic cells into induced pluripotent stem (iPS) cells, also implied that epigenetic derepression of silenced elements can occur [65]. By contrast, another group reported that silencing of TEs is stable during reprogramming of somatic cells to iPS cells [66].

We conclude that the distinct types of reporter expression and epigenetic regulation of new L1 insertions observed in mES cells or somatic tissues in vivo, in comparison with those in cultured somatic cells, could be related to the different cellular contexts or stages of differentiation in which L1 mobilization occurred initially (Fig. 8).

We inferred that the observed epigenetic marks at *de novo* L1 integrants could reflect the developmental timing at which they integrated (Fig. 8), by defining the epigenetic marks established and then maintained at new integrants. Notably, we measured reporter expression and epigenetic silencing long after the time of retrotransposition per se, i.e., after many cell divisions. This limitation was due to current technical constraints, since initially the L1 donor elements had retrotransposed in individual cells that are experimentally inaccessible except by single cell cloning and sequencing, or upon embryonic development and tissue differentiation.

The resulting, distinct forms of silencing at newly integrated L1 reporters observed in various contexts may have important implications for the expression of the L1s themselves, for the regulation of other genes neighboring the new insertions, and for chromosomal architecture. We conclude that *de novo* L1 retrotransposition can contribute to significant variability in epigenetic marks established in cellular genomes.

In contrast to our observations about a lack of de novo methylation at newly inserted L1 seuqences, previous experiments have demonstrated that newly inserted foreign DNA can undergo de novo methylation in cultured somatic cells [67, 68]. Therefore, as a control, we used PB DNA transposons to mobilize the same reporter genes, both in cultured cancer cells and in mES cells (Fig. 7). Although our L1 or PB reporters are comparable to transgenic insertions of foreign DNA in various cellular contexts, there are several fundamental differences. For example, the target site preferences of L1 retrotransposition vs. PB transposition vs. random integrants of transgenesis are all distinct. While multiple copies of a transgene frequently can recombine at a single genomic locus [69], individual copies of L1 reporter genes typically integrate at interspersed genomic sites. We observed two key differences in silencing of de novo L1 reporter integrants vs. PB integrants. First, we observed minimal or no variegation of PB reporter expression; instead, their expression appeared to be mostly "all or none". Second, the percentage of cells in which PB reporter integrants were silenced was much lower than that with variegated or silenced L1 integrants. These results also are consistent with a lack of PB integrant silencing observed in vivo [70, 71]. We speculate that these differences between L1 vs. PB reporter expression or silencing may reflect different genomic target site preferences of their mobilization. The differences also could be related to differences in the mechanisms of transposition by these elements. Thus, in comparison with the target sites of new L1 integrants, which are enriched slightly in intergenic genomic regions [72] or are distributed randomly [73], more than half of PB integrant sites are enriched inside expressed genes [74]. Alternatively, the PB-mobilized insertions would include PB inverted terminal repeats which could serve as insulators. However, a recent study showed that incorporation of bona fide insulator sequences flanking the transgenic PB reporter genes increased their expression [75]. This suggests that the PB sequences themselves may possess only minimal insulator activity.

In a recent study of epigenetic silencing of new L1 insertions in human embryonic carcinoma (EC) cells, histone deacetylation was identified as the silencing mechanism [31]. We note important similarities and differences between that study's results and our data. We confirmed that histone deacetylation occurs at new L1 insertions, but only in cultured cancer cells and not in mES cells. We found that new L1 insertions were densely methylated in mES cells (Fig. 5), whereas new insertions in hEC cells were not silenced by cytosine methylation [31]. The discrepancy could be attributed to differences in the host cells' species of origin; different epigenetic mechanisms operating in hEC vs. mES cells; distinct structures or sequences of the mobilized elements themselves; or various extents to which the cells

had differentiated in vitro. Notably, ES and EC cells represent very different stages of differentiation [76], despite similar levels of expression of "stem cell-like" markers such as OCT4. ES cells are derived from the inner cell mass of developing embryos, while EC cells are derived from germ cell tumors. Compared with ES cells, the EC cell line PA-1 [31] represents a later stage of embryogenesis. Mouse EC cells have been shown to express full length RNA and ORF1p from pre-existing L1 elements [77]. However, a possible relationship between derepressed chromatin in mES cells or hEC cells and the establishment of de novo epigenetic controls established at new L1 insertions is still unclear.

We also studied silencing of newly mobilized L1 insertions in vivo, both in differentiated tissues of pseudofounder mice and in their offspring. By contrast, the prior study did not include an analysis of silencing of new integrants in vivo [31]. In addition, the mobile genetic elements used as controls in the two studies to compare with L1 mobilization were very different. In the prior study, HIV-like retroviruses mobilized the transgenic reporter genes [31], whereas we used PB, a DNA transposon, to compare with L1 reporter silencing. These control vectors differ in their mechanisms of integration, genomic target sites, and the frequency of insertions generated per host genome. Each of these factors could play significant roles in shaping the downstream epigenetic silencing marks established at the new insertions.

We acknowledge potential limitations in our study. First, to investigate the state of expression and epigenetic regulation of newly inserted sequences that were mobilized by TEs, we artificially marked the L1 and PB donor elements using engineered, heterologous reporters including several different strong promoters and terminators. In comparison with native, unmarked elements, these reporter genes incorporated into donor TEs potentially could interfere with their mobilization. Moreover, upon integration they could trigger antisense transcripts [19] or otherwise artificially trigger or disrupt silencing by mimicking actively transcribed, protein-coding gene. A recent study noted that differences in promoters, reporter sequences or integration sites could influence reporter expression and thereby affect the study conclusions [78].

Second, we did not investigate L1 insertions that had newly integrated in germ line tissues. Extensive research has been conducted on the epigenetic control of pre-existing TEs in germ tissues during embryonic development. They appear to undergo a wave of demethylation followed by two distinct waves of de novo cytosine methylation [32, 79]. PIWI-interacting small RNAs (piRNAs), whose transcription is frequently initiated from TEs in germ tissues, mediate their regulation and silencing. Recent evidence also indicates that DNA methylation in hES cells is induced by PIWI/piRNA-mediated silencing. Small

RNAs also may play important roles in guiding establishment of *de novo* methylation in somatic tissues [39]. We currently are studying roles of possible small RNAs, e.g., including those generated from antisense transcripts [19], in targeting *de novo* methylation.

Third, new insertions were not identified immediately after their integration in single cells. This approach has become technically possible recently, so we could identify and characterize new insertions in individual cells or very small subclonal populations within a few cell divisions of integration. Their minimal allelic fractions would require use of ultra-deep sequencing, making further analysis technically difficult but feasible for the first time.

Fourth, we did not perform chromatin immunoprecipitation (ChIP) experiments to assess enrichment of particular histone modfiications at the new L1 integrants, although we measured cytosine methylation in detail. However, in a recent paper describing L1 reporter silencing by histone deacetylation in hEC cells, confirmatory ChIP data [31] were strongly consistent with our HDAC drug inhibition studies (Fig. 4, Additional file 1: Figure S6).

The donor TE constructs used here did not include a means to terminate conditionally their capacity for ongoing retrotransposition. Future designs will incorporate this feature, to avoid the potential for ongoing genetic variability, e.g., after subcloning individual cells.

In some cases, rather than assay for reporter expression, we monitored retrotransposition by using PCR-based assays to identify genomic insertions of newly transposed reporter genes. As was the case with reporter expression assays used here, we did not select positively or negatively on cells harboring such inserted sequences to preferentially include expressed reporters or exclude silenced reporters.

And finally, although in vivo mouse models harboring control *PB* donor elements have been developed, whereby we could compare their silencing during development and in diverse tissues, such strains were unavailable to us. However, published studies have indicated that *PB* insertions typically are not silenced even when selection is not imposed [70, 80].

In summary, we showed here that the expression and silencing of newly integrated sequences mobilized by L1 retrotransposition appear to be associated with the cellular, developmental and genomic contexts of their integration. We hypothesize that the distinct epigenetic marks set up at new TE insertions that integrated in different cellular or developmental contexts may have various downstream consequences. For example, recent evidence suggests that most new somatic L1 insertions mobilized during human cancer development can mediate only minimal, if any, impacts on neighboring gene expression, unless they cause direct insertional mutagenesis of coding exons [11, 25]. By contrast, new insertions occurring early in development could more significantly disrupt the expression of neighboring genes, in part because their allelic fraction would be higher. In addition, the epigenetic silencing including cytosine methylation that is established at them would be expected to be more repressive and stable (Fig. 8). Thus we speculate that somatic TE that integrated early in development would exert stronger disruptive effects on neighboring genes, because more repressive epigenetic controls including methylcytosine marks would be established at them. These findings can be compared with those of Doerfler et al., who characterized de novo methylation established at foreign DNA introduced into mammalian cell genomes [67].

We propose that these findings may have important practical implications for evaluation and understanding of new TE insertions in various biological contexts. For example, they may facilitate a novel experimental assay of transposition timing, i.e., to identify when the elements mobilized in vivo. Thus we would expect to find dense cytosine methylation strongly repressing newly integrated, polymorphic L1 insertions that were mobilized early in development or passed through the germ line (Fig. 8). Such integrants might be present at a high allele fraction (e.g., 50%, in heterozygosity). By contrast, a somatic L1 polymorphism occurring later in development or differentiation would be expected to be mosaic, so it would be present at a much lower allelic fraction in one tissue and not another, such as in a tumor but not in matched normal tissues. This type of new insertion would be silenced more dynamically and reversibly by histone deacetylation. These epigenetic characteristics would suggest that its mobilization occurred in differentiated somatic cells.

Conclusions

Analysis of newly inserted genomic sequences retrotransposed by L1 in various somatic cells and tissues revealed distinct patterns of expression and epigenetic regulation. In cancer cell lines, the newly retrotransposed integrants typically underwent rapid transcriptional gene silencing, but they lacked cytosine methylation, and their reporter expression was reversible and oscillated frequently. Silenced or variegated reporter expression was strongly and uniformly reactivated by treatment with inhibitors of histone deacetylation. By contrast, newly inserted sequences retrotransposed by L1 in mouse embryonic stem (ES) cells underwent rapid silencing by dense cytosine methylation. Similarly, *de novo* cytosine methylation at new integrants also was observed in several distinct somatic tissues of adult pseudofounder mice. We conclude that the host cellular and developmental contexts of retrotransposition are significant determinants of reporter expression and epigenetic silencing at newly integrated sequences mobilized by L1 retrotransposition. We have proposed a model whereby reporter expression of somatic TE integrants reflects the timing, molecular mechanism, epigenetic controls and the genomic, cellular and developmental contexts of their mobilization.

Methods

Plasmid constructs

To construct an L1 donor launched from an episomal plasmid, i.e., pDES46 (Fig. 1), where human L1 was marked by the *TEM1*-AI beta-lactamase/artificial intron (AI) reporter cassette inserted into its 3′ UTR, we first deleted the BamHI site in ORF2 of human L1.3 as present in pJM101/L1.3 (kindly provided by Dr. John V. Moran, Univ. Michigan) using site-directed mutagenesis. The L1.3/NeoR-AI promoter cassette was then moved into pBSII-KS using NotI and BamHI. We then introduced double-stranded oligonucleotides (containing Bst1107I -HSVTKpolyA-MluI) to replace the NeoR-AI and promoter fragments from pJM101/L1.3. BseRI sites were introduced to flank both *TEM1* (sequences obtained from the vector pBLAK-b which were then codon optimized) and beta-globin AI, using fusion PCR. Both of these constructs were cut out using BseRI and ligated together seamlessly. The resulting L1.3/TEM1-AI construct was then excised from pBSII-KS using NotI and BamHI and ligated into pCEP4 (Invitrogen), yielding pDES46.

As previously described, a conditionally activated mouse synthetic L1 donor, i.e., pJH435 (Fig. 5), was constructed by introducing a loxP-stop-loxP cassette between a strong composite promoter and the ORFeus L1 donor [54]. This construct consisted of a composite CMV immediate early enhancer and modified chicken beta-actin promoter; a floxed beta-geo/stop cassette comprised of a hybrid beta-galactosidase/neomycin phosphotransferase fusion gene and triple tandem copies of the SV40 late polyadenylation signal; L1 ORFeus ORF1 and ORF2 [56]; a GFP-based retrotransposition indicator cassette with its own promoter and polyadenylation signal; and beta-globin polyadenylation signal.

The constitutively active mouse synthetic L1 donor, i.e., pJL5, was prepared by cloning the *TEM1*-AI reporter cassette into the 3′ UTR of L1 ORFeus. Similar to pDES46, pJL5 also was cloned in the episomal plasmid pCEP4 backbone.

Cultured cells

HCT116 (human colorectal carcinoma) cells, kindly provided by Drs. Ina Rhee, Christoph Lengauer and Bert Vogelstein (Johns Hopkins University), were cultured in McCoy's 5A modified medium (Gibco, Life Technologies) supplemented with 10% heat inactivated fetal bovine serum (FBS) and 1% penicillin-streptomycin (P/S), at 37 °C and 5% CO_2 in a humidified chamber. HeLa.JVM (a subclone of human cervical carcinoma) cells, provided by Dr. John V. Moran (University of Michigan), were cultured in Dulbecco's Modified Eagle Medium (DMEM) with the same supplements as HCT116 cells. E14Tg2a.4 mouse embryonic stem (ES) cells, derived from 129P2 ES cells, were provided by Dr.

Allan Bradley (Sanger Institute) and the BayGenomics resource. They were cultured without feeder cells in Glasgow's Modified Eagle's Medium (Sigma Aldrich) supplemented with 10% ES cell-qualified FBS, 1% non-essential amino acids, 1% L-glutamine, 1% sodium pyruvate, 0.1 mM 2-mercaptoethanol (2ME) and ESGRO (Millipore) at 1000U/mL in a 7% CO_2 atmosphere.

Fugene 6 (Roche Applied Science, USA) was used to transfect the cultured cancer cell lines. Cells were transfected with pDES46 and were selected on Hygromycin (Invitrogen) at 0.3 mg/ml for 2 weeks. Resulting HygroR cells were cloned by limiting dilution into 96-well plates and screened for single colonies. Colonies were assayed for TEM1p beta-lactamase expression using the CCF2-AM assay [19]. Certain colonies were picked for further downstream analysis.

Mouse ES cells including Bruce4 cells (provided by Dr. Colin Stewart, NCI; [53]), and the resulting Truck_305 cells containing a mouse synthetic L1 ORFeus donor transgene (purchased from Ozgene [54]), were grown in high-glucose DMEM, supplemented with 15% ES-cell-qualified FBS, 1% L-glutamine, 1% non-essential amino acids, 0.1 mM 2ME, 50 U/ml of P/S and ESGRO in 10% CO_2. For feeder cells, *Neo2*-expressing mouse embryonic fibroblasts (MEFs) were cultured in DMEM supplemented with 10% FBS and 1% P/S, in 5% CO_2, arrested using Mitomycin C or gamma irradiation, and seeded in dishes. The mES cells were added 2–3 days after feeder cell seeding.

To activate L1 retrotransposition in these cells, an adenoviral vector encoding Cre recombinase was introduced to excise the *lacZ* LSL cassette. Briefly, one million Truck_305 cells were incubated for 30 min with the adenoviral Cre vector (Viral Technology Laboratory, NCI Frederick) in a 7% CO2 incubator, at various multiplicities of infection (MOIs) ranging from 10 to 200 per cell, and then plated into a well of a 6-well dish that was pre-seeded with mouse feeder (PMEF) cells. Cells were evaluated for cell death by light microscopy after 18 h incubation. Based on the extent of cell death and colony morphology, MOIs 25 and 50 were found to be optimal. Cell clones exposed to these MOIs were expanded. After at least 12 d in culture, with periodic changes of culture medium, cells were stained with crystal violet and X-gal, to visualize the colonies and assay for presence of the *lacZ* LSL cassette. ES cell clones that did not stain blue with X-gal, which indicated activated smL1 expression and potential retrotransposition, were propagated.

Reporter assays

To quantify beta-lactamase activity and protein expression, cells were stained with CCF2-AM substrate (Life Technologies) [43] by replacing culture medium with 1 mL loading solution (2 µL of a 1 mM CCF2-AM solution, 16 µl of Solution B, 10 µl of 250 mM Probenicid

(Sigma) and 972 µl Hanks' Balanced Salt Solution, HBSS) per 9.6 cm2 well. Cells were incubated in the dark at room temperature for one hr with gentle shaking, washed with HBSS, and visualized using an Axiovert 200 M inverted fluorescence microscope (Zeiss) equipped with blue/aqua and beta-lactamase ratio filter sets (Chroma Technology Corp.) and either an ORCA-ER high resolution digital camera (Hamamatsu Photonics) using Openlab software (version 4.0.2, Improvision), or a Zeiss camera using AxioVision software. Flow cytometric analysis [47] was performed using a BD LSR II flow cytometer with a 405 nm violet laser, 440/40 nm (blue) and 530/30 nm (green) filters, and FACSDiva software (BD Biosciences). Ratios of blue to green intensities were collected as a linear parameter. Each flow cytometry session included positive and negative controls to normalize outputs.

In vivo mouse models

Transgenic founder and pseudofounder mice were generated by pronuclear injection of linearized, marked smL1 donor cassette into fertilized eggs as described previously [55]. The unspliced L1 donor construct as well as spliced new genomic L1 insertions were identified by PCR amplification of the reporter gene, into and/or across the AI, respectively. Pseudofounder mice were defined by the presence of new spliced L1 insertions and the absence of the unspliced donor element. Their progeny were generated by backcrossing the founder and pseudofounder mice with wildtype mice. Certain *de novo* L1 integrants were transmitted in heterozygosity.

Genomic DNA isolation and recovery of TE integrants

To extract genomic DNA, we added 600 µL lysis buffer with 420 µg/mL of proteinase K to cells growing in a 48 well plate. Lysis buffer consisted of 50 mM Tris-Cl, pH 8.0, 100 mM EDTA, 100 mM NaCl and 1% SDS. Next, 200 uL saturated NaCl was added, followed by an equal volume of isopropanol. After centrifugation, DNA pellets were washed with 70% ethanol, dried and resuspended in TE buffer.

Southern blotting was performed to identify lengths and amounts of genomic fragments harboring new TE insertions. Genomic DNA was electrophoresed, blotted, and probed for the *TEM1* reporter using standard methods.

For recovery of TE integrants in cultured cancer cells using inverse PCR (iPCR), genomic DNA isolated from clonal populations of cells was digested with a restriction enzyme (RE) such as XbaI, EcoRI or HindIII. Upon heat inactivation of the RE, digested products were diluted to 1 ng/µL in a total of 500 µL, and incubated with T4 DNA ligase overnight at 16 °C for intra-molecular ligation. After ethanol precipitation, DNA was resuspended and used in iPCR reactions using primers DES682 and DES209 which annealed to the 3′ end of

the retrotransposed cassette. PCR reactions consisted of 40 cycles at 94 °C for 30 s, 55 °C for 30 s and 72 °C for 2 m 20 s. Each of the several bands observed by gel electropheresis after PCR were cloned into pCR2.1 using TOPO cloning kit (Invitrogen) and transformed into bacteria. Colony PCR using M13 forward and reverse primers was used to identify bacterial colonies containing cloned insert. PCR products were cleaned up and sequenced. To map *de novo* insertion sites, Sanger sequence reads were aligned against the reference human genome (hg19) using Blat.

To map new L1 integrants in mES cells using linear amplification-mediated (LAM-) PCR, we chose three ES cell clones (i.e., 1B6, 1C6 and 2B2) in which the spliced (i.e., retrotransposed) reporter gene had been identified by PCR assays. LAM-PCR reactions were set up with 50 ng gDNA from each ES cell clone, 2 nM dNTPs, 5 nM 5′-biotinylated primers DES3171 or DES3174, 1 uL of Advantage 2 enzyme in 1× buffer (Clontech), for 50 cycles (20 s at 95C, 45 s at 60C, and 90 s at 68C). Streptavidin-coated magnetic beads (200 mcg) were washed twice in 100 µL of binding buffer (1 M NaCl, 5 mM Tris, pH 7.5, 0.5 mM EDTA) using a magnetic separation stand, resuspended in 50 µL of 2× binding buffer, and mixed with the linear PCR reaction. The suspension was incubated for 60 min at RT under constant agitation, and then washed three times in 200 µL of wash buffer (10 mM NaCl, 5 mM Tris pH7.5, 0.5 mM EDTA and 0.01% Triton X-100). For second-strand synthesis, the matrix-bound DNA was resuspended in 20 µL of a reaction mixture containing 500 nM dNTPs, 100 ng/µL random hexamers, 5 U Klenow enzyme (New England Biolabs) and 1× NEB Buffer 2, and incubated at 37C for 60 min. After washing first with wash buffer and then twice in 1× reaction buffer, dsDNA was restricted using HaeIII or Sau3AI (NEB) at 37C for 2 hr, washed again in wash buffer followed by twice in 1× ligation buffer, and ligated with either HaeIII adapter (DES3177 and DES3178) or Sau3AI adapter (DES3177 and DES3179) using T4 DNA ligase at 16C overnight. To elute products, beads were resuspended in 5 µL of 0.1 N NaOH and incubated at RT for 10 min. The eluate was separated from the matrix using the magnetic stand, and was neutralized by adding 5 µL of Tris-Cl, pH 7.0.

To perform nested PCR, 1 µL of the eluate or from a 1/100 dilution from the first round of PCR was added as template in a second (nested) PCR reaction in 50 µL. Primers for the adapter (DES3181) and nested adapter (DES3182) were paired with DES3172 and nested primer DES3173 (HaeIII), or DES3175 and nested primer (Sau3AI), respectively, in the donor plasmid. Products were cloned into the Topo-TA pCR2.1 backbone (Invitrogen). To map integration junctions, Sanger sequencing was performed.

Bisulfite sequencing

Bisulfite sequencing was performed using either the EZ DNA Methylation kit (Zymo Research) or the Qiagen Epitect kit, following the manufacturers' instructions, respectively. Alternatively, we prepared fresh sodium hydroxide, sodium bisulfite and hydroquinone solutions to denature and treat DNA samples, which were then purified using Microcon-30 centrifugal spin columns (Amicon, Millipore) [81].

Histone deacetylase (HDAC) inhibitors

The culture medium for cells growing at 50–70% confluence was replaced by medium containing one of HDAC inhibitors including TsA, scriptaid, apicidin, butyrate and nicotinamide (Sigma Chemical Co.) at the specified concentrations. After treatment for a specific time (12–24 h), the cells stained using the CCF2-AM assay and observed using a fluorescence microscope.

Additional file

Additional file 1: Figure S1. Schematic of beta-lactamase reporter assay. **Figure S2.** Schematics of newly integrated sequences retrotransposed by L1. **Figure S3.** Standardized blue and green cell populations as controls for flow cytometry analysis. **Figure S4.** Dense maintenance methylation of pre-existing L1 retrotransposons in cultured human colorectal cancer (HCT116) cells is reduced dramatically in *DNMT1* and *DNMT3b* methyltransferase double knockout cells. **Figure S5.** Lack of cytosine methylation at silenced, de novo L1 reporter insertions in cultured HeLa cells. **Figure S6.** Variable L1 reporter expression in cultured cancer cells is associated with changes in histone acetylation. **Table T1.** Oligonucleotides used in this study. **Table T2.** De novo L1 integrant features. **Table T3.** Expression status of predicted SAGE tags from consensus human L1 template sequence in sense and antisense orientation. (PDF 18544 kb)

Abbreviations

2ME: 2-mercaptoethanol; AI: Artificial intron; ChIP: Chromatin immunoprecipitation; DKO: Double knockout; EC: Embryonic carcinoma; ERV: Endogenous retrovirus; ES: Embryonic stem; FBS: Fetal bovine serum; FRET: Fluorescence resonance energy transfer; GFP: Green fluorescent protein; GFPuv: Green fluorescent protein derivative tuned to UV light excitation; HDAC: Histone deacetylase; hES: Human embryonic stem; iPCR: Inverse polymerase chain reaction; LAM-PCR: Linear amplication-mediated polymerase chain reaction; MEFs: Mouse embryonic fibroblasts; mES: Mouse embryonic stem; NF: Not found; ORF: Open reading frame; P/S: Penicillin/streptomycin; PB: PiggyBac; piRNA: PIWI-interacting small RNA; RE: Restriction endonuclease (enzyme); RIFT: Retrotransposon initiated fusion transcript; RT: Room temperature; TE: Transposable element; TGS: Transcriptional gene silencing; TsA: Trichostatin A; TSD: Target site duplication; UTR: Untranslated region; WT: Wildtype

Acknowledgments

We thank Drs. Colin Stewart (previously at NCI-Frederick), John V. Moran (Univ. Michigan), Jeffrey S. Han (Tulane Univ.), Ina Rhee, Christoph Lengauer and Bert Vogelstein (Johns Hopkins Univ.), Allan Bradley (Sanger Institute) for providing cell lines and reagents; Anthony S. Baker (Ohio State University) for expert help in preparing figures; Jennifer Malloy and Jonathan Cherry (Zeiss) for assistance in setting up filters in fluorescence microscopy; Kathleen Noer (NCI-Frederick) for assistance with flow cytometry experiments; and members of the Symer laboratory for helpful discussions.

Funding

Funding for this study was provided by 1Z01BC010628, Intramural Research Program, National Cancer Institute, National Institutes of Health (D.E.S.); and by the Ohio State University Comprehensive Cancer Center (D.E.S.).

Authors' contributions

MK, JL, SEF, KEH, JCS, TLS, PKT, WA and DES generated reagents and/or experimental data; MK and DES analyzed data and helped to draft and edit the manuscript; MK, JL, WA, JDB and DES participated in various stages of the design of the study; DES conceived of the study; and DES supervised experiments, analyzed data, and wrote and edited the final manuscript. All authors read and approved the final manuscript.

Competing interests

The authors declare that they have no competing interests.

Author details

[1]Department of Biological Sciences, Birla Institute of Technology and Science Pilani, Pilani 333031, Rajasthan, India. [2]Laboratory of Immunobiology, Mouse Cancer Genetics Program and Basic Research Laboratory, Center for Cancer Research, National Cancer Institute, Frederick, MD 21702, USA. [3]Present Address: Birla Institute of Technology and Science, Pilani, Dubai campus, Dubai, United Arab Emirates. [4]Department of Cancer Biology and Genetics, The Ohio State University, Columbus, OH, USA. [5]Department of Internal Medicine, The Ohio State University, Columbus, OH, USA. [6]Biomedical Sciences Graduate Program, The Ohio State University, Columbus, OH, USA. [7]Present Address: National Heart, Lung and Blood Institute, National Institutes of Health, Bethesda, MD, USA. [8]Present Address: Aventiv Research, Inc., Columbus, OH, USA. [9]Present Address: Drug Safety Research and Development, Pfizer, Inc., Groton, CT, USA. [10]Present Address: Biocon, Bangalore, India. [11]Department of Molecular Biology and Genetics, The Johns Hopkins University School of Medicine, Baltimore, MD, USA. [12]Present Address: Department of Pharmaceutical Sciences, South Dakota State University, Brookings, SD, USA. [13]Present Address: Institute for Systems Genetics, New York University Langone Medical Center, New York, NY, USA. [14]Human Cancer Genetics Program, and Department of Biomedical Informatics, The Ohio State University, Columbus, OH, USA. [15]Human Cancer Genetics Program, Department of Cancer Biology and Genetics, and Department of Biomedical Informatics, The Ohio State University, Tzagournis Research Facility, Room 440, 420 West 12th Ave, Columbus, OH 43210, USA.

References

1. Lander ES, Linton LM, Birren B, Nusbaum C, Zody MC, Baldwin J, Devon K, Dewar K, Doyle M, FitzHugh W, et al. Initial sequencing and analysis of the human genome. Nature. 2001;409(6822):860–921.
2. Waterston RH, Lindblad-Toh K, Birney E, Rogers J, Abril JF, Agarwal P, Agarwala R, Ainscough R, Alexandersson M, An P, et al. Initial sequencing and comparative analysis of the mouse genome. Nature. 2002;420(6915): 520–62.
3. Levin HL, Moran JV. Dynamic interactions between transposable elements and their hosts. Nat Rev Genet. 2011;12(9):615–27.
4. Akagi K, Li J, Symer DE. How do mammalian transposons induce genetic variation? A conceptual framework. BioEssays. 2013;35(4):397–407.
5. van den Hurk JA, Meij IC, Seleme MC, Kano H, Nikopoulos K, Hoefsloot LH, Sistermans EA, de Wijs IJ, Mukhopadhyay A, Plomp AS, et al. L1 retrotransposition can occur early in human embryonic development. Hum Mol Genet. 2007;16(13):1587–92.
6. Muotri AR, Chu VT, Marchetto MC, Deng W, Moran JV, Gage FH. Somatic mosaicism in neuronal precursor cells mediated by L1 retrotransposition. Nature. 2005;435(7044):903–10.
7. Coufal NG, Garcia-Perez JL, Peng GE, Yeo GW, Mu Y, Lovci MT, Morell M, O'Shea KS, Moran JV, Gage FH. L1 retrotransposition in human neural progenitor cells. Nature. 2009;460(7259):1127–31.
8. Baillie JK, Barnett MW, Upton KR, Gerhardt DJ, Richmond TA, De Sapio F, Brennan PM, Rizzu P, Smith S, Fell M, et al. Somatic retrotransposition alters the genetic landscape of the human brain. Nature. 2011;479(7374):534–7.
9. Iskow RC, McCabe MT, Mills RE, Torene S, Pittard WS, Neuwald AF, Van Meir EG, Vertino PM, Devine SE. Natural mutagenesis of human genomes by endogenous retrotransposons. Cell. 2010;141(7):1253–61.

10. Evrony GD, Cai X, Lee E, Hills LB, Elhosary PC, Lehmann HS, Parker JJ, Atabay KD, Gilmore EC, Poduri A, et al. Single-neuron sequencing analysis of l1 retrotransposition and somatic mutation in the human brain. Cell. 2012;151(3):483–96.

11. Tubio JM, Li Y, Ju YS, Martincorena I, Cooke SL, Tojo M, Gundem G, Pipinikas CP, Zamora J, Raine K, et al. Mobile DNA in cancer. Extensive transduction of nonrepetitive DNA mediated by L1 retrotransposition in cancer genomes. Science. 2014;345(6196):1251343.

12. Moran JV, DeBerardinis RJ, Kazazian Jr HH. Exon shuffling by L1 retrotransposition. Science. 1999;283(5407):1530–4.

13. Symer DE, Connelly C, Szak ST, Caputo EM, Cost GJ, Parmigiani G, Boeke JD. Human L1 retrotransposition is associated with genetic instability in vivo. Cell. 2002;110(3):327–38.

14. Gilbert N, Lutz S, Morrish TA, Moran JV. Multiple fates of L1 retrotransposition intermediates in cultured human cells. Mol Cell Biol. 2005;25(17):7780–95.

15. Akagi K, Li J, Stephens RM, Volfovsky N, Symer DE. Extensive variation between inbred mouse strains due to endogenous L1 retrotransposition. Genome Res. 2008;18(6):869–80.

16. Speek M. Antisense promoter of human L1 retrotransposon drives transcription of adjacent cellular genes. Mol Cell Biol. 2001;21(6):1973–85.

17. Han JS, Szak ST, Boeke JD. Transcriptional disruption by the L1 retrotransposon and implications for mammalian transcriptomes. Nature. 2004;429(6989):268–74.

18. Belancio VP, Roy-Engel AM, Deininger P. The impact of multiple splice sites in human L1 elements. Gene. 2008;411(1–2):38–45.

19. Li J, Kannan M, Trivett AL, Liao H, Wu X, Akagi K, Symer DE. An antisense promoter in mouse L1 retrotransposon open reading frame-1 initiates expression of diverse fusion transcripts and limits retrotransposition. Nucleic Acids Res. 2014;42(7):4546–62.

20. Druker R, Bruxner TJ, Lehrbach NJ, Whitelaw E. Complex patterns of transcription at the insertion site of a retrotransposon in the mouse. Nucleic Acids Res. 2004;32(19):5800–8.

21. Kaer K, Branovets J, Hallikma A, Nigumann P, Speek M. Intronic L1 retrotransposons and nested genes cause transcriptional interference by inducing intron retention, exonization and cryptic polyadenylation. PLoS One. 2011;6(10):e26099.

22. Li J, Akagi K, Hu Y, Trivett AL, Hlynialuk CJ, Swing DA, Volfovsky N, Morgan TC, Golubeva Y, Stephens RM, et al. Mouse endogenous retroviruses can trigger premature transcriptional termination at a distance. Genome Res. 2012;22(5):870–84.

23. Chen JM, Stenson PD, Cooper DN, Ferec C. A systematic analysis of LINE-1 endonuclease-dependent retrotranspositional events causing human genetic disease. Hum Genet. 2005;117(5):411–27.

24. Callinan PA, Batzer MA. Retrotransposable elements and human disease. Genome Dyn. 2006;1:104–15.

25. Scott EC, Gardner EJ, Masood A, Chuang NT, Vertino PM, Devine SE. A hot L1 retrotransposon evades somatic repression and initiates human colorectal cancer. Genome Res. 2016;26(6):745–55.

26. Yoder JA, Walsh CP, Bestor TH. Cytosine methylation and the ecology of intragenomic parasites. Trends Genet. 1997;13(8):335–40.

27. Jaenisch R, Bird A. Epigenetic regulation of gene expression: how the genome integrates intrinsic and environmental signals. Nat Genet. 2003;33(Suppl):245–54.

28. Alves G, Tatro A, Fanning T. Differential methylation of human LINE-1 retrotransposons in malignant cells. Gene. 1996;176(1–2):39–44.

29. Florl AR, Lower R, Schmitz-Drager BJ, Schulz WA. DNA methylation and expression of LINE-1 and HERV-K provirus sequences in urothelial and renal cell carcinomas. Br J Cancer. 1999;80(9):1312–21.

30. Suter CM, Martin DI, Ward RL. Hypomethylation of L1 retrotransposons in colorectal cancer and adjacent normal tissue. Int J Colorectal Dis. 2004;19(2):95–101.

31. Garcia-Perez JL, Morell M, Scheys JO, Kulpa DA, Morell S, Carter CC, Hammer GD, Collins KL, O'Shea KS, Menendez P, et al. Epigenetic silencing of engineered L1 retrotransposition events in human embryonic carcinoma cells. Nature. 2010;466(7307):769–73.

32. Grandi FC, Rosser JM, Newkirk SJ, Yin J, Jiang X, Xing Z, Whitmore L, Bashir S, Ivics Z, Izsvak Z, et al. Retrotransposition creates sloping shores: a graded influence of hypomethylated CpG islands on flanking CpG sites. Genome Res. 2015;25(8):1135–46.

33. Pannell D, Osborne CS, Yao S, Sukonnik T, Pasceri P, Karaiskakis A, Okano M, Li E, Lipshitz HD, Ellis J. Retrovirus vector silencing is de novo methylase independent and marked by a repressive histone code. EMBO J. 2000;19(21):5884–94.

34. Rival-Gervier S, Lo MY, Khattak S, Pasceri P, Lorincz MC, Ellis J. Kinetics and epigenetics of retroviral silencing in mouse embryonic stem cells defined by deletion of the D4Z4 element. Mol Ther. 2013;21(8):1536–50.

35. Morgan HD, Sutherland HG, Martin DI, Whitelaw E. Epigenetic inheritance at the agouti locus in the mouse. Nat Genet. 1999;23(3):314–8.

36. Ekram MB, Kang K, Kim H, Kim J. Retrotransposons as a major source of epigenetic variations in the mammalian genome. Epigenetics. 2012;7(4):370–82.

37. Oey H, Isbel L, Hickey P, Ebaid B, Whitelaw E. Genetic and epigenetic variation among inbred mouse littermates: identification of inter-individual differentially methylated regions. Epigenetics Chromatin. 2015;8:54.

38. Rakyan VK, Blewitt ME, Druker R, Preis JI, Whitelaw E. Metastable epialleles in mammals. Trends Genet. 2002;18(7):348–51.

39. Cowley M, Oakey RJ. Transposable elements re-wire and fine-tune the transcriptome. PLoS Genet. 2013;9(1):e1003234.

40. Moran JV, Holmes SE, Naas TP, DeBerardinis RJ, Boeke JD, Kazazian Jr HH. High frequency retrotransposition in cultured mammalian cells. Cell. 1996;87(5):917–27.

41. Ostertag EM, Prak ET, DeBerardinis RJ, Moran JV, Kazazian Jr HH. Determination of L1 retrotransposition kinetics in cultured cells. Nucleic Acids Res. 2000;28(6):1418–23.

42. Xie Y, Rosser JM, Thompson TL, Boeke JD, An W. Characterization of L1 retrotransposition with high-throughput dual-luciferase assays. Nucleic Acids Res. 2011;39(3):e16.

43. Zlokarnik G, Negulescu PA, Knapp TE, Mere L, Burres N, Feng L, Whitney M, Roemer K, Tsien RY. Quantitation of transcription and clonal selection of single living cells with beta-lactamase as reporter. Science. 1998;279(5347):84–8.

44. Pornthanakasem W, Mutirangura A. LINE-1 insertion dimorphisms identification by PCR. Biotechniques. 2004;37(5):750–2.

45. Ostertag EM, Kazazian Jr HH. Twin priming: a proposed mechanism for the creation of inversions in L1 retrotransposition. Genome Res. 2001;11(12):2059–65.

46. Lee E, Iskow R, Yang L, Gokcumen O, Haseley P, Luquette LJ, 3rd, Lohr JG, Harris CC, Ding L, Wilson RK, et al. Landscape of somatic retrotransposition in human cancers. Science. 2012;337(6097):967–71.

47. Knapp T, Hare E, Feng L, Zlokarnik G, Negulescu P. Detection of beta-lactamase reporter gene expression by flow cytometry. Cytometry A. 2003;51(2):68–78.

48. Barnes DE, Lindahl T. Repair and genetic consequences of endogenous DNA base damage in mammalian cells. Annu Rev Genet. 2004;38:445–76.

49. Alexandrov LB, Nik-Zainal S, Wedge DC, Aparicio SA, Behjati S, Biankin AV, Bignell GR, Bolli N, Borg A, Borresen-Dale AL, et al. Signatures of mutational processes in human cancer. Nature. 2013;500(7463):415–21.

50. Rhee I, Bachman KE, Park BH, Jair KW, Yen RW, Schuebel KE, Cui H, Feinberg AP, Lengauer C, Kinzler KW, et al. DNMT1 and DNMT3b cooperate to silence genes in human cancer cells. Nature. 2002;416(6880):552–6.

51. Saha S, Sparks AB, Rago C, Akmaev V, Wang CJ, Vogelstein B, Kinzler KW, Velculescu VE. Using the transcriptome to annotate the genome. Nat Biotechnol. 2002;20(5):508–12.

52. Pezic D, Manakov SA, Sachidanandam R, Aravin AA. piRNA pathway targets active LINE1 elements to establish the repressive H3K9me3 mark in germ cells. Genes Dev. 2014;28(13):1410–28.

53. Kontgen F, Suss G, Stewart C, Steinmetz M, Bluethmann H. Targeted disruption of the MHC class II Aa gene in C57BL/6 mice. Int Immunol. 1993;5(8):957–64.

54. An W, Han JS, Schrum CM, Maitra A, Koentgen F, Boeke JD. Conditional activation of a single-copy L1 transgene in mice by Cre. Genesis. 2008;46(7):373–83.

55. An W, Han JS, Wheelan SJ, Davis ES, Coombes CE, Ye P, Triplett C, Boeke JD. Active retrotransposition by a synthetic L1 element in mice. Proc Natl Acad Sci U S A. 2006;103(49):18662–7.

56. Han JS, Boeke JD. A highly active synthetic mammalian retrotransposon. Nature. 2004;429(6989):314–8.

57. Ostertag EM, Kazazian Jr HH. Biology of mammalian L1 retrotransposons. Annu Rev Genet. 2001;35:501–38.

58. Symer DE, Boeke JD. An everlasting war dance between retrotransposons and their metazoan hosts. In: Kurth RBN, editor. Retroviruses: molecular biology, genomics and pathogenesis. Norwich: Caister Academic Press; 2010. p. 1–33.

59. Whitelaw E, Martin DI. Retrotransposons as epigenetic mediators of phenotypic variation in mammals. Nat Genet. 2001;27(4):361–5.

60. Okano M, Xie S, Li E. Cloning and characterization of a family of novel mammalian DNA (cytosine-5) methyltransferases. Nat Genet. 1998;19(3):219–20.

61. Janicki SM, Tsukamoto T, Salghetti SE, Tansey WP, Sachidanandam R, Prasanth KV, Ried T, Shav-Tal Y, Bertrand E, Singer RH, et al. From silencing to gene expression: real-time analysis in single cells. Cell. 2004;116(5):683–98.

62. Henikoff S, McKittrick E, Ahmad K. Epigenetics, histone H3 variants, and the inheritance of chromatin states. Cold Spring Harb Symp Quant Biol. 2004;69:235–43.

63. Chen T, Ueda Y, Dodge JE, Wang Z, Li E. Establishment and maintenance of genomic methylation patterns in mouse embryonic stem cells by Dnmt3a and Dnmt3b. Mol Cell Biol. 2003;23(16):5594–605.

64. Macia A, Munoz-Lopez M, Cortes JL, Hastings RK, Morell S, Lucena-Aguilar G, Marchal JA, Badge RM, Garcia-Perez JL. Epigenetic control of retrotransposon expression in human embryonic stem cells. Mol Cell Biol. 2011;31(2):300–16.

65. Wissing S, Munoz-Lopez M, Macia A, Yang Z, Montano M, Collins W, Garcia-Perez JL, Moran JV, Greene WC. Reprogramming somatic cells into iPS cells activates LINE-1 retroelement mobility. Hum Mol Genet. 2012;21(1):208–18.

66. Quinlan AR, Boland MJ, Leibowitz ML, Shumilina S, Pehrson SM, Baldwin KK, Hall IM. Genome sequencing of mouse induced pluripotent stem cells reveals retroelement stability and infrequent DNA rearrangement during reprogramming. Cell Stem Cell. 2011;9(4):366–73.

67. Orend G, Knoblauch M, Kammer C, Tjia ST, Schmitz B, Linkwitz A, Meyer G, Maas J, Doerfler W. The initiation of de novo methylation of foreign DNA integrated into a mammalian genome is not exclusively targeted by nucleotide sequence. J Virol. 1995;69(2):1226–42.

68. Doerfler W. Epigenetic consequences of foreign DNA insertions: de novo methylation and global alterations of methylation patterns in recipient genomes. Rev Med Virol. 2011;21(6):336–46.

69. Garrick D, Fiering S, Martin DI, Whitelaw E. Repeat-induced gene silencing in mammals. Nat Genet. 1998;18(1):56–9.

70. Ding S, Wu X, Li G, Han M, Zhuang Y, Xu T. Efficient transposition of the piggyBac (PB) transposon in mammalian cells and mice. Cell. 2005;122(3):473–83.

71. Nakanishi H, Higuchi Y, Kawakami S, Yamashita F, Hashida M. piggyBac transposon-mediated long-term gene expression in mice. Mol Ther. 2010;18(4):707–14.

72. Babushok DV, Ostertag EM, Courtney CE, Choi JM, Kazazian Jr HH. L1 integration in a transgenic mouse model. Genome Res. 2006;16(2):240–50.

73. Gasior SL, Preston G, Hedges DJ, Gilbert N, Moran JV, Deininger PL. Characterization of pre-insertion loci of de novo L1 insertions. Gene. 2007;390(1–2):190–8.

74. Meir YJ, Weirauch MT, Yang HS, Chung PC, Yu RK, Wu SC. Genome-wide target profiling of piggyBac and Tol2 in HEK 293: pros and cons for gene discovery and gene therapy. BMC Biotechnol. 2011;11:28.

75. Mossine VV, Waters JK, Hannink M, Mawhinney TP. piggyBac transposon plus insulators overcome epigenetic silencing to provide for stable signaling pathway reporter cell lines. PLoS One. 2013;8(12):e85494.

76. Andrews PW, Matin MM, Bahrami AR, Damjanov I, Gokhale P, Draper JS. Embryonic stem (ES) cells and embryonal carcinoma (EC) cells: opposite sides of the same coin. Biochem Soc Trans. 2005;33(Pt 6):1526–30.

77. Martin SL, Branciforte D. Synchronous expression of LINE-1 RNA and protein in mouse embryonal carcinoma cells. Mol Cell Biol. 1993;13(9):5383–92.

78. Cook PR, Tabor GT. Deciphering fact from artifact when using reporter assays to investigate the roles of host factors on L1 retrotransposition. Mob DNA. 2016;7:23.

79. Molaro A, Falciatori I, Hodges E, Aravin AA, Marran K, Rafii S, McCombie WR, Smith AD, Hannon GJ. Two waves of de novo methylation during mouse germ cell development. Genes Dev. 2014;28(14):1544–9.

80. Wen S, Zhang H, Li Y, Wang N, Zhang W, Yang K, Wu N, Chen X, Deng F, Liao Z, et al. Characterization of constitutive promoters for piggyBac transposon-mediated stable transgene expression in mesenchymal stem cells (MSCs). PLoS One. 2014;9(4):e94397.

81. Cheng RY, Hockman T, Crawford E, Anderson LM, Shiao YH. Epigenetic and gene expression changes related to transgenerational carcinogenesis. Mol Carcinog. 2004;40(1):1–11.

10

Discovery of rare, diagnostic *Alu*Yb8/9 elements in diverse human populations

Julie Feusier[*] (iD), David J. Witherspoon, W. Scott Watkins, Clément Goubert, Thomas A. Sasani and Lynn B. Jorde

Abstract

Background: Polymorphic human *Alu* elements are excellent tools for assessing population structure, and new retrotransposition events can contribute to disease. Next-generation sequencing has greatly increased the potential to discover *Alu* elements in human populations, and various sequencing and bioinformatics methods have been designed to tackle the problem of detecting these highly repetitive elements. However, current techniques for *Alu* discovery may miss rare, polymorphic *Alu* elements. Combining multiple discovery approaches may provide a better profile of the polymorphic *Alu* mobilome. *Alu*Yb8/9 elements have been a focus of our recent studies as they are young subfamilies (~2.3 million years old) that contribute ~30% of recent polymorphic *Alu* retrotransposition events. Here, we update our ME-Scan methods for detecting *Alu* elements and apply these methods to discover new insertions in a large set of individuals with diverse ancestral backgrounds.

Results: We identified 5,288 putative *Alu* insertion events, including several hundred novel *Alu*Yb8/9 elements from 213 individuals from 18 diverse human populations. Hundreds of these loci were specific to continental populations, and 23 non-reference population-specific loci were validated by PCR. We provide high-quality sequence information for 68 rare *Alu*Yb8/9 elements, of which 11 have hallmarks of an active source element. Our subfamily distribution of rare *Alu*Yb8/9 elements is consistent with previous datasets, and may be representative of rare loci. We also find that while ME-Scan and low-coverage, whole-genome sequencing (WGS) detect different *Alu* elements in 41 1000 Genomes individuals, the two methods yield similar population structure results.

Conclusion: Current *in-silico* methods for *Alu* discovery may miss rare, polymorphic *Alu* elements. Therefore, using multiple techniques can provide a more accurate profile of *Alu* elements in individuals and populations. We improved our false-negative rate as an indicator of sample quality for future ME-Scan experiments. In conclusion, we demonstrate that ME-Scan is a good supplement for next-generation sequencing methods and is well-suited for population-level analyses.

Keywords: Retrotransposon, Mobilome, Polymorphism, Population genetics, Human ancestry, Ancestry informative markers

Background

With >1.1 million copies, *Alu* elements are the most abundant and active retrotransposons in the human genome [1–3]. *Alu* elements are members of the SINE family of elements and utilize the LINE-1 endonuclease and reverse-transcriptase for retrotransposition [4]. This process inserts the *Alu* element, including its constitutive poly(A) tail and a target site duplication (TSD) sequence, into the genome [4, 5]. These hallmarks provide evidence of a retrotransposition event rather than a duplication or rearrangement.

While the vast majority of *Alu* elements in the human genome expanded during primate evolution and are no longer active, there are at least 42 retrotranspositionally active subfamilies today [6–10]. Furthermore, an active element with unique mutations has the potential to establish a new subfamily through retrotransposition in the genome [8, 10–12]. Recently retrotransposed *Alu* elements in some of these subfamilies are polymorphic for their presence or absence in the genome and are therefore useful for population and forensic analyses [7, 13–20]. *Alu* elements also contribute to the variation and regulation of the human genome [16, 18, 21–23], thus highlighting the importance of characterizing rare, ancestrally informative loci.

* Correspondence: jfeusier@genetics.utah.edu
Department of Human Genetics, University of Utah School of Medicine, Salt Lake City, UT, USA

Detecting all polymorphic *Alu* elements in humans has been challenging for several reasons. First, the typical output from high-throughput sequencing are 100 bp paired end reads and do not completely cover the length of the 300-bp *Alu* element nor the flanking region necessary for proper mapping [13]. Second, *Alu* elements are commonly found within repetitive regions, which cause alignment errors and inaccurate mapping [18, 24, 25]. Third, the datasets analyzed thus far had insufficient coverage (e.g. 1000 Genomes Project has on average only ~7× per sample) to accurately assemble all *Alu* elements [13, 16, 18, 21]. Finally, different bioinformatics tools report different mobile element sets, and it appears that multiple tools are necessary to detect the whole mobilome [26, 27].

Mobile element scanning (ME-Scan), a method developed for mobile element discovery, attempts to addresses the mapping problem by allowing for high-coverage sequencing of the 5′ flank of the *Alu* breakpoint, the junction between the (unique) genomic sequence and the *Alu* element [28, 29]. ME-Scan can be modified to target specific mobile elements or subfamilies and can be applied in a wide variety of organisms [28–31]. In our study, ME-Scan targets the 7 bp insertion in *Alu*Yb8/9 elements, allowing subfamily-specific amplification and insertion detection using high-throughput sequencing protocols [28, 29, 32, 33]. The *Alu*Yb8/9 subfamilies are particularly interesting as they are young (~2.3 million years old) [34, 35] and active elements. Specifically, ~28% of the polymorphic *Alu* elements in a recent study [6] and ~33% of characterized disease-causing de novo *Alu* elements [36] are members of the Yb8 or Yb9 subfamilies. Here, we present an analysis of 213 individuals (including 43 from the 1000 Genomes (1KG) Project) from 18 diverse populations (Additional file 1: Table S1) using an updated protocol of ME-Scan. This refined examination allows us to characterize new rare *Alu*Yb8/9 elements, to analyze subfamilies, and to discover ancestrally informative markers. We also compare our detection of *Alu* elements to low-coverage, whole-genome sequenced datasets.

Results
Replicate and false positive analysis
We updated ME-Scan with standard Illumina primers to better facilitate library preparation and sequencing (Additional file 2: Supplemental Methods). Eleven independent replicates of an African Pygmy individual, AFP20, were sequenced via ME-Scan to assess run-to-run consistency and library quality. We performed locus-specific PCR to validate 22 non-reference insertions that were present in at least one AFP20 replicate but absent from the rest of the dataset (singletons) (Additional file 1: Tables S2, S3). Eight single-replicate insertions and two insertions with low read counts within SVA_D (SINE/VNTR/ *Alu*) elements did not have an *Alu* element when detected

by PCR. All remaining positions except one, which was located within a segmental duplication on chromosome 17, contained an *Alu* insertion. We also tested nine insertions found in AFP20 and in one other individual (doubletons), and all nine insertions were confirmed by PCR (Additional file 1: Tables S2, S3). We conclude that sequencing replicates may reduce false positives and improve the detection rate for singleton mobile element insertion events, the most difficult class of *Alu* elements to detect.

For assessment of sample quality, we filtered our previous set of presumably fixed *Alu* elements to 1601 elements that were not located within segmental duplications and highly likely to be fixed in the human genome (Additional file 1: Table S4). These loci should be easily detected by ME-Scan (Additional file 1: Table S4). We found that there is a linear inverse relationship between the false negative rate of these presumably fixed elements and the detection rate of the rare insertions in the AFP20 replicates (Additional file 3: Figure S1). Specifically, the replicates with less than a 10% false-negative rate had the highest (> = 75%) detection rate of the rare loci. Therefore, samples showing a false negative rate of more than 10% for these 1601 fixed loci are very likely to be of low quality.

Since most individuals in the study did not have replicates, it was necessary to establish a true positive threshold for all singletons and doubletons in the dataset. We PCR-validated 60 singleton (Additional file 3: Figure S2) and six doubleton loci (Additional file 1: Tables S2, S3). Building from our past studies [28, 29], the number of unique reads, instead of total read count, was the best indicator of a true-positive *Alu* insertion (Additional file 3: Figure S2). Based on these validation studies, a threshold of at least eight unique reads was required to call putative singleton and doubleton insertions. This resulted in a list of 5288 loci that were either previously established elements (Repbase *Alu*Y8/9) [37] or non-reference loci with at least eight unique reads in an individual.

We used principal components analysis (PCA) to examine the consistency between the population structure obtained with our updated protocol versus previously published protocols (Additional file 3: Figure S3) [28]. The Brahmin, YRI, and TSI samples were sequenced using different primers (*Alu*SPv2) than the rest of the samples (*Alu*SPv3) in the previously published dataset [28]. The second largest principal component (Additional file 3: Figure S3) separates the samples processed with different primers; however, there appears to be good consistency among the two datasets given this difference in primers. Therefore, we are confident in the updated protocol and our new criterion of sample quality.

Identification of population-specific *Alu* elements
Ancestry-informative *Alu* elements can complement single nucleotide polymorphisms in detecting admixture or

population structure [13, 15, 16]. In our cohort of 213 individuals, 30.03% of the 5288 loci (See Methods) are specific to one regional group (Fig. 1). We then sought to identify and characterize rare, population-specific *Alu* elements for ancestry studies. From the initial 5288 elements, loci found in our published analysis as well as the datasets we previously examined were removed (see Methods) [15, 28, 38–40]. To minimize false-positives due to mapping error, insertions that were within 50 bp of a reference *Alu* element or a simple repeat were also removed (BEDTools v2.19.1 Quinlan and Hall, 2010 [41]). This resulted in a list of 323 presumably population-specific loci: 117 from Africa, 103 from East Asia, 33 from India, and 70 from Europe (Additional file 1: Table S5, S6). We randomly selected 50 insertions and were able to design PCR primers and had sufficient DNA to test 30 candidate loci (Additional file 1: Tables S2, S3, S6). One primer set failed to amplify the predicted reference band, and the predicted reference element from the reference sample but no *Alu* fragment was amplified for 12 candidate loci. In total, 17/29 loci were true polymorphic *Alu* elements, including a novel *Alu* element on the Y chromosome (chrY:9,992,131 [hg19]) that was found in an East Asian individual. As expected, the validation rate (58%) for these very low frequency loci and singletons was lower than previous validation rates for common loci by ME-Scan [28, 29]. Because of our sample sizes, these insertions may not be truly population-specific but may be present at a higher copy frequency in one regional group than others.

The presence of a set of very rare *Alu* elements may be sufficient to classify an individual into a specific population. To identify diagnostic population-specific *Alu* elements, we genotyped six additional population-specific insertions with varying allele frequencies (0.006–0.087 in

the specific population) (Additional file 1: Table S2, S3, S7) on a panel of 95 individuals (24 African, 24 European, 24 Indian, 23 East Asian) (Additional file 1: Table S8). Four of the loci were not detected in the population panel. An element (chr9:114,889,844 [hg19]) that was validated in two East Asian individuals from ME-Scan was detected in two additional East Asian individuals in the panel and absent in the other populations (Additional file 1: Tables S7, S8). This element may be more common than our analysis suggests (0.0303 with ME-Scan and 0.0357 with ME-Scan and Panel in East Asian individuals) because DNA was unavailable for three of the seven individuals detected for this locus by ME-Scan. Another element (chr9:114,940,676 [hg19]) was detected in nine copies in Africans via ME-Scan and was also present as a heterozygote in two African individuals and absent from the other populations in the panel (Additional file 1: Tables S7, S8). The minor copy frequency (0.0723) of this *Alu* element is statistically significantly different in the African population than the other populations (Wilson binomial 95% CI (0.0409–0.1250) for African, Wilson binomial 95% CI (0.0000–0.0332) for East Asian, the population with the lowest number of haploid genomes at this locus). These *Alu* elements are rare within one population group, may be absent or present at a very low copy number frequency in other populations, and add to a growing number of markers useful for ancestry studies.

Discovery of *Alu* elements in exonic regions

Alu insertions inside exons are rare and often deleterious in humans [15, 32, 36], so we investigated non-reference exonic insertions in our dataset. We annotated candidate insertions by their presence or absence in noncoding and coding exonic regions [28]. We detected 17 loci within noncoding exonic regions and validated 3/3 polymorphic *Alu* elements within UTRs via PCR (Additional file 1: Tables S2, S3, S9). We also detected and designed primers for two candidate coding exonic loci (Additional file 1: Tables S10). Both primers amplified the expected reference band, and an *Alu* element was detected in one locus.

We detected a heterozygous *Alu* insertion in exon 3 of *METTL20* (methyltransferase like 20) in the East Asian individual, 92–40-6 (Fig. 2a). *METTL20* was the first reported mitochondrial lysine methyltransferase characterized in animals and is thought to methylate non-histone proteins [42, 43]. Specifically, the *Alu*Yb8 element inserted near the start of the exon and duplicated the last seven nucleotides of the intron, including the AG splice acceptor site, as well as the first seven nucleotides of the exon as part of the TSD (Fig. 2b). This *Alu* element was also detected in the recent 1KG structural variation dataset [18] and appears to be present at very low frequencies (0.015 minor copy frequency in ME-Scan

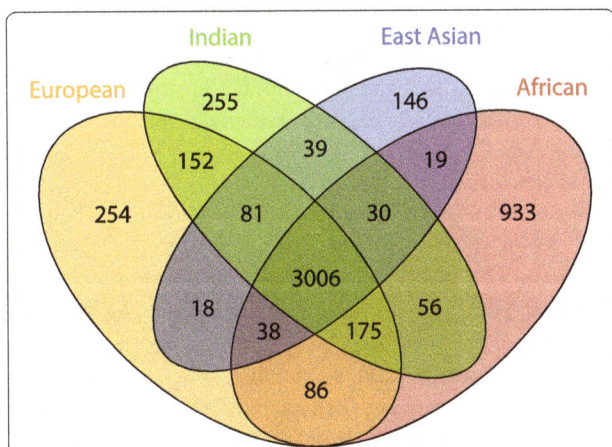

Fig. 1 Venn diagram of *Alu* elements among 4 regional populations. Each individual was placed into one of four regional groups. Every putative locus per individual (5288 total loci) was added into the particular regional group

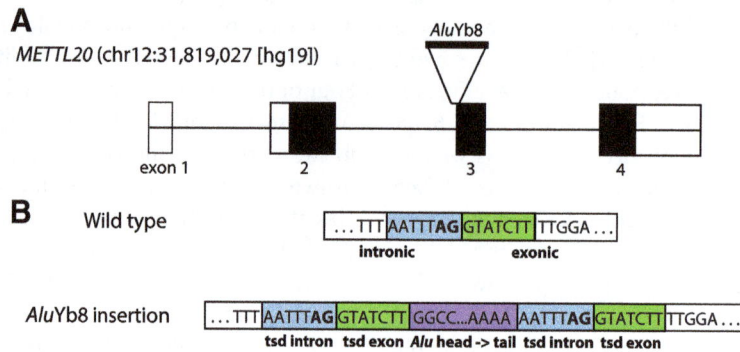

Fig. 2 Diagram of an identified heterozygous *Alu* insertion in *METTL20*. **a**: Diagram of *AluY*b8 insertion in *METTL20*. Open boxes indicate untranslated regions, closed boxes indicate coding regions, and lines indicate intronic regions. **b**: Diagram of the WT and the *AluY*b8 insertion sequences in exon 3 of *METTL20*. The light blue indicates intronic TSD region, green indicates exonic TSD region, and purple is the *Alu* sequence. The insertion of the *Alu* element duplicated the AG splice acceptor site, indicated in bold font

and 0.00097 minor copy frequency in 1KG) in the East Asian population. Further examination of the *METTL20* transcript will be required to determine if the *Alu* element is exonized through alternative slicing of the TSD AG splice acceptor site, thus potentially altering the function of this protein in some populations.

Comparison of Yb8/9 elements detected by ME-Scan and WGS in individuals from the 1000 Genomes Project

Tens of thousands of polymorphic *Alu* elements have been discovered through the HapMap and 1KG consortiums [3, 18, 21, 44]. To assess consistency across platforms, we compared *Alu* elements found by ME-Scan to *Alu* (including non-Yb8/9) elements from the Phase3 1KG dataset in 41 1KG high-quality samples present in both datasets [18]. We performed PCA of these 41 1KG individuals using polymorphic *Alu* elements that were detected in either the ME-Scan or Phase3 datasets [18] (Fig. 3a). A PCA of 191 shared loci for both datasets reveals consistency between the two approaches (Fig. 3b).

Some *Alu* elements may have been missed in the low-coverage (~7×) WGS datasets [3, 18]. We examined the number of loci shared between the datasets to assess the concordance of the methods. The Phase3 dataset contains only polymorphic loci, so loci present in Repbase or reference build hg19 were removed from both datasets to attempt to address this bias. Each method found hundreds of unique loci in the 411KG individuals, as shown in Fig. 3c. Over 99.6% of the shared elements were either classified as *Alu*Yb elements or were not classified in the Phase3 dataset (Additional file 3: Figure S4). It is not surprising that there are thousands of unique loci in the Phase3 dataset compared to ME-Scan, given that the 1KG analysis did not target specific *Alu* subfamilies.

Next, we sought to determine whether ME-Scan detects novel *Alu* elements not detected by WGS in 1KG individuals. After comparison of multiple published

datasets (see Methods) and filtering out false positive loci, 313 presumably novel *Alu* element insertions were identified in ME-Scan that had at least eight unique reads in at least one individual (Additional file 1: Tables S11, S12). Of these 313 presumably novel loci, 174 were detected in the 43 1KG individuals that were sequenced by ME-Scan (NA07346 and NA20515 were not in the comparison analyses) (Additional file 1: Table S12). Furthermore, a novel, validated population-specific *Alu* element (chr8:116,728,191 [hg19]) was found in TSI individual NA20518.

Characterization of PCR-validated *Alu*Yb8/9 elements and identification of potential source elements

We performed Sanger sequencing and alignments of 68 validated rare *Alu*Yb8 (N = 58) and *Alu*Yb9 (N = 10) elements from the loci validated by PCR (Additional file 1: Tables S2, S13). Five *Alu* elements had a 5′ truncation of up to 20 bp, but the truncations did not impact subfamily identification (Additional file 1: Table S13). All elements had been correctly mapped to within 1 bp, after adjustment of 5′ modifications, of the predicted junction location (Additional file 1: Table S13). Fourteen and four of our loci were exact matches to the Yb8 and Yb9 consensus sequences, respectively (Additional file 1: Table S13). Nine of the 15 Yb8b1 elements were an exact match to Yb8b1 (a subfamily of Yb8) [6], and all three Yb11 elements (a subfamily of Yb9) were an exact match to Yb11 [35]. Because we targeted *Alu* elements with the 7 bp insertion that is diagnostic of many *Alu*Yb subfamilies, it was not surprising that eight of the elements belonged to other Yb subfamilies. The elements diverged from their respective consensus subfamily by an average of 0.431% (+/− 0.635 s.d.), and 45.5% of the elements were full-length and an exact match to the consensus sequence based on BLAST+ analysis (Additional File 1: Table S12, see distribution in Additional

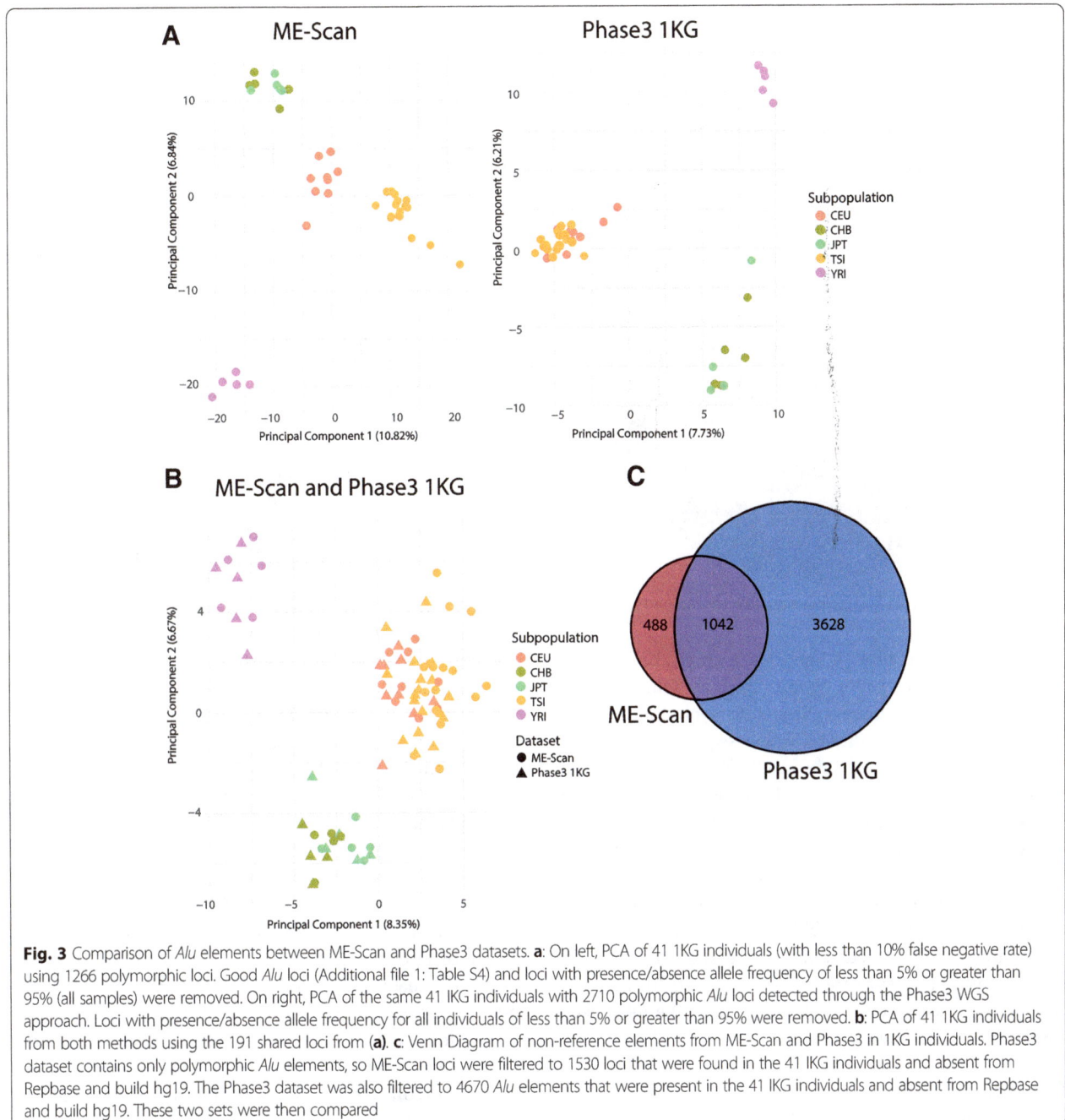

Fig. 3 Comparison of *Alu* elements between ME-Scan and Phase3 datasets. **a**: On left, PCA of 41 1KG individuals (with less than 10% false negative rate) using 1266 polymorphic loci. Good *Alu* loci (Additional file 1: Table S4) and loci with presence/absence allele frequency of less than 5% or greater than 95% (all samples) were removed. On right, PCA of the same 41 1KG individuals with 2710 polymorphic *Alu* loci detected through the Phase3 WGS approach. Loci with presence/absence allele frequency for all individuals of less than 5% or greater than 95% were removed. **b**: PCA of 41 1KG individuals from both methods using the 191 shared loci from (**a**). **c**: Venn Diagram of non-reference elements from ME-Scan and Phase3 in 1KG individuals. Phase3 dataset contains only polymorphic *Alu* elements, so ME-Scan loci were filtered to 1530 loci that were found in the 41 1KG individuals and absent from Repbase and build hg19. The Phase3 dataset was also filtered to 4670 *Alu* elements that were present in the 41 1KG individuals and absent from Repbase and build hg19. These two sets were then compared

file 3: Figure S5) [45]. The *Alu* and flanking sequence for each locus is presented in Additional file 4.

We examined the distribution of our Sanger-sequenced elements among the *Alu*Yb subfamilies (Fig. 4) and compared this to a previous *Alu*Yb subfamily distribution analysis [6]. Notably, we detected more elements that belong to the recently characterized *Alu*Yb8 subfamily, *Alu*Yb8b1, than in [6]. However, the proportion of Yb8b1 elements between the datasets was not significantly different (Fisher exact test, $P > 0.186$). Furthermore, the difference in the proportion of *Alu*Yb8 elements (the only other subfamily

that possibly differed) between the datasets was also not statistically significant (Fisher exact test, $P > 0.318$). Therefore, we conclude that this distribution is similar to the Yb8/9 subfamily distribution in [6].

Active *Alu* elements have the potential to retrotranspose in the genome; these "source" elements have at least four characteristic hallmarks: intact box A and B internal RNA Polymerase III (pol III) promoters [10, 11, 46], intact SRP9/14 sites [11], a poly(A) tail at least 20 bases long (preferably uninterrupted As) [12], and a pol III termination sequence, TTTT, within 15 bp of the TSD downstream of the poly(A)

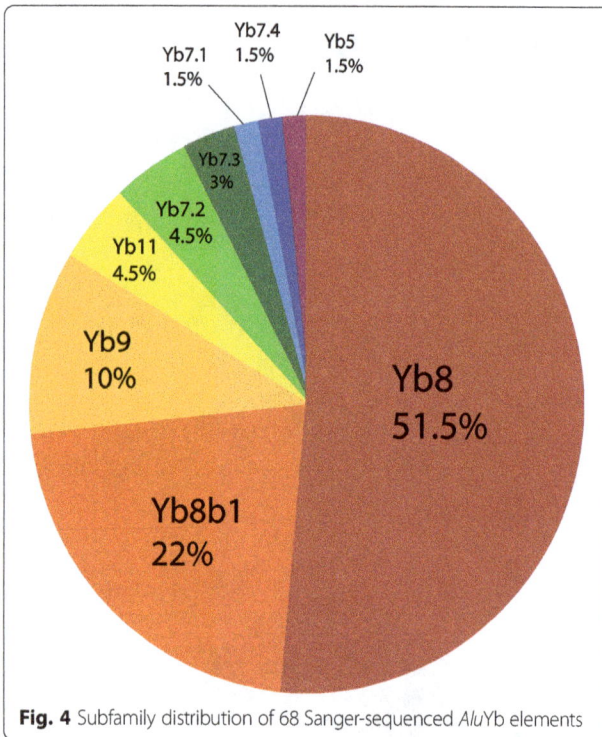

Fig. 4 Subfamily distribution of 68 Sanger-sequenced *Alu*Yb elements

tail [46]. Sixty-two of the 68 Sanger-sequenced *Alu* elements had enough sequence information via Sanger sequencing for analysis of these hallmarks. Sequences typically terminate within the poly(A) tail during Sanger sequencing; however, we estimated an approximate length for each poly(A) tail (Additional file 1: Table S13, Additional file 4). Only one element had an A-tail shorter than 20 bp, and another element had evidence of substitutions within the poly(A) tail. Overall, we detected 11 elements across five *Alu*Yb subfamilies that contained the hallmarks of potential source elements (Additional file 1: Table S13). Potential source elements are difficult to accurately detect, and other factors like the 5′ flanking sequence are important for pol III transcription [47, 48]; however, these 11 elements are the most likely candidates in this dataset.

Discussion

In this analysis, we present and utilize an updated version of a recently developed mobile element scanning technique, ME-Scan [28, 29], to examine *Alu*Yb8/9 elements in worldwide human populations. Our updated method is consistent with the previous ME-Scan protocol (Additional file 3: Figure S3) [28] and standardizes the entire ME-scan protocol for use on the Illumina HiSeq2000 without instrument adjustments (Additional file 2). We also present sequence information for 68 rare *Alu*Yb8/9 elements (Additional file 4), including 23 presumably population-specific loci and 11 elements with hallmarks of active source elements. Furthermore, ME-Scan is able to detect hundreds of *Alu* insertions previously not found by non-targeted high-throughput sequencing

methods, thus demonstrating a clear utility for multiple approaches to fully characterize the mobilome.

Discovery of rare, polymorphic *Alu* elements can be useful for distinguishing human ancestral identity. One key limiting factor, particularly with population-specific loci, is the number of new individuals being studied. With 213 individuals in this dataset, 74% and 13% of the population-specific loci were singletons and doubletons, respectively. This indicates that there may be hundreds, and potentially thousands, of unidentified *Alu* elements present at low minor allele frequencies in the human population, and potentially private mutations through de novo retrotransposition (the current expected de novo mutation rate is ~1:20 births [49]). Additionally, all of the Sanger-sequenced *Alu*Yb8/9 elements were present as heterozygotes in ME-Scan individuals, with the exception of *Alu*Yb8 at position chr9:114,889,844 [hg19], which was found as a homozygote in one individual and a heterozygote in seven individuals (Additional file 1: Table S9, S10). The preliminary findings from testing six rare, population-specific loci on a PCR panel of 95 individuals revealed that these loci may be diagnostic of specific populations, as they were present at a low allele frequency in one population (0.0049–0.0724) and absent from the rest. This finding also highlights the sensitivity of our method for detecting rare *Alu* insertions. Further examination of large cohorts will reveal additional diagnostic loci, as the majority of high-frequency *Alu* elements in the human genome have already been identified.

Alu discovery is challenging due to mapping/alignment errors and low sequencing depth of repetitive regions [26, 27]. The majority of the tested loci identified in these analyses were located in repetitive regions, and the false-positive loci may be due to alignment or PCR artifacts. Additionally, six of the 12 false-positive population-specific loci were singletons in individuals with >10% false negative rates. This helps to validate our criterion that >10% false negative rate indicates a poor-quality sample. Another two of the 12 false-positive loci were also detected in different individuals in the Phase3 1KG dataset [18]. This could be due to sample identity error, technical, mapping, or PCR artifacts, but it underscores the fact that PCR is still an important validation component of next-generation sequencing approaches.

Alu subfamily classification is an active field, and at least six new subfamilies have been classified in recent years [6, 35]. One goal of this project was to characterize the subfamily distribution of rare *Alu* elements and potentially identify very young subfamilies. Notably, 15 of our 68 elements belong to the *Alu*Yb8b1 subfamily, adding support to the classification of this new subfamily [6]. Another interesting discovery was that our subfamily distribution of polymorphic Yb8/9 elements recapitulated the distribution from a previous study [6]. Thus, we conclude that ME-Scan does not appear to be biased within subfamilies with the 7 bp insertion.

In search of new subfamilies, we identified preliminary evidence of a novel subfamily in the AluYb8 lineage. We identified four loci that differed from the Yb8 reference sequence with C to T substitutions at CpG sites on positions 207, 213, and 258. A BLAT search of this sequence in both [hg19] and human genome build [38] revealed one exact match, and alignments to a previously published dataset [6] revealed two additional loci (Alu 161 and Alu 356 from [6]). The sequence's location in the [hg19] reference genome (chr9:98,266,017 [hg19]) has the hallmark of an active source element, indicating that this subfamily may be currently active. However, CpG sites mutate at a rate six to ten times faster than non-CpG sites, and these mutations may have occurred after retrotransposition [9, 50, 51]. Classification of active subfamilies through de novo or somatic retrotransposition events rather than from sequence information would help to answer this question, as this would eliminate mutations that occur after retrotransposition. Further evidence will be needed to determine whether these three CpG mutations are diagnostic of a novel subfamily, AluYb8c3, or a collection of independent, random events.

For population genomics analyses, we demonstrate that PCA results based on ME-Scan compare almost perfectly to those of WGS approaches (Fig. 3a, b) [16, 18]. Platform differences did not seem to be involved in the first two principal components of the PCA of the 1KG in the ME-Scan and Phase3 datasets (Fig. 3b). However, TSI does not cluster with CEU in the AluYb8/9 loci from ME-Scan, whereas TSI and CEU cluster together using loci from different Alu subfamilies in the Phase3 dataset (Fig. 3a). This is likely due to a library bias in ME-Scan, as the TSI were sequenced in a different library than the rest of the 1KG individuals. We also found that there are hundreds of unique loci in the 41 1KG individuals in either dataset (Fig. 3c). These results demonstrate that complementary methods, such as WGS and ME-Scan, provide a more complete genomic assessment of the Alu mobilome than either method alone.

Conclusions

Here we demonstrate that ME-Scan detection is consistent with WGS approaches and is an independent complementary method for AluYb8/9 discovery. The updated protocol and threshold criteria allow for future studies to be performed with relative ease. Even as the cost of WGS continues to decrease, we conclude that ME-Scan provides alternate options in the field of transposable element population genomics and is scalable from pilot experiments to much larger projects involving the analysis of polymorphisms in hundreds of individuals.

Methods
DNA samples and ME-Scan protocol
The ME-Scan protocol was standardized to the Illumina HiSeq 2000 platform. A detailed report of the protocol including primers is provided in the Additional file 4: Supplemental Methods. Data were mapped to hg19 using bwa align (bwa version 0.7.9a) [52] and uploaded to SQL developer for analysis. Read set processing was the same as described in [28].

Two hundred thirty-three samples (213 unique individuals) were sequenced using the ME-Scan protocol and Illumina sequencing. These individuals were sampled from 21 groups, including 18 geographical ancestry groups: 6 Nande, 5 YRI (Yoruba in Ibadan, Nigeria), 16 Hema, and 24 Pygmy from sub-Saharan Africa; 22 Brahmin, 2 Irula, 2 Kapu, 2 Khonda Dora, 20 Madiga, 26 Mala, 2 Relli and 2 Yadava from south India; 18 TSI (Toscani in Italy), 10 CEU (CEPH samples from Utah) and 23 Europeans from west Europe; 5 CHB (Han Chinese in Beijing, China), 5 JPT (Japanese in Tokyo, Japan), 8 Japanese, 4 Vietnamese, an individual from Taiwan, and 10 individuals of mixed Asian ancestry from east Asia. DNA from 43 individuals (TSI, CEU, CHB, JPT, and YRI) was obtained from transformed lymphoblast cells lines from the HapMap Project [53]. DNA was obtained from whole blood for the remaining individuals (including the PCR panel), who have been described previously [54–56]. DNA for the TSI 1000 Genomes individuals and non-1000 Genomes individuals were available for PCR validation. Most of the indexed individuals were combined into 9 pooled libraries of ~25 individuals per library, designated AFHAFN, ASIAN, BRA, CAUC, HapMap, MADIGA, MALA, PYG, and TSI. Twenty-two indexed samples were combined into 5 pooled libraries that contained samples that were not part of this study, with ~53 individuals per library. These libraries were arbitrarily named Library 10–14 for this study.

PCR validation and oligonucleotide primer design
Each locus was viewed on the hg19 build on the UCSC genome browser [57]. The DNA sequence was obtained with 500 bp of flanking sequence upstream and downstream of the potential breakpoint and was entered into primer-Blast and verified by in silico PCR from the UCSC genome browser [58]. In cases where primer-Blast was unable to create a primer set, the sequence was entered into Primer3 and the primer set was verified in primer-Blast [58, 59].

PCR amplifications of ~25 ng of template DNA were performed in 25 μl reactions according to the Phusion HotStart Flex DNA Polymerase protocol (using 5× GC buffer), with the exception that the quantity of 10 μM primers was reduced to 1 μl each. The thermocycler conditions were: initial denaturation at 98C for 20s, 34 cycles of denaturation at 98C for 20s, optimal annealing temperature (58–62) at 30s, extension at 72C for 30s, and a final extension at 72C for 7 min. Every primer set reaction was performed on the individual(s) with the candidate *Alu* element, a positive control (an individual not expected to contain the *Alu* element) and H$_2$O. PCR amplicons of 24 µl were run on a 2% gel containing 0.12 mg/ml ethidium bromide for 60 min at 160 V. Gels were imaged using a Fotodyne Analyst Investigator Eclipse machine.

Sanger sequencing

PCR amplicons of 20 µl per loci were run on a 2% agarose gel. The band that was shifted ~300 bp above the wildtype band was cut out and purified for sequencing using the Qiaquick gel extraction kit (Qiagen). When the candidate *Alu* element was present in multiple individuals, the DNA was pooled prior to purification. A total of 9.5 µl of purified DNA and 0.5 µl of 10 µM primer were used for Sanger sequencing. Each *Alu* element was also sequenced using an internal *Alu*Yb primer (ACGGAGTCTCGCTCTGTCG) that starts near the poly(A) tail and continues through the head of the *Alu* into the flanking region for double coverage of most of the *Alu* element. All Sanger sequencing reads were analyzed using Sequencher [60].

Matching ME-Scan reads to reference genomes and published datasets

We matched the ME-Scan sequenced loci to the RepeatMasker-annotated hg19 reference genome [38], as in [28]. The positions were not corrected for possible 5′ truncations. Therefore, we added a 30 bp buffer upstream of the breakpoint was on the correct strand. We also compared the loci to dbRIP and two datasets for PCR validation [15, 21, 40]. The exonic regions were annotated as in [28]. We did not remove previously published ME-Scan-identified loci, as those had not been validated, with the exception of discovering population-specific loci. A putative list was made of loci that matched "*Alu*Yb8" or "*Alu*Yb9" by RepeatMasker or had at least eight unique reads in an individual (Additional file 1: Table S5).

After PCR validation, we further compared our results with recently published datasets to identify unpublished *Alu*Yb8/9 elements (Additional file 1: Tables S11, S12) [6, 13, 18, 28, 35, 61]. We extended the reference range to within 30 bp on either side of the ME-Scan breakpoint position for comparison with non-Repbase datasets. Additionally, we used the liftOver tool [57] in the UCSC genome browser to compare the build [hg38] *Alu* elements with these loci. Matches for all datasets are reported in Additional file 1: Table S10 and the novel loci are reported in Additional file 1: Table S11.

1KG *Alu* elements were downloaded from the Phase3 data release of the 1000 Genomes Project [3, 18, 62] (ftp://ftp.1000genomes.ebi.ac.uk/vol1/withdrawn/phase3/integrated_sv_map/ALL.wgs.integrated_sv_map_v2.20130502.svs.genotypes.vcf.gz).

Additional files

Additional file 1: Tables S1-S12. This file contains supplementary Tables S1-S13 as well as a table of contents with table names (XLSX 1249 kb)

Additional file 2: Supplementary Methods. This file contains supplementary methods that contain the improved ME-Scan protocol (DOCX 29 kb)

Additional file 3: This file contains Figures S1-S5 and figure legends. (PDF 4383 kb)

Additional file 4: FASTA sequences of 68 *Alu* elements. This file contains high-quality sequence from Sanger sequencing of 68 *Alu* elements. The nucleotides are color-coded for *Alu*, TSD, A and B boxes, SRP9/14 sites, and pol III termination signals (DOCX 34 kb)

Abbreviations

1KG: 1000 Genomes project; pol III: RNA Polymerase III; SINE: Short interspersed nuclear element; TSD: Target site duplication

Acknowledgments

We thank Brian Dalley of the Microarray and Genomic Analysis Core Facility at the Huntsman Cancer Institute for advice and technical assistance. We thank Cedric Feschotte and Jinchuan Xing for advice during the early stages of manuscript preparation. For helpful comments, we gratefully acknowledge Justin Tackney and Jonathan Downie.

Funding

LBJ was supported by NIH grants GM59290, GM104390, and GM118335.

Authors' contributions

DW and WSW conceived of the study and designed the pre-sequencing analysis, including the improvement of ME-Scan. JF designed and carried out the post-sequencing analysis with advice from all authors, and particularly WSW and LBJ. DW was a designer and contributor in the exonic and singleton analyses. CG matched the loci to the reference datasets and performed the analysis in Fig. 3c. TAS was a major contributor in the population-specific PCR analysis. JF wrote the manuscript and all authors revised the manuscript. All authors read and approved the final manuscript.

Competing interests

The authors declare that they have no competing interests.

References

1. Hasler J, Strub K. Alu elements as regulators of gene expression. Nucleic Acids Res. 2006;34:5491–7.
2. The International Human Genome Mapping Consortium. A physical map of the human genome. Nature 2001;409:934–41.
3. The 1000 Genomes Project Consortium. A global reference for human genetic variation. Nature. 2015;526:68–74.
4. Dewannieux M, Esnault C, Heidmann T. LINE-mediated retrotransposition of marked Alu sequences. Nat Genet. 2003;35:41–8.
5. Christensen SM, Eickbush TH. R2 target-primed reverse transcription: ordered cleavage and polymerization steps by protein subunits asymmetrically bound to the target DNA. Mol Cell Biol. 2005;25:6617–28.
6. Konkel MK, Walker JA, Hotard AB, Ranck MC, Fontenot CC, Storer J, et al. Sequence Analysis and Characterization of Active Human Alu Subfamilies Based on the 1000 Genomes Pilot Project. Genome Biol Evol. 2015;7:2608–22.
7. Batzer MA, Deininger PL. Alu Repeats and Human Genomic Diversity. Nat Rev Genet. 2002;3:370–9.
8. Han K, Xing J, Wang H, Hedges DJ, Garber RK, Cordaux R, et al. Under the genomic radar: The Stealth model of Alu amplification. Genome Res. 2005;15:655–64.
9. Xing J, Hedges DJ, Han K, Wang H, Cordaux R, Batzer M A. Alu element mutation spectra: molecular clocks and the effect of DNA methylation. J Mol Biol. 2004;344:675–82.

10. Mills RE, Bennett EA, Iskow RC, Devine SE. Which transposable elements are active in the human genome? Trends Genet. 2007;23:183–91.

11. Bennett EA, Keller H, Mills RE, Schmidt S, Moran JV, Weichenrieder O, et al. Active Alu retrotransposons in the human genome. Genome Res. 2008;18:1875–83.

12. Dewannieux M, Heidmann T. Role of poly(A) tail length in Alu retrotransposition. Genomics. 2005;86:378–81.

13. Wildschutte JH, Baron AA, Diroff NM, Kidd JM. Discovery and characterization of Alu repeat sequences via precise local read assembly. Nucleic Acids Res. 2015;43:10292–307.

14. Tajnik M, Vigilante A, Braun S, Hänel H, Luscombe NM, Ule J, et al. Intergenic *Alu* exonisation facilitates the evolution of tissue-specific transcript ends. Nucleic Acids Res. 2015;43:gkv956.

15. Hormozdiari F, Alkan C, Ventura M, Hajirasouliha I, Malig M, Hach F, et al. Alu repeat discovery and characterization within human genomes. Genome Res. 2011:840–9.

16. Rishishwar L, Tellez Villa CE, Jordan IK. Transposable element polymorphisms recapitulate human evolution. Mobile DNA. 2015;6:21.

17. Salem AH, Kilroy GE, Watkins WS, Jorde LB, Batzer MA. Recently integrated Alu elements and human genomic diversity. Mol Biol Evol. 2003;20:1349–61.

18. Sudmant PH, Rausch T, Gardner EJ, Handsaker RE, Abyzov A, Huddleston J, et al. An integrated map of structural variation in 2,504 human genomes. Nature. 2015;526:75–81.

19. Gu Z, Jin K, Crabbe MJC, Zhang Y, Liu X, Huang Y, et al. Enrichment analysis of Alu elements with different spatial chromatin proximity in the human genome. Protein Cell. Higher Education Press. 2016;7:250–66.

20. Wang L, Rishishwar L, Mariño-Ramírez L, Jordan IK. Human population-specific gene expression and transcriptional network modification with polymorphic transposable elements. Nucleic Acids Res. 2017;45:2318–28.

21. Stewart C, Kural D, Strömberg MP, Walker JA, Konkel MK, Stütz AM, et al. A comprehensive map of mobile element insertion polymorphisms in humans. PLoS Genet. 2011;7:e1002236.

22. Witherspoon DJ, Watkins WS, Zhang Y, Xing J, Tolpinrud WL, Hedges DJ, et al. Alu repeats increase local recombination rates. BMC Genomics. 2009;10:530.

23. Xing J, Zhang Y, Han K, Xing J, Zhang Y, Han K, et al. Mobile elements create structural variation: Analysis of a complete human genome. Genome Res. 2009:1516–26.

24. Thung DT, de Ligt J, Vissers LEM, Steehouwer M, Kroon M, de Vries P, et al. Mobster: accurate detection of mobile element insertions in next generation sequencing data. Genome Biol. 2014;15:488.

25. Wu J, Lee W-P, Ward A, Walker JA, Konkel MK, Batzer MA, et al. Tangram: a comprehensive toolbox for mobile element insertion detection. BMC Genomics. 2014;15:795.

26. Ewing AD. Transposable element detection from whole genome sequence data. Mobile DNA. 2015;6:24.

27. Rishishwar L, Mariño-Ramírez L, Jordan IK. Benchmarking computational tools for polymorphic transposable element detection. Brief Bioinform. 2016;bbw072:1–11.

28. Witherspoon DJ, Zhang YH, Xing JC, Watkins WS, Ha H, Batzer MA, et al. Mobile element scanning (ME-Scan) identifies thousands of novel Alu insertions in diverse human populations. Genome Res. 2013;23:1170–81.

29. Witherspoon DJ, Xing J, Zhang Y, Watkins WS, Batzer MA, Jorde LB. Mobile element scanning (ME-Scan) by targeted high-throughput sequencing. BMC Genomics. 2010;11:410.

30. Platt RN, Zhang Y, Witherspoon DJ, Xing J, Suh A, Keith MS, et al. Targeted Capture of Phylogenetically Informative Ves SINE Insertions in Genus Myotis. Genome Biol Evol. 2015;7:1664–75.

31. Ha H, Loh JW, Xing J. Identification of polymorphic SVA retrotransposons using a mobile element scanning method for SVA (ME-Scan-SVA). Mobile DNA. 2016;7:15.

32. Xing J, Witherspoon DJ, Jorde LB. Mobile element biology: new possibilities with high-throughput sequencing. Trends Genet Elsevier Ltd. 2013;29:280–9.

33. Ha H, Wang N, Xing J. Library Construction for High-Throughput Mobile Element Identification and Genotyping. Methods Mol Biol. Totowa, NJ: Humana Press; 2015. p. 1–15.

34. Carter AB, Salem A, Hedges DJ, Keegan CN, Kimball B, Walker JA, et al. Genome-wide analysis of the human Yb-lineage. Hum Genomics. 2004;1:167–78.

35. Ahmed M, Li W, Liang P. Identification of three new Alu Yb subfamilies by source tracking of recently integrated Alu Yb elements. Mobile DNA. 2013;4:25.

36. Hancks DC, Kazazian HH. Roles for retrotransposon insertions in human disease. Mobile DNA. 2016;7:9.

37. Jurka J, Kapitonov V, Pavlicek A, Klonowski P, Kohany O, Walichiewicz J. Repbase Update, a database of eukaryotic repetitive elements. Cytogenet Genome Res. 2005;110:462–7.

38. Smit A, Hubley R, Green P. RepeatMasker Open-3.0. 2010.

39. Stewart C, Kural D, Strömberg MP, Walker JA, Konkel MK, Stütz AM, et al. A comprehensive map of mobile element insertion polymorphisms in humans. PLoS Genet. 2011;7

40. Wang J, Song L, Grover D, Azrak S, Batzer MA, Liang P. dbRIP: A Highly Integrated Database of Retrotransposon Insertion Polymorphisms in Humans. Hum Mutat. 2006;27:323–9.

41. Quinlan AR, Hall IM. BEDTools: a flexible suite of utilities for comparing genomic features. Bioinformatics. 2010;26:841–2.

42. Cloutier P, Lavallée-Adam M, Faubert D, Blanchette M, Coulombe B. A Newly Uncovered Group of Distantly Related Lysine Methyltransferases Preferentially Interact with Molecular Chaperones to Regulate Their Activity. PLoS Genet. 2013;9:e1003210.

43. Małecki J, Ho AYY, Moen A, Dahl H-A, Falnes PØ. Human METTL20 is a mitochondrial lysine methyltransferase that targets the β subunit of electron transfer flavoprotein (ETFβ) and modulates its activity. J Biol Chem. 2015;290:423–34.

44. The 1000 Genomes Project Consortium. A map of human genome variation from population-scale sequencing. Nature. 2010;467:1061–73.

45. Camacho C, Coulouris G, Avagyan V, Ma N, Papadopoulos J, Bealer K, et al. BLAST+: architecture and applications. BMC Bioinformatics. 2009;10:421.

46. Comeaux MS, Roy-Engel AM, Hedges DJ, Deininger PL. Diverse cis factors controlling Alu retrotransposition: what causes Alu elements to die? Genome Res. 2009;19:545–55.

47. Parekh RB, Dwek R, Sutton B. Upstream sequences modulate the internal promoter of the human 7SL RNA gene. Nature. 1985;318:452–7.

48. Chesnokov I, Schmid CW. Flanking sequences of an Alu source stimulate transcription in vitro by interacting with sequence-specific transcription factors. J Mol Evol. 1996;42:30–6.

49. Cordaux R, Hedges DJ, Herke SW, Batzer MA. Estimating the retrotransposition rate of human Alu elements. Gene. 2006;373:134–7.

50. Batzer MA, Kilroy GE, Richard PE, Shaikh TH, Desselle TD, Hoppens CL, et al. Structure and variability of recently inserted Alu family members.[erratum appears in Nucleic Acids Res 1991 Feb 11;19(3):698–9]. Nucleic Acids Res. 1990;18:6793–8.

51. Labuda D, Striker G. Sequence conservation in Alu evolution. Nucleic Acids Res. 1989;17:2477–91.

52. Li H, Durbin R. Fast and accurate short read alignment with Burrows-Wheeler transform. Bioinformatics. 2009;25:1754–60.

53. The International HapMap 3 Consortium, Principal investigators, Altshuler DM, Gibbs RA, Peltonen L, Project coordination leaders, et al. Integrating common and rare genetic variation in diverse human populations. Nature. 2010;467:52–8.

54. Bamshad MJ, Watkins WS, Dixon ME, Jorde LB, Rao BB, Naidu JM, et al. Female gene flow stratifies Hindu castes. Nature. 1998;395:651.

55. Jorde LB, Bamshad MJ, Watkins WS, Zenger R, Fraley AE, Krakowiak PA, et al. Origins and affinities of modern humans: a comparison of mitochondrial and nuclear genetic data. Am J Hum Genet. 1995;57:523–38.

56. Watkins WS, Bamshad M, Dixon ME, Bhaskara Rao B, Naidu JM, Reddy PG, et al. Multiple Origins of the mtDNA 9-bp Deletion in Populations of South India. Am J Phys Anthropol. 1999;109(147–15):147–58.

57. Kent WJ, Sugnet CW, Furey TS, Roskin KM, Pringle TH, Zahler AM, et al. The Human Genome Browser at UCSC The Human Genome Browser at UCSC. Genome Res. 2002;12:996–1006.

58. Ye J, Coulouris G, Zaretskaya I, Cutcutache I, Rozen S, Madden TL. Primer-BLAST: A tool to design target-specific primers for polymerase chain reaction. BMC Bioinformatics. 2012;13:134.

59. Untergasser A, Cutcutache I, Koressaar T, Ye J, Faircloth BC, Remm M, et al. Primer3-new capabilities and interfaces. Nucleic Acids Res. 2012;40:1–12.

60. Corporation GC. Sequencher. Ann Arbor: Gene Codes Corporation; 2015.

61. David M, Mustafa H, Brudno M. Detecting Alu insertions from high-throughput sequencing data. Nucleic Acids Res. 2013;41:1–13.

62. Durbin RM, Altshuler DL, Durbin RM, Abecasis GR, Bentley DR, Chakravarti A, et al. A map of human genome variation from population-scale sequencing. Nature. 2010;467:1061–73.

11

Identification of a novel HERV-K(HML10): comprehensive characterization and comparative analysis in non-human primates provide insights about HML10 proviruses structure and diffusion

Nicole Grandi[1†], Marta Cadeddu[1†], Maria Paola Pisano[1], Francesca Esposito[1], Jonas Blomberg[2] and Enzo Tramontano[1,3*]

Abstract

Background: About half of the human genome is constituted of transposable elements, including human endogenous retroviruses (HERV). HERV sequences represent the 8% of our genetic material, deriving from exogenous infections occurred millions of years ago in the germ line cells and being inherited by the offspring in a Mendelian fashion. HERV-K elements (classified as HML1–10) are among the most studied HERV groups, especially due to their possible correlation with human diseases. In particular, the HML10 group was reported to be upregulated in persistent HIV-1 infected cells as well as in tumor cells and samples, and proposed to have a role in the control of host genes expression. An individual HERV-K(HML10) member within the major histocompatibility complex C4 gene has even been studied for its possible contribution to type 1 diabetes susceptibility. Following a first characterization of the HML10 group at the genomic level, performed with the innovative software RetroTector, we have characterized in detail the 8 previously identified HML10 sequences present in the human genome, and an additional HML10 partial provirus in chromosome 1p22.2 that is reported here for the first time.

Results: Using a combined approach based on RetroTector software and a traditional Genome Browser Blat search, we identified a novel HERV-K(HML10) sequence in addition to the eight previously reported in the human genome GRCh37/hg19 assembly. We fully characterized the nine HML10 sequences at the genomic level, including their classification in two types based on both structural and phylogenetic characteristics, a detailed analysis of each HML10 nucleotide sequence, the first description of the presence of an Env Rec domain in the type II HML10, the estimated time of integration of individual members and the comparative map of the HML10 proviruses in non-human primates.

Conclusions: We performed an unambiguous and exhaustive analysis of the nine HML10 sequences present in GRCh37/hg19 assembly, useful to increase the knowledge of the group's contribution to the human genome and laying the foundation for a better understanding of the potential physiological effects and the tentative correlation of these sequences with human pathogenesis.

Keywords: Human endogenous retroviruses, Herv, HML10, Herv-k(C4), RetroTector, Cancer, Autoimmune diseases

* Correspondence: tramon@unica.it
†Equal contributors
[1]Department of Life and Environmental Sciences, University of Cagliari, Cagliari, Italy
[3]Istituto di Ricerca Genetica e Biomedica, Consiglio Nazionale delle Ricerche (CNR), Monserrato, Cagliari, Italy
Full list of author information is available at the end of the article

Background

The human genome is formed in small proportion by coding sequences (~2%), while it is constituted for about half of repeated elements, among which the human endogenous retroviruses (HERV) account for ~8% of it. HERVs have been acquired as the consequence of ancient retroviral infections affecting the germ line cells over several million years [1], and consequently transmitted to the offspring in a Mendelian way [2]. In the course of evolution, HERV sequences have hoarded abundant mutations, causing loss of virulence and contributing to their actual composition [3]. Despite the accumulation of substitutions, insertions and deletions, a number of HERV genes have maintained functional Open Reading Frames (ORF) and some HERV proteins are known to be involved in important physiological functions. The main examples are Syncytin-1 and -2, two Env proteins encoded by a HERV-W [4, 5] and a HERV-FRD provirus [6], respectively, providing essential fusogenic and immunosuppressive functions to human placenta [6–9]. To explain their persistence in the human genome, it has been proposed that HERVs could be neutral sequences, thus not negatively selected and removed during evolution (parasitic theory), or, conversely, they could be involved in important cellular functions leading to their positive selection over time (symbiotic theory) [10]. However, the former theory does not exclude the latter, being possible that, after the initial acquisition, the random accumulation of mutations by the viral DNA could led to the synthesis of divergent proteins that acquired a role for the host, enabling HERVs symbiotic persistence in our DNA [10, 11]. HERVs are currently divided into three main classes according to their similarity to exogenous elements: I (*Gammaretrovirus*- and *Epsilonretrovirus*-like), II (*Betaretrovirus*-like) and III (*Spumaretrovirus*-like). The further classification of HERV groups is currently based mainly on *pol* gene phylogeny, even if the taxonomy has been for a long time based on discordant criteria, such as the human tRNA complementary to the Primer Binding Site (PBS) of each group [12]. In this way, individual HERV groups have been identified based on the amino acid associated to the tRNA putatively priming the reverse transcription, i.e. tryptophan (W) for HERV-W sequences and lysine (K) for HERV-K supergroup. Among class II elements, the HERV-K sequences were originally identified due to their similarity to the Mouse Mammary Tumor Virus (MMTV, *Betaretroviruses)* [13], and are in fact classified accordingly in 10 so-called human MMTV-like clades (HML1–10) [3]. The HERV-K elements are currently highly investigated due to their possible association with human diseases, especially regarding cancer and autoimmunity. One of the most interesting HERV-K clade is the HML10 one, initially identified due to a full-length provirus integrated in anti-sense orientation within the ninth intron of the fourth component of human complement gene (*C4A*) in the class III region of the major histocompatibility complex (MHC) on chromosome 6 short arm [14]. This HML10 provirus was subsequently named HERV-K(C4), and showed a typical retroviral structure with 5'- and 3'Long Terminal Repeats (LTR) flanking *gag*, *pol* and *env* genes. The human *C4* gene is part of the so-called RCCX cassette, a genetic module composed by four genes: *STK19* (serine/threonine nuclear protein kinase), *C4* (either in an acid *C4A* form or a basic *C4B* form), *CYP21* (steroid 21-hydroxylase) and *TXN* (tenascin) [15]. Remarkably, *CYP21A2* contains a recombination site leading to the presence, in the human population, of polymorphic monomodular (69%), bimodular (17%) and trimodular (14%) RCCX cassettes, containing one, two, and three *C4* functional copies, respectively [16]. Interestingly, HERV-K(C4) presence or absence determines a dichotomous *C4* gene size polymorphism, showing a long (22,5 kb) or a short (16 kb) form, respectively [14, 17, 18]. About three quarters of *C4* genes belong to the long variant, including the HERV-K(C4) integration that could be present in 1 to 3 copies according to the *C4* harboring gene copy number. For European-diploid genome, the most common *C4* copy number is of four copies: two *C4A* and two *C4B* [16]. Subsequently, in the human genome assembly reference sequence, HERV-K(C4) provirus is present in two copies, one inserted in *C4A* and one in *C4B*, thought to be evolved from a *C4* duplication event in a non-human primate ancestor [15] and leading to the presence of two identical proviral insertions separated by ~26 Kb. Based on time of insertion calculation, HERV-K(C4) provirus integration has been estimated to be occurred between 10 and 23 million years ago (mya) [19]. Of note, MHC is the genome region being associated with more disorders than any other one, especially concerning autoimmune and infectious diseases [20].

Cell-culture studies on HERV-K(C4) expression pointed out that i) HERV-K(C4) is expressed in various human cell lines and tissues, including cells playing an important role in the immune system [18]; ii) HERV-K(C4) antisense transcripts are present in cells constitutively expressing C4, while there is no evidence of HERV-K(C4) sense transcripts [18, 21], iii) the expression of retroviral-like constructs is significantly downregulated in C4 expressing cells [21], and iv) this downregulation is dose-dependently modulated following interferon-gamma stimulation of C4 expression [18, 21]. These evidences suggested a role of HERV-K(C4) in the control of homologous genes expression through antisense inhibition as a plausible defense strategy against exogenous retroviral infections [21]. The latter could also be able to influence HML10 group expression, as shown by the enhancement of HML10 transcription in persistently (but not de novo) HIV-1 infected cells [22]. With regards to autoimmune diseases, a recent study proposed an association between HERV-K(C4) copy number and type 1 diabetes, reporting that affected individuals have significantly fewer copies of HERV-K(C4), which could be also linked to some disease-associated MHC II alleles [23]. Therefore, it has been speculated that this HML10 copy number could be a novel marker of type 1 diabetes susceptibility, and that the insertion of other HML10 elements may contribute to the protection against this disease by antisense transcripts expression [23]. However, no final proof of this has been shown yet, while a previous study analyzing the transmission of HERV-K(C4) in type-1 diabetes patients refuted its role as a potential susceptibility marker for diabetes [24], suggesting that HERV-K(C4) could just be a passive partner in human genetic reshuffling.

Overall, besides the possible role of the well studied HERV-K(C4) provirus, also other HML10 copies integrated within the human genome can be involved in the antisense control of homologous gene expression, possibly having a role in human pathogenesis. Thus, the comprehensive characterization of the HML10 group at the genomic level could provide a reliable background for understanding the specific origin, regulatory mechanisms, structure and physio-pathological effects of the transcripts reported in human cells, especially in the presence of exogenous infections, cancer and autoimmunity.

In the light of this, aiming to have a complete map of HML10 and other HERV sequences present in the human genome, we previously analyzed GRCh37/hg19 assembly, reporting a comprehensive map of 3173 conserved HERV insertions [3]. To this purpose we used the RetroTector software (ReTe), which allows the identification of full retroviral integrations through the detection of conserved retroviral motifs are their connection into chains, reconstructing the original sequence [25]. A multi-step classification approach allowed the exhaustive characterization of 39 "canonical" HERV groups, and 31 additional "non canonical" clades showing mosaicism as the consequence of recombination and secondary integrations [3]. Starting from this unique dataset, we focused on the deeper genetic analysis of individual HERV groups, which still remains a major bioinformatics goal [26], starting from the ones supposedly to be involved in human pathogenesis.

Using ReTe, we performed the first global analysis of the HML10 group presence in the human GRCh37/hg19 genome assembly, identifying a total of eight sequences that have been classified as HML10 [3]. More recently, seven of these eight HML10 elements have been further described as non-randomly distributed among chromosomes, but preferentially found nearby human genes, with a strong prevalence of intronic localization and

antisense orientation with respect to the surrounding gene [27]. In the same work, three HML10 proviruses integrated in reverse orientation within human introns were investigated in cell-culture models for their promoter capacity showing, for all three, a transcriptional activity in at least one LTR [27]. Authors suggested the potential antisense negative regulation of encompassing genes that, in the case of the HML10 provirus within human pro-apoptotic DAP3 (Death-associated protein 3) gene (HML10(DAP3)), was found to be efficiently suppressed by interferon γ [27]. Interestingly, the inactivation of this HML10 provirus resulted in an increase of DAP3 expression, triggering cell death and supporting the functional relevance of these retroviral transcripts in suppressing DAP3 mediated apoptosis [27]. Considering that the HML10 group was previously reported to be expressed in various cancer cell lines [28–31], the upregulation of HML10(DAP3), as well as other HML10 proviruses, could possibly be involved in the apoptotic-resistant phenotype of human malignancies [27].

Hence, also considering that the above mentioned study [27] included a lower number of HML10 proviral elements as compared to our previously reported dataset [3], we decided to provide a complete characterization of the group at the genomic level, reporting additional information about the HML10 single members phylogeny, structure and dynamics of entry and colonization of the primate lineages, and identifying a HML10 locus not previously reported.

Results
Localization and characterization of HERV-K(HML10) sequences

Following the report of a duplicated HML10 integration in the C4 genes [32], in our previous analysis performed through the bioinformatics tool ReTe, a total of eight HML10 sequences were identified, seven of which were reported for the first time [3] (Table 1). Seven of these were then used in a subsequent study that did not include the HML10 provirus in locus 19p13.2 [27], possibly relying on its misleading annotation by RepeatMasker. 19p13.2 HML10 provirus, in fact, is indeed ~550 nucleotides shorter as compared to the relative annotation in Genome Browser, which improperly associated to this HML10 locus an additional 5′ portion that is albeit not part of the HML10 proviral structure, being instead an HML9 LTR (LTR14C) that probably belongs to a surrounding HML9 proviral sequence. Thus, this HML10 provirus actually lacks both LTRs and represents a secondary proviral insertion separating a pre-existent HML9 provirus 5′LTR (flanking the HML10 provirus in 5′) from the rest of its internal sequence (flanking the HML10 provirus in 3′).

Table 1 HML10 proviral sequences localized in the human genome GRCh37/hg19 assembly

Locus	Coordinates [a]	Length	First reference	RVNR [b]	Genomic context	Secondary integrations
1p36.13	1:20,253,380–20,259,203 (−)	5824	Vargiu 2016	5836	intergenic	–
1p22.2	1:89,551,973–89,554,309	2337	this study	–	intergenic	–
1q22	1:155,661,620–155,669,312 (−)	7693	Vargiu 2016	6073	DAP3 (+) L1 MB7 (−)	AluSp 155,663,467–155,663,784 (+) MER11 155,667,171–155,668,256 (−)
6p22.1	6:27,155,300–27,164,058 (+)	8759	Vargiu 2016	2101	intergenic	AluY 27,158,573–27,158,903 (+) AluYc 27,158,904–27,159,195 (+) AluY 27,159,341–27,159,663 (+) AluY 27,159,784–27,160,001 (−) LTR13A 27,162,010–27,163,209 (−)
6p21.33	a) 6:31,952,469–31,958,829 (−)	6361	Tassabehji 1994	2116	C4A (+)	–
	b) 6:31,985,207–31,991,567 (−)	6361	Tassabehji 1994	2115	C4B (+)	–
6q22.31	6:122,825,990–122,833,238 (−)	7249	Vargiu 2016	2320	PKIB (+)	AluY 122,827,840–122,828,145 (−) AluY 122,829,905–122,830,202 (−) AluY 122,830,590–122,830,893 (−)
19p13.2	19:7,860,947–7,865,932 (−)	4986	Vargiu 2016	4599	intergenic	AluY 7,861,800–7,862,107 (−) AluY 7,862,886–7,863,179 (+) AluY 7,863,787–7,864,090 (−) AluY 7,865,512–7,865,832 (+)
19q13.41	19:52,964,148–52,969,750 (−)	5458	Vargiu 2016	4762	ZNF578 (+)	–
Yq11.221	Y:15,105,784–15,113,006 (−)	7223	Vargiu 2016	5104	L1M3f (−)	LTR2B 15,106,449–15,106,924 (−) AluY 15,111,205–15,111,507 (−)

[a]Chromosome: start-end (strand). Positions are referred to the human genome sequence, assembly GRCh37/hg19
[b]Individual sequences identifiers in the first reference study (Vargiu et al. 2016, [3])

Regarding the previous identification of HML10 genomic loci, it should be considered that ReTe uses a collection of generic conserved motifs for HERV sequences recognition, which can be mutated or lost in defective proviruses [3], possibly constituting a "bias" responsible for the missed detection of less conserved HERV group members. Hence, as previously described for the HERV-W group [33], to complete the HML10 sequences identification the human genome we also performed a traditional BLAT search in Genome Browser using the RepBase HERV-K(C4) provirus reference sequence (assembled as LTR14-HERVKC4-LTR14) [34] as a query. This approach confirmed the presence of the eight HML10 proviruses previously identified by ReTe [3] and revealed the presence of an additional HML10 provirus in locus 1p22.2, with an overall number of nine HERV-K(HML10) sequences in the human genome (Table 1).

In agreement with the previously adopted nomenclature [35], we indicated the HML10 sequences using their unique chromosomal position and, if more sequences were present in the same locus, we used consecutive letters ("a" and "b") to univocally indicate each of them (Table 1). Overall, HML10 proviral sequences were present in chromosomes 1, 6, 19 and Y. Particularly, chromosome 6 held 3 integrations (including the duplicated proviral sequence in locus 6p21.33), chromosomes 1 and 19 showed 3 and 2 sequences, respectively, and 1 element was found in chromosome Y. The number of HML10 elements found in each chromosome, including

the previously reported solitary LTR relics [27], was compared to the expected number of integrations based on the single chromosomes size (Fig. 1), considering that the current solitary LTRs are ancestral proviral insertions that underwent LTR-LTR homologous recombination. Results showed that the number of HML10 integration events observed is often discordant with respect to the expected amounts, suggesting a nonrandomly integration pattern of the group in the various chromosomes. In particular, most of human chromosomes showed a number of HML10 insertions lower than expected, with the exception of chromosomes 6, 9, 17, 21, 22, X and Y that held around twice the number of expected insertions, reaching a 9-fold increase in chromosome 19. For some of these chromosomes, such as 17 and 19 ones, an enrichment in HML10 insertions could be expected considering their particularly high gene density, as the HML10 proviruses are known to show prevalent integration in intronic regions [3, 27], as observed also for other HERV groups preferentially inserted in proximity to human genes [36]. In chromosomes with low recombination rate, such as chromosome Y, the relative abundance of HERV may instead be due to the absence of major recent rearrangements [36], or to an higher rate of HERV fixation in the male germ line, favoring HERV persistence [37]. To verify the nonrandomness of HML10 integrations distribution in human chromosomes, we compared the actual number of HML10 loci with the expected one with a random

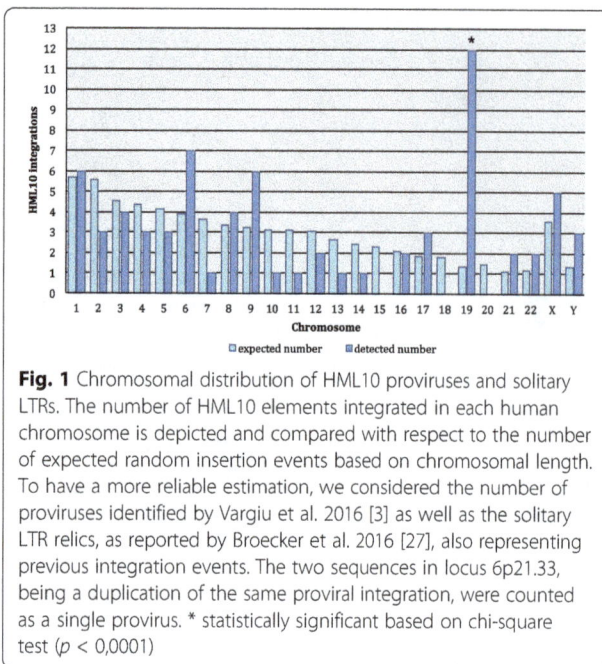

Fig. 1 Chromosomal distribution of HML10 proviruses and solitary LTRs. The number of HML10 elements integrated in each human chromosome is depicted and compared with respect to the number of expected random insertion events based on chromosomal length. To have a more reliable estimation, we considered the number of proviruses identified by Vargiu et al. 2016 [3] as well as the solitary LTR relics, as reported by Broecker et al. 2016 [27], also representing previous integration events. The two sequences in locus 6p21.33, being a duplication of the same proviral integration, were counted as a single provirus. * statistically significant based on chi-square test ($p < 0,0001$)

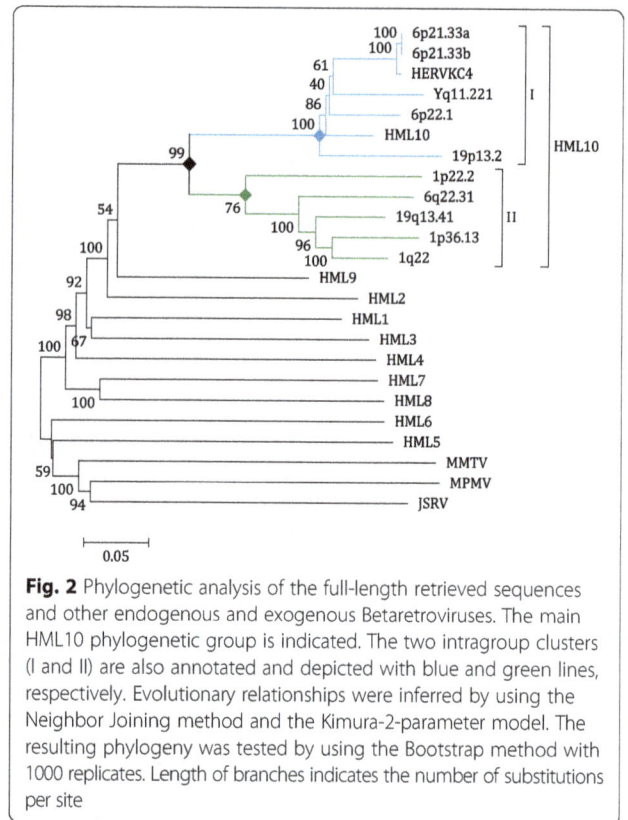

Fig. 2 Phylogenetic analysis of the full-length retrieved sequences and other endogenous and exogenous Betaretroviruses. The main HML10 phylogenetic group is indicated. The two intragroup clusters (I and II) are also annotated and depicted with blue and green lines, respectively. Evolutionary relationships were inferred by using the Neighbor Joining method and the Kimura-2-parameter model. The resulting phylogeny was tested by using the Bootstrap method with 1000 replicates. Length of branches indicates the number of substitutions per site

integration pattern through a chi-square (χ^2) test. Results rejected the null hypothesis that HML10 sequences are randomly distributed in the human genome, supporting an overall non-random integration pattern through an highly significant p value ($p < 0,0001$). However, when applied to the individual chromosomes, the same test showed that the variation between observed and expected number of HML10 integration was not statistically significant (mean p value = 0,4) except for chromosome 19, which was confirmed to be significantly enriched in HML10 sequences (p < 0,0001) making hence the overall statistics significant (Fig. 1).

In order to confirm the belonging of the newly identified sequence to the HML10 group, we performed a Neighbor Joining (NJ) phylogenetic analysis of the full-length proviruses, including the HML1–10 RepBase reference sequences [34] assembled as LTR-internal portion-LTR from Dfam database [38] as well as the main representative exogenous Betaretroviruses (MMTV; Mason-Pfizer Monkey Virus, MPMV and Jaagsiekte sheep retrovirus, JSRV) (Fig. 2). The phylogenetic analysis confirmed that the newly identified partial proviral sequence in locus 1p22.2 belongs to the HML10 group, clustering with the previously identified HML10 elements and with the Dfam and RepBase HML10 HERV-K(C4) proviral reference sequences with a 99 bootstrap support. Overall, this phylogenetic group is clearly separated from the other endogenous and exogenous Betaretroviruses, even if sharing higher similarity with the HML9 and HML2 references. Interestingly, within this main phylogenetic group we observed two different clusters, that we named type I and II, which were statistically supported by bootstrap

values (100 and 76, respectively) (Fig. 2). Type I HML10 sequences (blue lines) include both the Dfam HML10 reference and the HERV-K(C4) representative provirus, corresponding to the duplicated integrations in locus 6p21.33. Type II elements (green lines) showed a more divergent structure with respect to the group references, especially regarding the proviral locus 1p22.2 that is also less related to the other cluster II members.

HML10 proviruses structural characterization

Considering that the phylogeny of the HML10 full-length proviruses revealed the clear presence of type I and II sequences, we analyzed in detail the nucleotide structure of the individual members to gain a comprehensive knowledge of the uniqueness of each HML10 locus and to characterize the main differences between the two types. To this aim, we aligned all the HML10 proviruses nucleotide sequences to the RepBase reference LTR14-HERVKC4-LTR14, namely HERV-K(C4), corresponding to the two duplicated proviral insertions in locus 6p21.33. For each HML10 provirus, we annotated all insertions and deletions up to 1 nucleotide as well as the presence of the main structural and regulatory features, as referred to the LTR14-HERVKC4-LTR14 RepBase sequence (Fig. 3). Particularly, we verified the conservation of LTR motifs relevant for retroviral expression, i.e. a Tata box (TATAAA, nucleotides

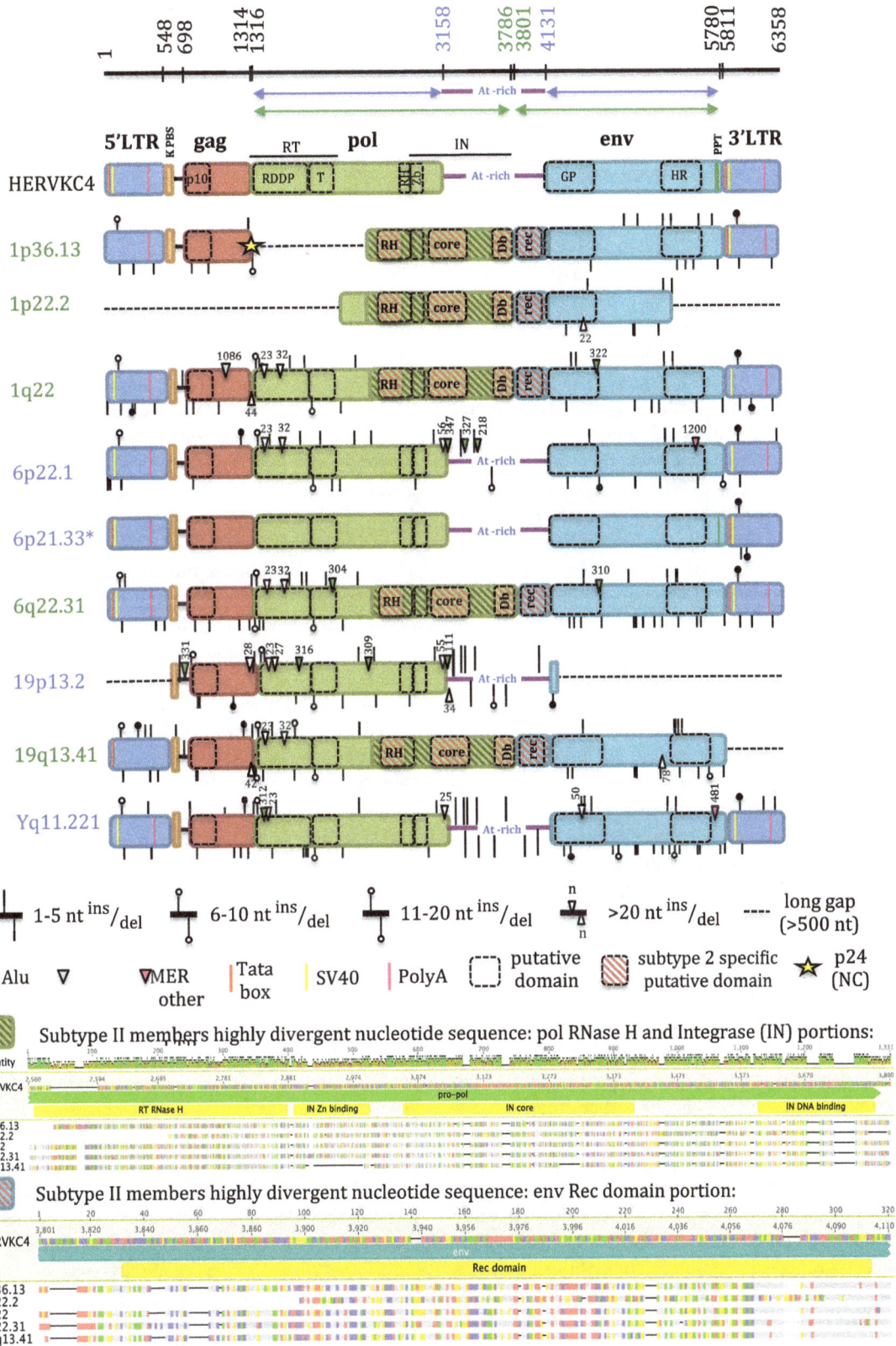

Fig. 3 (See legend on next page.)

30–35 and 5840–5845), a SV40 enhancer (GTGGAAAG, nucleotides 65–72 and 5875–5882) and a PolyA signal (AATAAA, nucleotides 384–389 and 6194–6199), as well as the conservation of the PBS sequence (nucleotides 552–569) and the polypurine tract (PPT, nucleotides 5786–5798). We also analyzed the presence of functional domains in the retroviral genes, as predicted by the NCBI tool for conserved domains search [39] (Fig. 3). In addition, we assessed whether the ~830 nucleotides A/T-rich stretch previously reported between the *pol* and *env* genes of HERV-K(C4) proviral insertion (from nucleotide 3159 to nucleotide 3189) [14] was present in any other HML10 sequence. Interestingly, a correspondent portion with a comparable enrichment in A/T nucleotides (ranging from about 67% to 73%) was identified in type I proviruses only, being present also in all the members other than HERV-K(C4) (data not shown). Overall, the HML10 proviruses showed a complete retroviral structure, and the analysis allowed us to better define the location of the main retroviral genes with respect to what has been previously reported in RepBase database (Fig. 3). The majority of HML10 proviruses retained two LTRs (nucleotides 1–548 and 5811–6358) flanking the *gag* (698–1314), *pol* (1316–3786) and *env* (3801–5780) genes. Some HML10 proviral sequences, however, were defective for at least one retroviral element: loci 1p22.2 and 19p13.2 lack, for example, both LTRs, a portion of the *env* gene and, in the case of 1p22.2, the PBS sequence and the whole *gag* gene. Locus 19q13.41 lacks the 3′LTR, while locus 1p36.13 lacks the 5′portion of *pol* gene but, remarkably, it present indeed the *gag* p24 nucleocapsid region, which resulted instead absent in all the other analyzed sequences. Regarding the LTR regulatory sites (Tata box, SV40 and PolyA), all the HML10 proviruses LTRs showed nucleotide changes in at least one motif, except for locus 6q22.31 that showed conserved nucleotide sequences for all the considered features in both LTRs, in line with its reported promoter activity in cell cultures [27] (Fig. 3). Moreover, the presence of the above-mentioned A/T-rich stretch in type I HML10 sequences constitutes a variation in the *pol* and *env* genic structure, because this portion has traditionally been considered as not included in the sequence of these two genes in HERV-K(C4) [14] and, actually, its presence in type I

sequences corresponds to the absence of any putative Pol and Env functional domains. Thus, while the *pol* gene start position and the *env* gene terminal position are common to both types members, type I *pol* and *env* genes appear to end before (*pol*, nucleotide 3158), and start after (*env*, nucleotide 4131), the correspondent genes in type II HML10 sequences, respectively (Fig. 3). The NCBI search for conserved domains predicted the presence of some functional features shared by all the group members retaining the harboring gene portion: a Gag p10 domain (core region), Pol Reverse Transcriptase (RT) RNA Dependent DNA Polymerase (RDDP) and thumb domains, a Pol Integrase (IN) Zinc binding site, and Env Glycoprotein and Heptad Repeats regions. None of the HML10 elements retained instead any domain that could suggest the presence of a *pro* gene, which seems to be defective for the whole group. In addition, it is interesting to note that some other predicted domains were identified only in a subset of HML10 elements, all belonging to type II sequences (Fig. 3). The latter showed, in fact, a highly divergent nucleotide structure when compared to the HERV-K(C4) reference, in *pol* Ribonuclease H (RNase H) and IN portions, as well as in the 5′ region of *env* gene. Of note, these peculiar genic regions of type II proviral sequences correspond, in sequence positions, to the above-mentioned A/T-rich stretch found exclusively for HML10 type I elements, further confirming the high nucleotide divergence of such element with respect to the type II *pol* 3′ and *env* 5′ portions (Fig. 3). The search for conserved motifs in such regions revealed the peculiar presence, in type II HML sequences, of i) a longer putative Pol RNase H domain; ii) an IN core domain, iii) an IN DNA binding site and iv) an Env Rec domain, which were contrarily not found in any of the HML10 type I proviruses. Particularly, the presence of a putative Rec domain was unexpected, since such accessory protein has been reported to be present in the HERV-K(HML2) proviruses only [40–42], where its expression has been tentatively linked to cancer development. Thus, we characterized in more detail such HML10 Rec domain through the bioinformatics analysis of the correspondent putative proteins and their comparison to the already characterized HML2 Rec proteins present in UniProt database [43].

Characteristics of the newly identified HML10 Rec putative proteins

In order to characterize in more detail the Rec coding region in HML10 subtype II elements, we built a NJ phylogenetic tree of the five subtype II proviruses Rec sequences after their bioinformatics translation in the correspondent putative proteins (puteins) (Fig. 4). The amino acids sequences of nine previously published HERV-K(HML2) Rec proteins as well as the analogues Human Immunodeficiency Virus 1 (HIV-1) Rev and Human T Lymphotropic Virus 1 (HTLV-1) and Simian T Lymphotropic Virus 1 (STLV-1) Rex proteins were included as references (see Methods). As shown in Fig. 4, 1p22.2 Rec putein showed the highest relation to the HERV-K(HML2) Rec proteins, with a 99 bootstrap value. This cluster was itself related to the other four HML10 Rec puteins, supported by a 93 bootstrap value. Differently, the putein obtained from the translation of the correspondent nucleotide portion of HERV-K(C4), used as representative for type I HML10 elements, did not show remarkable phylogenetic similarity to any Rec sequence, as suggested by the presence of the A/T-rich stretch in this region.

To further investigate the possible relevance of the five Rec puteins identified in type II HML10 sequences, we analyzed the occurrence of premature internal stop codons and frameshifts as compared to UniProt HML2 Rec proteins (Fig. 5). Remarkably, two of the five HML10 Rec ORFs (locus 1q22 and 1p22.2) showed an intact structure devoid of premature stop codons and frameshifts, theoretically encoding for 76 and 72 amino acids puteins, respectively (Fig. 5). 1p36.13 Rec putein

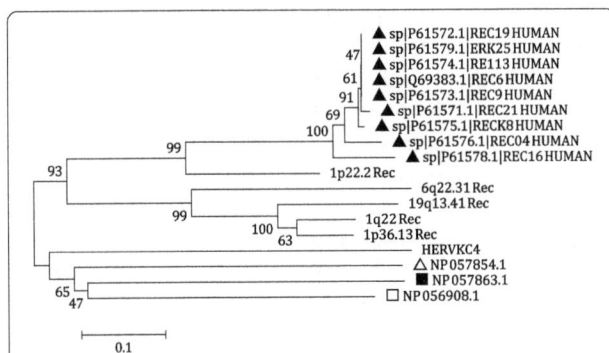

Fig. 4 Phylogenetic analysis of the HML10 subtype II Rec putative proteins. The HML10 subtype II proviruses nucleotide sequences corresponding to a predicted Rec domain were translated and the obtained putative proteins (puteins) were analyzed in a NJ tree including previously reported HERV-K HML2 Rec proteins (black triangles) and the analogues HIV-1 Rev. (white triangle), HTLV-1 Rex (black square) and STLV Rex (white square) proteins. Evolutionary relationships were inferred by using the Neighbor Joining method and the p-distance model. The resulting phylogeny was tested by using the Bootstrap method with 1000 replicates. Length of branches indicates the number of substitutions per site

showed instead a single internal stop codon at residue 24, whose reversion could theoretically lead to the production of a full-length putein. The Rec puteins in HML10 loci 6q22.31 and 19q13.41 show a more defective structure, being affected by 3 premature stop codons (6q22.31, positions 24, 29 and 49) and one internal frameshift (19q13.41, between residues 17 and 18), respectively. Thus, we focused our attention on the two HML10 Rec puteins with potentially intact ORFs (locus 1q22 and 1p22.2), evaluating the preservation of important functional domains as described for HERV-K(HML2) Rec proteins (Fig. 5). The latter present, in fact, two motifs needed for nuclear localization and export (NLS and NES, respectively) [44]. The analysis showed that, while all HML10 Rec puteins apparently lack the NLS portion, both 1q22 and 1p22.2 Rec puteins present a recognizable putative NES domain (Fig. 5).

Estimated time of integration

A special property of proviral sequences is that their LTRs are identical at the time of integration, so that their divergence (D) after endogenization depends on the genome random mutation rate per million years, allowing to estimate the time of integration (T) of each provirus [45]. Even if this method has been widely used to calculate the HERV sequences approximate age, it is affected by important limitations, as previously reported [33]. Firstly, it is not applicable to those proviruses lacking one or both LTRs and, secondly, it may underestimate T values, as it has been shown comparing the T values to the presence in non human primates of the HERV proviruses orthologous sequences [33]. For these reasons, we estimated the HML10 proviruses age through a multiple approach of T calculation, based on the D percentage value between i) the 5′ and 3′ LTRs of the same provirus (LTR vs LTR, possible for 7/9 HML10 sequences); ii) each LTR and a generated LTR consensus sequence; and iii) the *gag*, *pol* and *env* genes and a generated consensus sequence. Both consensus sequences have been generated following the majority-rule by the multiple alignments of all HML10 proviruses. Briefly, for each approach, the T value has been estimated by the relation $T = D\%/0,2\%$, where $0,2\%$ represents the human genome random mutation rate expressed in substitutions/nucleotide/million years [46–48]. With regards to the D between the two LTRs of the same provirus, the obtained T value has been further divided for a factor of 2, considering that after endogenization each LTR accumulates random substitutions independently. For each provirus, the final T value has been calculated as the average of the T values obtained with the different approaches. Noteworthy, the final T value has also been validated by the identification of the Oldest Common Ancestor (O.C.A., i.e. the most distantly related primate species

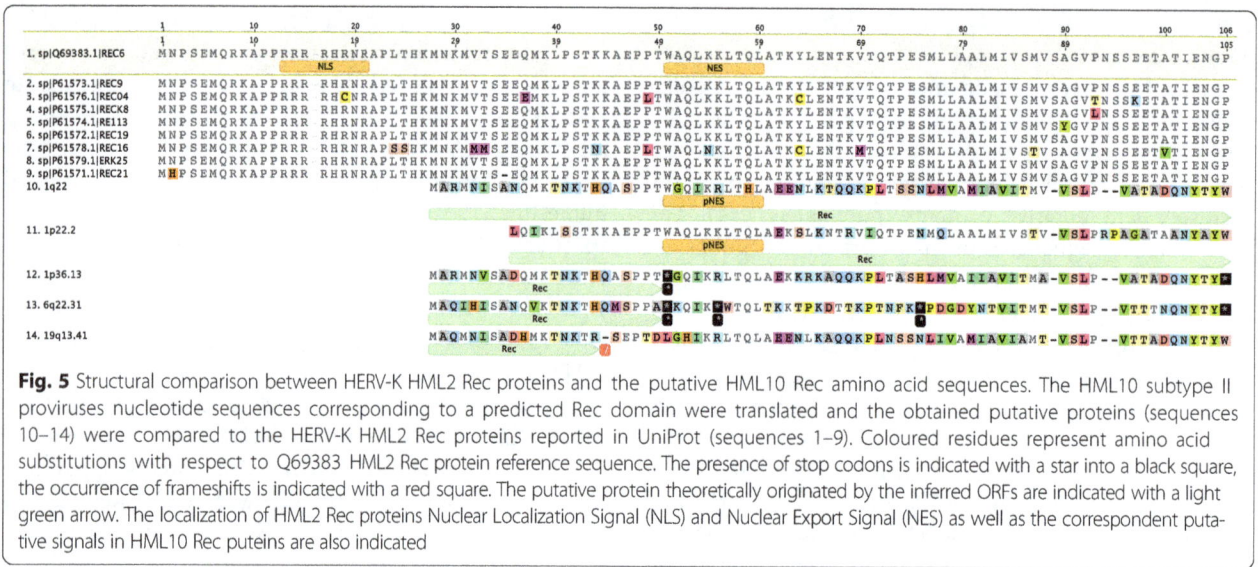

Fig. 5 Structural comparison between HERV-K HML2 Rec proteins and the putative HML10 Rec amino acid sequences. The HML10 subtype II proviruses nucleotide sequences corresponding to a predicted Rec domain were translated and the obtained putative proteins (sequences 10–14) were compared to the HERV-K HML2 Rec proteins reported in UniProt (sequences 1–9). Coloured residues represent amino acid substitutions with respect to Q69383 HML2 Rec protein reference sequence. The presence of stop codons is indicated with a star into a black square, the occurrence of frameshifts is indicated with a red square. The putative protein theoretically originated by the inferred ORFs are indicated with a light green arrow. The localization of HML2 Rec proteins Nuclear Localization Signal (NLS) and Nuclear Export Signal (NES) as well as the correspondent putative signals in HML10 Rec puteins are also indicated

presenting the correspondent orthologous insertion), which also provides details on the period of proviruses formation (Table 2 and Fig. 6).

In general, the HML10 group spreading in the primate lineages occurred between 40 and 20 mya, after the divergence between New World Monkeys and Old World Monkeys, with the majority of proviral insertions occurring in Rhesus macaque (Table 2 and Fig. 6). It is interesting to note that, as previously observed [33], the LTR vs LTR method gave significantly lower T values than the consensus based approaches ($p < 0,001$), showing, in fact, a D value average of 3,6% versus the 6% D average obtained with the consensus based methods. Thus, it can be concluded that T values obtained with the sole traditional LTR vs LTR approach could generally led to some underestimation, possibly indicating an earlier integration period instead of the actual one, which was also confirmed by the proviruses O.C.A.. A similar underestimation, even if with lower confidence ($p < 0,05$), was observed in the genes vs consensus method when comparing the T value calculated with the *pol* gene to the ones calculated for the *gag* and *env* genes, possibly suggesting a lower variability of the *pol* region, that is in fact known to be generally the most conserved retroviral portion (Table 2 and Fig. 6). Moreover, in the specific case of the duplicated sequence in locus 6p21.33, the presence of a low T value could possibly be biased by the fact that these sequences are located within an important genic region, presenting an overall lower substitution rate, and, for sequence 6p21.33b, the fact that has been recently created by a large gene duplication. It is worth to note that the apparent loss of both 6p21.33 proviral copies in different evolutionarily intermediate primates species, as already reported [32], is another confounding factor for the accurate T estimation of these elements.

Finally, it is interesting to note that HML10 type II sequences are older than HML10 type I insertions, showing an average estimated time of integration of 35,5 mya ago with respect to a medium age of 25, 9 mya calculated for type I elements.

Comparative identification of orthologous insertions in non-human primates

Most HERVs entered into the primates lineages between 10 and 50 mya, during primates evolutionarily speciation. The most ancient HERV-K HML group, the HML-5 one, has been estimated to have integrated before the separation of New and Old World Monkeys, occurred about 43 mya, while the other HMLs appeared later on in several subsequent waves of colonization of the *Catarrhini* parvorder only (Old World Monkeys and Hominoids). Hence, in order to gain more details on the HML10 diffusion in the various primate species, we searched the HML10 sequences orthologous to each provirus retrieved in the human genome in the genome assemblies of one New World Monkey (Marmoset; *Platyrrhini* parvorder), one Old World Monkey (Rhesus macaque; *Catarrhini* parvorder) and 4 Hominoids (Gibbon, Orangutan, Gorilla and Chimpanzee; *Catarrhini* parvorder). As shown in Table 3, six of the nine HML10 proviruses found in the human genome have corresponding orthologous sequences in all the analyzed *Catarrhini* species, from Chimpanzee to Rhesus, confirming an approximate main period of HML10 group diffusion between 43 and 30 mya. 1p22.2 partial provirus is also present from human to Rhesus, but its orthologous insertion in the Gorilla genome is missing, possibly due to a deletion event. With regards to the provirus integrated in locus 6p21.33, the two identical

Table 2 HML10 sequences estimated time of integration

	LTR vs LTR	LTR vs consensus	*gag* vs consensus	*pol* vs consensus [a]	*env* vs consensus [b]	*Average*	O.C.A. [c]
1p36.13	14.1	21.0	22.5	no *pol* (62 nt only)	31.9	*22.4*	rhesus
1p22.2	no 5' and 3'LTRs	no 5' and 3'LTRs	no *gag*	no *pol*	45.0	*45.0*	rhesus
1q22	14.7	44.1	35.7	28.9	32.7	*31.2*	rhesus
6p22.1	12.7	36.5	43.0	18.9	32.8	*28.8*	rhesus
6p21.33a	22.9	18.0	25.2	21.3	21.3	*21.7*	rhesus[d]
6p21.33b	22.9	18.0	25.2	21.3	21.3	*21.7*	orangutan[d]
6q22.31	17.2	38.8	38.9	44.8	35.1	*35.0*	rhesus
19p13.2	no 5' and 3'LTRs	no 5' and 3'LTRs	[e]	20.8	no *env* (48 nt only)	*20.8*	rhesus
19q13.41	no 3'LTR	46.0	37.4	27.2	45.9	*39.1*	rhesus
Yq11.221	20.8	45.2	41.5	30.4	44.7	*36.5*	rhesus
Average	*17.9*	*33.5*	*33.7*	*26.7*	*34.5*	*28,58*	

[a]partial sequence: nucleotides 1277–2571 in LTR14-HERVKC4-LTR14
[b]partial sequence: nucleotides 4103–5810 in LTR14-HERVKC4-LTR14
[c]Oldest Common Ancestor
[d]Provirus loss in various intermediate species: chimpanzee, gorilla, orangutan and gibbon (6p21.33a); chimpanzee, gorilla, gibbon and rhesus (6p21.33b)
[e]sequence showing an highly divergent gag sequence, giving an estimated T of 165,7 that was not taken into account for the final T calculation

copies are localized in the human complement C4A and C4B genes, known to reside on duplicated segments of DNA. In particular, the C4 genes of some *Catarrhini* primates exhibit a long/short dichotomous size variation due to the presence/absence of these HML10 integrations, while chimpanzee and gorilla only contain short C4 genes [19, 32]. In line with this, 6p21.33a and 6p21.33b orthologous HML10 insertions were localized in Rhesus and Orangutan genome sequences, respectively, but are absent in the other analyzed species (Table 3). Finally, the

orthologous HML10 provirus in locus Yq11.221 could be localized in Chimpanzee genome only, because no comparative information are available for the Y chromosome of the other primate species (Table 3).

In addition to the non-human primates HML10 sequences orthologous to human loci, we wanted also to assess whether the group period of proliferation activity could have also determined species-specific insertions outside of the human evolutionary lineage. Thus, we performed BLAT searches in the above mentioned non-human primates genome sequences using the HML10 group LTR14-HERVKC4-LTR14 RepBase sequence [34] from Dfam database [38] as a query. The analysis showed that no additional species-specific HML10 integrations are present in Chimpanzee, Gorilla, Orangutan and Rhesus genome sequences (data not shown), while a HML10 provirus apparently lacking orthologous loci in the other primate species was found in Gibbon assembly chr5:62,078,165–62,086,762. This provirus was in part recognized as HML9 sequence based on RepeatMasker annotation track, but its inclusion in a NJ phylogenetic tree with all the 10 HML groups reference sequences confirmed its belonging to the HML10 group (data not shown).

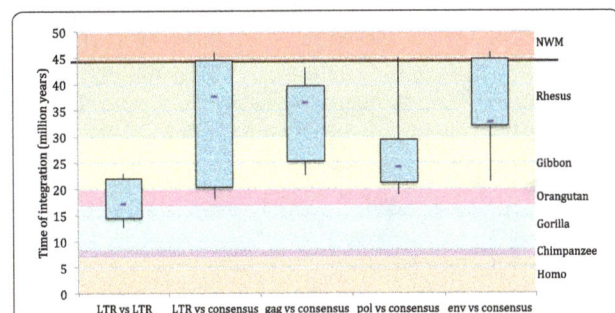

Fig. 6 Overview of HML10 group colonization of primate lineages. Boxplot representations of HML10 group period of entry in primate lineages. The estimated age (in million years) was calculated considering the divergence values between i) the 5' and 3' LTRs of the same provirus; ii) each LTR and a generated consensus; iii) *gag*, *pol* and *env* genes and a generated consensus. The approximate period of evolutionarily separation of the different primate species are also indicated and have been retrieved from Steiper et al. 2006 [70] and Perelman et al. 2011 [71]. Boxes represent the main period of HML10 group diffusion in primates based on the different approaches of calculation, including from 25 to 75 percentiles and showing the mean value as a blue dash. Whiskers indicate the minimum and maximum estimated age

Retroviral features analysis

Beside these major determinants, the various HERV genera share some specific features, which are also valuable for taxonomic purposes [49]. Particularly, it is known that Class II Betaretrovirus-like HERVs, including the HERV-K HML1–10 groups, commonly present a PBS sequence putatively recognizing a Lysine (K) tRNA. The human tRNA supposed to prime the retrotranscription process, in fact, has been used for a long time for HERV

Table 3 HML10 sequences orthologous loci in non-human primates genome

Human locus	Chimpanzee	Gorilla	Orangutan	Gibbon	Rhesus	Marmoset
1p36.13 (−)	1:19,897,252–19,903,183 (−)	1:20,573,241–20,579,060 (−)	1:210,407,411–210,413,307 (+)	24:19,115,921–19,117,286 (−)	1:22,729,037–22,740,752 (−)	x
1p22.2 (−)	1:89,883,243–89,885,583(−)	x	1:139,752,930–139,755,294 (+)	12:87,503,425–87,505,758	1:92,543,319–92,545,983 (−)	x
1q22 (−)	1:133,941,236–133,948,931 (−)	1:134,686,645–134,687,185 (−)	1:95,817,622–95,818,162 (+)	assembly gap	1:134,772,475–134,779,343 (−)	x
6p22.1 (+)	6:27,446,871–27,456,058 (+)	6:28,001,913–28,010,233 (+)	6:28,071,758–28,078,582 (+)	1a:72,438,487–72,447,474 (+)	4:27,112,448–27,121,339 (+)	x
6p21.33a (−)	x	x	x	x	4:32,223,558–32,230,572 (−)	x
6p21.33b (−)	x	x	6:32,500,019–32,506,424 (−)	x	x	x
6q22.31 (−)	6:123,707,066–123,714,005 (−)	6:122,872,935–122,879,489 (−)	6:125,032,218–125,039,364 (−)	3:109,711,272–109,718,216 (−)	4:143,675,558–143,676,403 (−)	x
19p13.2 (−)	19:7,923,717–7,929,241 (−)	19:8,020,313–8,024,861 (−)	19:7,962,003–7,966,295 (−)	10:66,445,268–66,447,647 (+)	19:8,140,869–8,144,331 (+)	x
19q13.41 (−)	19:57,389,749–57,395,370 (−)	19:49,869,509–49,875,109 (−)	19:53,964,824–53,970,559 (−)	10:72,725,038–72,730,734 (−)	19:58,261,760–58,267,798 (−)	x
Yq11.221 (−)	Y:20,496,417–20,503,728 (−)	–	–	–	–	–

For each human HML10 locus (for precise start and end positions, see Table 1), chromosome coordinates and strand of orthologous loci are given for the other regarded non-human *Catarrhini* primate reference genome sequences. Apparent absence of a HML10 sequence in the orthologous genome position is indicated by "x". Regarding the HML10 locus on the human chromosome Y, comparative information is available for chimpanzee genome sequence only (see main text)

nomenclature and, even if now it is considered poorly reliable for taxonomic classification, it remains a characteristic feature of the different HERV groups. Among the nine HML10 proviruses analyzed, eight conserve a PBS sequence, while locus 1p22.2 provirus is defective for a big 5′ retroviral portion and lacks 5′LTR and *gag* gene. As expected, when present, the PBS sequence is located 3 residues downstream the 5′LTR and is 18 nucleotide in length, except for 19q13.41 provirus that has a single nucleotide insertion between residues 10 and 11 (Fig. 7). All the analyzed PBS were predicted to recognize a Lysine tRNA and show a conserved nucleotide composition, as indicated in the logo generated from the PBS sequences alignment (Fig. 7).

Other common features of Class II Betaretrovirus-like HERV groups are i) a Pro C-terminal G-patch motif, ii) a Pro N-terminal dUTPase, and iii) two Gag NC Zinc finger motifs [3, 49]. In the case of the HML10 sequences, however, these features are not present due to the absence of the harboring retroviral genome portions. As described, in fact, all HML10 proviruses lack the entire *pro* gene and, with the exception of locus 1p36.13, the *gag* NC portion (Fig. 3). However, the analysis of HML10 locus 1p36.13 revealed also in this provirus the partial deletion of the gene 3′ terminal portion, i.e. the one normally including both the Zinc finger motifs.

Finally, the HML10 group is known to be biased for the Adenine (A) content, showing around the 34% of A and only the 17% of Guanine (G) nucleotides in the canonical

sequences [3]. Such G to A hypermutation could be due to host RNA editing systems, as commonly observed with APOBEC3G enzymes in *Lentiviruses* [50]. The analysis of our complete dataset nucleotide frequencies confirmed a bias for A, showing in average a 33% of A (maximum = 36%, minimum = 31%, standard deviation = 2) and a 18% of G (maximum = 21%, minimum = 15%, standard deviation = 2). In addition to this skewed purine composition, we observed a weak bias in pyrimidine amount, with 28% of Thymine (T) (maximum = 28%, minimum = 27%, standard deviation = 1) and 21% of Cytosine (C) (maximum = 22%, minimum = 19%, standard deviation = 1).

Phylogenetic analyses

To gain more insights into the HML10 group phylogeny, we analyzed all identified HML10 proviruses using the nucleotide sequences of *gag*, *pol* and *env* genes to generate NJ trees, including also the reference sequences of all Dfam HERV-K groups (HML-1 to 10) and of some representative exogenous Betaretroviruses (MMTV, MPMV and JSRV) (see Methods) (Fig. 8). The presence of two types of HML10 proviruses, was confirmed in the NJ trees of both *pol* and *env* genes, but not in the *gag* gene (Fig. 8), in agreement with the HML10 individual loci structural characterization, which already pointed out that the major differences between type I and type II elements are located in the *pol* RNase H and IN portions

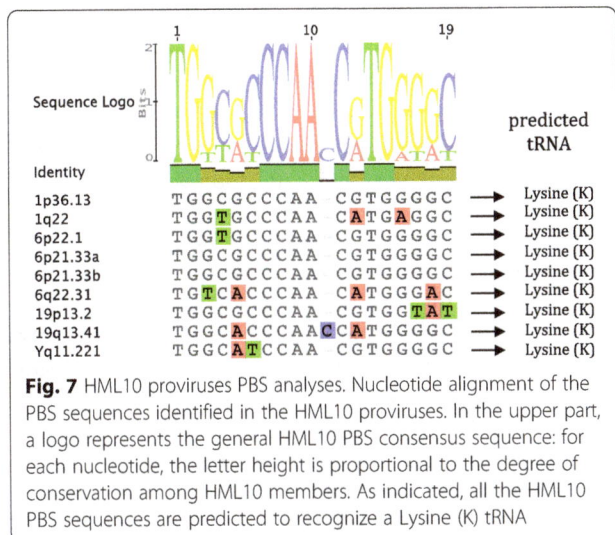

Fig. 7 HML10 proviruses PBS analyses. Nucleotide alignment of the PBS sequences identified in the HML10 proviruses. In the upper part, a logo represents the general HML10 PBS consensus sequence: for each nucleotide, the letter height is proportional to the degree of conservation among HML10 members. As indicated, all the HML10 PBS sequences are predicted to recognize a Lysine (K) tRNA

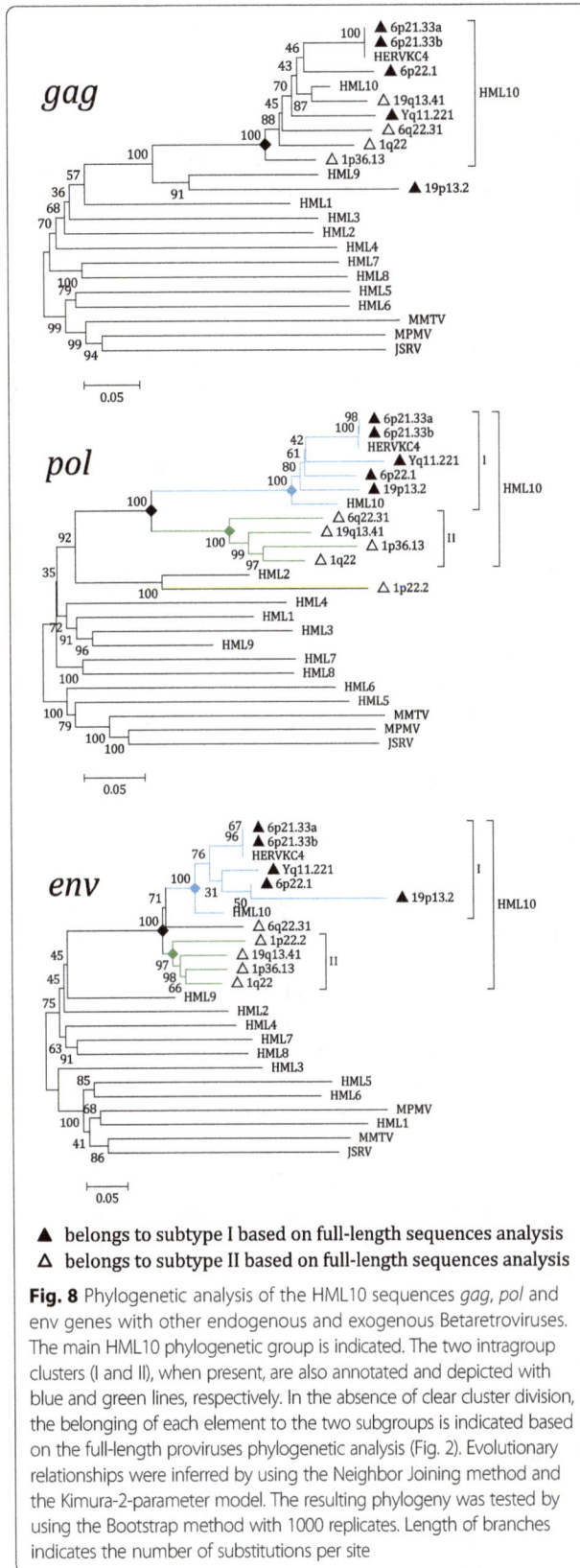

▲ belongs to subtype I based on full-length sequences analysis
△ belongs to subtype II based on full-length sequences analysis

Fig. 8 Phylogenetic analysis of the HML10 sequences *gag*, *pol* and env genes with other endogenous and exogenous Betaretroviruses. The main HML10 phylogenetic group is indicated. The two intragroup clusters (I and II), when present, are also annotated and depicted with blue and green lines, respectively. In the absence of clear cluster division, the belonging of each element to the two subgroups is indicated based on the full-length proviruses phylogenetic analysis (Fig. 2). Evolutionary relationships were inferred by using the Neighbor Joining method and the Kimura-2-parameter model. The resulting phylogeny was tested by using the Bootstrap method with 1000 replicates. Length of branches indicates the number of substitutions per site

and in the *env* 5′ region. More in details, the *gag* gene phylogenetic analysis revealed that all HML10 sequences group together with 100 bootstrap support, except for 19p13.2 provirus, which was related instead to the HML9 reference sequence. Due to the fact that this HML10 provirus has been inserted as a secondary integration within a pre-existing HML9 proviral sequence, a part of the flanking HML9 element could have been erroneously associated to the encompassed HML10 element. To assess this possibility, we analyzed 19p13.2 HML10 with respect to both HML10 and HML9 Dfam references with Recco software [51], detecting eventual recombination events among aligned sequences (data not shown). Indeed, an internal portion of the 19p13.2 provirus (from nucleotide 755 to nucleotide 1384, 15% of the total length) is effectively more similar to HML9 reference, being albeit included in a "true" HML10 proviral sequence (nt 1–754 and 1285–4986, 85% of the total length) and suggesting the previous occurrence of a recombination event involving the gag gene and leading to a HML10 mosaic form (data not shown).

Differently, in *pol* tree the phylogenetic clusters of type I and II proviruses were supported by the maximum bootstrap value (100), including all the respective proviruses as already classified based on the full length nucleotide sequence, except for locus 1p22.2. The latter *pol* sequence, similarly to what observed for locus 19p13.2 *gag* gene, showed instead higher similarity to the HML2 group reference sequence. The same type I and II phylogenetic clusters have been observed in *env* gene phylogenetic analysis, showing also in this case a high bootstrap support (100 and 98, respectively). In this tree, subtype II sequence in locus 6q22.31 showed an intermediate position, sharing some high similarities with type I cluster also.

For completeness, we analyzed the phylogeny of the HML10 proviral 5′ and 3′LTR also, including the LTR references for HML1 to 10 groups and for the exogenous Betaretroviruses MMTV, MPMV and JSRV. As expected, all the HML10 proviruses 5′ and 3′LTR sequences grouped together with the group reference LTR14, supported by a 100 bootstrap value (Additional file 1: Figure S1). Within this phylogenetic group, both LTRs of the same proviral element were generally coupled with bootstrap values ranging from 91 to 100, but no clusters dividing the LTRs of type I and type II HML proviruses were observed, confirming an overall common LTR sequence for both subgroups (Additional file 1: Figure S1).

Discussion

Initially identified due to the presence of an integrated proviral sequences in the human C4 gene [32], the HML10 group expression has been proposed to affect a number of biological processes. The HERV-K(C4) prototype sequence is, in fact, normally expressed in various human cells, almost exclusively producing antisense transcripts [18, 21] that have been hypothesize to act as i) regulators of homologous genes expression through antisense inhibition, ii) possible defense mechanism against exogenous infections, iii) potential contributor to autoimmune diseases involving the complement components [21]. Recently, some HML10 proviruses, other than HERV-K(C4) and originally reported by Vargiu et al. [3], have been investigated for their promoter capacity and expression, further supporting their possible role as antisense regulators of host genes [27]. This is of particular interest, considering that most HML10 elements are located within human introns in antisense orientation, and many of them, in addition to the well studied HERV-K(C4) insertions, can potentially influence host functions. Interestingly, the antisense expression of HML10 provirus in locus 1q22 downregulated the encompassing gene DAP3 in cell culture, leading to an apoptotic-resistant cell phenotype [27]. These findings, together with the reported generic group expression in various tumor cell lines, could suggest a contribution of some HML10 loci to human malignancies, potentially through to the loss of apoptosis cell control. Overall, while these findings made the HML10 group one of the most interesting HERV groups, the lack of the complete identification of the HML10 integrations and the lack of a comprehensive investigation of the single HML10 loci impeded the assessment of their specific contribution to human transcriptome and to human pathogenesis [52].

In the present work, we completed the identification of the HML10 proviruses, reporting for the first time an additional HML10 sequence in locus 1p22.2. The latter, even if characterized by a defective structure, being 2337 nucleotides in length and showing the *pol* and *env* genes

only, constitutes a partial but "true" HML10 provirus based on structural and phylogenetic analyses. Hence, given the HML10 proviruses reported in our previous study [3], there are nine HML10 sequences in the human genome. In addition, we analyzed and characterized in great detail the structure, phylogeny and estimated period of diffusion of these ten HML10 proviruses providing, to our knowledge, the most complete representation of the HML10 group up to date. The chromosomal distribution of these proviruses and the HML10 solitary LTR relics revealed a non-random integration pattern, showing clusters of sequences with a number of integration higher than expected, especially in chromosomes 6, 9, 19, X and Y. This bias, in the case of gene-rich chromosomes such as 17 and 19 ones, is probably linked to the strong preference of HML10 elements to be inserted in proximity or within human gene introns [3, 27], while for the Y chromosome, showing a lower recombination rate, it could be linked to a greater rate of HERV fixation [37]. The phylogenetic analysis of the full length proviral nucleotide sequences revealed the presence of two well supported clusters, identified here as type I and II and including 4 and 5 members, respectively, and further confirmed by the phylogenetic analysis of both *pol* and *env* genes. Interestingly, the structural analysis of such regions showed that both types of HML10 sequences have some specific domains, being present in all the same-type members but not found in the correspondent portion of the other-type sequences. In the case of type I sequences, we found that the A/T-rich stretch previously reported between the *pol* and *env* genes of HERV-K(C4) provirus [14] is present also in the other 3 type I elements. Similar A/T-rich regions have been reported also in other HERV LTRs [53, 54] as well as in the *env* gene of a HML2 provirus in locus 5q33.2 [42], but the function of such portion in these sequences as well as in HML10 type I elements is still unknown. In the case of type II HML10 elements, the portion corresponding to type I intergenic A/T-rich stretch presents instead putative functional domains of Pol and Env proteins not found in type I proviruses, such as the RNase H 5′ portion, the IN core and DNA binding domains and, of further note, an Env Rec domain, whose presence has been confirmed also through the phylogenetic analysis of the five type II HML10 proviruses Rec puteins. Until now, Rec was considered to be exclusive of a subset of HERV-K(HML2) sequences [40–42]. HML2 Rec has been shown to be expressed in a wide range of tissues [55], interacting with a number of cellular proteins relevant for host physiological functions [56–59], and is currently highly investigated for its oncogenic potential (as reviewed in [60, 61]). Thus, the expression of a Rec analogue in HML10 sequences could contribute to human physiopathology and surely deserves to be further investigated, given that two of the five characterized HML10 Rec puteins did not harbor any premature stop codon or frameshift and presented a putatively functional NES. Other interesting structural peculiarities of HML10 group are the absence of *pro* gene and the presence of a shorter *gag* gene lacking the nucleocapsid portion, that was found only in 1p36.13 type II provirus. Apart from the possibility of an occasional loss of *pro* due to post-insertional mutations and deletions, such gene is usually present in HERV sequences, being often the most intact ORF [3]. Thus, to our knowledge, HML10 is the first HERV group systematically lacking the *pro* gene. While unlikely, it is hence possible to speculate that its original exogenous retroviruses could have evolved alternative mechanisms for protein cleavage, as observed for the coopted HERV-W Syncytin-1 Env, in which a peculiar four amino acids deletion made the protein constitutively fusogenic even in the absence of a functional viral Protease [62]. While such diffuse defective structure in *pro* and *gag* genes implied the absence of the relative Betaretroviruses characteristic features (Pro G-patch and dUTPase, Gag Zinc fingers), 8/9 HML10 sequences maintained the originally reported PBS sequence recognizing a K tRNA. Also the previously reported purine bias [3] was confirmed, showing an A frequency average of about 33%, and an unreported weak bias in pyrimidines amount, with an increase in T percentage (28%). The G to A bias could be explained by the action of host RNA APOBEC editing enzymes, as observed for HIV-1 [50] and HERV-K(HML2) [63] sequences, while the C to T hypermutation could be due to DNA methyltransferase methylation of CG dinucleotides, followed by the spontaneous deamination of methyl-C to T, as a potential silencing mechanism of retroelements. The time of integration estimation, performed for each HML10 sequence with a multiple and more reliable approach suggested that HML10 elements have been acquired by the primate lineages between 40 and 20 mya and mostly found in all the analyzed *Catarrhini* primates, but not in *Platyrrhini* species. This estimation was further corroborated by the identification of each human locus orthologous HML10 insertion in the genome assembly of 5 *Catarrhini* non-human primates species, providing the first comparative map of the group. This analysis also revealed a HML10 species-specific insertion in Gibbon chromosome 5, hence acquired after the evolutionary separation from subsequent species, i.e. less than 20 mya.

Conclusions

Besides the well studied HERV-K(C4) proviruses, also other HML10 sequences can be involved in the antisense control of homologous gene expression, possibly contributing to immune regulation and antiviral defense, as well as having a role in cancer development and auto-immunity. The present exhaustive characterization of all the HML10 sequences integrated in the human genome is thus the needed comprehensive background that is essential to assess the physio-pathological effects of HML10 expression.

Methods

HML10 sequences localization in human and non-human primates genomes

The HML10 sequences integrated in human genome assembly GRCh37/hg19 were identified based on the previous analysis of the latter with RetroTector software [3] combined with a UCSC Genome Browser [64, 65] BLAT search using the RepBase Update [34] assembled reference LTR14-HERVKC4-LTR14 as a query.

The HML10 loci orthologous to each human sequence have been identified through the comparative localization of the harboring genomic region for the following *Catarrhini* primate genome assemblies in UCSC Genome Browser:

- Chimpanzee (*Pan troglodytes*, assembly Feb. 2011 - CSAC 2.1.4/panTro4)
- Gorilla (*Gorilla gorilla gorilla*, assembly May 2011 - gorGor3.1/gorGor3)
- Orangutan (*Pongo pygmaeus abelii*, assembly July 2007 - WUGSC 2.0.2/ponAbe2)
- Gibbon (*Nomascus Leucogenys*, assembly Oct. 2012 - GGSC Nleu3.0/nomLeu3)
- Rhesus (*Macaca mulatta*, assembly Oct. 2010 - BGI CR_1.0/rheMac3)

while the search in Marmoset (*Platyrrhini* parvorder) genome sequence (*Callithrix jaccus*, assembly March 2009 - WUGSC 3.2/calJac3) gave negative results.

The eventual HML10 species specific insertion lacking an ortholog in humans have been searched in the same non human primates genome sequences through a UCSC Genome Browser [64, 65] BLAT search using the RepBase Update [34] assembled reference LTR14-HERVKC4-LTR14 as a query.

Analysis of HML10 chromosomal distribution

In order to estimate the expected number of integration events, each human chromosome length has been multiplied for the total number of HML10 insertions, including both proviruses and solitary LTR relics, and the obtained value has been divided for the total length of the human genome sequence. The number obtained, representing the expected proportion of HML10 insertion for each chromosome based on a random distribution principle, has been then compared to the actual amount of HML10 sequences.

HML10 proviral sequences alignment

Pairwise and multiple alignments of HML10 proviral nucleotide sequences were generated with Geneious bioinformatics software platform, version 8.1.4 [66] using MAFFT algorithm G-INS-i [67] with default parameters.

Pairwise and multiple alignments of HML10 puteins amino acid sequences were generated with Geneious bioinformatics software platform, version 8.1.4 [66] using MAFFT algorithm G-INS-i [67] with default parameters, after the bioinformatics translation of the correspondent gene portion.

All alignments have been visually inspected and, if necessary, manually corrected before further structural and phylogenetic analyses. The multiple alignment of the 9 HML10 proviral sequences with respect to LTR14-HERV-K(C4)-LTR14 reference is provided in fasta format as Additional file 2

Phylogenetic analyses

All phylogenetic trees were built from manually optimized multiple alignments generated by Geneious (see above) using Mega Software, version 6 [68] and NJ statistical method. Nucleotide and amino acid sequences NJ trees were built using the p-distance model and applying pairwise deletion option. Phylogenies were tested by the bootstrap method with 1000 replicates.

Beside HML10 proviral sequences, the trees included also the following reference sequences, as representative for endogenous and exogenous Betaretroviruses:

- HML10 prototype HERV-K(C4) RepBase [34] assembled nucleotide sequence (LTR14-HERVKC4-LTR14)
- HML1–10 Dfam [38] assembled nucleotide sequences: HML1 (LTR14A-HERVK14-LTR14A), HML2 (LTR5-HERVK-LTR5), HML3 (MER9B-HERVK9-MER9B), HML4 (LTR13-HERVK13-LTR13), HML5 (LTR22A-HERVK22-LTR22A), HML6 (LTR3-HERVK3-LTR3), HML7 (MER11D-HERVK11D-MER11D), HML8 (MER11A-HERVK11-MER11A), HML9 (LTR14C-HERVK14C-LTR14C) and HML10 (LTR14-HERVKC4-LTR14)
- MMTV nucleotide sequence (GenBank accession number: NC_001503.1)
- MPMV nucleotide sequence (GenBank accession number: NC_001550.1)
- JSRV nucleotide sequence(GenBank accession number: NC_001494.1)
- GenBank representative Rec proteins and their exogenous analogues amino acid sequences: HERV-

K HML2 (Q69383.1, P61573.1, P61576.1, P61575.1, P61574.1, P61572.1, P61578.1, P61579.1, P61571.1), HIV-1 Rev. (NP_057854), HTLV-1 Rex (NP_057863), STLV-1 Rex (NP_056908)

Structural analyses

The nucleotide sequence of each HML10 provirus has been aligned to the HML10 prototype HERV-K(C4) RepBase [34] assembled reference (LTR14-HERVKC4-LTR14) and all insertions and deletions ≥1 nucleotide as well as the main structural and regulatory features have been annotated in a graphical representation of the multiple alignment. The prediction of functionally relevant domains has been performed with the NCBI tool for conserved domains search [39] (https://www.ncbi.nlm.nih.gov/Structure/cdd/wrpsb.cgi)

PBS type and Betaretroviral features characterization

The PBS nucleotide sequence of each HML10 provirus has been aligned and compared with a library of 1171 known HERV PBS [3] to assign the most probably recognized tRNA. The general conservation of the PBS sequence among the HML10 proviruses has been represented by a logo generated at http://weblogo.berkeley.edu/logo.cgi [69] from the nucleotide alignment of all the HML10 PBS sequences.

The features known to be associated to Betaretroviruses, i.e. a Pro C-terminal G-patch motif (GYx2GxGLGx4GxnG), a Pro N-terminal dUTPase (DSDYxGEIQ), and two Gag NC Zinc finger motifs (CX2CX4HX4C) [3] were manually searched after the bioinformatics translation of the harboring genes (when present) in all the three possible reading frames with Geneious bioinformatics software platform, version 8.1.4 [66].

In order to individuate any bias in the HML10 sequences nucleotide composition, the relative frequencies of each nucleotide in the individual proviruses has been estimated by Geneious bioinformatics software platform, version 8.1.4 [66], after the manual removal of any eventual secondary integration. The final value for each nucleotide has been expressed as the average value obtained in the single HML10 proviruses.

Time of integration estimation

The time of integration of each HML10 provirus was estimated using a multiple approach of calculation, based on the percentage of divergent nucleotides (D%) between i) the two LTRs of each sequence, ii) each LTR of each sequence and a HML10 LTR consensus generated from our dataset alignment, and iii) the *gag*, *pol* and *env* genes of each sequence and a HML10 *gag*, *pol* and *env* consensus generated from our dataset alignment. Regarding *pol* and *env* genes, the nucleotides region showing high divergence between the two types of sequences were excluded, considering only the portions sharing a general identity comparable to the rest of the proviral structure (nucleotides 1277–2571 and 4103–5810 in LTR14-HERVKC4-LTR14 reference assembled reference, respectively). In particular, the pairwise D% between aligned nucleotide sequences was estimated, after removal of hypermutating CpG dinucleotides, by MEGA Software, version 6 [68], through a p-distance model with the pairwise deletion option applied. Variance was estimated by Neighbor Joining method with 1000 bootstrap replicates.

The estimated time of integration (T) was obtained according to the relation:

$$T = D\%/0,2\%$$

where 0.2% correspond to the neutral substitution rate acting on the human genome (percentage of mutation per nucleotide per million years). With regards to the D% between the two LTRs of the same provirus, which are known to be identical at time of integration, the T obtained was further divided by a factor of 2, considering that each LTR accumulates mutation independently.

For each HML10 provirus, the final T was expressed as the mean of the values obtained through the three approaches of D% calculation, after the exclusion of values with standard deviation >20%.

Additional files

Additional file 1: Figure S1. Phylogenetic analysis of the HML10 sequences 5'- and 3'LTRs with other endogenous and exogenous Betaretroviruses. The main HML10 phylogenetic group is indicated. In the absence of clear cluster division, the belonging of each element to the two subgroups is indicated based on the full-length proviruses phylogenetic analysis (Fig. 2). Evolutionary relationships were inferred by using the Neighbor Joining method and the Kimura-2-parameter model. The resulting phylogeny was tested by using the Bootstrap method with 1000 replicates. Length of branches indicates the number of substitutions per site. (PDF 12 kb)

Additional file 2: HML10 multiple alignment. FASTA multiple alignment of the 9 HML10 proviral sequences with respect to LTR14-HERV-K(C4)-LTR14 RepBase reference. (FASTA 149 kb)

Abbreviations
C4: fourth component of human complement gene; CYP21: steroid 21-hydroxylase; D: Divergence; DAP3: Death-associated protein 3; HERV: Human Endogenous Retroviruses; HIV-1: Human Immunodeficiency Virus 1; HML: Human MMTV-like; HTLV-1: Human T Lymphotropic Virus 1; IN: Integrase; JSRV: Jaagsiekte Sheep Retrovirus; LTR: Long Terminal Repeats; MHC: major histocompatibility complex; MMTV: Mouse Mammary Tumor Virus; MPMV: Mason-Pfizer Monkey Virus; mya: million years ago; NJ: Neighbor Joining; O.C.A.: Oldest Common Ancestor.; ORF: Open Reading Frame; PBS: Primer Binding Site; PPT: polypurine tract; puteins: putative proteins; RDDP: RNA Dependent DNA Polymerase; ReTe: RetroTector software; RNase H: Ribonuclease H; RP: serine/threonine nuclear protein kinase; RT: Reverse Transcriptase; STLV-1: Simian T Lymphotropic Virus 1; T: Time of integration; TNX: Tenascin extracellular matrix protein

Acknowledgements
Not applicable.

Funding
Not applicable.

Authors' contributions
NG and MC performed the analyses and wrote the manuscript. MPP and FE participated in the analysis and in the writing. JB and ET conceived and coordinated the study. All authors helped editing the manuscript and read and approved the final version.

Competing interests
The authors declare that they have no competing interests.

Author details
[1]Department of Life and Environmental Sciences, University of Cagliari, Cagliari, Italy. [2]Department of Medical Sciences, Uppsala University, Uppsala, Sweden. [3]Istituto di Ricerca Genetica e Biomedica, Consiglio Nazionale delle Ricerche (CNR), Monserrato, Cagliari, Italy.

References
1. Bannert N, Kurth R. The evolutionary dynamics of human endogenous retroviral families. Annu Rev Genomics Hum Genet. 2006;7:149–73.
2. Bock M, Stoye JP. Endogenous retroviruses and the human germline. Curr Opin Genet Dev. 2000;10:651–5.
3. Vargiu L, Rodriguez-Tomé P, Sperber GO, Cadeddu M, Grandi N, Blikstad V, et al. Classification and characterization of human endogenous retroviruses mosaic forms are common. Retrovirology. 2016;13
4. Blond JL, Beseme F, Duret L, Bouton O, Bedin F, Perron H, et al. Molecular characterization and placental expression of HERV-W, a new human endogenous retrovirus family. J Virol. 1999;73:1175–85.
5. Blond JL, Lavillette D, Cheynet V, Bouton O, Oriol G, Chapel-Fernandes S, et al. An envelope glycoprotein of the human endogenous retrovirus HERV-W is expressed in the human placenta and fuses cells expressing the type D mammalian retrovirus receptor. J Virol. 2000;74:3321–9.
6. Blaise S, de Parseval N, Bénit L, Heidmann T. Genomewide screening for fusogenic human endogenous retrovirus envelopes identifies syncytin 2, a gene conserved on primate evolution. Proc Natl Acad Sci U S A 2003;100: 13013–13018.
7. Mi S, Lee X, Li X, Veldman GM, Finnerty H, Racie L, et al. Syncytin is a captive retroviral envelope protein involved. Nature. 2000;403:785–9.
8. Mangeney M, Renard M, Schlecht-Louf G, Bouallaga I, Heidmann O, Letzelter C, et al. Placental syncytins: genetic disjunction between the fusogenic and immunosuppressive activity of retroviral envelope proteins. Proc Natl Acad Sci U S A. 2007;104:20534–9.
9. Tolosa JM, Schjenken JE, Clifton VL, Vargas A, Barbeau B, Lowry P, et al. The endogenous retroviral envelope protein syncytin-1 inhibits LPS/PHA-stimulated cytokine responses in human blood and is sorted into placental exosomes. Placenta. 2012;33:933–41.
10. Cegolon L, Salata C, Weiderpass E, Vineis P, Palù G, Mastrangelo G. Human endogenous retroviruses and cancer prevention: evidence and prospects. BMC Cancer. 2013;13:4.
11. Zeyl C, Bell G. Symbiotic DNA in eukaryotic genomes. Trends Ecol Evol. 1996;11:10–5.
12. Blomberg J, Benachenhou F, Blikstad V, Sperber G, Mayer J. Classification and nomenclature of endogenous retroviral sequences (ERVs): problems and recommendations. Gene. 2009;448:115–23.
13. Ono M, Yasunaga T, Miyata T, Ushikubo H. Nucleotide sequence of human endogenous retrovirus genome related to the mouse mammary tumor virus genome. J Virol. 1986;60:589–98.
14. Dangel AW, Mendoza AR, Menachery CD, Baker BJ, Daniel CM, Carroll MC, et al. The dichotomous size variation of human complement C4 genes is mediated by a novel family of endogenous retroviruses, which also establishes species-specific genomic patterns among old world primates. Immunogenetics. 1994;40:425–36.
15. Blanchong CA, Chung EK, Rupert KL, Yang Y, Yang Z, Zhou B, et al. Genetic, structural and functional diversities of human complement components C4A and C4B and their mouse homologues, Slp and C4. Int

Immunopharmacol. 2001;1:365–92.

16. Blanchong CA, Zhou B, Rupert KL, Chung EK, Jones KN, Sotos JF, et al. Deficiencies of human complement component C4A and C4B and heterozygosity in length variants of RP-C4-CYP21-TNX (RCCX) modules in caucasians. The load of RCCX genetic diversity on major histocompatibility complex-associated disease. J Exp Med. 2000;191:2183–96.

17. Chu X, Rittner C, Schneider PM. Length polymorphism of the human complement component C4 gene is due to an ancient retroviral integration. Exp Clin Immunogenet. 1995;12:74–81.

18. Mack M, Bender C, Schneider PM. Detection of retroviral antisense transcripts and promoter activity of the HERV-K(C4) insertion in the MHC III region. Immunogenetics. 2004;56:321–32.

19. Dangel AW, Baker BJ, Mendoza AR, Yu CY. Complement component C4 gene intron 9 as a phylogenetic marker for primates: long terminal repeats of the endogenous retrovirus ERV-K(C4) are a molecular clock of evolution. Immunogenetics. 1995;42:41–52.

20. Trowsdale J. Knight JC. Europe PMC Funders Group Major Histocompatibility Complex Genomics and Human Disease. 2015:301–23.

21. Schneider PM, Witzel-Schlömp K, Rittner C, Zhang L. The endogenous retroviral insertion in the human complement c4 gene modulates the expression of homologous genes by antisense inhibition. Immunogenetics. 2001;53:1–9.

22. Vincendeau M, Göttesdorfer I, Schreml JMH, Wetie AGN, Mayer J, Greenwood AD, et al. Modulation of human endogenous retrovirus (HERV) transcription during persistent and de novo HIV-1 infection. Retrovirology. 2015;12:27.

23. Mason MJ, Speake C, Gersuk VH, Nguyen Q-A, O'Brien KK, Odegard JM, et al. Low HERV-K(C4) copy number is associated with type 1 diabetes. Diabetes. 2014;63:1789–95.

24. Pani MA, Wood JP, Bieda K, Toenjes RR, Usadel KH, Badenhoop K. The variable endogenous retroviral insertion in the human complement C4 gene: a transmission study in type I diabetes mellitus. Hum Immunol. 2002;63:481–4.

25. Sperber G, Airola T, Jern P, Blomberg J. Automated recognition of retroviral sequences in genomic data–RetroTector. Nucleic Acids Res. 2007;35:4964–76.

26. Magiorkinis G, Belshaw R, Katzourakis A. "There and back again": revisiting the pathophysiological roles of human endogenous retroviruses in the post-genomic era. Philos Trans R Soc Lond Ser B Biol Sci. 2013;368(1626): 20120504.

27. Broecker F, Horton R, Heinrich J, Franz A, Schweiger M-R, Lehrach H, et al. The intron-enriched HERV-K(HML-10) family suppresses apoptosis, an indicator of malignant transformation. Mob DNA. 2016;7:25.

28. Schön U, Seifarth W, Baust C, Hohenadl C, Erfle V, Leib-Mösch C. Cell type-specific expression and promoter activity of human endogenous retroviral long terminal repeats. Virology. 2001;279:280–91.

29. Frank O, Jones-Brando L, Leib-Mosch C, Yolken R, Seifarth W. Altered transcriptional activity of human endogenous retroviruses in neuroepithelial cells after infection with Toxoplasma gondii. J Infect Dis. 2006;194:1447–9.

30. Diem O, Schäffner M, Seifarth W, Leib-Mösch C. Influence of antipsychotic drugs on human endogenous retrovirus (HERV) transcription in brain cells. PLoS One. 2012;7

31. Assinger A, Yaiw K-C, Göttesdorfer I, Leib-Mösch C, Söderberg-Nauclér C. Human cytomegalovirus (HCMV) induces human endogenous retrovirus (HERV) transcription. Retrovirology. 2013;10:132.

32. Tassabehji M, Strachan T, Anderson M, Campbell RD, Collier S, Lako M. Identification of a novel family of human endogenous retroviruses and characterization of one family member, HERV-K(C4), located in the complement C4 gene cluster. Nucleic Acids Res. 1994;22:5211–7.

33. Grandi N, Cadeddu M, Blomberg J, Tramontano E. Contribution of type W human endogenous retroviruses to the human genome : characterization of HERV - W proviral insertions and processed pseudogenes. Retrovirology BioMed Central. 2016:1–25.

34. Bao W, Kojima KK, Kohany O. Repbase update, a database of repetitive elements in eukaryotic genomes. Mob DNA. 2015;6:11.

35. Hughes JF, Coffin JM. Evidence for genomic rearrangements mediated by human endogenous retroviruses during primate evolution. Nat Genet. 2001; 29:487–9.

36. Medstrand P, van de Lagemaat LN, Mager DL. Retroelement distributions in the human Genome: variations associated with age and proximity to genes. Genome Res. 2002;12:1483–95.

37. Katzourakis A, Pereira V, Tristem M. Effects of recombination rate on human endogenous retrovirus fixation and persistence. J. Virol. 2007;81:10712–7.

38. Hubley R, Finn RD, Clements J, Eddy SR, Jones TA, Bao W, et al. The Dfam database of repetitive DNA families. Nucleic Acids Res. 2016;44:D81–9.

39. Marchler-Bauer A, Bo Y, Han L, He J, Lanczycki CJ, Lu S, et al. CDD/SPARCLE: functional classification of proteins via subfamily domain architectures. Nucleic Acids Res. 2017;45:D200–3.

40. Magin C, Löwer R, Löwer J. cORF and RcRE, the rev/Rex and RRE/RxRE homologues of the human endogenous retrovirus family HTDV/HERV-K. J Virol. 1999;73:9496–507.

41. Magin-Lachmann C, Hahn S, Strobel H, Held U, Löwer J, Löwer R. Rec (formerly Corf) function requires interaction with a complex, folded RNA structure within its responsive element rather than binding to a discrete specific binding site. J Virol. 2001;75:10359–71.

42. Subramanian RP, Wildschutte JH, Russo C, Coffin JM. Identification, characterization, and comparative genomic distribution of the HERV-K (HML-2) group of human endogenous retroviruses. Retrovirology. 2011;8:90.

43. Bateman A, Martin MJ, O'Donovan C, Magrane M, Alpi E, Antunes R, et al. UniProt: the universal protein knowledgebase. Nucleic Acids Res. 2017;45: D158–69.

44. Boese A, Sauter M, Mueller-Lantzsch N. A rev-like NES mediates cytoplasmic localization of HERV-K cORF. FEBS Lett. 2000;468:65–7.

45. Lebedev YB, Belonovitch OS, Zybrova NV, Khil PP, Kurdyukov SG, Vinogradova TV, et al. Differences in HERV-K LTR insertions in orthologous loci of humans and great apes. Gene. 2000;247:265–77.

46. Johnson WE, Coffin JM. Constructing primate phylogenies from ancient retrovirus sequences. Proc Natl Acad Sci U S A. 1999;96:10254–60.

47. Nachman MW, Crowell SL. Estimate of the mutation rate per nucleotide in humans. Genetics. 2000;156:297–304.

48. Stoye JP. Endogenous retroviruses: still active after all these years? Curr Biol. 2001;11:914–6.

49. Jern P, Sperber GO, Blomberg J. Use of endogenous retroviral sequences (ERVs) and structural markers for retroviral phylogenetic inference and taxonomy. Retrovirology. 2005;2:50.

50. Mangeat B, Turelli P, Caron G, Friedli M, Perrin L, Trono D. Broad antiretroviral defence by human APOBEC3G through lethal editing of nascent reverse transcripts. Nature. 2003;424:99–103.

51. Maydt J, Lengauer T. Recco: recombination analysis using cost optimization. Bioinformatics. 2006;22:1064–71.

52. Grandi N, Tramontano E. Type W human endogenous retrovirus (HERV-W) integrations and their mobilization by L1 machinery: contribution to the human transcriptome and impact on the host physiopathology. Viruses. 2017;9

53. Kulski JK, Gaudieri S, Inoko H, Dawkins RL. Comparison between two human endogenous retrovirus (HERV)-rich regions within the major histocompatibility complex. J Mol Evol. 1999;48:675–83.

54. Benachenhou F, Sperber GO, Bongcam-Rudloff E, Andersson G, Boeke JD, Blomberg J. Conserved structure and inferred evolutionary history of long terminal repeats (LTRs). Mob DNA. 2013;4:5.

55. Schmitt K, Heyne K, Roemer K, Meese E, Mayer J. HERV-K(HML-2) rec and np9 transcripts not restricted to disease but present in many normal human tissues. Mob DNA. 2015;6:4.

56. Denne M, Sauter M, Armbruester V, Licht JD, Roemer K, Mueller-Lantzsch N. Physical and functional interactions of human endogenous retrovirus proteins Np9 and rec with the promyelocytic leukemia zinc finger protein. J Virol. 2007;81:5607–16.

57. Kaufmann S, Sauter M, Schmitt M, Baumert B, Best B, Boese A, et al. Human endogenous retrovirus protein Rec interacts with the testicular zinc-finger protein and androgen receptor. J Gen Virol. 2010;91:1494–502.

58. Hanke K, Chudak C, Kurth R, Bannert N. The Rec protein of HERV-K(HML-2) upregulates androgen receptor activity by binding to the human small glutamine-rich tetratricopeptide repeat protein (hSGT). Int J Cancer. 2013; 132:556–67.

59. Hanke K, Hohn O, Liedgens L, Fiddeke K, Wamara J, Kurth R, et al. Staufen-1 interacts with the human endogenous retrovirus family HERV-K(HML-2) rec and gag proteins and increases virion production. J Virol. 2013;87:11019–30.

60. Kassiotis G. Endogenous retroviruses and the development of cancer. J Immunol. 2014;192:1343–9.

61. Suntsova M, Garazha A, Ivanova A, Kaminsky D, Zhavoronkov A, Buzdin A. Molecular functions of human endogenous retroviruses in health and disease. Cell Mol Life Sci Springer Basel. 2015;72:3653–75.

62. Gimenez J, Mallet F. ERVWE1 (endogenous retroviral family W, Env(C7), member 1). Atlas Genet Cytogenet Oncol Haematol. 2008;12:134–48.

63. Lee YN, Malim MH, Bieniasz PD. Hypermutation of an ancient human retrovirus by APOBEC3G. J Virol. 2008;82:8762–70.
64. Kent W, Sugnet CW, Furey TS, Roskin KM, Pringle TH, Zahler AM, et al. The human genome browser at UCSC. Genome Res. 2002;12:996–1006.
65. Karolchik D, Barber GP, Casper J, Clawson H, Cline MS, Diekhans M, et al. The UCSC genome browser database: 2014 update. Nucleic Acids Res. 2014; 42:D764–70.
66. Kearse M, Moir R, Wilson A, Stones-Havas S, Cheung M, Sturrock S, et al. Geneious basic: an integrated and extendable desktop software platform for the organization and analysis of sequence data. Bioinformatics. 2012;28: 1647–9.
67. Katoh K, Standley DM. MAFFT multiple sequence alignment software version 7: improvements in performance and usability. Mol Biol Evol. 2013; 30:772–80.
68. Tamura K, Stecher G, Peterson D, Filipski A, Kumar S. MEGA6: molecular evolutionary genetics analysis version 6.0. Mol. Biol. Evolution. 2013;30: 2725–9.
69. Crooks GE, Hon G, Chandonia JM, Brenner SE. WebLogo: A sequence logo generator. Genome Res. 2004;14:1188–90.
70. Steiper ME, Young NM. Primate molecular divergence dates. Mol Phylogenet Evol. 2006;41:384–94.
71. Perelman P, Johnson WE, Roos C, Seuánez HN, Horvath JE. Moreira M a M, et al. a molecular phylogeny of living primates. PLoS Genet. 2011;7:1–17.

12

Transcription coupled repair and biased insertion of human retrotransposon L1 in transcribed genes

Geraldine Servant[1], Vincent A. Streva[1,2] and Prescott L. Deininger[1,3]*

Abstract

Background: L1 retrotransposons inserted within genes in the human genome show a strong bias against sense orientation with respect to the gene. One suggested explanation for this observation was the possibility that L1 inserted randomly, but that there was negative selection against sense-oriented insertions. However, multiple studies have now found that *de novo* and polymorphic L1 insertions, which have little opportunity for selection to act, also show the same bias.

Results: Here we show that the transcription-coupled sub-pathway of nucleotide excision repair does not affect the overall rate of insertion of L1 elements, which is in contrast with the regulation by the global sub-pathway of nucleotide excision repair. The transcription-coupled subpathway does cause a strong bias against insertion in the sense orientation relative to genes.

Conclusions: This suggests that a major portion of the L1 orientation bias might be generated during the process of insertion through the action of transcription-coupled nucleotide excision repair.

Keywords: L1 retrotransposon, Transcription-coupled repair, Target-primed reverse transcription, DNA repair, Mutagenesis

Background

Sequencing of the human genome revealed that transposable elements make up almost half of the genome [1, 2]. The long interspersed element L1 is the only active, autonomous retrotransposon in the human cells and constitutes 17% of the genome. L1 inserts are relatively randomly distributed in genic and intergenic regions, with the elements showing a genomic preference for AT-rich regions [3]. However, L1 copies within genes show a significant enrichment for the antisense orientation. It has been proposed that this orientation bias may be caused by a selection process limiting transcriptional interference with gene expression [3]. However, a similar trend is observed with published *de novo* inserts recovered in HeLa cells using an engineered L1 element, although there are insufficient data for that to

reach significance [4–6], and somatic L1 insertions identified in brain cells [7]. These latter findings would be expected to be subjected to much less selection and raise the possibility for an insertion-related mechanism controlling L1 insertion in actively transcribed genes, in a gene-orientated manner.

Transcription-coupled repair (TCR), a sub-pathway of nucleotide excision repair (NER), is a DNA repair pathway that excises helix distorting lesions. These lesions are typically caused by UV-light exposure or chemical compounds and they block the RNA polymerase II (RNAPII) processivity on the template strand of transcribing genes ([8] and Fig. 1). CSA and CSB (Cockayne Syndrome proteins A and B), the sensor proteins of the pathway, are recruited to the stalled RNAPII complex and initiate the excision process of the damaged strand. If the bulky DNA lesion is located on the coding strand of the gene or in an untranscribed genomic region, they do not interfere with the transcription process and are then subject to the slower, global genome repair (GGR) NER sub-pathway ([9, 10] and Fig. 1). After lesion recognition, the TCR and GGR mechanisms converge on a

* Correspondence: pdeinin@tulane.edu
[1]Tulane University, Tulane Cancer Center and the Department of Epidemiology, 1430 Tulane Ave, New Orleans, LA 70112, USA
[3]Tulane Cancer Center, SL66, Tulane University Health Sciences Center, 1430 Tulane Ave., New Orleans, LA 70112, USA
Full list of author information is available at the end of the article

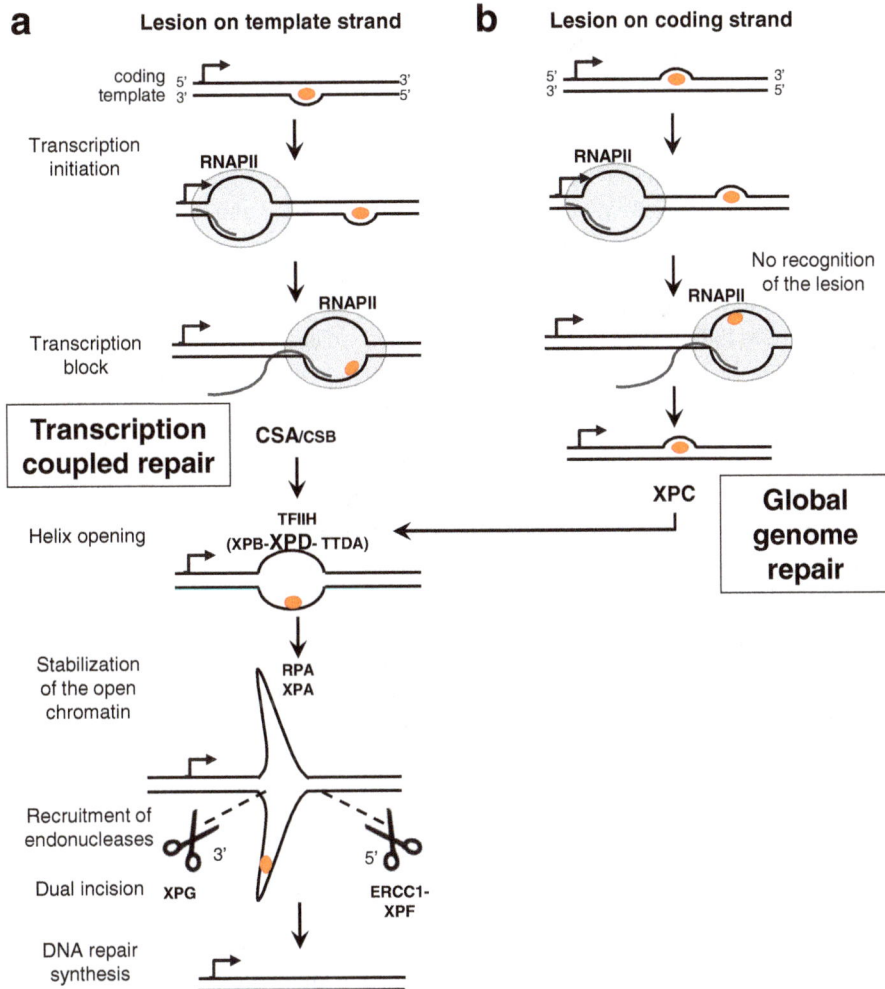

Fig. 1 The predicted influence of NER sub-pathways on the coding or template damaged strand of an actively transcribed gene. **a** and **b** Schematic representation of the repair of a bulky lesion located on the template strand (panel **a**) or on the coding strand (panel **b**) of an active gene. If the lesion is on the template strand (panel **a**), read by RNAPII during the transcription process, the lesion causes the RNAPII complex to stall. CSA and CSB proteins are the sensors of stalled RNAPII and recruit the transcription complex TFIIH to the site of the lesion. The helicase activities, XPB and XPD, of the TFIIH complex open the chromatin around the lesion. XPA and RPA stabilized the open structure of the chromatin. The endonucleases, ERCC1-XPF in 5' and XPG in 3' cleave the damaged strand. The gap is then filled by DNA repair polymerases and ligases. If the lesion is on the coding strand (panel **b**) and therefore not read by the RNAPII complex, the lesion does not interfere with the enzyme processivity and the gene is transcribed. The lesion can be later recognized by the XPC complex, the lesion binding proteins in the global genome repair (GGR), the second NER sub-pathway. After the lesion recognition step, both GGR and TCR are identical. In bold are the factors controlled for their impact on L1 insertion regulation in the present study

common series of steps. Briefly, the DNA helix is opened by the helicase proteins, XPD and XPB, of the TFIIH complex [11]. The open DNA structure is then stabilized by XPA-RPA proteins [12, 13]. ERCC1-XPF and XPG endonucleases cleave the damaged strand at 5' and 3' ends of the lesion [14, 15].

We have recently shown that several proteins of the NER pathway, notably two central proteins of the DNA repair, XPD and XPA [16], as well as the endonuclease ERCC1-XPF [17] and the lesion binding protein XPC of the GGR pathway, limits L1 retrotransposition [16]. In cells with mutations in these genes, the L1 retrotransposition rate increased and generated larger tandem site duplications(TSDs) at the insertion site that are abnormally large [4–6, 16, 18]. As the GGR pathway can inhibit L1 insertions, we hypothesize that the TCR sub-pathway may also serve the same role. The TCR sub-pathway is only active on the portion of the genome that is actively transcribing in any given cell ([8] and see Discussion) while the GGR sub-pathway activity would continue to protect the majority of the genome. Because of this, we expect that TCR would not greatly affect the overall rate of retrotransposition. However, we hypothesize that it might generate a strong bias against L1 insertions in the template strand of transcribed genes, thereby helping to explain the observed bias in orientation of L1 elements within genes.

Methods

Cell lines and culture conditions

HeLa cells (ATCC CCL2) were grown in eMEM supplemented with 10% Fetal Bovine Serum, 0.1 mM non-essential amino acids (Life Technologies) and 1 mM sodium pyruvate (Life Technologies) at 37° in a 5% carbon dioxide environment. The following cell lines were obtained from the Coriell Cell Repository: CSA-SV40 transformed fibroblasts (GM16094), XPC-SV40 transformed fibroblasts (GM15983), XPD- SV40 transformed fibroblasst (GM08207), the stably complemented version of XPD- cell line (XPD+) (GM15877). XPC-, XPD- and CSA- cell lines were grown in eMEM supplemented with 10% Fetal Bovine Serum, 0.1 mM non-essential amino acids (Life Technologies) at 37° in a 5% carbon dioxide environment. XPD+ cell line was grown in the DMEM supplemented with 10% Fetal Bovine Serum (Life Technologies). A stably complemented version of the CSA- cell line (CSA+) was generated in this study by transfecting CSA- cells with a CSA cDNA expression vector (# EX-S0507-M67, GeneCopoeia) along with a hygromycin selection vector to allow selection for integration. CSA+ cells are maintained in eMEM medium supplemented with 10% Fetal Bovine Serum (Life Technologies), 0.1 mM non-essential amino acids (Life Technologies) and 200 μg/mL hygromycin at 37° in a 5% carbon dioxide environment.

Plasmids

JM102/L1.3 contains the CMV promoter upstream of the L1.3 element deleted for the 5′ UTR and the *mneo* indicator cassette cloned in pCEP4 plasmid [19].

JM102/D702A/L1.3 derives from JM102/L1.3 and contains the reverse transcriptase deficient mutant of an L1.3 element and the *mneo* retrotransposition cassette cloned in pCEP4 vector [19].

TAM102/L1.3 contains the CMV promoter upstream of the L1.3 element deleted for the 5′ UTR and the *mblastI* indicator cassette cloned in pCEP4 vector [20].

TAM102/D702A/L1.3 derives from TAM/L1.3 and contains the reverse transcriptase deficient mutant of an L1.3 element and the *mblastI* indicator cassette cloned in pCEP4 vector [20].

TAM102/H230A/L1.3 derives from TAM102/L1.3 and contains the endonuclease deficient mutant of the L1.3 element and the *mblastI* indicator cassette cloned in pCEP4 vector [20].

EX-S0507-M67 (GeneCopoeia) contains the CSA cDNA driven by CMV promoter and a hygromycin resistance gene in pReceiverM67 vector.

The synL1_neo vector used for the recovery of *de novo* L1 inserts was previously described [21].

The pIRES2-EGFP vector (Clontech) contains a neomycin resistance gene expressed from a SV40 promoter.

The vector contains a multi-cloning site upstream of an IRES and eGFP marker. The cloned gene and eGFP marker are expressed from the CMV promoter on the same transcript.

All plasmid DNA were purified by Maxiprep kit (Qiagen). DNA quality was also evaluated by the visual assessment of ethidium bromide stained agarose gel electrophoresed aliquots.

Retrotransposition assays

Briefly, 5×10^6 CSA+ and CSA- cells were seeded in T75 flasks. Cells were transfected the next day at about 90% confluence using Lipofectamine 2000 (Life Technologies) following the manufacturer's protocol. Cells were transfected with 3 μg of L1.3 or L1.3-RT (–) construct tagged with the *mneo* retrotransposition cassette (JM102/L1.3 or JM102/D702A/L1.3) in T75 flasks. Two days after transfection, cells were selected for the transposition events in medium, containing 500 μg/mL Geneticin (Life Technologies). After 14 days, cells were fixed and stained with crystal violet solution (0.2% crystal violet in 5% acetic acid and 2.5% isopropanol) (Fig. 2b). Each assay was performed in triplicate. The number of neo[R] colonies was counted in each flask.

L1 toxicity and colony formation assay

L1 toxicity and colony formation assays were performed using the L1 episomal and the pIRES2-EGFP vectors. Briefly, 5×10^6 CSA+ and CSA- cells were seeded in T75 flasks. Cells were transfected the next day at about 90% confluence using Lipofectamine 2000 (Life Technologies) following the manufacturer's protocol. Cells were transfected with 3 μg of L1.3, or L1.3-EN (–) construct tagged with the *mblast* retrotransposition cassette (TAM102/L1.3, or TAM102/H230A/L1.3) and 0.5 μg of pIRES2-EGFP vectors (pIRES2-GFP was used because it contains a G418 resistance cassette). Cells were selected for the presence of the pIRES2-EGFP plasmid in selective medium containing 500 μg/ml geneticin (Life Technologies) for 14 days. The cells were then fixed and stained with crystal violet solution (0.2% crystal violet in 5% acetic acid and 2.5% isopropanol). The number of neo[R] colonies was counted in each flask.

RT-qPCR

Total RNA were extracted from a confluent T75 flask, using TRIzol Reagent (Life Technologies). We then carried out chloroform extraction and isopropanol precipitation. RNA was suspended in 100 μL of DEPC-treated water. The cDNA was synthetized using the Reverse Transcription System (Promega), following the manufacturer's protocol. Briefly 1 μg of total RNA was denatured at 65° for 5 min. The reverse transcription reaction was primed with Oligo(dT)$_{15}$ primers and incubated at 42°

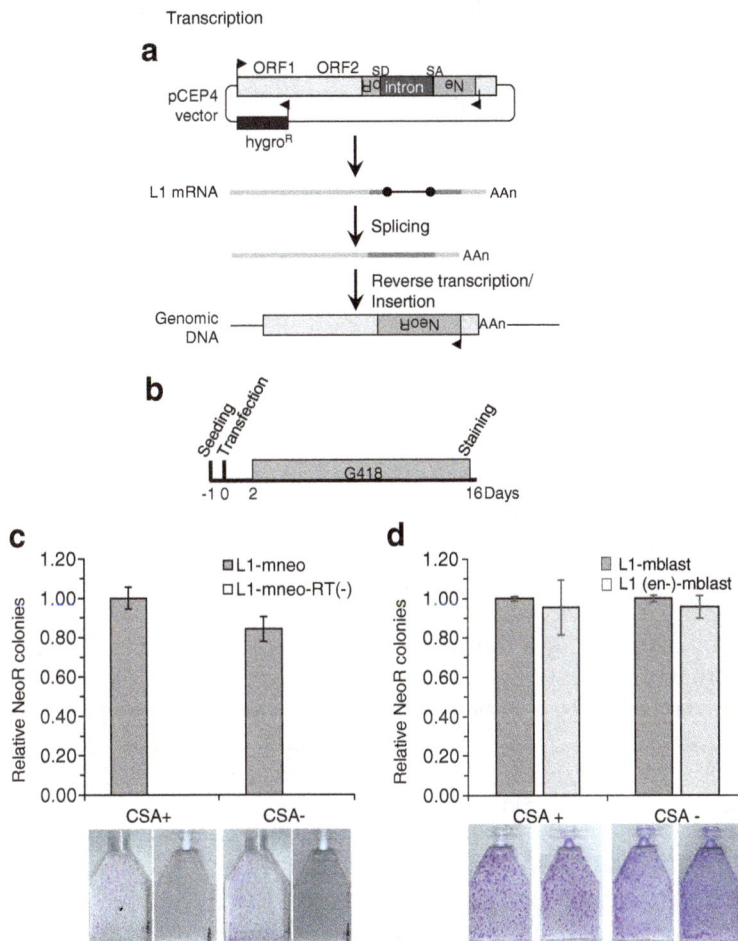

Fig. 2 L1 retrotransposition rate is not significantly different in CSA-deficient cells (CSA-) and in the stably complemented CSA-deficient cells (CSA+). **a** Schematic of L1 retrotransposition assay. The L1.3 element tagged at the 3' end with the *mneo* retrotransposition sensor is inserted in a pCEP4 vector (JM102/L1.3 vector). The retrotransposition cassette consists of a neomycin resistance (NeoR) gene in antisense orientation relative to the L1 element and expressed from its own promoter. The NeoR gene is not functional in the retrotransposition cassette because it is interrupted by an intron in L1 sense orientation. The NeoR gene becomes functional only after transcription, splicing, reverse transcription of L1 mRNA and insertion of L1 cDNA. **b** Schematic of the timing of the L1 retrotransposition assay. CSA- and CSA+ cells are seeded the day before transfection with JM102/L1.3 (L1.3-mneo-WT) or JM102/D702A/L1.3 (L1.3-mneo-RT(−)) expression vector. Two days after transfection, G418 selection is added to the growth medium and cells are kept under selection for 14 days. At the end of the assay, cells are fixed and stained and the number of NeoR colonies is determined. **c** CSA-deficient cells (CSA-) and the stably complemented version (CSA+) were transfected with JM102/L1.3 (L1-mneo) or JM102/D702A/L1.3 (L1-mneo-RT(−)) construct. Colony formation was assayed after two weeks under neomycin selection. The graph shows the relative colony number (average ± S.D.) of three independent experiments. Values are normalized to L1.3 WT vector. No significant differences ($p > 0.05$, two-tailed two sample Student's T-test) were observed between the L1-mneo expression constructs in the different CSA+ and CSA- cells. Representative examples of NeoR colony formation from L1 retrotransposition assay in CSA+ and CSA- cells were presented below the graph. No colonies were detected with the L1 element with a defective RT. **d** CSA-deficient cells (CSA-) and the complemented version (CSA+) were co-transfected with TAM102/L1.3 (L1 mblast), or TAM102/H230A/L1.3 (L1 (en-)-mblast) construct and pIRES2-EGFP vector, a vector carrying a constitutive NeoR expression cassette. Colony formation due to random integration of this transfected plasmid was assayed after two weeks under neomycin selection. The L1 expression constructs were only included as a functional L1 and a defective (en-) L1 so that the experiment can simultaneously test for differences in the CSA- and CSA+ cells for transfection, colony formation and potential toxicity from the L1. The graph shows the relative colony number (average ± S.D.) of three independent experiments. Values are normalized to L1.3 WT vector. No significant differences ($p > 0.05$, two-tailed two sample Student's T-test) were observed between the different L1 expression constructs in the different cell lines. Representative examples of NeoR colony formation from this L1 toxicity assay in CSA+ and CSA- cells are presented below the graph

for 1 h in a thermocycler (BioRad, C1000 Touch). The enzyme was then heat-inactivated at 85° for 5 min. The PCR amplification of CSA cDNA was performed using previously published primers [22]. Meanwhile, the PCR amplification of beta-actin cDNA was performed as a control of the assay. The PCR products were analyzed on a 1% agarose gel and the bands were gel extracted and cloned into TOPO-TA (Life Technologies). Cloned PCR products were Sanger sequenced using M13 forward and reverse primers. Samples were sent for Sanger sequencing to Elim Biopharmaceuticals, Inc., Hayward, California. Lasergene 10 SeqBuilder software was utilized for sequence analysis and the sequences were compared to the reference cDNA using BLAST software (website: https://blast.ncbi.nlm.nih.gov/Blast.cgi).

Recovery of *de novo* L1 insertions

De novo L1 insertion recovery was performed as previously described [16]. Briefly, 5×10^6 CSA- and CSA+ cells were transfected with 3 μg synL1_neo rescue vector [21] using Lipofectamine 2000 reagent (Life Technologies). Cells were selected with 500 μg/mL of Geneticin (Life Technologies) for 14 days to allow for colony formation. Neo^R cells were harvested by trypsinization and genomic DNA was extracted using a Qiagen DNeasy Blood and Tissue kit. Genomic DNA was digested with 100 U of *Hin*dIII (NEB) overnight at 37°. The following day, digested genomic DNA was self-ligated using 1200 U T4 DNA ligase (NEB) in a volume of 1 mL overnight at room temperature. DNA was purified and concentrated using centrifugal filters (Amicon Ultra, 0.5 mL, 50 K, Millipore). Purified DNA was transformed by electroporation into competent DH5α *E. coli* (Life Technologies). Individual kanamycin-resistant colonies were grown and plasmid DNA was harvested using SV Wizard miniprep kit (Promega). The 5′ end of the *de novo* L1 insertion was sequenced using primers specific to the L1 rescue plasmid and primer walking until the 5′ end of the insert was recovered as described in [20]. Because sequencing through a long adenosine tract at the 3′ end of the L1 insertions is not effective, the 3′ flanking genomic region was sequenced by ligation mediated PCR based on [23, 24]. Briefly, a pool of five to six L1 rescue vectors was digested with *Stu*I (NEB) to relax supercoils, and then sheared by sonication using a Bioruptor (Diagenode, high, 30 s on, 90 s off, for 12 min). Sheared plasmid DNA was primer extended using an oligo specific to the 3′ end of the synL1_neo rescue plasmid (3′_rescue_1: 5′ ATATATGAG TAACCTGAGGC 3′ or 3′_rescue_1_secondpA: 5′ GTGGGCATTCTGTCTTGTTC 3′). Duplexed T-linkers were ligated using 10 U T4 DNA ligase and PCR was performed using the primers: linker specific (5′ ACACTCTTTCCCTACACGACGCTCTTCCGATCT 3′) and 3′_rescue_1 (or 3′_rescue_1_secondpA) primer. PCR was carried out with these steps: initial denaturation at 94°, 20 cycles of 94° for 30s, 60° for 1 min, 72° for 1 min, and a final extension for 10 min at 72°. PCR reactions were run on a 1% agarose gel and a light smear between 400 and 700 nt was gel extracted with the Qiaquick gel extraction kit (Qiagen). One μL of gel extracted DNA was subject to an additional 15 cycles of PCR amplification as described above using linker specific and nested 3′ rescue vector primers (3′_rescue_2: 5′ TGAGTAACCT GAGGCTATGCTG 3′ or 3′_rescue_2_secondpA: 5′

TTCTGTCTTGTTCCGGTTCTTAAT 3′). The nested PCR product was run on a 1% agarose gel and the resulting smear was gel extracted and cloned into TOPO-TA (Life Technologies). Cloned PCR products were Sanger sequenced using M13 forward and reverse primers to determine 3′ end junctions. Samples were sent for sequencing to Elim Biopharmaceuticals, Inc., Hayward, California. Lasergene 10 SeqBuilder software was utilized for sequence analysis. Flanking regions were mapped on the human reference genome hg19 (build 37) using Blat tool (https://genome.ucsc.edu/cgi-bin/hgBlat). The sequence data related to these insertions is included in Additional file 1: Table S3.

Immunoblot analysis

To evaluate expression of CSA protein in the cells, HeLa, XPC-, XPD-, XPD+, CSA- and CSA+ cells were haverested in 300 μl of lysis buffer (50 mM Tris, pH 7.2, 150 mM NaCl, 0.5% Triton X-100, 10 mM EDTA, 0.5% SDS). After 10 min of sonication (Bioruptor, Diagenode, manufacturer's recommended settings), lysates were clarified by centrifugation for 15 min at 4° at 13,000 rpm and the protein concentration was determined by Bradford assay (Biorad). 40 μg of protein was fractionated on a 4–12% bis-tris polyacrylamide gel (Life Technologies). Proteins were transferred to a nitrocellulose membrane using the iBlot gel transfer system from Life Technologies (manufacturer's settings). The membrane was blocked for 1 h at room temperature in PBS (pH 7.4), 0.1% Tween 20 (Sigma), 5% skim milk powder (OXOID) and then incubated overnight at 4° with an anti-CSA monoclonal antibody (D-2, sc-376,981, Santa Cruz Biotechnology) diluted at 1:500 and an anti-GAPDH antibody (FL-335, sc-25,778, Santa Cruz Biotechnology) diluted at 1:1000 in PBS, 0.1% Tween 20, 3% non-fat dry milk. The membrane was then incubated for 1 h at room temperature with the secondary goat anti-mouse or donkey anti-rabbit HRP-conjugated antibody (sc-2005, sc-2313, Santa Cruz Biotechnology) diluted at 1:100,000 in PBS, 0.1% Tween 20, 3% non-fat milk. Signals were detected using Super Signal West Femto Chemiluminescent Substrate (Pierce).

UV sensitivity assay

The protocol was adapted from [25]. Briefly 5×10^5 cells were seeded in 6-cm plates and grown in growth medium for 24 h. The growth medium was removed and the cells were irradiated in the presence of 1 mL of 1X phosphate buffer saline (PBS) with a bactericidal UVC lamp (254 nm, 1.57 J/m^2/s) at 0, 3, 6, 9 and 12 J/m^2 UVC dose. The PBS was removed and replaced with growth medium. After 4 days, cells were counted with a hemocytometer to determine cell survival. Cell survival was calculated as the percent of live cells in the irradiated sample relative to the untreated sample.

RNA-Seq analysis of HeLa gene expression

RNA was isolated from HeLa cells as described for RT-PCR. 5 μg of RNA was submitted to the University of Wisconsin Biotechnology Center (http://www.biotech. wisc.edu/services/dnaseq/services/Illumina) for polyA selection and strand-specific 2 × 100 bp RNA sequencing on an Illumina HiSeq2000. Approximately 40 million reads were subjected to RSEM analysis [26] on the human GR38 reference genome and output calculated for all of the ENCODE coding gene alignments in FPKM (fragments per kilobase per million reads).

Results

CSA protein does not control the rate of L1 retrotransposition

In GGR-deficient cells, we have observed an increase of 3–10-fold in L1 retrotransposition rate in comparison to the complemented cell lines, suggesting that the NER repair pathway limits L1 insertion to the genome [16]. We therefore wondered if the L1 retrotransposition rate would also increase in TCR-deficient cells. SV40-transformed, CSA-deficient (CSA-) skin fibroblasts were obtained from Coriell Cell Repository from a patient suffering from cockayne syndrome (see materials and methods). These cells express a truncated CSA mRNA that does not produce functional CSA protein and the cells are remarkably sensitive to UV light exposure ([22] and Additional file 2: Figure S1). We stably complemented the cells by transfection with a CSA cDNA expression vector under selection and controlled for the efficiency of the complementation with a functional UV sensitivity assay (Materials and Methods and [25]). The data revealed that the stably complemented (CSA+) cells are less sensitive to UV light exposure (Additional file 2: Figure S1A). RT-PCR and immunoblot assays confirmed the overexpression of CSA mRNA and protein in the stably complemented cells (Additional file 2: Figs. S1A and S1B).

To test the activity level of the L1 retrotransposon in CSA-deficient and complemented cells, we performed an L1 retrotransposition assay by transfecting the cells with the JM102/L1.3 vector expressing the L1.3 element tagged at the 3′end with *mneoI* retrotransposition cassette [19]. The retrotransposition cassette contains an antisense neomycin resistant gene, interrupted by a sense oriented intron that is spliced only in L1 mRNA (Fig. 2a). Therefore, the neo[R] gene becomes expressed and functional only after retrotransposition. The assay allows for an estimation of L1 retrotransposition rate by counting Neo[R] colonies 14 days after selection (Fig. 2b). In contrast to the results obtained in GGR-deficient cells, the retrotransposition assays do not show a rate increase in CSA- cells in comparison to isogenic CSA+ cells (Fig. 2c and Additional file 2: Figs. S2A-C). There were also no measurable differences in L1-caused toxicity in the cells or cell growth as shown in Fig. 2d and Additional file 2: Figs. S2D-F. This study suggests that if there is a difference of L1 retrotransposition rate in these cells, it is relatively minor, as we would have predicted based on the relatively small portion of the genome under surveillance by the TCR-NER pathway at any one time ([8] and see RNA-Seq gene expression data (Additional file 2: Figure S3)).

de novo L1 inserts do not generate large duplications at the target site in CSA-deficient cells

In GGR-deficient cells, we also observed that abnormally large duplications (over 1 kb) were formed at the L1 insertion site [16]. We therefore decided to investigate the features of L1 *de novo* insertions in CSA-deficient and complemented cells (Additional file 1: Tables S1 and S2). We have recovered 60 and 75 L1 *de novo* insertions from CSA-deficient and complemented cells, respectively (Additional file 1: Tables S1 and S2), using the synL1_neo rescue vector and the previously published method (Materials and Methods section and [16, 20, 27]). Surprisingly, the characteristics of L1 *de novo* insertions were very similar in CSA- and CSA+ cells. No chromosome was specifically targeted by L1 *de novo* insertions. No significant difference was identified in the median length of the inserts in CSA+ and CSA- cells (3401 and 3642 bp respectively) (Fig. 3a). Additionally, we found about 21% of L1 *de novo* insertions were full length in both cells lines, consistent with 10% - 30% observed in previous studies [1, 4, 28–30]. Except for one recovered insert in CSA- cells, all L1 *de novo* insertions had a poly-A tail and their target site sequences were T-rich, close to the TTTT/A consensus sequence (Additional file 1: Tables S1 and S2; [4, 6, 20]). Deletions (2 to 2000 bp) at the target site of L1 *de novo* insertion were identified in 19 out of 60 insertions (31,6%) in CSA- deficient and in 21 out of 77 insertions (27%) in the complemented cells (Additional file 1: Tables S1 and S2). A high rate of genomic deletions was also reported in XPD+ and HeLa cells (47% and 26%, respectively) [4, 16]. Typical target-site duplications (TSDs) duplications were primarily observed at the target site of L1 *de novo* insertions recovered from CSA- and CSA+ cells (Additional file 1: Tables S1 and S2). The TSD size ranged from 1 to 29,902 bp in CSA- cells and from 1 to 3450 bp in CSA+ cells with a median length of 13 and 12 bp in CSA+ and CSA- cells, respectively (Fig. 3b). These data corresponded to the typical observations reported in HeLa cells or complemented NER cells (15 bp on average) [4, 16, 18] and were very different to the abnormally large TSDs (over 1 kb on average) observed in the other GGR-deficient cells [16].

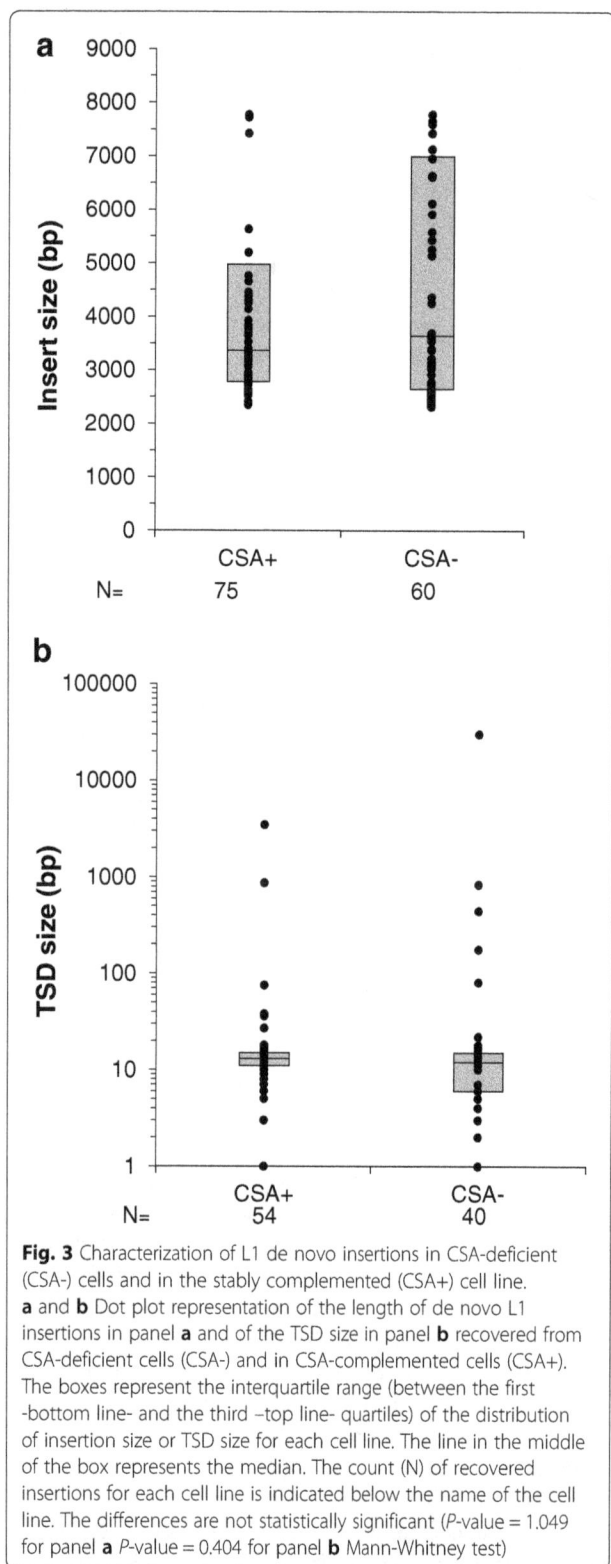

Fig. 3 Characterization of L1 de novo insertions in CSA-deficient (CSA-) cells and in the stably complemented (CSA+) cell line. **a** and **b** Dot plot representation of the length of de novo L1 insertions in panel **a** and of the TSD size in panel **b** recovered from CSA-deficient cells (CSA-) and in CSA-complemented cells (CSA+). The boxes represent the interquartile range (between the first -bottom line- and the third –top line- quartiles) of the distribution of insertion size or TSD size for each cell line. The line in the middle of the box represents the median. The count (N) of recovered insertions for each cell line is indicated below the name of the cell line. The differences are not statistically significant (P-value = 1.049 for panel **a** P-value = 0.404 for panel **b** Mann-Whitney test)

Does TCR-NER influence the insertional bias of L1 elements in genes?

We then investigated the distribution of L1 *de novo* insertions from our tagged vector in the genomes of

CSA-deficient and complemented cells. Because the TCR pathway specifically excises the DNA lesion that interrupts the transcription process, it seems likely that a nascent L1 insert in the template strand would block transcription, and possibly trigger TCR to remove the inhibiting L1 retrotransposition event.

As observed in the reference genome and in many cell lines (Additional file 2: Figure S4A and [3]), L1 *de novo* insertions were almost equally dispersed in genic and intergenic regions of the genome of both CSA- and CSA+ cells (Additional file 1: Tables S1 and S2 and Fig. 4a). Nevertheless, when L1 *de novo* insertions were integrated within genes in the complemented CSA+ cells, we characterized twice as many antisense-oriented as sense-oriented insertions (62.1% and 37.9% respectively) (Fig. 4b). This observation agreed with the previously reported trends for the genomic orientation of L1 elements in genes [4, 21, 31] (see Additional file 2: Figure S3), L1 *de novo* insertions in HeLa cells (see Additional file 1: Table S4) and brain cells [4, 7]. In contrast, L1 *de novo* insertions showed no significant bias in sense versus antisense orientation in CSA-deficient cells (Fig. 4b).

We reasoned that if the TCR sub-pathway would influence L1 orientation in genes, any steps in the pathway downstream from the sensor (CSA) would influence similarly L1 *de novo* insertions, while the sensor for the GGR sub-pathway (XPC) would not have the same effect. XPC-deficient cells showed a similar orientation bias for L1 *de novo* insertions to those seen in other TCR-proficient cells (Additional file 2: Figure S4B). However, L1 *de novo* insertions were equally sense and antisense oriented in XPD-deficient cells, which are defective for the downstream NER pathway factor that affects both TCR and GGR (Additional file 2: Figure S4B). In XPD+ cells, the complemented version of XPD- cells, the orientation bias was again observed for L1 *de novo* insertions (Additional file 2: Figure S4B).

In conclusion, our results revealed that L1 *de novo* insertions were preferentially antisense oriented in cells proficient for the TCR pathway (TCR+, Fig. 4c), such as HeLa, CSA+, XPD+ cells as well as XPC- cells. In TCR-deficient cells (TCR-, Fig. 4c), such as CSA- and XPD- cells, the orientation of L1 *de novo* insertions within genes was random (Fig. 4c).

Expression of genes in which L1 inserted in HeLa cells

Because TCR is only active when a transcription complex hits a DNA lesion on the template strand, we predicted that sense-strand L1 insertions (occurring in the template strand) would be depleted in actively transcribing genes relative to the antisense-oriented insertions that would not be predicted to be affected by TCR (see Additional file 2: Figure S5). We therefore carried out a quantitation of gene expression for the

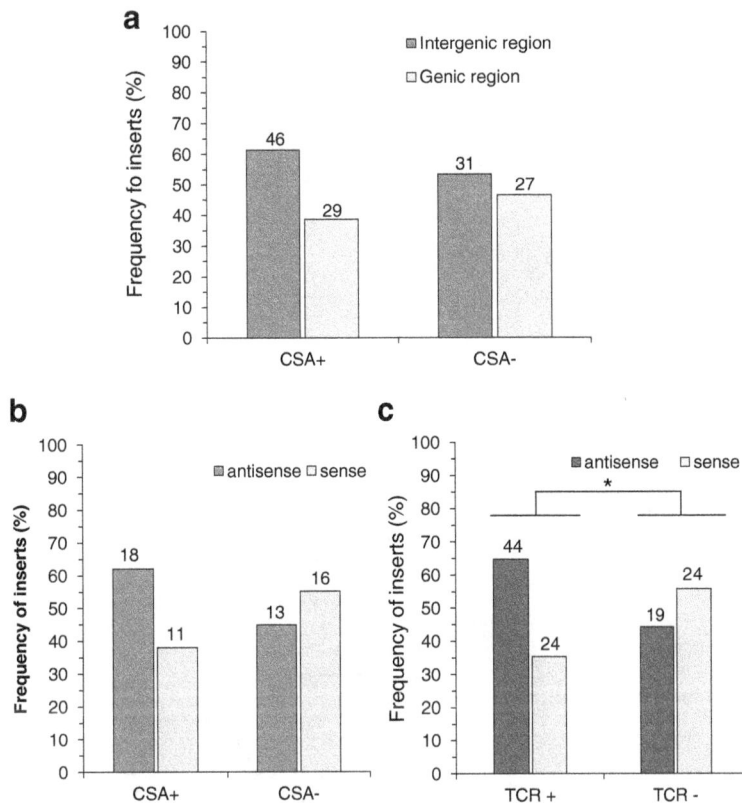

Fig. 4 The tendency of L1 de novo insertions in the antisense orientation within genes decreased in the cells deficient in the TCR pathway (CSA- and XPD- cells). **a** Analysis of the distribution of L1 de novo insertions in the genome of CSA- and CSA+ cells. Bars represent the frequency of de novo L1 insertions in genic and intergenic regions of the genome. The numbers over the bars represent the counts of recovered insertions for each condition. The difference in the genomic repartition of the insertions between CSA+ and CSA- cells is not statistically significant ($p = 0.227$; Fisher exact test). **b** Analysis of the orientation of L1 de novo insertions within genes in CSA+ and CSA- cells. Bars represent the frequency of de novo L1 insertions in sense and antisense orientation within genes. The numbers over the bars represent the counts of recovered insertions for each condition. The difference in the counts of sense-oriented vs antisense-oriented recovered L1 insertions between CSA+ and CSA- cells is not statistically significant ($p = 0.292$; Fisher exact test). **c** Analysis of the orientation of L1 de novo insertions within genes in TCR proficient (TCR+: HeLa, CSA+, XPD+, XPC-) and deficient (TCR-: XPD-, CSA-) cell lines. Bars represent the frequency of de novo L1 inserts in sense and antisense orientation within genes in TCR+ and TCR- cell lines. The numbers over the bars represent the combinations of the counts of recovered L1 insertions in HeLa, CSA+, XPD+ and XPC- cells (TRC+ cell lines) and in XPD- and CSA- cells (TRC- cell lines). The difference in the counts of sense-oriented vs antisense-oriented recovered L1 insertions between TCR+ and TCR- cell lines is statistically significant ($p = 0.049$; Fisher exact test)

ENCODE coding sequences in the human genome from HeLa cells. HeLa cells were chosen because they have an intact TCR pathway and because there is more available data on *de novo* inserts in HeLa cells than any other cell line.

In this study, approximately 80% of the cellular genes had little or no transcription (Additional file 2: Figure S3) confirming that they would be unlikely targets for TCR. Many of the expressed genes had expression levels less than 1 % the level of GAPDH, suggesting that they might be less subject to TCR than more actively transcribed genes.

When we examined HeLa *de novo* inserts analyzed with the rescue approach utilized in this manuscript from Gilbert et al. [4] and from this study, we see 39 inserts in the antisense orientation relative to ENCODE

genes and 17 in the sense orientation (Additional file 1: Table S1). This ratio of antisense to sense is very similar to the ratio seen in the genome [3]. When we look at the expression levels from those genes, we see that the genes with antisense inserts have an average FPKM expression value of almost 25, while the sense inserts are in genes with less than 10 FPKM. This is significant at the 0.04 level in a two-tailed T-test. Furthermore, given that the majority of ENCODE genes have no measured expression, it is interesting that even though the genes in which the insertions occurred are not highly expressed, there is also a depletion of insertions in non-expressed genes. We are not sure if this represents a preferred target for insertion or the requirement for open chromatin to allow the selectable marker in the L1 element to express.

Discussion

Although L1 retrotransposons are inserted throughout the human genome, these autonomous mobile elements have been found to be located with a strong antisense bias within genes [3]. This orientation bias is a characteristic of referenced and established L1 elements as well as polymorphic and *de novo* insertions (Additional file 1: Table S1) [4, 7]. Although it has been suggested that the bias may be the result of selection eliminating the insertions in the sense orientation that might be more disruptive of gene expression [32], this seems unlikely to have a strong influence on the *de novo* insertions. Thus, it is worth considering whether there is a specific mechanism limiting sense insertion in genes, possibly limiting the mutagenic impact of these insertion events. In the present study, we have demonstrated that recovered L1 *de novo* insertions are equally sense and antisense oriented within active genes in CSA- and XPD- deficient cells, both defective in the TCR

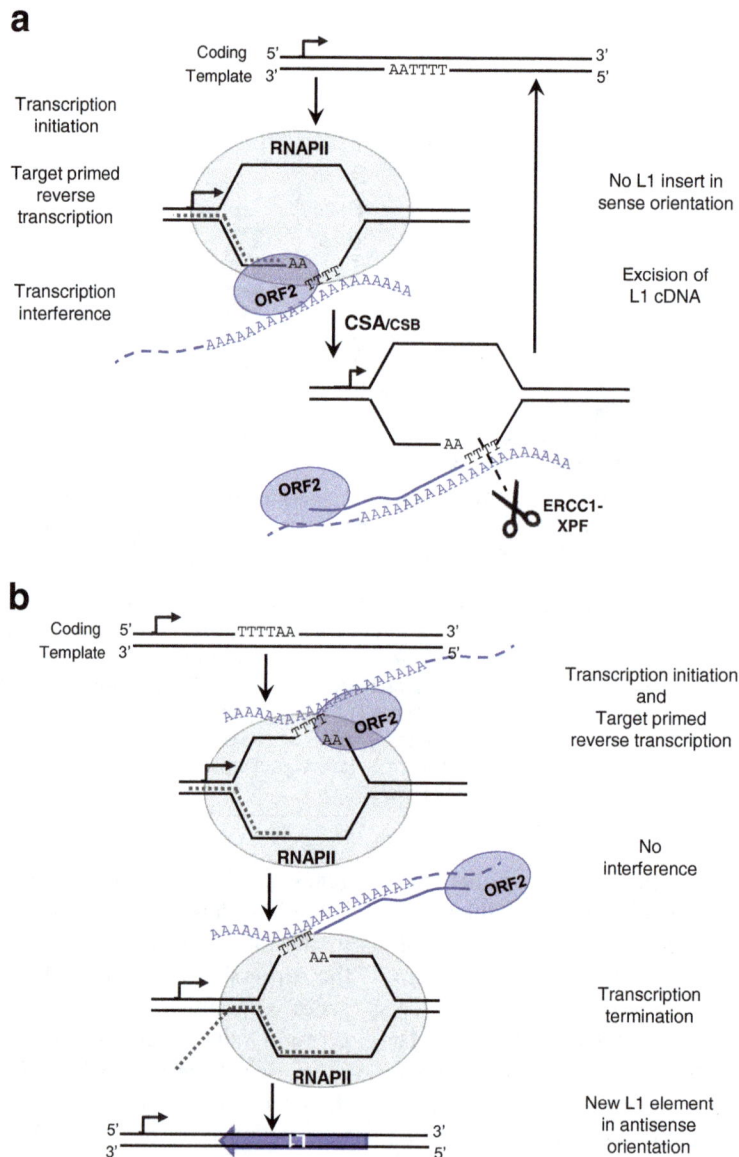

Fig. 5 The TCR pathway prevents the L1 retrotransposon from inserting into the template strand during the transcription of active genes. **a** If an L1 insertion targets the template strand during the transcription of the gene, it can interfere with the RNAPII complex and stop transcription. CSA and CSB proteins detect the stalled RNAPII and recruit the other proteins of the NER pathway. The elongating L1 cDNA is then cleaved and the original DNA sequence is restored allowing completion of the gene transcription. This process results in the depletion of sense L1 insertions in actively transcribed genes. **b** If the L1 machinery targets the coding strand during the gene transcription, it does not interfere with the RNAPII complex. The transcription as well as the L1 insertion process can be completed. Thus, a de novo L1 insert ends up in antisense orientation within the active gene

pathway. These results suggest that the TCR pathway is responsible for much of the orientation bias of L1 elements in the human genome, although we cannot rule out some post-insertional selection influences as well. This demonstrates that in addition to the influence of GGR on L1 retrotransposition rate, the TCR subpathway also influences the distribution of inserts.

In cells proficient for TCR, the pathway is recruited at stalled RNAPII complex and excises DNA lesions blocking the RNAPII processivity on the template strand (Fig. 1). After the repair, the transcription process is re-initiated. If L1 elements insert in the template strand of a gene, they would end up in the same orientation as the gene [33]. Insertions in the coding strand that would result in antisense insertions would not be expected to stall RNAPII and induce TCR (Fig. 5a). The data presented in our study suggest that the TCR pathway may prevent the insertion of L1 elements in the template strand of actively transcribed genes, but not in the coding strand, leading to the observed orientation bias of L1 inserts in the genome. This is supported by both the ratios of sense to antisense inserts (Fig. 4), as well as the tendency for sense inserts to be present in less expressed genes (Additional file 1: Table S1) than antisense inserts in HeLa cells with active TCR. Conversely, if the L1 machinery targets the coding strand, there would be no interference with the RNAPII complex and a L1 de novo insertion would be able to occur (Fig. 5b). The de novo insertion would be in antisense orientation within the gene.

Our data are consistent with the model that the TCR pathway may minimize interference of gene expression by new L1 retrotransposition events. We did not observe a strong effect of the TCR regulation on the overall L1 retrotransposition rate because only a small part of the genome is actively and efficiently transcribed at any given time in a cell (see Additional file 2: Figure S3) and the rest can still be protected from de novo L1 insertion by the GGR pathway (Additional file 2: Figure S5). The TCR pathway, which is essential for the protection of gene expression, represents a unique mechanism in the regulation of L1 retrotransposition especially during embryonic development when L1 activity is high [34] and L1-caused mutations could be detrimental for cell survival.

Although L1 elements are distributed throughout the genome, there are likely to be multiple factors that influence their distribution. L1 elements preferentially insert into a locally A + T-rich target sequence [6, 35, 36]. Thus, it is likely that the relative density of such A + T-rich target sequences may influence the rate of insertion in those regions. In addition, insertion of L1 sequences into genes may provide various signals that either fully or partially disrupt expression of the gene [37–39] resulting in negative selective pressure that will eventually lead to depletion of genes in which L1 insertions

have occurred [40–43] This is likely a contributor to the relative paucity of L1 within genes that increases over evolutionary time [3, 41]. The insertion of L1 sequences may be more disruptive in one orientation relative to another [3] which could also lead to selection for more L1 elements in one orientation relative to another within genes. However, our finding that TCR can contribute strongly to such an insertion bias provides a mechanism that may establish such a bias immediately, without requiring time for selective pressure to alter the frequency.

Conclusions

This work shows that the previously observed bias against sense-oriented L1 elements in genes is primarily due to transcription-coupled nucleotide excision repair being able to block sense insertions, rather than principally being due to selection post insertion. This would serve to minimize the negative impact of L1 insertions on gene expression.

Additional files

Additional file 1: Table S1. Characteristics of recovered de novo L1 inserts in CSA-deficient cells. This table describes the general characteristics of the L1 inserts isolated from the CSA-minus cells. **Table S2.** Characteristics of recovered de novo L1 inserts in stably complemented CSA + cells. This table describes the general characteristics of the L1 inserts isolated from the cells that have been complemented to be CSA+. **Table S3A&B.** DNA sequences flanking rescued L1 inserts. S3A has the sequence data from the L1 insertion rescues for the CSA-minus cells, while S3B has similar data for the complemented cells that are now CSA plus. **Table S4.** FPKM values for de novo L1 inserts in HeLa cells that inserted within genes. (ZIP 130 kb)

Additional file 2: Figure S1. Control for the efficiency of the complementation of CSA-deficient cells. **Figure S2.** L1 retrotransposition rate is not significantly different in CSA-deficient cells (CSA-) and in the stably complemented CSA-deficient cells (CSA+). **Figure S3.** FPKM counts for Encode genes expressed in HeLa. **Figure S4.** The tendency of de novo L1 elements to insert in the antisense orientation within genes is lost in the cells deficient in the TCR pathway (CSA- and XPD- cells). **Figure S5.** Model of regulation of L1 insertion in genes by the TCR pathway. (ZIP 241 kb)

Abbreviations

bp: Base pairs; CMV: Cytomegalovirus; CSA: Cockayne syndrome protein A; CSB: Cockayne syndrome protein B; eGFP: Enhanced green fluorescent protein; ERCC1: Excision repair 1; FPKM: Fragments per kilobase per million reads; GGR: Global genome repair sub-pathway of NER; IRES: Internal ribosome entry site; J/m^2/s: Joules per meter squared per second; L1: LINE-1 or Long, INterspersed Element-1; mblast: Blasticidin resistance cassette; mneo: Neomycin resistance cassette; Neo^R: Neomycin or geneticin resistance; NER: Nucleotide excision repair; nt: Nucleotide; PBS: Phosphate-buffered saline; PCR: Polymerase chain reaction; RNAPII: RNA polymerase II; RNA-Seq: Next generation sequencing protocol for RNA; RT-PCR: Reverse transcription – PCR; SV40: Simian virus 40; TCR: Transcription-coupled repair; TSD: Target site duplication; UV: Ultraviolet; XPA: Xeroderma pigmentosum protein A; XPC: Xeroderma pigmentosum protein C; XPD: Xeroderma pigmentosum protein D; XPF: Xeroderma pigmentosum protein F

Acknowledgements

We wish to thank other members of the Deininger laboratory and the Consortium for Mobile Elements at Tulane for their constructive criticism of these studies, including manuscript editing by Tiffany Kaul. We wish to acknowledge the expert help of Melody Baddoo and the Cancer Crusaders Bioinformatics Core.

Funding

This work was supported by USPHS grant # R01 GM045668 and R01 GM121812.

Author's contributions

GS carried out and did the preliminary analysis for most of the experiments as well as preparing the first draft of the manuscript. VS helped with some of the rescue experiments and participated in manuscript revisions. PD provided guidance, and oversight of experimental design as well as some data analysis and editing of the manuscript. All authors read and approved the final manuscript.

Competing interests

The authors declare that they have no competing interests.

Author details

[1]Tulane University, Tulane Cancer Center and the Department of Epidemiology, 1430 Tulane Ave, New Orleans, LA 70112, USA. [2]Present Address: Division of Infectious Diseases, Boston Children's Hospital and Harvard Medical School, 300 Longwood Ave, Boston, MA 02115, USA. [3]Tulane Cancer Center, SL66, Tulane University Health Sciences Center, 1430 Tulane Ave., New Orleans, LA 70112, USA.

References

1. Lander ES, Linton LM, Birren B, Nusbaum C, Zody MC, Baldwin J, Devon K, Dewar K, Doyle M, FitzHugh W, et al. Initial sequencing and analysis of the human genome. Nature. 2001;409(6822):860–921.
2. de Koning AP, Gu W, Castoe TA, Batzer MA, Pollock DD. Repetitive elements may comprise over two-thirds of the human genome. PLoS Genet. 2011; 7(12):e1002384.
3. Medstrand P, van de Lagemaat LN, Mager DL. Retroelement distributions in the human genome: variations associated with age and proximity to genes. Genome Res. 2002;12(10):1483–95.
4. Gilbert N, Lutz S, Morrish TA, Moran JV. Multiple fates of L1 retrotransposition intermediates in cultured human cells. Mol Cell Biol. 2005; 25(17):7780–95.
5. Gilbert N, Lutz-Prigge S, Moran JV. Genomic deletions created upon LINE-1 retrotransposition. Cell. 2002;110(3):315–25.
6. Symer DE, Connelly C, Szak ST, Caputo EM, Cost GJ, Parmigiani G, Boeke JD. Human l1 retrotransposition is associated with genetic instability in vivo. Cell. 2002;110(3):327–38.
7. Upton KR, Gerhardt DJ, Jesuadian JS, Richardson SR, Sanchez-Luque FJ, Bodea GO, Ewing AD, Salvador-Palomeque C, van der Knaap MS, Brennan PM, et al. Ubiquitous L1 mosaicism in hippocampal neurons. Cell. 2015; 161(2):228–39.
8. Mellon I, Spivak G, Hanawalt PC. Selective removal of transcription-blocking DNA damage from the transcribed strand of the mammalian DHFR gene. Cell. 1987;51(2):241–9.
9. Sugasawa K, Ng JM, Masutani C, Iwai S, van der Spek PJ, Eker AP, Hanaoka F, Bootsma D, Hoeijmakers JH. Xeroderma pigmentosum group C protein complex is the initiator of global genome nucleotide excision repair. Mol Cell. 1998;2(2):223–32.
10. Riedl T, Hanaoka F, Egly JM. The comings and goings of nucleotide excision repair factors on damaged DNA. EMBO J. 2003;22(19):5293–303.
11. Coin F, Marinoni JC, Rodolfo C, Fribourg S, Pedrini AM, Egly JM. Mutations in the XPD helicase gene result in XP and TTD phenotypes, preventing interaction between XPD and the p44 subunit of TFIIH. Nat Genet. 1998; 20(2):184–8.
12. Vasquez KM, Christensen J, Li L, Finch RA, Glazer PM. Human XPA and RPA DNA repair proteins participate in specific recognition of triplex-induced helical distortions. Proc Natl Acad Sci U S A. 2002;99(9):5848–53.
13. Wang M, Mahrenholz A, Lee SH. RPA stabilizes the XPA-damaged DNA complex through protein-protein interaction. Biochemistry. 2000;39(21):6433–9.
14. Sijbers AM, van der Spek PJ, Odijk H, van den Berg J, van Duin M, Westerveld A, Jaspers NG, Bootsma D, Hoeijmakers JH. Mutational analysis of the human nucleotide excision repair gene ERCC1. Nucleic Acids Res. 1996;24(17):3370–80.
15. O'Donovan A, Davies AA, Moggs JG, West SC, Wood RD. XPG endonuclease makes the 3' incision in human DNA nucleotide excision repair. Nature. 1994;371(6496):432–5.
16. Servant G, Streva VA, Derbes RS, Wijetunge MI, Neeland M, White TB, Belancio VP, Roy-Engel AM, Deininger PL. The nucleotide excision repair pathway limits L1 Retrotransposition. Genetics. 2017;205(1):139–53.
17. Gasior SL, Roy-Engel AM, Deininger PL. ERCC1/XPF limits L1 retrotransposition. DNA Repair (Amst). 2008;7(6):983–9.
18. Ichiyanagi K, Okada N. Mobility pathways for vertebrate L1, L2, CR1, and RTE clade retrotransposons. Mol Biol Evol. 2008;25(6):1148–57.
19. Moran JV, Holmes SE, Naas TP, DeBerardinis RJ, Boeke JD, Kazazian HH Jr. High frequency retrotransposition in cultured mammalian cells. Cell. 1996;87(5):917–27.
20. Morrish TA, Gilbert N, Myers JS, Vincent BJ, Stamato TD, Taccioli GE, Batzer MA, Moran JV. DNA repair mediated by endonuclease-independent LINE-1 retrotransposition. Nat Genet. 2002;31(2):159–65.
21. Gasior SL, Preston G, Hedges DJ, Gilbert N, Moran JV, Deininger PL. Characterization of pre-insertion loci of de novo L1 insertions. Gene. 2007; 390(1–2):190–8.
22. Ridley AJ, Colley J, Wynford-Thomas D, Jones CJ. Characterisation of novel mutations in Cockayne syndrome type a and xeroderma pigmentosum group C subjects. J Hum Genet. 2005;50(3):151–4.
23. Streva VA, Jordan VE, Linker S, Hedges DJ, Batzer MA, Deininger PL. Sequencing, identification and mapping of primed L1 elements (SIMPLE) reveals significant variation in full length L1 elements between individuals. BMC Genomics. 2015;16:220.
24. Yuanxin Y, Chengcai A, Li L, Jiayu G, Guihong T, Zhangliang C. T-linker-specific ligation PCR (T-linker PCR): an advanced PCR technique for chromosome walking or for isolation of tagged DNA ends. Nucleic Acids Res. 2003;31(12):e68.
25. Emmert S, Kobayashi N, Khan SG, Kraemer KH. The xeroderma pigmentosum group C gene leads to selective repair of cyclobutane pyrimidine dimers rather than 6-4 photoproducts. Proc Natl Acad Sci U S A. 2000;97(5):2151–6.
26. Li B, Dewey CN. RSEM: accurate transcript quantification from RNA-Seq data with or without a reference genome. BMC Bioinformatics. 2011;12:323.
27. El-Sawy M, Kale SP, Dugan C, Nguyen TQ, Belancio V, Bruch H, Roy-Engel AM, Deininger PL. Nickel stimulates L1 retrotransposition by a post-transcriptional mechanism. J Mol Biol. 2005;354(2):246–57.
28. Beck CR, Collier P, Macfarlane C, Malig M, Kidd JM, Eichler EE, Badge RM, Moran JV. LINE-1 retrotransposition activity in human genomes. Cell. 2010; 141(7):1159–70.
29. Brouha B, Schustak J, Badge RM, Lutz-Prigge S, Farley AH, Moran JV, Kazazian HH Jr. Hot L1s account for the bulk of retrotransposition in the human population. Proc Natl Acad Sci U S A. 2003;100(9):5280–5.
30. Iskow RC, McCabe MT, Mills RE, Torene S, Pittard WS, Neuwald AF, Van Meir EG, Vertino PM, Devine SE. Natural mutagenesis of human genomes by endogenous retrotransposons. Cell. 2010;141(7):1253–61.
31. Ponnaluri VK, Ehrlich KC, Zhang G, Lacey M, Johnston D, Pradhan S, Ehrlich M. Association of 5-hydroxymethylation and 5-methylation of DNA cytosine with tissue-specific gene expression. Epigenetics. 2017;12(2):123–38.
32. Han JS, Boeke JD. LINE-1 retrotransposons: modulators of quantity and quality of mammalian gene expression? BioEssays. 2005;27(8):775–84.
33. Wheelan SJ, Aizawa Y, Han JS, Boeke JD. Gene-breaking: a new paradigm for human retrotransposon-mediated gene evolution. Genome Res. 2005; 15(8):1073–8.
34. Kano H, Godoy I, Courtney C, Vetter MR, Gerton GL, Ostertag EM, Kazazian HH Jr. L1 retrotransposition occurs mainly in embryogenesis and creates somatic mosaicism. Genes Dev. 2009;23(11):1303–12.
35. Monot C, Kuciak M, Viollet S, Mir AA, Gabus C, Darlix JL, Cristofari G. The specificity and flexibility of l1 reverse transcription priming at imperfect T-tracts. PLoS Genet. 2013;9(5):e1003499.
36. Daniels GR, Deininger PL. Integration site preferences of the Alu family and similar repetitive DNA sequences. Nucleic Acids Res. 1985;13(24):8939–54.
37. Belancio VP, Hedges DJ, Deininger P. LINE-1 RNA splicing and influences on mammalian gene expression. Nucleic Acids Res. 2006;34(5):1512–21.
38. Perepelitsa-Belancio V, Deininger P. RNA truncation by premature polyadenylation attenuates human mobile element activity. Nat Genet. 2003;35(4):363–6.
39. Han JS, Szak ST, Boeke JD. Transcriptional disruption by the L1 retrotransposon and implications for mammalian transcriptomes. Nature. 2004;429(6989):268–74.

Detection of the LINE-1 retrotransposon RNA-binding protein ORF1p in different anatomical regions of the human brain

Debpali Sur[1], Raj Kishor Kustwar[1†], Savita Budania[1†], Anita Mahadevan[3], Dustin C. Hancks[4,5], Vijay Yadav[2], S. K. Shankar[3] and Prabhat K. Mandal[1*]

Abstract

Background: Recent reports indicate that retrotransposons – a type of mobile DNA – can contribute to neuronal genetic diversity in mammals. Retrotransposons are genetic elements that mobilize via an RNA intermediate by a "copy-and-paste" mechanism termed retrotransposition. Long Interspersed Element-1 (LINE-1 or L1) is the only active autonomous retrotransposon in humans and its activity is responsible for ~ 30% of genomic mass. Historically, L1 retrotransposition was thought to be restricted to the germline; however, new data indicate L1 s are active in somatic tissue with certain regions of the brain being highly permissive. The functional implications of L1 insertional activity in the brain and how host cells regulate it are incomplete. While deep sequencing and qPCR analysis have shown that L1 copy number is much higher in certain parts of the human brain, direct in vivo studies regarding detection of L1-encoded proteins is lacking due to ineffective reagents.

Results: Using a polyclonal antibody we generated against the RNA-binding (RRM) domain of L1 ORF1p, we observe widespread ORF1p expression in post-mortem human brain samples including the hippocampus which has known elevated rates of retrotransposition. In addition, we find that two brains from different individuals of different ages display very different expression of ORF1p, especially in the frontal cortex.

Conclusions: We hypothesize that discordance of ORF1p expression in parts of the brain reported to display elevated levels of retrotransposition may suggest the existence of factors mediating post-translational regulation of L1 activity in the human brain. Furthermore, this antibody reagent will be useful as a complementary means to confirm findings related to retrotransposon biology and activity in the brain and other tissues in vivo.

Keywords: Retrotransposon, LINE-1, ORF1p antibody, Active retrotransposition in human brain, Somatic mosaicism

Background

Historically, the genome was thought to be identical in every cell throughout an organism except immune cells and germ cells. Notably, the discovery of transposable elements and their mobilization in somatic and germ cells indicates that genomes within an organism are by no means static [1, 2]. Since initial characterization by Barbara McClintock, transposons have been considered as insertional mutagens; in other words transposon activity may result in single-gene disease [1, 3, 4].

Typically considered "selfish" parasitic sequences, recent findings dispute this traditional model and have demonstrated the multifaceted potential of transposons. Indeed, transposons are now appreciated as major players in genome evolution in most organisms including mammals [5].

Along with being widespread across mammalian genomes, Long Interspersed Element –1 (LINE-1 or L1) is the only active autonomous retrotransposon in the modern human genome [6]. In addition, L1 is the most abundant retrotransposon by sequence mass accounting for 17% of the human genome (~500,000 copies) [7, 8]. L1 mobilizes from one genomic location to another using RNA as an intermediate via a process referred to

* Correspondence: mandal.prabhat@gmail.com; pkm31fbt@iitr.ac.in
†Equal contributors
[1]Department of Biotechnology, IIT Roorkee, Roorkee, Uttarakhand, India
Full list of author information is available at the end of the article

as retrotransposition. Thus, these types of elements are referred to as retrotransposons [6]. Although most of the L1s are inactive due to point mutations, 5′-truncations and other rearrangements including inversions, around 80–100 L1s are active in any given human [9].

An active, full-length L1 is ~6 kb in length. It encodes an internal promoter within a ~900 base pair (bp) 5′-UTR, two open-reading frames termed ORF1 and ORF2 separated by a small inter-ORF spacer and a 3′-UTR (~205 bp). Genomic insertions end in a polyA sequence derived from the mRNA polyA tail (~40-120 bp) and are flanked by direct repeats of varying length known as target-site duplications (TSD, typically 4–20 bp in length) at the site of insertion [4, 6, 10]. ORF1 encodes a protein (ORF1p) with single-stranded nucleic acid binding activity [11, 12], whereas ORF2-encodes a protein (ORF2p) with demonstrated reverse transcriptase (RT) [13] and endonuclease (EN) activities [14]. Both proteins are required for retrotransposition in *cis* [15]. Notably, along with mobilizing its own RNA, L1 activity is responsible for dispersing eight thousand processed pseudogene insertions [16–20], more than 1.2 million Alu elements – a type of SINE – and ~2700 SINE-R/VNTR/Alu (SVA) elements throughout the human genome [7, 21–25].

Although ORF1p is critical for retrotransposition its roles in *cis* and *trans*-mobilization are incomplete [26]. While human ORF1p does not display amino acid sequence similarity to other known proteins [27]; recent studies have revealed that the 40 kDa ORF1p has three distinct domains: coiled-coil (CC), RNA recognition motif (RRM) and carboxy terminal domain (CTD) [28, 29]. In-vitro studies have demonstrated that both human and mouse ORF1p are non-sequence specific single stranded RNA and DNA binding proteins with nucleic acid chaperone activity [12, 28, 29]. Studies on the localization of these proteins in cell lines suggest that ORF1p is mainly cytoplasmic and may concentrate in certain regions of the cytoplasm resulting in the formation of cytoplasmic foci [30–34]. Interestingly, a fraction of ORF1p is observed in stress granules and in the nucleus of cells [32, 33, 35]. Furthermore, detection of ORF1p and its cytoplasmic and nuclear localisation has also been reported in healthy and cancer human tissues [36–38].

Although L1 retrotransposition in most somatic cells is generally silenced by a variety of defence mechanisms and host factors, such as the APOBEC3 proteins, [39–42] presumably to limit insertional mutagenesis, transgenic animal models and deep-sequencing studies have shown that L1 is highly active in certain regions of the brain (e.g. *hippocampus*) [2, 43–45]. The pioneering work of Muotri et al. [2], which involved insertion analysis of transgenic mice carrying an engineered L1 that upon retrotransposition

expresses green fluorescent protein (GFP) [46], unexpectedly identified retrotransposition-competent cells in many regions of the brain including cortex, hypothalamus, cerebellum and hippocampus. While an increase in L1 copy-number has been observed using qPCR or next-generation sequencing in certain brain disorders like ATM deficiency [47, 48], Rett Syndrome, schizophrenia [49] and autism [50], the biology of L1 proteins in the human brain of healthy individuals that were not diagnosed with any neurodegenerative disease is poorly understood.

Here, we describe a novel polyclonal antibody against the RRM domain of human L1-ORF1p which we generated to investigate L1 protein expression in the human brain. Using this antibody, we detect ORF1p in various parts of the human brain derived from post-mortem samples. Interestingly, we observe differential expression of ORF1p when comparing the same brain region of samples from different ages. Together, these data provide in vivo evidence for L1 protein expression in the human brain and describe a new antibody available to the community.

Methods

Cloning and purification of RRM domain from L1-ORF1p and generation of polyclonal antibody against human L1-ORF1p (RRM)

The RNA Recognition Motif (RRM) domain of human ORF1 from L1-RP (Accession number -AF148856.1) [51] was selected (Fig. 1a and Additional file 1) as the epitope to immunize a rabbit for antibody generation. The RRM domain was isolated from ORF1-RRMF (RRM domain cloned at *EcoRI-NotI* sites of pcDNA6/ myc-HisB) [20, 52] using *EcoRI* and *NotI*. The ORF1-RRM fragment was cloned into *EcoRI* and *NotI* of pET-28a vector (EMD Biosciences) for protein expression in bacteria. The expressed protein and corresponding nucleotide sequence are provided in Additional file 1. The His-tagged L1-ORF1-RRM protein was expressed in *E. coli* strain BL21 and purified on nickel-NTA Agarose (Qiagen) according to the manufacturer's protocol. Purified human ORF1 RRM domain, with molecular mass of approximately 15 kDa (vector sequence plus RRM, details in Additional file 1), was used to immunize rabbit (Immunization protocol: Additional file 1). Immunized whole serum from the rabbit without further purification was used for the experiments described in this study to detect ORF1p in cell and tissue lysates.

Cell culture and Transfection

HEK293T (human embryonic kidney), HeLa (cervical carcinoma), MCF-7 (Breast cancer), DU145 (Prostate cancer), and NIH3T3 (Mouse fibroblast) cells were maintained in a tissue culture incubator at 5% CO_2, 37 °C in high glucose Dulbecco's modified Eagle medium

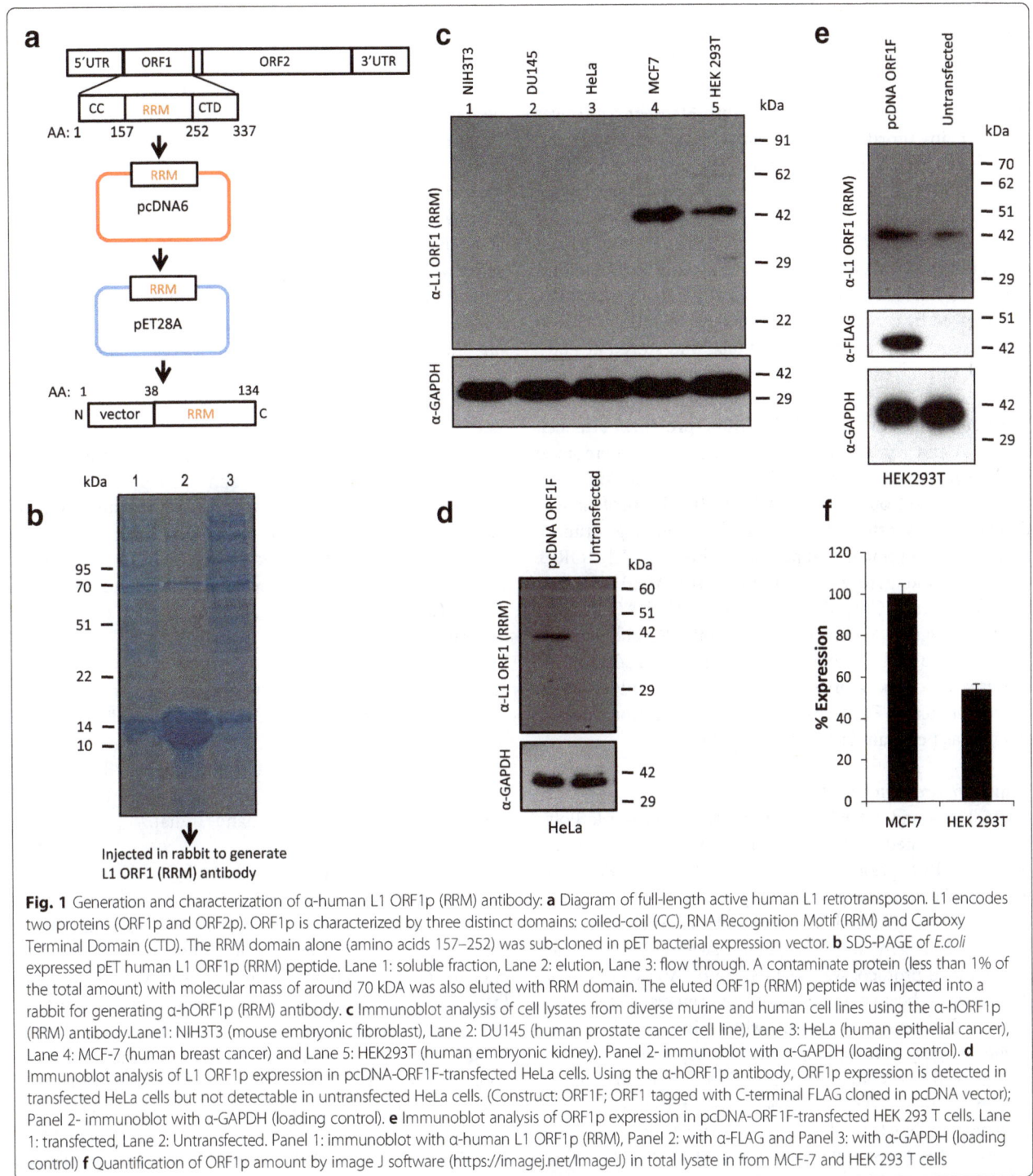

Fig. 1 Generation and characterization of α-human L1 ORF1p (RRM) antibody: **a** Diagram of full-length active human L1 retrotransposon. L1 encodes two proteins (ORF1p and ORF2p). ORF1p is characterized by three distinct domains: coiled-coil (CC), RNA Recognition Motif (RRM) and Carboxy Terminal Domain (CTD). The RRM domain alone (amino acids 157–252) was sub-cloned in pET bacterial expression vector. **b** SDS-PAGE of *E.coli* expressed pET human L1 ORF1p (RRM) peptide. Lane 1: soluble fraction, Lane 2: elution, Lane 3: flow through. A contaminate protein (less than 1% of the total amount) with molecular mass of around 70 kDA was also eluted with RRM domain. The eluted ORF1p (RRM) peptide was injected into a rabbit for generating α-hORF1p (RRM) antibody. **c** Immunoblot analysis of cell lysates from diverse murine and human cell lines using the α-hORF1p (RRM) antibody.Lane1: NIH3T3 (mouse embryonic fibroblast), Lane 2: DU145 (human prostate cancer cell line), Lane 3: HeLa (human epithelial cancer), Lane 4: MCF-7 (human breast cancer) and Lane 5: HEK293T (human embryonic kidney). Panel 2- immunoblot with α-GAPDH (loading control). **d** Immunoblot analysis of L1 ORF1p expression in pcDNA-ORF1F-transfected HeLa cells. Using the α-hORF1p antibody, ORF1p expression is detected in transfected HeLa cells but not detectable in untransfected HeLa cells. (Construct: ORF1F; ORF1 tagged with C-terminal FLAG cloned in pcDNA vector); Panel 2- immunoblot with α-GAPDH (loading control). **e** Immunoblot analysis of ORF1p expression in pcDNA-ORF1F-transfected HEK 293 T cells. Lane 1: transfected, Lane 2: Untransfected. Panel 1: immunoblot with α-human L1 ORF1p (RRM), Panel 2: with α-FLAG and Panel 3: with α-GAPDH (loading control) **f** Quantification of ORF1p amount by image J software (https://imagej.net/ImageJ) in total lysate in from MCF-7 and HEK 293 T cells

(DMEM) supplemented with 10% fetal bovine calf serum (Gibco, Thermo Fisher Scientific) and 100 U/ml penicillin-streptomycin (Gibco, Thermo Fisher Scientific). For transfections, cell lines were seeded into a 35 mm plate to achieve 30–50% confluency within 8–12 h prior to transfection. Using Fugene 6 (Promega), 1–1.5 μg of plasmid DNA was transfected into the cell lines according

to manufacturer's instructions. Transfected cells were incubated for an additional 48 h before proceeding to any experiment.

Protein extraction and immunoblots

Whole cell lysates from cell lines were prepared using lysis buffer A [composition: 20 mM Tris-Cl pH 7.8,137 mM

NaCl and 1% NP-40 supplemented with 1X protease inhibitor cocktail (Roche)]. The lysate was cleared by centrifugation at 2500×g, 5 min at 4 °C. To prepare brain tissue lysate, around 150–200 mg frozen brain tissue (post mortem frontal cortex tissue from 80 year old) was crushed in mortar pestle using liquid nitrogen and transferred to 1.5 ml tube containing 250 μl cold RIPA buffer [150 mM NaCl, 1% NP-40, 0.5% Na-deoxycholate, 0.1% SDS, 50 mM Tris-Cl pH -8.0 with protease inhibitor cocktail (Roche)]. The crushed tissue was then passed through an 18 gauge needle 5–8 times and incubated on ice for 45 min with intermittent mixing. Finally, the lysate was cleared by centrifugation (12,000×g, 10 min, 4 °C) and supernatant transferred to a new tube and stored at –80 °C until further use. The Bradford reagent (Bio-Rad) was used to estimate the protein concentration. The proteins were separated by SDS-PAGE (Mini protein Tetra cell Bio-Rad) and wet transferred to nitrocellulose membrane (Millipore) by applying 100 V for 75 min (Bio-Rad mini trans blot electrophoretic transfer cell). The protein was detected by Western blot using the following primary antibody: polyclonal rabbit anti human L1 ORF1 (1:33,000, and 1:20,000 dilution), anti-GAPDH (1:6000 dilution) (Santa Cruz Biotechnology, anti-FLAG (1,3000 dilution) (Sigma). Secondary anti- rabbit HRP and secondary anti-mouse HRP were purchased from Jackson ImmunoResearch Laboratories, USA. Western blots were developed using ECL western blotting detection reagent (Pierce) as per manufacturer's instructions.

Immunofluorescence analysis

2×10^5 HeLa and MCF-7 cells were seeded on sterile Poly-L- lysine coated cover slips in 35 mm tissue culture plates 12–18 h prior transfection. The following day, cells were transfected using 1 μg of plasmid DNA (prepared using GeneJet Plasmid miniprep Kit, Thermo Scientific) and 3 μl of Fugene 6 Transfection Reagent. The immunofluorescence protocol was adapted from "Abcam protocol" (http://www.abcam.com/protocols/immunocytochemistry-immunofluorescence-protocol) with minor modifications.One day post-transfection, media was aspirated, cells were washed with ice cold 1XPBS and fixed by incubating the cells in 100% chilled (–20 °C) methanol for 10 min. at room temperature. Next, fixed cells were washed three times for 10 min with immunofluorescence wash buffer (composition: 0.05% sodium azide, 0.1% BSA, 0.75% glycine, 0.04% Tween-20, and 0.2% Triton X-100 in 1X PBS) using gentle agitation. Permeabilization of fixed cells was performed by incubating cells in permeabilization buffer (1X PBS containing 0.5% Triton X-10) for 3–5 min. at room temperature. Afterwards, cells were rinsed with immunofluorescence wash buffer three times and each time cells were allowed to sit in the wash buffer for 5 min for better quenching. The fixed and permeabilized cells were blocked

for an hour in room temperature by incubating in blocking solution (1% BSA in 1XPBST). Subsequently, cells were incubated with human α-ORF1p (RRM) primary antibody (1:500 diluted in blocking solution) at 4 °C overnight. The next day, cells were washed three times with immunofluorescence wash buffer as stated above followed by incubation with secondary antibody [Alexa fluor 488; Jackson Immuno Research laboratories (1:300 diluted in blocking solution)] for one hour at room temperature in a dark room. Immediately after this, cells were rinsed twice in immune fluorescence wash buffer for five minutes at room temperature. After washing, cells were counterstained with Hochst 33342 for 10 min at room temperature and mounted on slides with DPX mounting media. Samples were then analysed with appropriate fluorescent filters on confocal laser scanning microscope (LSM 780, Carl Zeiss, Germany).

Tissue Specimens: Brain tissue samples were collected in the form of formalin-fixed paraffin embedded (FFPE) sections on slides and frozen tissue from the Human Tissue Repository for Neurobiological Studies (HBTR), Human Brain Bank, Department of Neuropathology, National Institute of Mental Health and Neurosciences, (Bangalore, India). Following proper consent, all the samples were collected from victims of road traffic accident. The tissues were taken from zones distal from the site of injury. All investigations were conducted in accordance with ethical principles embodied in the declaration of tissue request and material transfer agreement [IHEC No. BT/IHEC-IITR/2017/6673; Institute Human Ethics Committee (IHEC), Indian Institute of Technology Roorkee, Utarakhand, India].

Immunohistochemistry (IHC)

Paraffin-embedded brain tissue sections on glass slides were de-paraffinized rehydrated in descending grade of ethanol solutions before proceeding for antigen retrieval. The antigen retrieval step was adapted from "Abcam protocol" (http://www.abcam.com/protocols/immunocytochemistry-immunofluorescence-protocol). The process was performed in a common household vegetable steamer (pressure cooker) using Tris-EDTA antigen retrieval buffer (10 mM Tris base, 1 mM EDTA solution, 0.05% Tween 20, pH -9.0). Next, slides were washed 2 X 5 min each in TBST (1X TBS containing 0.025% Triton-X100) and then blocked in blocking solution (1% BSA in 1X TBST) for 1 h at room temperature. Thereafter, slides were incubated with polyclonal rabbit α-ORF1p (RRM) antibody (1:500 diluted in blocking reagent) at 4 °C overnight in humid chamber. The next day, slides were washed with 1X TBST and treated with 0.3% hydrogen peroxide to quench any peroxidise present within the tissue. Slides were then incubated with secondary antibody [Goat α-Rabbit HRP 1: 500 dilution (Jackson ImmunoResearch)] for an hour at room temperature. The slides were washed 3 × 10 min at room temperature with gentle agitation. Signals were visualised by adding 3–3'- Dia amino benzidinetetrahydrochloride (DAB substrate) solution to the slides and were counterstained with haematoxylin, (Himedia) dehydrated with ascending order of ethanol and mounted with DPX mounting media. Images were captured using a light microscope (Leica Microsystems) equipped with a camera. Intensity of DAB stained regions was measured with ImageRatio software [53] and plotted as a percentage of expression. α-Neurofilament (NE14) (Biogenex) raised in mouse was used as neuronal marker (1:500 dilution) that preferentially stained the neurons.

Results

Characterization α-human ORF1p (RRM) antibody by immunoblotting and immunofluorescence

Human L1 ORF1p is a 338 amino acids (L1RP, Accession number: AF148856.1) protein with a predicted mass of 40 kDa [11, 51] with RNA binding and nucleic acid chaperone activity [12].ORF1p is characterized by three distinct domains: Coiled Coil (CC) (AA: 52–153 relative to L1RP Accession number AF148856.1), RNA Recognition Motif (RRM) (AA: 157–252) and Carboxy Terminal Domain (CTD) (AA: 264–323) (Fig. 1a) [29]. Although much has been learned from cell culture and genomic studies about L1 biology, our understanding of retrotransposition in vivo is far from complete. Here we sought to generate an additional tool to investigate L1 activity, namely an antibody reactive to ORF1p that would be useful for detecting the native protein. To generate ORF1p antibody we selected the RRM domain as the epitope of interest for three reasons: 1) a previous study [29] showed high expression of this domain, 2) the same study showed that the expressed protein was retained in the soluble fraction (native form) in a bacterial expression system and 3) the RRM domain is easier to handle due to its smaller size (MW 12 kDa) relative to full-length ORF1p (MW 40 kDa). The RRM domain from human (h) ORF1 was cloned into a bacterial expression vector (Fig. 1a and Additional file 1: Figure S1). Following expression in bacteria, the RRM domain was purified using Ni- agarose chromatography. Analysis of the purified protein by SDS-PAGE and Coomassie staining revealed a distinct band of ~15 kDa and a minor contaminant protein (less than 1% compared to the RRM band) at ~70 kDa (Fig. 1b).

To generate a hORF1p specific antibody, we injected the purified RRM domain into a New Zealand rabbit. Following isolation, serum was assayed for α-hORF1p (RRM) by Western blot analysis on protein lysates generated from human cancer cell lines known to express varying levels of ORF1p (U87, MCF7, HeLa, Du145 and HEK293T) (Fig. 1c). Robust expression of a 40 kDa protein- approximate mass of ORF1p- was detected in MCF-7 and HEK293T cell lines [20, 38, 54] while DU145 and HeLa cells lacked detectable expression (Fig. 1c). Loading the same amount of total protein lysate followed by western blot analysis using increased serum concentration (1:20,000 instead of 1:33,000 dilution) (Additional file 1: Figure S2a) revealed an extra band of lower molecular weight (~25 kDa) only in samples containing the 40 kDa band. To assess sensitivity of α-hORF1p (RRM), we carried out western blot analysis using increasing amounts of total lysate from HEK293T cells (10 μg, 20 μg and 40 μg) which revealed a distinct single band at 40 kDa when the primary antibody was used at a 1:33,000 dilution (Additional file 1:Figure S2b, Panel 2); while a similar experiment with the same amount of protein lysate but more concentrated serum (1: 20,000) detected a smaller ~25 kDa band in the lane loaded with 40 μg and 20 μg of protein lysate (Additional file 1: Figure S2b, Panel 1). Furthermore, we performed western analysis using total lysate from the E.coli expression cells (pET30b induced in E.coli BL-21 strain). The absence of other non-specific bands suggested that the small fraction (less than 1% of RRM band) of unknown 70 kDa bacterial protein which co-purified with the RRM peptide was not immunogenic in rabbit (Additional file 1: Figure S2c).

Quantification of band intensity by densitometry revealed that ORF1p expression in MCF-7 cells is almost twice the amount detected in HEK293T cells (Fig. 1C, lane 4 and lane 5; Fig. 1f).Consistent with species-specificity, serum failed to detect any band in cell lysate obtained from a mouse cell line [Fig. 1C, lane 1 (NIH-3T3)]. To further characterize specificity, we assayed reactivity of serum on protein lysates from HeLa and HEK293T cells transfected with a construct containing L1-ORF1 sequence tagged by a FLAG-epitope at the C-terminus of ORF1 (pcDNA-ORF1F) [20] (Fig. 1d, Panel 1), (Fig. 1e, Panel 1). α-FLAG and α-GAPDH served as controls (Fig. 1d, Panel 2), (Fig. 1e, Panel 2 and Panel 3). Along with demonstrating that the serum isolated from the rabbit injected with the hORF1p (RRM) peptide contains an antibody reactive and specific to human L1-ORF1p, these data indicate that our antibody is capable of detecting endogenous denatured L1 protein.

To determine whether α-hORF1p (RRM) can detect endogenous hORF1p in its native conformation, we performed immunofluorescence (IF) on cultured MCF-7 and HeLa cells characterized for the presence or absence of L1 ORF1p by immunoblot analyses (Fig. 2). Consistent with our Western blot data, no fluorescence was detected in untransfected HeLa cells (Fig. 2b, Panel 1); however, transfection with pcDNA-ORF1F revealed cytoplasmic staining (Fig. 2b, Panel 2). Indeed, cytoplasmic localization

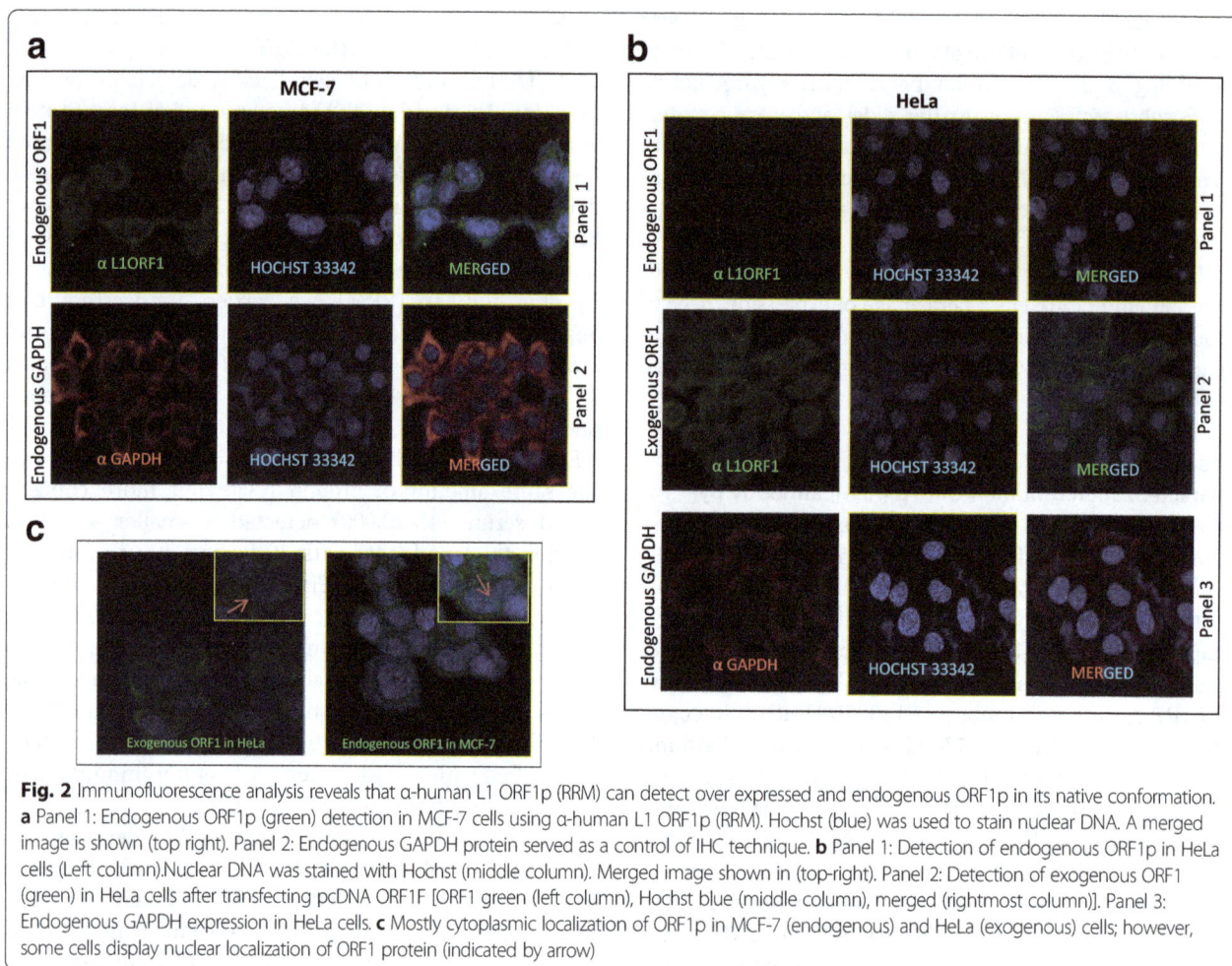

Fig. 2 Immunofluorescence analysis reveals that α-human L1 ORF1p (RRM) can detect over expressed and endogenous ORF1p in its native conformation. **a** Panel 1: Endogenous ORF1p (green) detection in MCF-7 cells using α-human L1 ORF1p (RRM). Hochst (blue) was used to stain nuclear DNA. A merged image is shown (top right). Panel 2: Endogenous GAPDH protein served as a control of IHC technique. **b** Panel 1: Detection of endogenous ORF1p in HeLa cells (Left column).Nuclear DNA was stained with Hochst (middle column). Merged image shown in (top-right). Panel 2: Detection of exogenous ORF1 (green) in HeLa cells after transfecting pcDNA ORF1F [ORF1 green (left column), Hochst blue (middle column), merged (rightmost column)]. Panel 3: Endogenous GAPDH expression in HeLa cells. **c** Mostly cytoplasmic localization of ORF1p in MCF-7 (endogenous) and HeLa (exogenous) cells; however, some cells display nuclear localization of ORF1 protein (indicated by arrow)

of exogenous ORF1p has been reported for a variety of cell lines including HeLa [31, 33, 35]. These data indicate that α-hORF1p (RRM) can detect over-expressed ORF1p in its native conformation in fixed HeLa cells by IF.

To assay whether α-hORF1p (RRM) can detect endogenous hORF1p, we carried out IF on MCF-7 cells (Fig. 2a). In agreement with our hypothesis, we detected endogenous ORF1p using α-hORF1p (RRM) (Fig. 2a) and the localization mirrored that of the exogenously transfected ORF1p in HeLa cells (Fig. 2b); specifically, we observed mainly cytoplasmic staining of ORF1p in MCF-7 cells by immunofluorescence (Fig. 2a, Panel 1). GAPDH serves as an internal control (Fig. 2a, Panel 2). Notably, few cells (less than 5%) both in MCF-7 (endogenous ORF1p) and HeLa (exogenous ORF1p) showed nuclear localisation of ORF1p (Fig. 2c). The data demonstrate that α-hORF1p (RRM) is able to detect endogenous ORF1p in a cancer cell line (Fig. 1c and 2a) and that minimal background fluorescence is observed in cell lines lacking ORF1p expression by Western blot analysis using our antibody (Fig. 2b).

Detection of endogenous ORF1p in human tissues using α-hORF1p (RRM)

While significant progress has been made recently regarding our understanding of endogenous L1 activity primarily using next-generation sequencing technology for insertion analysis in disease states [49, 50] and animal models [2] such as mouse, less is known about human L1 retrotransposition and protein functions in vivo in somatic tissues. To this end, we carried out immunohistochemistry using α-hORF1p (RRM) on a variety of human samples, including brain tissues [36, 38] where L1 insertional activity is known to be increased [2, 43, 44].

To assay for ORF1p expression in different regions of human brain, we performed immunohistochemistry (IHC) on formalin fixed paraffin embedded brain sections. We first examined L1-ORF1p expression in three different regions - frontal cortex, hippocampus and basal ganglia of a brain from a post-mortem 55 years old female (victim of a traffic accident) with no known neurological or psychiatric illness. All three regions show significant staining in neurons with α-hORF1p (RRM)

(Fig. 3). A hippocampus section exposed to the secondary antibody alone (Fig. 3, panel 4 middle) did not exhibit any specific immunostaining (negative control).To assay specificity of α-hORF1p (RRM) in IHC, we utilized an additional three controls: 1) primary α-hORF1p (RRM) (raised in rabbit) followed by secondary α-mouse (Fig. 3, Panel 4 rightmost), 2) primary α-His (raised in mouse) and secondary α-mouse (Additional file 1: Figure S2d) and 3) primary non-immune sera (rabbit) and secondary α-rabbit (Additional file 1: Figure S2e); in all the instances, no signal was detected. In another control, total lysate from 80 year old frontal cortex tissue and MCF-7 cells probed with non-immune rabbit sera by immunoblotting didn't show any signal (Additional file 1: Figure S2f). To determine whether the cells which stained with α-hORF1p (RRM) are neurons, we performed IHC using α-Neurofilament (NE-14) a neuronal marker in a hippocampus section

from 55 year-old brain (Additional file 1: Figure S3b). These data demonstrate that NE-14 stained neurons show morphological similarities with α-hORF1p (RRM) staining cells. To account for age or sex bias potentially associated with L1-ORF1p expression, we stained post-mortem samples from a 15-year-old male and an 80-year-old female. These data show that ORF1p levels are noticeably lower for the 15-year-old sample across all three regions tested (frontal cortex, hippocampus and basal ganglia) when compared to stained samples from the 55-year-old and 80- year-old individuals (Fig. 3). Surprisingly, we observed very high ORF1p expression in the 80-year-old frontal cortex (Fig. 3, Panel 3 leftmost). Staining another frontal cortex section from 77-year-old brain showed similar very high expression of ORF1p (Fig. 3, Panel 4 leftmost). Quantification of DAB signal (e.g. ORF1p positive cells) using ImmunoRatio software [53] indicate that ORF1p expression in the

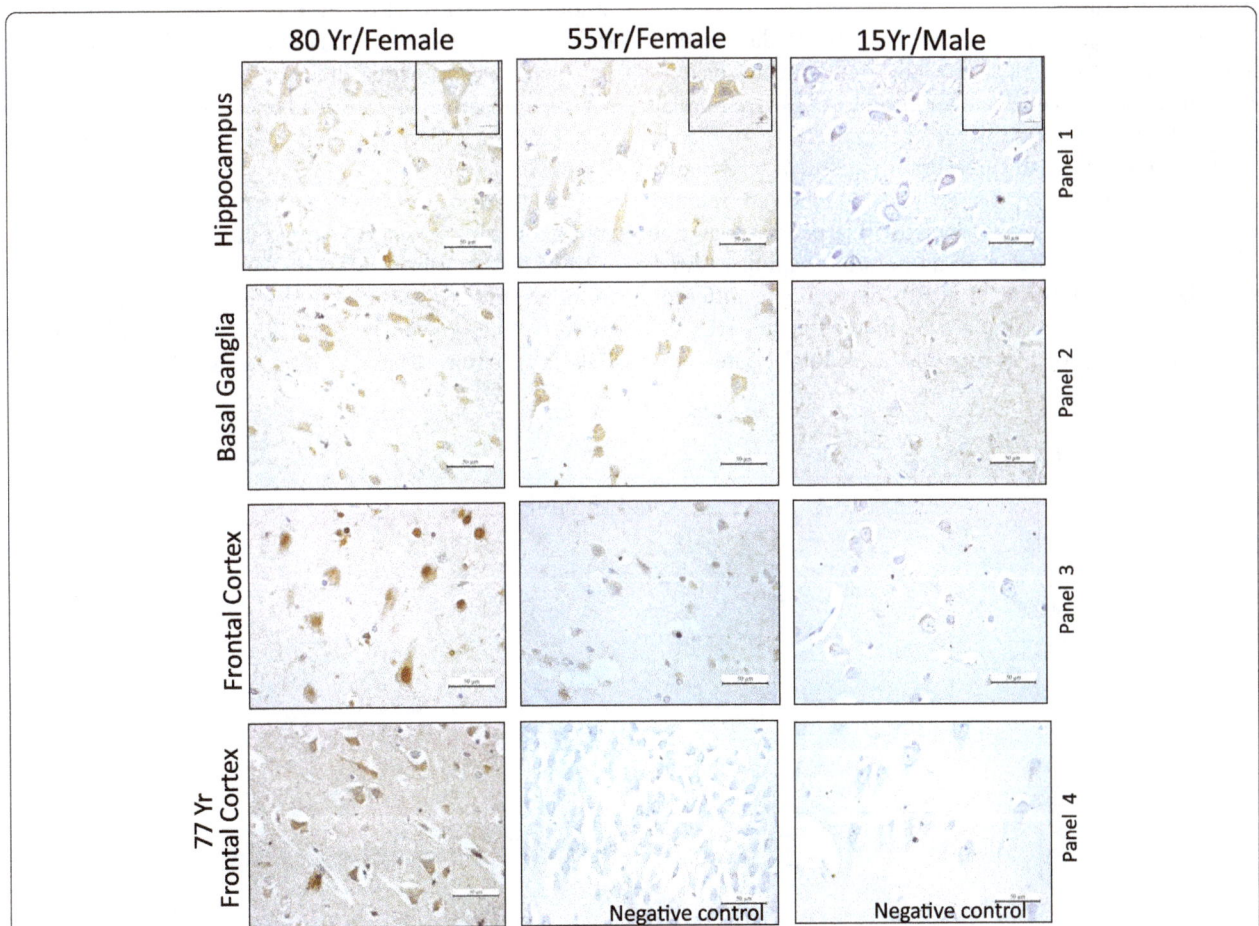

Fig. 3 Immunoperoxidase detection of endogenous L1-ORF1p in different regions of sections of the normal human brain. Human α-L1 ORF1p (RRM) was used to detect the expression L1 activity in human three different brain tissues: hippocampus, basal ganglia and frontal cortex. Samples from three different aged brains (80 year, 55 year and 15 year old) were analyzed. Images were taken at 40X magnifications. Panel 4, leftmost: Immunoperoxidase staining of frontal cortex from 77 year old individual. Panel 4 middle: As a negative control, the immunostaining procedure was performed without primary antibody on a hippocampus section from a 55-year old individual. Panel 4 rightmost: In another negative control, IHC was performed on the same section using primary α-hORF1p (RRM) (raised in rabbit) followed by secondary α-mouse

hippocampus and basal ganglia samples from the 55-year-old and 88-year-old individuals is similar (Additional file 1: Figure S3a, Panel 1 and Panel 3). In contrast, ORF1p expression in samples from the 80-year-old frontal cortex (Fig. 3, Panel 3 leftmost; Additional file 1: Figure S3A, Panel 3) are approximately 3-fold more intense relative to the frontal cortex sample from the 55-year-old. Furthermore, the expression of ORF1p observed in the 15-year-old frontal cortex is less than 5% of that observed in the sample from the 80-year-old (Additional file 1: Figure S3a, Panel 3).

While quantification of ORF1p expression in the basal ganglia samples from the three individuals showed similar levels (Additional file 1: Figure S3, Panel 2), the signal intensity of stained cells coming from the 15-year-old individual (Fig. 3, Panel 2 rightmost) was significantly less. We speculate that the increased value for the basal ganglia for the 15-year-old is due to increased tissue matrix staining, a technical problem we were unable to circumvent.

Along with gaining insight into the tissue distribution and abundance of L1-ORF1p, IHC can provide insights regarding the sub-cellular distribution of ORF1p. Similar to our IF analysis (Fig. 2), we observe ORF1p primarily in the cytoplasm of all three brain regions. Interestingly, the frontal cortex of the 80-year-old showed intense staining of ORF1p in the nucleus; this pattern was not observed for the other regions or other samples tested (Fig. 4a). To complement ORF1p detection in different parts of post-mortem human brain section by IHC, we performed Western blot analysis using total lysate from

the 80 year old frontal cortex. Using total lysate from MCF-7 cells as a control, we were able to detect ORF1p in the 80 year old frontal cortex tissue (Fig. 4b, Panel 1). The GAPDH immunoblotting was used as an internal control (Fig. 4b, Panel 2). These data further support that L1 ORF1p is present in different anatomical regions of human brain including robust levels in the frontal cortex region.

To further our interrogation of ORF1p expression in the human brain, we carried out IHC on additional sections derived from the following regions: medulla oblongata, midbrain, thalamus and spinal cord (Fig. 5). Notably, we did not have access to any of these tissues in the case of the 55-years old individual, and there was no spinal cord tissue available from the 15-years old. For tissues derived from the 80-year-old individual, we detected ORF1p positive cells for the medulla oblongata, midbrain and thalamus but not the spinal cord (Fig. 5, Panel 1 rightmost). Consistent with barely detectable amounts of L1 ORF1p in the 15-year-old individual's (Fig. 3) frontal cortex, basal ganglia and hippocampus, our IHC experiments did not provide any evidence for the presence of L1 ORF1p in thalamus, midbrain, and medulla oblongata of the same individual (Fig. 5, Panel 2).

In addition to the brain, L1 may also be expressed in other somatic tissues in vivo although at least one other study suggests otherwise [38]. To test if L1 is expressed in other human tissues besides the brain, we assayed three tissues previously tested for ORF1p expression by IHC [38] kidney, liver and lung along with heart tissues

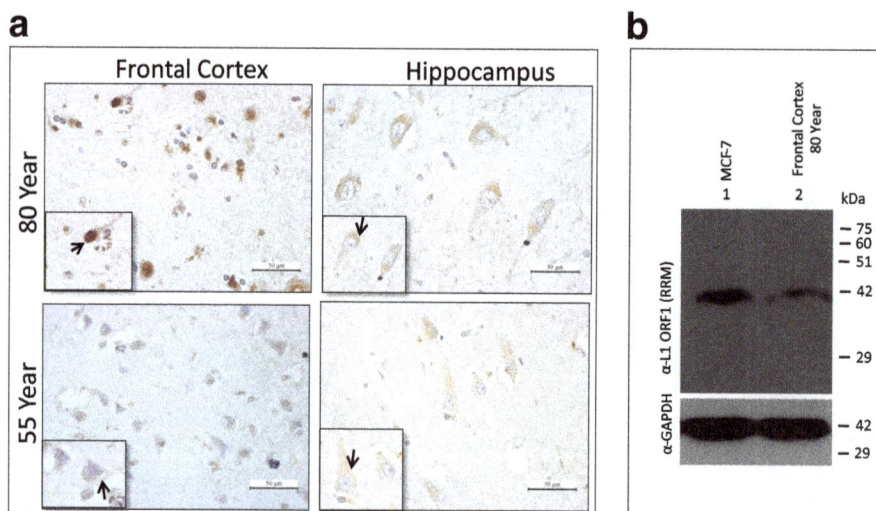

Fig. 4 Nuclear-cytoplasmic localization of endogenous ORF1p in frontal cortex and hippocampal sections obtained from 55 year- and 80 year-old brain. **a** Differences in the abundance of nuclear-localized ORF1p between frontal cortex relative to hippocampal section from 80 year-old. Immunohistochemistry analysis was carried as described in methods using α-L1 ORF1p (RRM). **b** Detection of L1 ORF1p in total lysate from 80 year old frontal cortex by Western blotting. α-human L1 ORF1p (RRM) detects ORF1p (~40 kDa) in total lysate from 80 year old frontal cortex (lane 2). Total lysate from MCF-7 cells was used as control (lane 1)

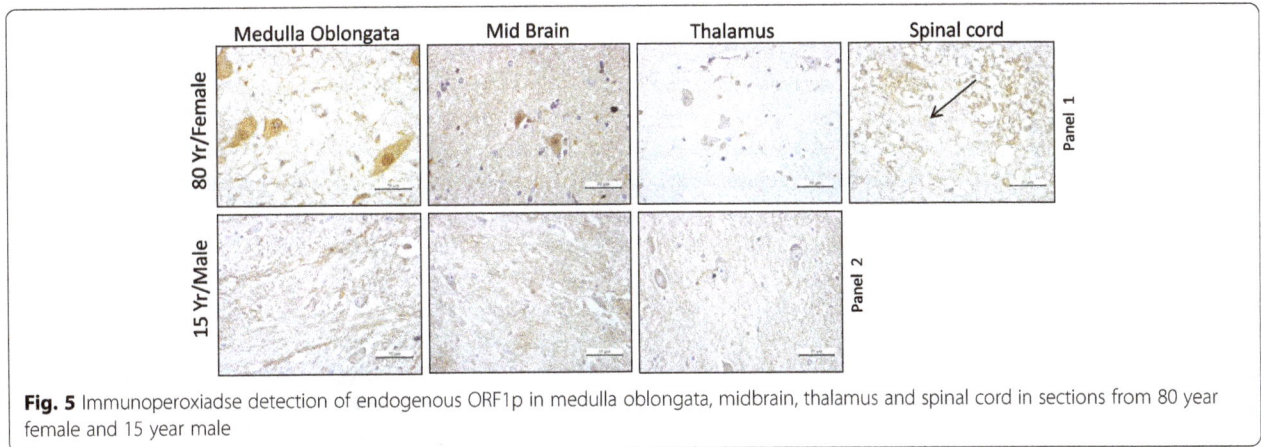

Fig. 5 Immunoperoxiadse detection of endogenous ORF1p in medulla oblongata, midbrain, thalamus and spinal cord in sections from 80 year female and 15 year male

using α-hORF1p (RRM). Consistent with Rodic et al. data, our immunohistochemistry analysis also suggests little to no ORF1p expression in these tissues (Fig. 6a, Panel 1). IHC for GAPDH expression showed very high expression in all four tissues tested (Fig. 6a, Panel 2) as well as in brain sections (data not shown). Quantitation using ImmunoRatio software [53] showed GAPDH expression is comparable in all four tissues whereas expression of ORF1p is more abundant in heart tissue compared to other non-brain samples (Fig. 6b).

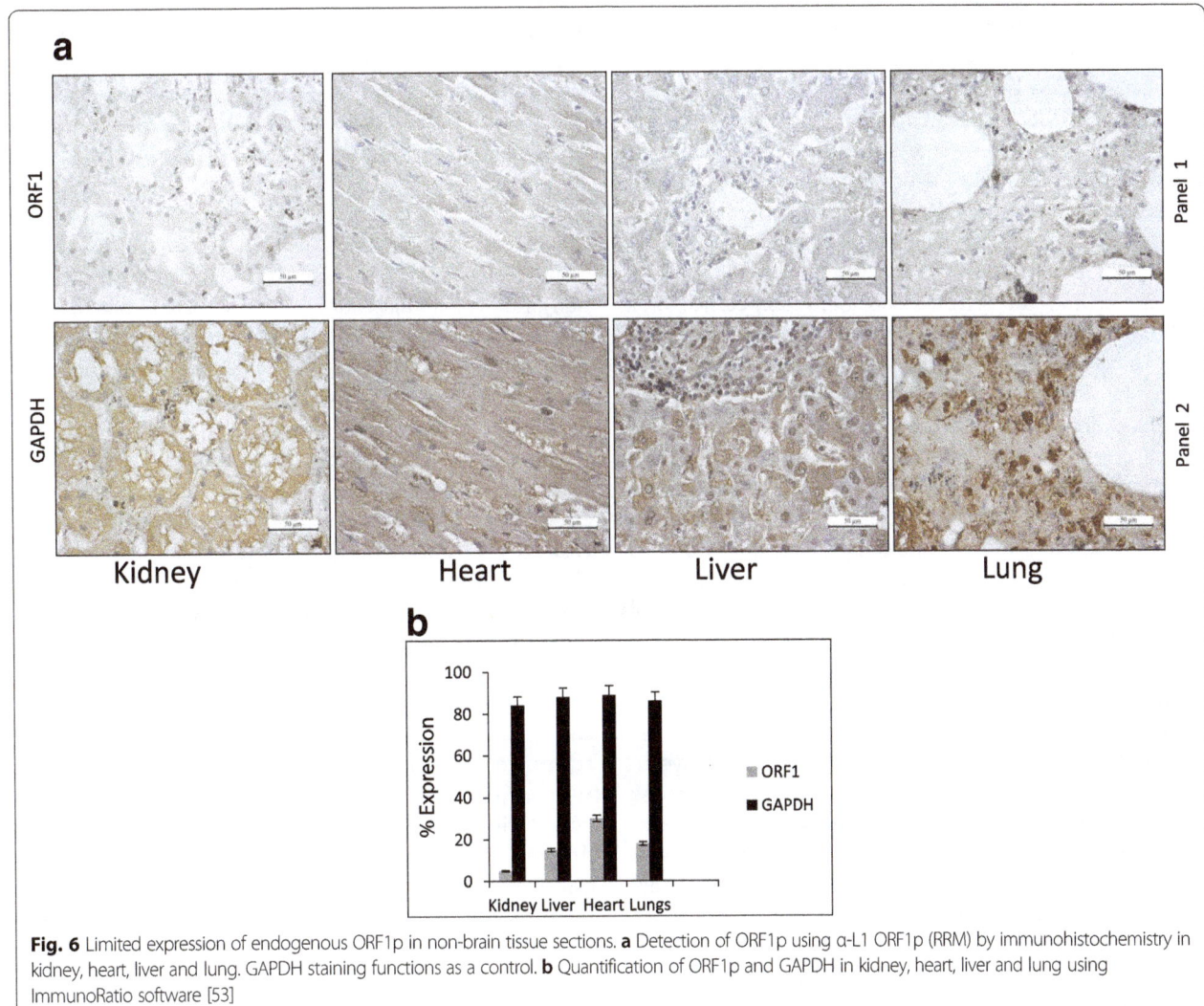

Fig. 6 Limited expression of endogenous ORF1p in non-brain tissue sections. **a** Detection of ORF1p using α-L1 ORF1p (RRM) by immunohistochemistry in kidney, heart, liver and lung. GAPDH staining functions as a control. **b** Quantification of ORF1p and GAPDH in kidney, heart, liver and lung using ImmunoRatio software [53]

Discussion

ORF1 protein expression in the brain is widespread

Recently, the role of retrotransposon activity in brain function and neuronal plasticity has gained significant interest. Although several studies to date have reported an increase in L1 insertions in certain brain regions such as the hippocampus [2, 43, 45], our understanding of L1 protein expression in the brain is limited. In this study, we report the first in vivo detection of L1 protein expression in sections from multiple distinct regions of the post-mortem human brain. (Figs. 3, 4, and 5).

ORF1p is one of two proteins encoded by LINE-1 retrotransposons; both of which are required for retrotransposition *in cis* [15]. While significant insights pertaining to ORF1p biology have been gained from cell culture, biochemical, genetic, and structural studies, less is known regarding its function in vivo. Here, we set out to establish a new reagent - α-hORF1p (RRM) – that would be useful to complement existing tools and studies [24, 37, 38, 54, 55] including insertion analysis in brain tissues using next-gen sequencing.

Using four different cancer cell lines (MCF-7, HeLa, DU145, HEK293T) we observe very high levels of endogenous ORF1p in MCF-7 and HEK293T cells, but are unable to detect ORF1p in lysates from HeLa and DU145 cells by Western analysis. These data are consistent with previous studies which have also observed high levels of endogenous ORF1p in many breast cancer tumors [37] and breast cancer cell lines (T47D, SKBr3, BT-20, MCF-7, Hs578T) [54]. Even though L1 protein expression and the retrotransposition-competence of a cell (e.g., new insertions) is known to vary across cancers and cancer cell lines [38, 56–58] perhaps MCF-7 cells might be useful in identifying factors important for ORF1p expression and stability.

A potential association between LINE-1 ORF1p expression and aging

The detection of ORF1p by immunohistochemistry and its quantification in several distinct regions of the human brain derived from different individuals provides additional support that L1 is indeed active in this organ. We observe differences in the intensity and abundance of ORF1p in the brain tissue samples across individuals. Specifically, tissues derived from two individuals older than 50-years of age displayed markedly increased levels of ORF1p relative to samples derived from a 15-year-old male (Fig. 3). These data are supported by previous studies which have identified variation in L1 copy number using qPCR-based assays across different brain regions and individuals; but, the exact relationship between endogenous L1 protein expression and insertion frequency remains incomplete [43]. Importantly, the data here assaying seven different brain regions (frontal cortex, hippocampus and basal ganglia, thalamus,

midbrain, medulla oblongata and spinal cord) along with four other organs (liver, lung, kidney and heart) from three different individuals increase our understanding of human tissues that permit endogenous ORF1p expression.

Consistent with our detection of robust endogenous ORF1p expression in the hippocampus and frontal cortex, an elevated insertion frequency has been observed in these tissues by single-cell analysis and deep-sequencing [45]. Our quantification of staining indicates significantly higher expression of ORF1p in basal ganglia, hippocampus and frontal cortex when compared to other brain regions tested in samples derived from the 55-year old. Interestingly, our data indicate that samples originating from even older brains (e.g. 77- and 80-years-old) display more detectable expression of ORF1p. Perhaps the most striking finding from IHC analysis of brain tissues is the near absence of ORF1p staining in samples derived from a 15-year-old brain in light of our ability to easily detect ORF1p in samples from older brains.

Importantly, the frequency and impact of new insertions in brain tissues is still being debated. For instance, single neuron sequencing performed by [59] to assay rates of retrotransposition in the frontal cortex and caudate nucleus calculated less than 0.1 insertions per neuron. Their data suggested that an increase in ORF1p expression within a particular region of the brain might serve some other function and does not correlate with the number of L1 insertions in that region of the brain. Consistently, our data assessing ORF1p expression in non-brain tissues like kidney, heart, liver and lung did not detect expression of ORF1p in agreement with Rodic et al. [38]. In contrast, IHC analysis of adult testis, another non-brain tissue have demonstrated significant expression of ORF1p and ORF2p by IHC analysis [36].

Although we acknowledge that the sample size is very small in this study, it is tempting to speculate that endogenous ORF1p expression increases with age. However, at this time we cannot rule out that the observed ORF1p expression differences seen in this study may be due to inter-individual variation in the number of "hot" L1s each person inherits [60]. Relatedly, the longevity regulating protein Sirtuin 6 (SIRT6) has been reported to suppress L1 retrotransposition. Specifically, SIRT6 enforces silencing of L1 by establishing transcriptionally repressive heterochromatin at L1 genomic sequences [61]. With aging, SIRT6 activity is depleted allowing the activation of silenced L1 elements [61]. Future studies similar to this report, which include ORF2p reverse transcriptase assays and deep-sequencing analysis for the genomic L1 insertion content, will likely resolve whether older brains display elevated rates of L1 retrotransposition (e.g., an increase in L1 insertions) relative to younger brains and any associated biological impact.

Detection of the LINE-1 retrotransposon RNA-binding protein ORF1p in different anatomical regions...

169

Conclusions

Our findings show elevated expression of L1ORF1p in different parts of post-mortem human brain compared to other body parts like kidney, heart, liver and lung. We have seen individuals of different ages display very different expression of L1ORF1p, especially in the frontal cortex. Overall, our data show ORF1p levels in brain tissues vary from person to person where age might have some influence on L1 retrotransposition.

Abbreviation

BSA: Bovine Serum Albumin; CC: Coiled Coil; CTD: Carboxy Terminal Domain; DAB: 3-3'- Diaaminobenzidinetetrahydrochloride (DAB substrate); DMEM: Dulbecco's Modified Eagle medium; DNA: Deoxyribonucleic Acid; ECL: Enhanced Chemiluminescence; EDTA: Ethylene-Diamine Tetra-Acetic Acid; FFPE: Formalin-Fixed Paraffin Embedded; GAPDH: Glyceraldehyde 3-Phosphate Dehydrogenase; HBTR: Human Brain Tissue Repository; HEK: Human Embryonic Kidney; HRP: Horseradish Peroxidase; IF: Immunofluorescence; IHC: Immunohistochemistry; kDa: kilo Dalton; LINE: Long INterpersed Element; MCF-7: Michigan Cancer Foundation-7; mm: Millimeter; mM: Millimolar; MW: Molecular Weight; NaCl: Sodium Chloride; °C: degree centigrade; ORF: Open Reading Frame; PBS: Phosphate Buffered Saline; PBS-T: Phosphate Buffered Saline-Tween; qPCR: quantitative Polymerase Chain Reaction; RNA: Ribonucleic Acid; RRM: RNA Recognition Motif; SDS-PAGE: Sodium Dodecyl Sulphate-Polyacrylamide Gel Electrophoresis; SINE: Short Interpersed Element; SIRT6: Sirtuin 6; SVA: (SINE-R/VNTR/Alu); TBST: Tris Buffered Saline-Tween; TBST: Tris Buffered Saline-Tween; UTR: Untranslated region; V: Volt; x g: Times gravity; µg: microgram

Acknowledgements

We thank Dr. Biplob Bhattacharya and Dr. Jayita, Earth Science IIT Roorkee for helping with microscopy. We thank Dr. Sudha Bhattacharya (School of Environmental Sciences, Jawaharlal Nehru University, and New Delhi, India) for helping with reagents and chemicals required in this study. We thank two anonymous reviewers for critically reviewing the manuscript.

Funding

This work was supported by grants to P.K.M from Department of Science and Technology (DST), India (EMR/2014/000167) and Faculty Initiative Grant IIT Roorkee (FIG100638). D.C.H. is funded by a K99/R00 Pathway to Independence Award from the National Institutes of Health (NIGMS), U.S.A.

Author's contributions

DS conducted all the experiments and helped to write the manuscript. RKK cloned, expressed, and purified L1-ORF1-RRM domain using a bacterial expression system. SV helped in generating antibody. AM provided the brain sections on slides and frozen brain tissues. DCH helped in writing and editing the manuscript. VY generated the L1-ORF1-RRM antibody. SKS provided brain sections on slides, analysed data and edited the manuscript. PKM conceived of the study, supervised experiments, analysed data and wrote the manuscript. All authors read and approved the final manuscript.

Competing interests

The authors declare that they have no competing interest.

Author details

[1]Department of Biotechnology, IIT Roorkee, Roorkee, Uttarakhand, India. [2]School of Environmental Sciences, Jawaharlal Nehru University, New Delhi, India. [3]Human Brain Tissue Repository (HBTR), Neurobiology Research Centre, NIMHANS, Bangalore 560 029, India. [4]Department of Human Genetics, University of Utah, Salt Lake City, UT, USA. [5]Present address: Department of Immunology, UT South-western Medical Centre, Dallas, TX, USA.

References

1. McClintock B. Chromosome organisation and genic expression. Cold Spring Harbor Symp Quant Biol. 1951;16:13–47.
2. Muotri AR, Chu VT, Marchetto MC, Deng W, Moran JV, Gage FH. Somatic mosaicism in neuronal precursor cells mediated by L1 retrotransposition. Nature. 2005;435(7044):903–10.
3. Kazazian Jr HH, Wong C, Youssoufian H, Scott AF, Phillips DG, et al. Haemophilia a resulting from de novo insertion of L1 sequences represents a novel mechanism for mutation in man. Nature. 1988;332(6160):164–6.
4. Hancks DC, Kazazian HH Jr. Roles for retrotransposon insertions in human disease. Mob DNA. 2016; https://doi.org/10.1186/s13100-016-0065-9.
5. Biemont CA. Brief history of the status of transposable elements: from junk DNA to major players in evolution. Genetics. 2010;186(4):1085–93.
6. Richardson SR, Doucet AJ, Kopera HC, Moldovan JB, Garcia-Perez JL, Moran JV. The influence of LINE-1 and SINE Retrotransposons on mammalian genomes. Microbiol Spectr. 2014; https://doi.org/10.1128/microbiolspec. MDNA3-0061-2014.
7. Lander ES, Linton LM, Birren B, Nusbaum C, Zody MC, Baldwin J, Devon K, Dewar K, et al. Initial sequencing and analysis of the human genome. Nature. 2001;409(6822):860–21.
8. Mandal PK, Kazazian HH Jr. SnapShot: Vertebrate transposons. Cell. 2008; 135(1):192.
9. Brouha B, Schustak J, Badge RM, Lutz-Prigge S, Farley AH, Moran JV, Kazazian HH Jr. Hot L1s account for the bulk of retrotransposition in the human population. Proc Natl Acad Sci U S A. 2003;100(9):5280–5.
10. Scott AF, Schmeckpeper BJ, Abdelrazik M, Comey CT, O'Hara B, Rossiter JP, Cooley T, Heath P, Smith KD, Margolet L. Origin of the human L1 elements: proposed progenitor genes deduced from a consensus DNA sequence. Genomics. 1987;1(2):113–25.
11. Holmes SE, Singer MF, Swergold GD. Studies on p40, the leucine zipper motif-containing protein encoded by the first open reading frame of an active human LINE-1 transposable element. J Biol Chem. 1992;267(28):19765–8.
12. Martin SL, Bushman FD. Nucleic acid chaperone activity of the ORF1 protein from the mouse LINE-1 retrotransposon. Mol Cell Biol. 2001;21(2):467–75.
13. Mathias SL, Scott AF, Kazazian Jr HH, Boeke JD, Gabriel A. Reverse transcriptase encoded by a human transposable element. Science. 1991;254(5039):1800–10.
14. Feng Q, Moran JV, Kazazian Jr HH, Boeke JD. Human L1 retrotransposon encodes a conserved endonuclease required for retrotransposition. Cell. 1996;87(5):905–16.
15. Moran JV, Holmes SE, Naas TP, DeBerardinis RJ, Boeke JD, Kazazian HH Jr. High frequency retrotransposition in cultured mammalian cells. Cell. 1996;87(5):917–27.
16. Esnault C, Maestre J, Heidmann T, Human LINE. retrotransposons generate processed pseudogenes. Nat Genet. 2000;24(4):363–7.
17. Pei B, Sisu C, Frankish A, Howald C, Habegger L, XJ M, Harte R, Balasubramanian S, Tanzer A, Diekhans M, Reymond A, Hubbard TJ, Harrow J, Gerstein MB. The GENCODE pseudogene resource. Genome Biol. 2012; https://doi.org/10.1186/gb-2012-13-9-r51.
18. Zhang Z, Harrison PM, Liu Y, Gerstein M. Millions of years of evolution preserved: a comprehensive catalog of the processed pseudogenes in the human genome. Genome Res. 2003;13(12):2541–58.
19. Karro JE, Yan Y, Zheng D, Zhang Z, Carriero N, Cayting P, Harrrison P, Gerstein M. Pseudogene.org: a comprehensive database and comparison platform for pseudogene annotation. Nucleic Acids Res. 2007;35(Database issue):D55–60.
20. Mandal PK, Ewing AD, Hancks DC, Kazazian Jr HH. Enrichment of processed pseudogene transcripts in L1-ribonucleoprotein particles. Hum Mol Genet. 2013;22(18):3730–48.
21. Dewannieux M, Esnault C, Heidmann T. LINE-mediated retrotransposition of marked Alu sequences. Nat Genet. 2003;35(1):41–8.
22. Ostertag EM, Goodier JL, Zhang Y, Kazazian HH Jr. SVA elements are nonautonomous retrotransposons that cause disease in humans. Am J Hum Genet. 2003;73(6):1444–51.
23. Hancks DC, Goodier JL, Mandal PK, Cheung LE, Kazazian Jr HH. Retrotransposition of marked SVA elements by human L1s in cultured cells. Hum Mol Genet. 2011; 20(17):3386–400.
24. Raiz J, Damert A, Chira S, Held U, Klawitter S, Hamdorf M, Löwer J, Strätling WH, Löwer R, Schumann GG. The non-autonomous retrotransposon SVA is trans-mobilized by the human LINE-1 protein machinery. Nucleic Acids Res. 2012;40(4):1666–83.
25. Wang H, Xing J, Grover D, Hedges DJ, Han K, Walker JA, Batzer MASVA. Elements: a hominid-specific retroposon family. J Mol Biol. 2005;354(4):994–07.

26. Martin SL. Nucleic acid chaperone properties of ORF1p from the non-LTR retrotransposon, LINE-1. RNA Biol. 2010;7(6):706–11.

27. Martin SL. The ORF1 protein encoded by LINE-1: structure and function during L1 retrotransposition. J Biomed Biotechnol. 2006;1:45621.

28. Khazina E, Weichenrieder O. Non-LTR retrotransposons encode noncanonical RRM domains in their first open reading frame. Proc Natl Acad Sci U S A. 2009;106(3):731–6.

29. Khazina E, Truffault V, Büttner R, Schmidt S, Coles M, Weichenrieder O. Trimeric structure and flexibility of the L1ORF1 protein in human L1 retrotransposition. Nat Struct Mol Biol. 2011;18(9):1006–14.

30. Goodier JL, Ostertag EM, Engleka KA, Seleme MC, Kazazian HH Jr. A potential role for the nucleolus in L1 retrotransposition. Hum Mol Genet. 2004;13(10):1041–8.

31. Goodier JL, Zhang L, Vetter MR, Kazazian HH Jr. LINE-1 ORF1 protein localizes in stress granules with other RNA-binding proteins, including components of RNA interference RNA-induced silencing complex. Mol Cell Biol. 2007;27(18):6469–83.

32. Goodier JL, Mandal PK, Zhang L, Kazazian Jr HH. Discrete subcellular partitioning of human retrotransposon RNAs despite a common mechanism of genome insertion. Hum Mol Genet. 2010;19(9):1712–25.

33. Doucet AJ, Hulme AE, Sahinovic E, Kulpa DA, Moldovan JB, Kopera HC, Athanikar JN, Hasnaoui M, Bucheton A, Moran JV. Characterization of LINE-1 ribonucleoprotein particles. PLoS Genet. 2010;6(10) https://doi.org/10.1371/journal.pgen.1001150.

34. Horn AV, Klawitter S, Held U, Berger A, Vasudevan AA, Bock A, et al. Human LINE-1 restriction by APOBEC3C is deaminase independent and mediated by an ORF1p interaction that affects LINE reverse transcriptase activity. Nucleic Acids Res. 2014;42(1):396–416.

35. Kirilyuk A, Tolstonog GV, Damert A, Held U, Hahn S, et al. Functional endogenous LINE-1 retrotransposons are expressed and mobilized in rat chloroleukemia cells. Nucleic Acids Res. 2008;36(2):648–65.

36. Ergün S, Buschmann C, Heukeshoven J, Dammann K, Schnieders F, Lauke H, Chalajour F, Kilic N, Strätling WH, Schumann GG. Cell type-specific expression of LINE-1 open reading frames 1 and 2 in fetal and adult human tissues. J Biol Chem. 2004;279(26):27753–63.

37. Harris CR, Normart R, Yang Q, Stevenson E, Haffty BG, Ganesan S, Cordon-Cardo C, Levine AJ, Tang LH. Association of nuclear localization of a long interspersed nuclear element-1 protein in breast tumors with poor prognostic outcomes. Genes Cancer. 2010;1(2):115–24.

38. Rodić N, Sharma R, Sharma R, Zampella J, Dai L, Taylor MS, Hruban RH, Iacobuzio-Donahue CA, et al. Long interspersed element-1 protein expression is a hallmark of many human cancers. Am J Pathol. 2014;184(5):1280–6.

39. Chiu YL, Greene WC. The APOBEC3 cytidine deaminases: an innate defensive network opposing exogenous retroviruses and endogenous retroelements. Annu Rev Immunol. 2008;26:317–53.

40. Schumann GG, Gogvadze EV, Osanai-Futahashi M, Kuroki A, Münk C, et al. Unique functions of repetitive transcriptomes. Int Rev Cell Mol Biol. 2010; 285:115–88.

41. Pizarro JG, Cristofari G. Post-transcriptional control of LINE-1 Retrotransposition by cellular host factors in somatic cells. Front Cell Dev Biol. 2016;4:14. https://doi.org/10.3389/fcell.2016.00014.

42. Goodier JL. Restricting retrotransposons: a review. Mob DNA. 2016;7:16. https://doi.org/10.1186/s13100-016-0070-z.

43. Coufal NG, Garcia-Perez JL, Peng GE, Yeo GW, Mu Y, Lovci MT, Morell M, O'Shea KS, Moran JV, Gage FH. L1 retrotransposition in human neural progenitor cells. Nature. 2009;460(7259):1127–31.

44. Baillie JK, Barnett MW, Upton KR, Gerhardt DJ, Richmond TA, et al. Somatic retrotransposition alters the genetic landscape of the human brain. Nature. 2011;479(7374):534–7.

45. Upton KR, Gerhardt DJ, Jesuadian JS, Richardson SR, Sánchez-Luque FJ, Bodea GO, Ewing AD, Salvador-Palomeque C, et al. Ubiquitous L1 mosaicism in hippocampal neurons. Cell. 2015;161(2):228–39.

46. Ostertag EM, Prak ET, DeBerardinis RJ, Moran JV, Kazazian JHH. Determination of L1 retrotransposition kinetics in cultured cells. Nucleic Acids Res. 2000;28(6):1418–23.

47. Coufal NG, Garcia-Perez JL, Peng GE, Marchetto MC, Muotri AR, Mu Y, Carson C, Macia A, Moran JV, Gage FH. Ataxia telangiectasia mutated (ATM) modulates long interspersed element-1 (L1) retrotransposition in human neural stem cells. Proc Natl Acad Sci U S A. 2011;108(51):20382–7.

48. Muotri AR, Marchetto MC, Coufal NG, Oefner R, Yeo G, Nakashima K, Gage FH. L1 retrotransposition in neurons is modulated by MeCP2. Nature. 2010; 468(7322):443–6.

49. Bundo M, Toyoshima M, Okada Y, Akamatsu W, Ueda J, Nemoto-Miyauchi T, Sunaga F, et al. Increased l1 retrotransposition in the neuronal genome in schizophrenia. Neuron. 2014;81(2):306–13.

50. Shpyleva S, Melnyk S, Pavliv O, Pogribny I, Jill James S. Overexpression of LINE-1 Retrotransposons in autism brain. Mol Neurobiol. 2017; https://doi.org/10.1007/s12035-017-0421-x.

51. Kimberland ML, Divoky V, Prchal J, Schwahn U, Berger W, Kazazian Jr HH. Full-length human L1 insertions retain the capacity for high frequency retrotransposition in cultured cells. Hum Mol Genet. 1999;8(8):1557–60.

52. Mandal PK, Kazazian Jr HH. Purification of L1-Ribonucleoprotein particles (L1-RNPs) from cultured human cells. Methods Mol Biol. 2016;1400:299–310. https://doi.org/10.1007/978-1-4939-3372-3_19.

53. Tuominen VJ, Ruotoistenmäki S, Viitanen A, Jumppanen M, Isola J. ImmunoRatio: a publicly available web application for quantitative image analysis of estrogen receptor (ER), progesterone receptor (PR) and Ki-67. Breast Cancer Res. 2010;12(4):R56. https://doi.org/10.1186/bcr2615.

54. Chen L, Dahlstrom JE, Chandra A, Board P, Rangasamy D. Prognostic value of LINE-1 retrotransposon expression and its subcellular localization in breast cancer. Breast Cancer Res Treat. 2012;136(1):129–42.

55. Bratthauer GL, Fanning TG. Active LINE-1 retrotransposons in human testicular cancer. Oncogene. 1992;7(3):507–10.

56. Garcia-Perez JL, Morell M, Scheys JO, Kulpa DA, Morell S, et al. Epigenetic silencing of engineered L1 retrotransposition events in human embryonic carcinoma cells. Nature. 2010;466(7307):769–73.

57. Philippe C, Vargas-Landin DB, Doucet AJ, van Essen D, Vera-Otarola J, et al. Activation of individual L1 retrotransposon instances is restricted to cell-type dependent permissive loci. elife. 2016;e13926 https://doi.org/10.7554/eLife.13926.

58. Lee E, Iskow R, Yang L, Gokcumen O, Haseley P, Luquette LJ 3rd, Lohr JG, et al. Cancer genome atlas research network. Landscape of somatic retrotransposition in human cancers. Science. 2012;337(6097):967–71.

59. Evrony GD, Lee E, Park PJ, Walsh CA, et al. Resolving rates of mutation in the brain using single-neuron genomics. elife. 2016;5 https://doi.org/10.7554/eLife.12966.

60. Beck CR, Collier P, Macfarlane C, Malig M, Kidd JM, Eichler EE, Badge RM, Moran JV. LINE-1 retrotransposition activity in human genomes. Cell. 2010; 141(7):1159–70.

61. Van Meter M, Kashyap M, Rezazadeh S, Geneva AJ, Morello TD, Seluanov A, Gorbunova V. SIRT6 represses LINE1 retrotransposons by ribosylating KAP1 but this repression fails with stress and age. Nat Commun. 2014;23(5):5011. https://doi.org/10.1038/ncomms6011.

14

Transposable element polymorphisms recapitulate human evolution

Lavanya Rishishwar[1,2,3], Carlos E. Tellez Villa[2,4] and I. King Jordan[1,2,3]*

Abstract

Background: The human genome contains several active families of transposable elements (TE): Alu, L1 and SVA. Germline transposition of these elements can lead to polymorphic TE (polyTE) loci that differ between individuals with respect to the presence/absence of TE insertions. Limited sets of such polyTE loci have proven to be useful as markers of ancestry in human population genetic studies, but until this time it has not been possible to analyze the full genomic complement of TE polymorphisms in this way.

Results: For the first time here, we have performed a human population genetic analysis based on a genome-wide polyTE data set consisting of 16,192 loci genotyped in 2,504 individuals across 26 human populations. PolyTEs are found at very low frequencies, > 93 % of loci show < 5 % allele frequency, consistent with the deleteriousness of TE insertions. Nevertheless, polyTEs do show substantial geographic differentiation, with numerous group-specific polymorphic insertions. African populations have the highest numbers of polyTEs and show the highest levels of polyTE genetic diversity; Alu is the most numerous and the most diverse polyTE family. PolyTE genotypes were used to compute allele sharing distances between individuals and to relate them within and between human populations. Populations and continental groups show high coherence based on individuals' polyTE genotypes, and human evolutionary relationships revealed by these genotypes are consistent with those seen for SNP-based genetic distances. The patterns of genetic diversity encoded by TE polymorphisms recapitulate broad patterns of human evolution and migration over the last 60–100,000 years. The utility of polyTEs as ancestry informative markers is further underscored by their ability to accurately predict both ancestry and admixture at the continental level. A genome-wide list of polyTE loci, along with their population group-specific allele frequencies and F_{ST} values, is provided as a resource for investigators who wish to develop panels of TE-based ancestry markers.

Conclusions: The genetic diversity represented by TE polymorphisms reflects known patterns of human evolution, and ensembles of polyTE loci are suitable for both ancestry and admixture analyses. The patterns of polyTE allelic diversity suggest the possibility that there may be a connection between TE-based genetic divergence and population-specific phenotypic differences.

Keywords: Transposable elements, Polymorphism, Population genetics, Human ancestry, Admixture, Ancestry informative markers, Phylogenetics, Alu, L1, SVA

Background

Much of the human genome sequence, anywhere from ~50 to 70 % depending on estimates [1, 2], is derived from transposable elements (TE). The vast majority of TE-derived sequences in the genome are remnants of ancient insertion events, which are no longer capable of transposition. Nevertheless, there remain a few families of actively transposing human TEs [3]; the active families of human TEs include Alu [4, 5], L1 [6, 7] and SVA [8, 9] elements. Alu elements are 7SL RNA-derived short interspersed nuclear elements (SINEs) [10, 11], L1s are a family of long interspersed nuclear elements (LINEs) [12, 13], and SVA elements are composite TEs that are made up of human endogenous retrovirus sequence, simple sequence repeats and Alu sequence [14, 15]. All three of these active families of human TEs are retrotransposons that transpose via reverse transcription of an RNA intermediate. L1s are

* Correspondence: king.jordan@biology.gatech.edu
[1]School of Biology, Georgia Institute of Technology, 310 Ferst Drive, Atlanta, GA 30332-0230, USA
[2]PanAmerican Bioinformatics Institute, Cali, Valle del Cauca, Colombia
Full list of author information is available at the end of the article

autonomous retrotransposons that encode the enzymatic machinery necessary to catalyze their own retrotransposition [16], whereas Alu and SVA elements are transposed in *trans* by the L1 machinery [17, 18].

If members of these active TE families transpose in the germline, they can create novel insertions that are capable of being inherited, thereby generating human-specific polymorphisms. Such polymorphic TE (polyTE) insertion sites have been shown to be valuable genetic markers for studies of human ancestry and evolution. PolyTEs provide a number of advantages for such population genetic studies [3, 19]. First, the presence of a polyTE insertion site shared by two or more individuals nearly always represents identity by descent [19, 20]. This is because there are so many possible insertion sites genome-wide, and transposition rates are so low, that the probability of independent insertion at the same site in two individuals is negligible. Second, since newly inserted TEs rarely undergo deletion they are highly stable polymorphisms. These two characteristics underscore the fact that polyTE markers are completely free of homoplasies, i.e. identical states that do not represent shared ancestry, which are far more common for single nucleotide polymorphisms (SNPs). Another useful feature of polyTEs for population genetic studies is the fact that the ancestral state of polyTE loci is known to be absence of the insertion [21, 22]. Finally, polyTEs are practically useful markers since they can be rapidly and accurately typed via PCR-based assays.

A number of previous studies have leveraged TE polymorphisms for the analysis of human ancestry and evolution [3, 18, 19, 21–27]. Most of these studies have focused on Alu elements; there have been far fewer human population genetic studies using L1 markers and to our knowledge no such studies using polymorphic SVA elements. Alus are particularly advantageous for these types of studies because their small size allows them to be readily PCR amplified; furthermore, both the presence and absence of Alu insertions can yield amplification products from a single PCR. Ancestry studies that use TE polymorphisms have relied on a number of selection criteria in order to try and define the most useful polyTE loci for human population differentiation. For instance, polyTE loci have often been identified via literature surveys of specific gene mutations caused by TE insertions. Analysis of the human genome sequence has also been used to identify intact members of the youngest (i.e. recently active) subfamilies of Alus and L1s in order to try and predict potentially mobile sequences. Once potential polyTE marker loci are chosen using these methods, they need to be empirically evaluated with respect to their levels of polymorphism within and between populations. These approaches, while somewhat *ad hoc* and laborious, have in fact proven to be useful

for the identification of polyTE loci that serve as ancestry informative markers (AIMs).

The most recent data release from the 1000 Genome Project (Phase3 November 2014) includes, for the first time, a comprehensive genome-wide data set of polyTE sites. There are a total of 16,192 such polyTE loci reported for 2,504 individuals across 26 human populations. These newly available data provide an unprecedented level of depth and resolution for polyTE-based studies of human ancestry and evolution. With these data, it is now possible to evaluate the relationship between TE polymorphism and human evolution in a systematic and unbiased way. In addition, individual polyTE loci genome-wide can be evaluated with respect to their utility as AIMs as well as their applicability to ancestry studies for specific population groups. Such an analysis could provide a useful resource for investigators interested in conducting their own targeted studies on specific populations. With such a comprehensive, genome-wide polyTE data set, it is also possible to evaluate the marker utility of previously underutilized L1 and SVA sequences. For this study, we have conducted a genome-wide population genetic analysis of human TE polymorphisms in order to address precisely these kinds of issues. This work represents the most comprehensive study of human polyTEs to date.

Results
Human population genomics of polyTEs
There are three families of polymorphic transposable elements (polyTEs) that show variation in presence/absence patterns at individual insertion sites across human genome sequences; these are Alu (SINE), L1 (LINE) and chimeric SVA elements. The Phase3 data release (November 2014) of the 1000 Genomes Project provides the most complete catalog of human transposable element insertion site polymorphisms available to date. Presence/absence genotypes for these human polyTEs are available for 2,504 individuals from 26 human populations across 16,192 genomic sites.

We characterized the frequencies and distributions of human polyTEs for the 26 populations organized into 5 continental groups: African, Asian, European, Indian and American (Table 1). The vast majority of human polyTEs are found at low frequencies within and between human populations; 15,141 (93.5 %) of polyTE loci show < 5 % overall allele frequencies (Fig. 1a). Nevertheless, there is substantial variability of individual polyTE allele frequencies among populations from different continental groups (Fig. 1b). Accordingly, there are higher numbers of polyTEs with continental group-specific allele frequencies > 5 % (Fig. 1c), and numerous individual polyTE loci are exclusively present within a single continental group (Fig. 1d). On average, ~25 % of individual polyTE loci are exclusive to a specific continental

Table 1 Human populations analyzed in this study

	Color	Short	Full Description	n
African (n=504)		ESN	Esan in Nigeria	99
		GWD	Gambian in Western Division, The Gambia	113
		LWK	Luhya in Webuye, Kenya	99
		MSL	Mende in Sierra Leone	85
		YRI	Yoruba in Ibadan, Nigeria	108
Asian (n=504)		CDX	Chinese Dai in Xishuangbanna, China	93
		CHB	Han Chinese in Bejing, China	103
		CHS	Southern Han Chinese, China	105
		JPT	Japanese in Tokyo, Japan	104
		KHV	Kinh in Ho Chi Minh City, Vietnam	99
European (n=503)		CEU	Utah residents with Northern and Western European ancestry	99
		FIN	Finnish in Finland	99
		GBR	British in England and Scotland	91
		IBS	Iberian populations in Spain	107
		TSI	Toscani in Italy	107
Indian (n=489)		BEB	Bengali in Bangladesh	86
		GIH	Gujarati Indian in Houston,TX	103
		ITU	Indian Telugu in the UK	102
		PJL	Punjabi in Lahore,Pakistan	96
		STU	Sri Lankan Tamil in the UK	102
American (n=504)		ACB	African Caribbean in Barbados	96
		ASW	African Ancestry in Southwest US	61
		CLM	Colombian in Medellin, Colombia	94
		MXL	Mexican Ancestry in Los Angeles, California	64
		PEL	Peruvian in Lima, Peru	85
		PUR	Puerto Rican in Puerto Rico	104

Populations are organized into five continental groups, and the number of individuals in each population is shown. The same population-specific color codes are used throughout the manuscript

group. These results are consistent with the possibility that polyTE genotypes may serve as useful markers of genomic ancestry. Results of the same analyses are shown for individual polyTEs families in Additional file 1: Figure S1. Alu is by far the most abundant family of polyTEs followed by L1 and SVA. All three polyTE families show similar levels of continental group-specific insertions.

PolyTE genotypes were analyzed in order to evaluate the polyTE genetic diversity levels for different continental groups and for different TE families. To do this, presence/absence patterns at all polyTE loci were used to genotype individual human genomes and pairwise allele sharing distances between individuals were computed based on these polyTE genotypes (see Methods). African populations have the highest levels of polyTE genetic diversity and Asian populations show the lowest diversity (Fig. 2a). These data are similar to what has been shown in previous studies of polyTEs [27] and for SNP-based genetic diversity [28]. All of the differences in median genetic diversity levels between pairs of population groups are highly statistically significant ($0 \leq P \leq 8.5 \times 10^{-56}$ Wilcoxon ranked sum test). African populations also have the highest levels of variation in polyTE genetic diversity for any of the non-admixed groups, consistent with human origins in Africa and the bottleneck experienced by other population groups during their migrations out of Africa [29, 30]. The overall effect of recent admixture in

the Americas is revealed by the broad distribution of polyTE genetic diversity among the American populations, and African admixture among these same populations probably accounts for the fact that this group has the second highest level of median diversity seen for all continental groups (Fig. 2a). For polyTE families, Alu has the highest diversity followed by SVA and L1 (Fig. 2b). The relative levels of continental group polyTE genetic diversity are the same for all three families of polyTEs (Fig. 2c–d).

Human evolutionary relationships based on polyTEs

The distributions of polyTE genotypes among individuals were analyzed in an effort to reconstruct the evolutionary relationships among human individuals and populations. To do this, PolyTE genotype allele sharing distances were used to generate multi-dimensional scaling (MDS) plots showing the genetic relationships among all individuals (Fig. 3a) and the average genetic relationships between individual populations (Fig. 3b). Phylogenetic reconstruction was also used to show the average polyTE genotype-based relationships between populations (Fig. 3c). The evolutionary relationships revealed by this analysis are entirely consistent with previous analyses based on individual nucleotide level variation assessed via SNP-based genotypes [31], and very similar to what has previously been seen based on Alu polymorphisms [23]. African, Asian

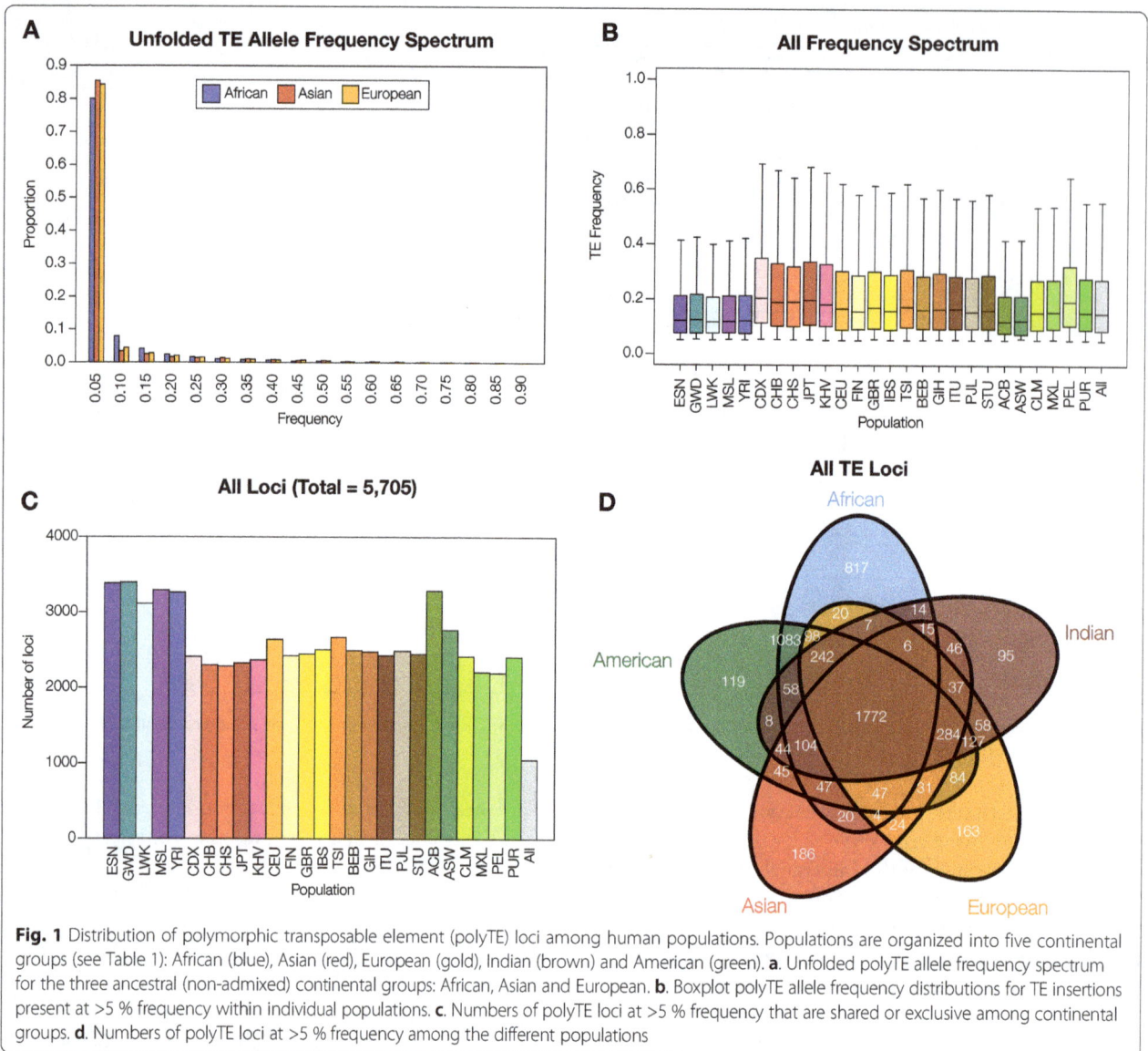

Fig. 1 Distribution of polymorphic transposable element (polyTE) loci among human populations. Populations are organized into five continental groups (see Table 1): African (blue), Asian (red), European (gold), Indian (brown) and American (green). **a**. Unfolded polyTE allele frequency spectrum for the three ancestral (non-admixed) continental groups: African, Asian and European. **b**. Boxplot polyTE allele frequency distributions for TE insertions present at >5 % frequency within individual populations. **c**. Numbers of polyTE loci at >5 % frequency that are shared or exclusive among continental groups. **d**. Numbers of polyTE loci at >5 % frequency among the different populations

and European continental groups represent the three poles of human genomic variation with the more ancient admixed Indian group and more recent admixed American group in between. In the phylogenetic analysis, the African populations are the most basal with the European and Asian populations being derived.

One of the advantages of using TE polymorphisms for ancestry inference is that the ancestral state for any polyTE loci can be confidently taken to be the absence of an insertion [21, 22]. This property allows for the creation of a hypothetical ancestral genome characterized by the absence of insertions across all polyTE loci. When such a hypothetical ancestor is included in the polyTE-based reconstruction of human evolutionary relationships, it maps near the center of the MDS plots closer to the African populations (Fig. 3a and b), and it maps closest to the root of the phylogeny between the

African and non-African lineages (Fig. 3c). These results confirm that polyTE insertions are derived allelic states.

For the most part, there is high coherence of polyTE genotypes within both individual populations and for continental groups. The only exception seen is for the admixed American continental group, which has two distinct subgroups, a Latino subgroup (PEL, MXL, CLM and PUR) with primarily European and Asian admixture and an African-American subgroup (ACB and ASW) with primarily African and European admixture (Fig. 3d). The relative admixture levels seen for these populations are consistent with previous nucleotide level SNP-based analysis [32, 33]. The apparent Asian admixture of the Latino subgroup reflects Native American ancestry owing to the fact that Native Americans are relatively recently derived from East Asian populations [34]. As there are no Native American samples in the 1000 Genomes Project

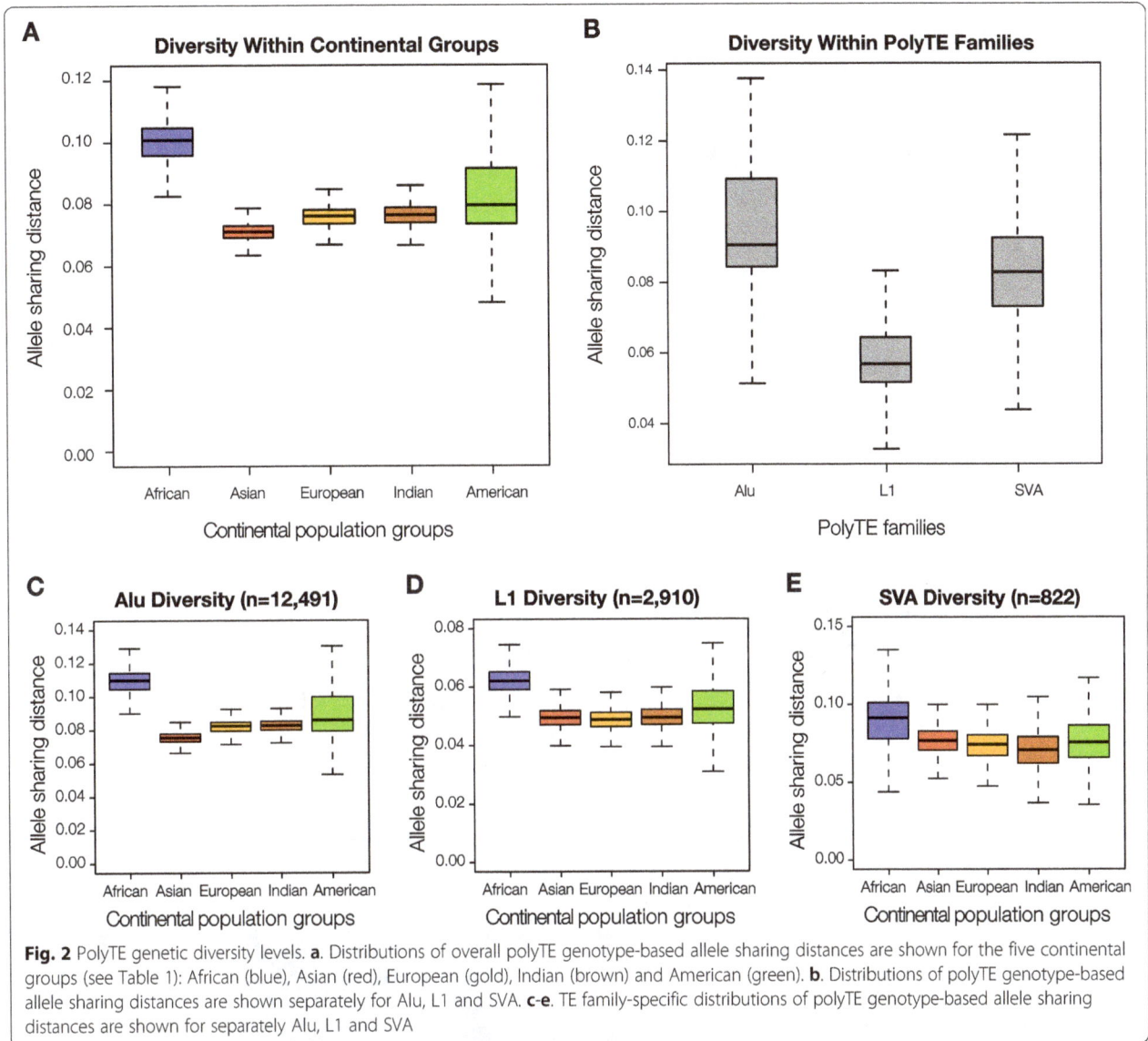

Fig. 2 PolyTE genetic diversity levels. **a**. Distributions of overall polyTE genotype-based allele sharing distances are shown for the five continental groups (see Table 1): African (blue), Asian (red), European (gold), Indian (brown) and American (green). **b**. Distributions of polyTE genotype-based allele sharing distances are shown separately for Alu, L1 and SVA. **c-e**. TE family-specific distributions of polyTE genotype-based allele sharing distances are shown for separately Alu, L1 and SVA

Data [28, 35], the East Asian genome sequences appear as most closely related to the Latino subgroup. CLM and PUR show relatively higher levels of European, and to a lesser extent African, admixture than seen for PEL and MXL (Fig. 3d). We also attempted to infer Native American ancestry in admixed American populations by imputing polyTE genotypes for Native American populations from the Human Genome Diversity Project based on the 1000 Genome Project imputation panels. The ancestry contribution fractions for admixed American individuals are highly correlated between the observed Asian polyTE genotypes and the imputed Native American polyTE genotypes (Additional file 1: Figure S2).

Results of the same analyses are shown for individual polyTEs families in Additional file 1: Figures S3–S5. While the results are highly concordant for all three polyTE families, Alu ployTEs show the highest levels of

resolution for human evolutionary relationships owing to the far higher number of polymorphic Alu insertions available for analysis. Nevertheless, L1 and SVA elements also show the ability to differentiate human populations and continental groups suggesting that these previously under-utilized polyTEs may also serve as useful ancestry markers.

Ancestry prediction with polyTEs

Having established the overall ability of polyTE-based genotype analysis to capture known evolutionary relationships among human populations, we evaluated the ability of individual of polyTE loci to serve as useful markers for ancestry inference. To do this, levels of population differentiation for individual polyTE loci were assessed using the fixation index F_{ST} and the absolute allele frequency differences δ (see Methods). PolyTE loci-based F_{ST} and δ

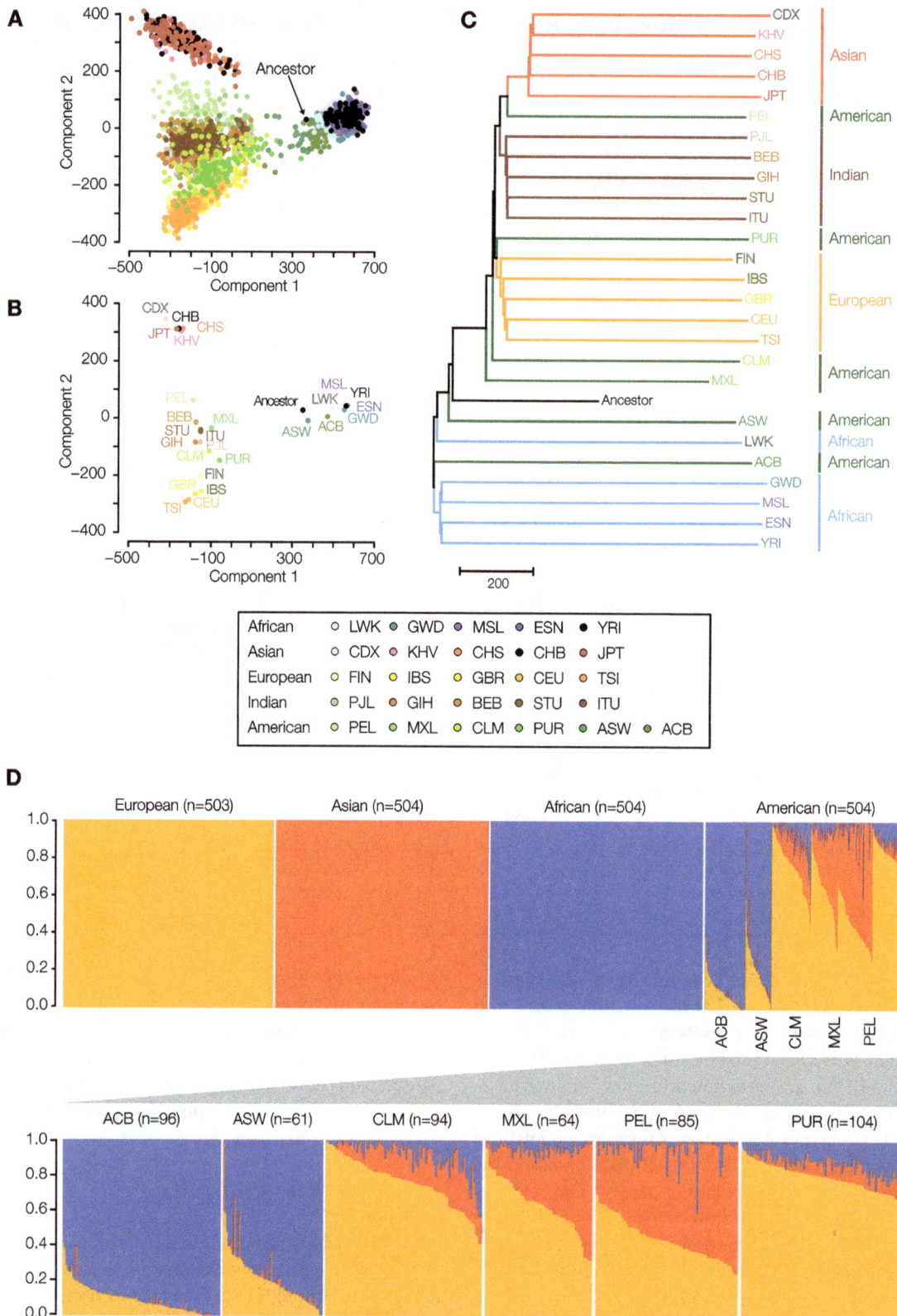

Fig. 3 (See legend on next page.)

distributions were computed for three-way comparisons between non-admixed continental groups (African, Asian and European) and for five-way comparisons between individual populations within the same non-admixed continental group (Additional file 1: Figures S6 and S7). As can be expected, individual polyTE loci show substantially higher levels of population differentiation (i.e. higher F_{ST} and δ values) for the between compared to the within continental group comparisons. This is consistent with the overall ability of polyTE genotypes to better distinguish between continental groups (Fig. 3) than within continental groups (Additional file 1: Figure S8). The same pattern has been observed for SNP-based AIMs [36]. Nevertheless, polyTE loci are able to provide some level of resolution for even closely related populations within continental groups. A comprehensive list of human polyTE loci along with their allele frequencies and F_{ST} and δ values, within and between populations, are provided in Additional file 2: Table S1 so that investigators can choose loci of interest as potential ancestry markers.

Interestingly, the overall levels of polyTE-based F_{ST} are fairly low even for the between continental group comparison (Additional file 1: Figure S6). F_{ST} levels ≥ 0.4 have previously been taken to indicate that a nucleotide SNP can serve as a useful ancestry informative marker (AIM) [36, 37]. There are no individual polyTE loci that conform to this AIM criteria; 0.39 is the highest polyTE F_{ST} value. This can be attributed to the overall low frequency of polymorphic TE insertions seen here (Fig. 1a) since low levels of within-group polyTE allele frequency will depress F_{ST} levels owing to high levels of within group heterozygosity. The values of δ appear to be somewhat more sensitive for the characterization of individual polyTE AIMs. Several different δ value thresholds have been proposed for AIM characterization over the years [36]: 0.3, 0.4 and 0.5. There are 371 (0.3), 79 (0.4) and 9 (0.5) polyTE loci with continental δ values that exceed these thresholds. Thus, individual polyTE loci appear to have moderate ability to differentiate human populations, whereas ensembles of polyTE loci can be used effectively to distinguish more closely and distantly related populations.

In light of the ability of individual polyTEs genotypes and overall polyTE genotype patterns to differentiate

human populations, we attempted to identify the smallest set of polyTE loci needed to accurately predict human ancestry. The accuracy of ancestry prediction was assessed for both non-admixed continental groups (African, Asian and European) and for individual populations within the African continental group. To do this for each comparison, the top 500 ancestry informative polyTE loci were ranked according to their F_{ST} levels and prediction accuracy was computed for sets of polyTE loci of sequentially decreasing size, going from 500 to 10 in steps of 10 (Fig. 4). Two measures of ancestry prediction, accuracy and error, were measured for each set of polyTE loci using the approach described in the Materials and Methods. When all polyTE loci are used, continental group ancestry prediction approaches 100 % accuracy with < 1 % error. As the number of polyTE loci used for ancestry prediction is steadily decreased from 500, the accuracy declines and the error increases. However, the changes in accuracy and error are relatively slight. For the top 100 polyTE loci, ancestry prediction is 86.9 % accurate with 0.3 % error. The smallest set of 10 polyTE loci yields 65.8 % accuracy and 2.7 % error. These results are similar to a previous report [27] that evaluated the minimum number of polymorphic Alu loci (~50) that would yield accurate genetic distances between human populations.

A similar approach was taken to evaluate the utility of polyTE genotypes for ancestry prediction within continental groups. Consistent with what is observed for the within continental group F_{ST} values (Additional file 1: Figure S6), polyTE genotypes have less power to discriminate ancestry for closely related populations from the same continental group (Fig. 4b). For the African populations, individual genotypes based on the entire set of polyTE loci yield an ancestry prediction accuracy of 48.3 % and an error of 6.7 %. Since there are five African populations, a random predictor would yield 20 % accuracy. Thus, the accuracy achieved by polyTE loci, while relatively low, is 2.4x greater than expected by chance alone. Accuracy does not change greatly with decreasing numbers of polyTE loci. 100 polyTE loci yields accuracy of 38.5 %, and the accuracy for 10 polyTE loci is 36.3 %. The error rate of prediction does steadily increase to 8.4 % for 100 polyTE loci and 21.3 % for 10 polyTE loci.

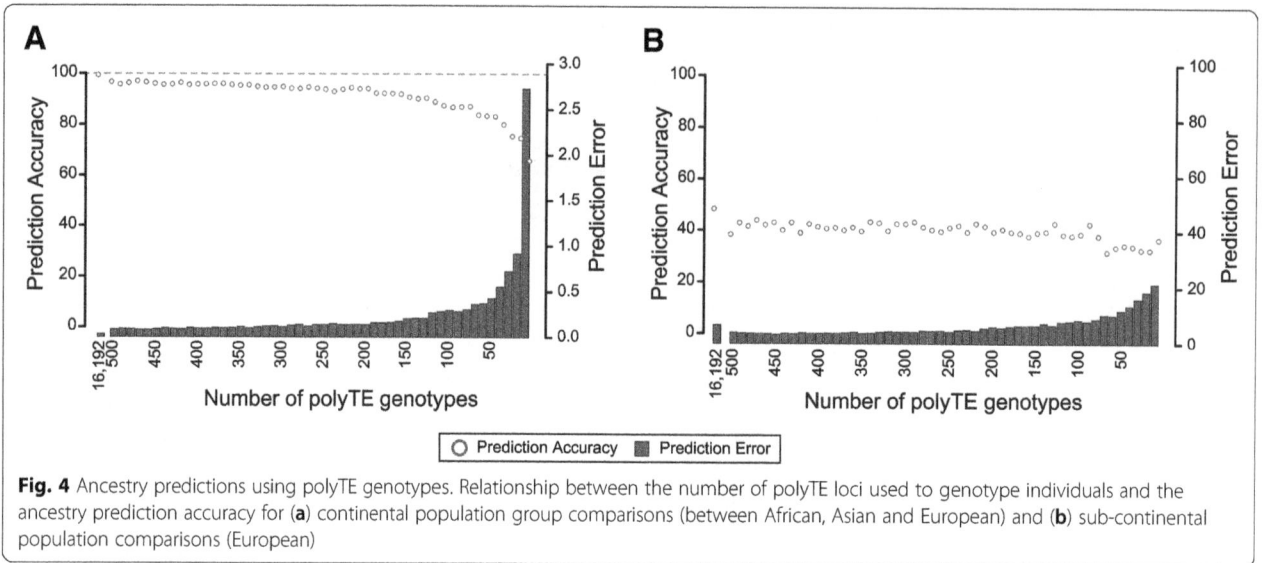

Fig. 4 Ancestry predictions using polyTE genotypes. Relationship between the number of polyTE loci used to genotype individuals and the ancestry prediction accuracy for (**a**) continental population group comparisons (between African, Asian and European) and (**b**) sub-continental population comparisons (European)

Admixture prediction with polyTEs

Having established the utility of small sets of polyTE loci to make ancestry inferences for non-admixed groups, we wished to similarly evaluate the ability of polyTE loci sets to allow for inferences about continental ancestry contributions to admixed populations. To do this, ancestral contributions from African and European populations to the admixed ASW American population were evaluated using sets of polyTE loci of decreasing size in a similar way as was done for ancestry prediction in non-admixed populations. In the case of admixture, prediction error levels were measured by comparing the ancestral admixture components computed from the entire set of 16,192 polyTE loci to those computed from the smaller polyTE loci sets (see Methods). As with ancestry prediction, error levels steadily increase with the use of decreasing numbers of polyTE loci (Fig. 5a). However, slightly larger numbers of polyTE loci are required to keep admixture inference error levels low; the use of 10 polyTE loci yields 3.4 % error, whereas a set of 50 polyTE loci reduces the error to 2.2 %. There is strong agreement in the results of continental ancestry contributions for this admixed population between analyses conducted with all polyTEs versus the top 50 polyTEs ($r = 0.62$; Fig. 5b).

Discussion

Human ancestry and admixture from polyTEs

Our analysis of a genome-wide set of human polyTE genotypes indicates that TE polymorphism patterns recapitulate the pattern of human evolution and migration over the last 60–100,000 years (Fig. 3 and Additional file 1: Figures S3–S5). While polyTEs considered as an ensemble provide substantial resolution for inferring ancestry and human relationships,

individual polyTE loci show moderate population differentiation levels (Additional file 1: Figure S6 and S7). This can be attributed to the fact that individual polyTE loci tend to be found at low allele frequencies (Fig. 1a). However, these same low frequency loci do show high levels of geographic differentiation, i.e. many of them are continental group or population specific (Fig. 1b). Therefore, when a relatively small set of these low frequency but highly geographically differentiated polyTE loci are used together, they do in fact provide substantial resolution for evolutionary analysis as well as ancestry and admixture inference (Figs. 4 and 5).

These results have important implications for the study of human evolution, ancestry and admixture by smaller labs that may not have access to the same level of resources as larger consortia or genome centers since analysis of a small set of polyTE loci (10–50 depending on the application) can prove to be quite informative. Given the size range of TEs insertions, in particular for Alus which are the most numerous family of polyTEs, element presence/absence patterns can be accurately characterized in a cost-effective way using (multiplex) PCR-based techniques. Protocols for PCR-based analysis of polyTEs are well established in a number of labs. The results of this study can be used to help investigators choose the specific TE loci of interest for their own evolutionary studies (see Additional file 2: Table S1 for a list of genomic locations of polyTEs and their allele frequencies and F_{ST} values).

Despite the overall utility of polyTEs as ancestry markers, results from this study suggest that they are not likely to be good markers for mapping by admixture linkage disequilibrium (MALD or admixture mapping)

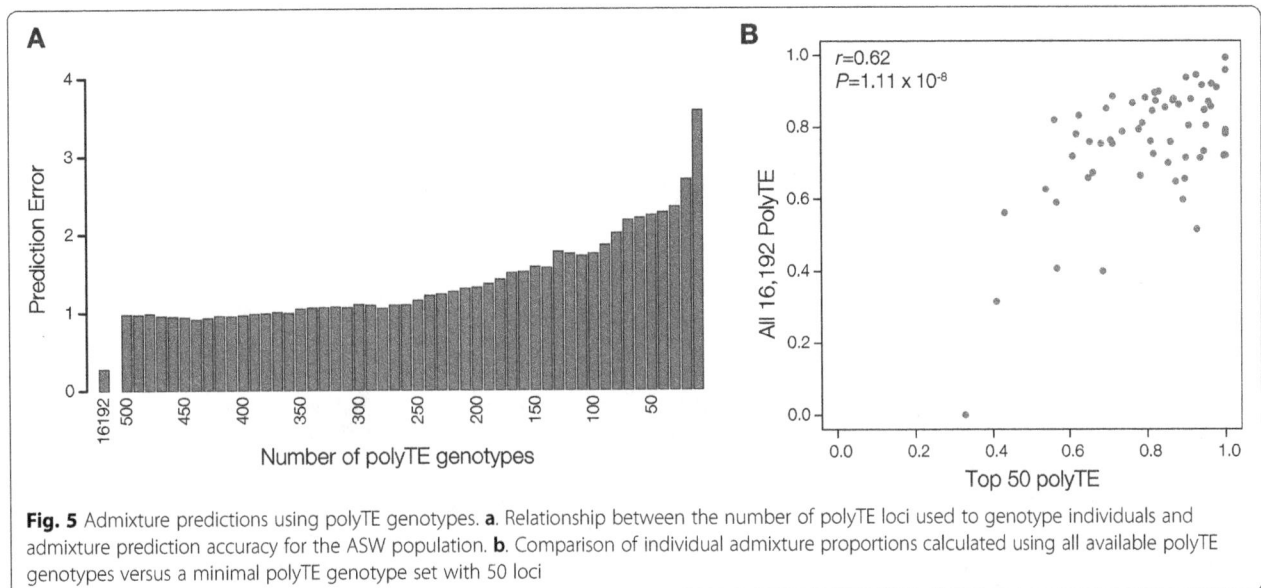

Fig. 5 Admixture predictions using polyTE genotypes. **a**. Relationship between the number of polyTE loci used to genotype individuals and admixture prediction accuracy for the ASW population. **b**. Comparison of individual admixture proportions calculated using all available polyTE genotypes versus a minimal polyTE genotype set with 50 loci

studies [38, 39]. These studies rely on detailed locus-specific assignments of ancestry across the genome in admixed individuals. In order to achieve this level of resolution, thousands of markers are needed and individual markers should have high levels of population differentiation (as measured by F_{ST} or other related metrics) [36]. Thus, SNPs would seem to remain the best choice of AIMs for MALD (admixture mapping) studies.

Deleteriousness and selection on polyTE insertions

Our initial analysis of human polyTEs within and between populations revealed that TE insertion polymorphisms are found at very low frequencies (Fig. 1a). This is consistent with the overall deleteriousness of TE insertions and accordingly their removal by purifying selection. The elimination of polyTEs by purifying selection is also underscored by the fact that polyTEs are vastly under-represented in genic and exonic regions (Additional file 1: Figure S9). Nevertheless, some polyTEs do rise to high allele frequencies and many also show high levels of geographic differentiation consistent with what has been seen for SNPs [28]. This differentiation is precisely what makes them good markers for ancestry inference, particularly when considered as an ensemble, but it also suggests the possibility that polyTE insertions may influence population specific phenotypes shaped by selection. Additional analysis on the effects of selection on TE polymorphisms, as well as the relationship between polymorphic TEs and potentially adaptive phenotypes, will be needed to test this assertion.

Conclusions

Polymorphic TE loci have long been used as markers in human population genetic studies, and they are known

to provide a number of advantages for such studies. The selection of which polyTE loci to use for population genetic studies has been largely *ad hoc*, based on a combination of literature and database surveys together with empirical evaluation on the suitability of individual loci as markers that can discriminate between populations. With the recent release of a genome-wide set of 16,192 TE polymorphisms by the 1000 Genomes Project [28, 35], genotyped across 2,504 individuals from 26 global populations, it is now possible to systematically evaluate the utility of polyTE loci for human population genetic and ancestry studies. We have leveraged these newly released data to conduct the first genome-scale analysis of polyTE genotypes for the study of human genetic ancestry. We show that the genetic diversity represented by TE polymorphisms reflects known patterns of human evolution, and define sub-sets of polyTE loci that can be used as ancestry informative markers. We provide ranked lists of the polyTE loci that can be used by researchers in the community for future ancestry and admixture analyses.

Methods
Transposable element polymorphisms

Human polymorphic transposable element (polyTE) genotypes were taken from the Phase3 data release (November 2014) of the 1000 Genomes Project [28, 35] (ftp://ftp-trace.ncbi.nih.gov/1000genomes/ftp/release/20130502/). These genotypes consist of phased presence/absence patterns of polyTE insertions at specific human genome sites for individual genomes, and they are characterized from human genome reference sequence mapped next-generation sequence data via 1) discordant read mapping for short paired-end reads and/or 2) split read mapping for longer reads as previously described [40]. PolyTE allele frequencies are calculated as the number of present TE insertions (*TEi*) normalized by the total number of sites in the population (*2n*): *TEi/2n*. The extent to which individual polyTE loci differentiate populations was computed using the fixation index F_{ST} with the Weir Cockerham method [41] implemented in VCFtools [42] and the δ parameter [36], which is defined as the absolute value of the difference in the allele frequencies between populations for TE polymorphisms.

Ancestry analysis

PolyTE-based allele sharing distances were computed for all pairs of human genomes by counting the total number of polyTE presence/absence alleles that differ between two individuals across all genomic insertion sites. Allele sharing distances computed in this way were projected in two-dimensional space using multi-dimensional scaling (MDS) implemented in R. This was done for pairwise distances computed between individual genomes and for average allele sharing distances among populations. Population average allele sharing distances were used to reconstruct a neighbor-joining [43] phylogenetic tree using the program MEGA6 [44].

Admixture analysis

The program ADMIXTURE was used to infer the proportion of ancestry contributions from ancestral populations to modern admixed populations from the Americas (ACB, ASW, CLM, MXL, PEL, PUR) based on polyTE genotypes. The program was first run in supervised mode with three ancestral clusters: African, Asian and European. Asian ancestry is taken here as a rough surrogate for Native American admixture in American populations given the relatively close evolutionary relationship between East Asian and Native American populations and the lack of Native American samples in the 1000 Genomes Project. PolyTE genotypes were then imputed for Native American genomes from the Human Genome Diversity Project [31, 45], using the impute panel from the 1000 Genomes Project with the program IMPUTE2 [46], and ADMIXTURE was run in supervised mode with the three ancestral clusters: "African, European and Native American. The ancestry contribution fractions for modern admixed populations from the Americas computed based on observed Asian polyTE genotypes and imputed Native American genotypes were correlated to check for consistency.

Ancestry and admixture prediction analyses

The program ADMIXTURE was used together with a cross-validation approach in order to predict the ancestry of individuals based on their polyTE genotypes. The cross-validation method relied on an 80 %/20 % split of the data, whereby 80 % of individual polyTE genotypes were used to

build a three-cluster ancestry model with ADMIXTURE. The remaining 20 % of individual polyTE genotypes were then tested against this model to predict their ancestry membership in one of the three groups. Group-specific ancestry was only assigned if the probability of group membership was calculated as $\geq$ 90 %. Accuracy is then defined as the number of correct ancestry predictions normalized by the total number of predictions made. Error is defined as the root-mean-square difference (*RMSD*) between the predicted and actual ancestry inference made with the complete data. *RMSD* values are reported as the average prediction error for all individuals. This process was done repeatedly across individual polyTE genotypes based on decreasing numbers of polyTE sites, from 500 to 10 in steps of 10. For each polyTE set, this 80/20 prediction process was repeated 100 times.

An analogous prediction approach was used to infer the continental ancestry contributions to an admixed American population (ASW) using ADMIXTURE. In this case, the training was done using individual polyTE genotypes from ancestral populations (African and European) and the testing was done using polyTE genotypes from admixed ASW individuals. This was done first using all 16,192 polyTE loci and then for individual polyTE genotypes based on decreasing numbers of polyTE sites, from 500 to 10 in steps of 10. The predicted ancestry contributions to admixed individuals were compared for results based on all polyTE loci and results based on reduced sets of polyTE loci using the root-mean-square difference (*RMSD*) for the African and European fractional ancestry contributions.

Abbreviations

TE: Transposable Elements; polyTE: Polymorphic Transposable Elements; LINE: Long Interspersed Nuclear Elements; SINE: Short Interspersed Nuclear Elements; SNP: Single Nucleotide Polymorphism; AIM: Ancestry Informative Marker; MDS: Multi-Dimensional Scaling; MALD: Mapping by Admixture Linkage Disequilibrium; RMSD: Root-Mean-Square Difference.

Competing interests

The authors declare that they have no competing interests.

Authors' contributions

LR and IKJ conceived of the study and designed the analysis. LR carried out the analysis. CETV did the imputation analysis. LR, CETV and IKJ wrote and revised the manuscript. All authors read and approved the final manuscript.

Acknowledgements

This work was funded the Georgia Institute of Technology Bioinformatics Graduate Program, IHRC-GIT Applied Bioinformatics Laboratory (ABiL) and

BIOS – the Colombian National Center of Bioinformatics and Computational Biology.

Author details

[1]School of Biology, Georgia Institute of Technology, 310 Ferst Drive, Atlanta, GA 30332-0230, USA. [2]PanAmerican Bioinformatics Institute, Cali, Valle del Cauca, Colombia. [3]BIOS Centro de Bioinformática y Biología Computacional, Manizales, Caldas, Colombia. [4]Escuela de Ingeniería de Sistemas y Computación, Universidad del Valle, Santiago de Cali, Colombia.

References

1. Lander ES, Linton LM, Birren B, Nusbaum C, Zody MC, Baldwin J, et al. Initial sequencing and analysis of the human genome. Nature. 2001;409(6822):860–921. doi:10.1038/35057062.
2. de Koning AP, Gu W, Castoe TA, Batzer MA, Pollock DD. Repetitive elements may comprise over two-thirds of the human genome. PLoS Genet. 2011;7(12):e1002384. doi:10.1371/journal.pgen.1002384.
3. Ray DA, Batzer MA. Reading TE leaves: new approaches to the identification of transposable element insertions. Genome Res. 2011;21(6):813–20. doi:10.1101/gr.110528.110.
4. Batzer MA, Gudi VA, Mena JC, Foltz DW, Herrera RJ, Deininger PL. Amplification dynamics of human-specific (HS) Alu family members. Nucleic Acids Res. 1991;19(13):3619–23.
5. Batzer MA, Deininger PL. A human-specific subfamily of Alu sequences. Genomics. 1991;9(3):481–7.
6. Brouha B, Schustak J, Badge RM, Lutz-Prigge S, Farley AH, Moran JV, et al. Hot L1s account for the bulk of retrotransposition in the human population. Proc Natl Acad Sci U S A. 2003;100(9):5280–5. doi:10.1073/pnas.0831042100.
7. Kazazian Jr HH, Wong C, Youssoufian H, Scott AF, Phillips DG, Antonarakis SE. Haemophilia A resulting from de novo insertion of L1 sequences represents a novel mechanism for mutation in man. Nature. 1988;332(6160):164–6. doi:10.1038/332164a0.
8. Wang H, Xing J, Grover D, Hedges DJ, Han K, Walker JA, et al. SVA elements: a hominid-specific retroposon family. J Mol Biol. 2005;354(4):994–1007. doi:10.1016/j.jmb.2005.09.085.
9. Ostertag EM, Goodier JL, Zhang Y, Kazazian Jr HH. SVA elements are nonautonomous retrotransposons that cause disease in humans. Am J Hum Genet. 2003;73(6):1444–51. doi:10.1086/380207.
10. Schmid CW, Deininger PL. Sequence organization of the human genome. Cell. 1975;6(3):345–58.
11. Ullu E, Tschudi C. Alu sequences are processed 7SL RNA genes. Nature. 1984;312(5990):171–2.
12. Fanning TG, Singer MF. LINE-1: a mammalian transposable element. Biochim Biophys Acta. 1987;910(3):203–12.
13. Burton FH, Loeb DD, Voliva CF, Martin SL, Edgell MH, Hutchison 3rd CA. Conservation throughout mammalia and extensive protein-encoding capacity of the highly repeated DNA long interspersed sequence one. J Mol Biol. 1986;187(2):291–304.
14. Ono M, Kawakami M, Takezawa T. A novel human nonviral retroposon derived from an endogenous retrovirus. Nucleic Acids Res. 1987;15(21):8725–37.
15. Shen L, Wu LC, Sanlioglu S, Chen R, Mendoza AR, Dangel AW, et al. Structure and genetics of the partially duplicated gene RP located immediately upstream of the complement C4A and the C4B genes in the HLA class III region. Molecular cloning, exon-intron structure, composite retroposon, and breakpoint of gene duplication. J Biol Chem. 1994;269(11):8466–76.
16. Moran JV, Holmes SE, Naas TP, DeBerardinis RJ, Boeke JD, Kazazian Jr HH. High frequency retrotransposition in cultured mammalian cells. Cell. 1996;87(5):917–27.
17. Dewannieux M, Esnault C, Heidmann T. LINE-mediated retrotransposition of marked Alu sequences. Nat Genet. 2003;35(1):41–8. doi:10.1038/ng1223.
18. Salem AH, Kilroy GE, Watkins WS, Jorde LB, Batzer MA. Recently integrated Alu elements and human genomic diversity. Mol Biol Evol. 2003;20(8):1349–61. doi:10.1093/molbev/msg150.
19. Batzer MA, Deininger PL. Alu repeats and human genomic diversity. Nat Rev Genet. 2002;3(5):370–9. doi:10.1038/nrg798.
20. Ray DA, Xing J, Salem AH, Batzer MA. SINEs of a nearly perfect character. Syst Biol. 2006;55(6):928–35.

21. Perna NT, Batzer MA, Deininger PL, Stoneking M. Alu insertion polymorphism: a new type of marker for human population studies. Hum Biol. 1992;64(5):641–8.
22. Batzer MA, Stoneking M, Alegria-Hartman M, Bazan H, Kass DH, Shaikh TH, et al. African origin of human-specific polymorphic Alu insertions. Proc Natl Acad Sci U S A. 1994;91(25):12288–92.
23. Stoneking M, Fontius JJ, Clifford SL, Soodyall H, Arcot SS, Saha N, et al. Alu insertion polymorphisms and human evolution: evidence for a larger population size in Africa. Genome Res. 1997;7(11):1061–71.
24. Witherspoon DJ, Marchani EE, Watkins WS, Ostler CT, Wooding SP, Anders BA, et al. Human population genetic structure and diversity inferred from polymorphic L1(LINE-1) and Alu insertions. Hum Hered. 2006;62(1):30–46. doi:10.1159/000095851.
25. Ray DA, Walker JA, Hall A, Llewellyn B, Ballantyne J, Christian AT, et al. Inference of human geographic origins using Alu insertion polymorphisms. Forensic Sci Int. 2005;153(2–3):117–24. doi:10.1016/j.forsciint.2004.10.017.
26. Terreros MC, Alfonso-Sanchez MA, Novick GE, Luis JR, Lacau H, Lowery RK, et al. Insights on human evolution: an analysis of Alu insertion polymorphisms. J Hum Genet. 2009;54(10):603–11. doi:10.1038/jhg.2009.86.
27. Watkins WS, Rogers AR, Ostler CT, Wooding S, Bamshad MJ, Brassington AM, et al. Genetic variation among world populations: inferences from 100 Alu insertion polymorphisms. Genome Res. 2003;13(7):1607–18. doi:10.1101/gr.894603.
28. Genomes Project C, Abecasis GR, Auton A, Brooks LD, DePristo MA, Durbin RM, et al. An integrated map of genetic variation from 1,092 human genomes. Nature. 2012;491(7422):56–65. doi:10.1038/nature11632.
29. Jakobsson M, Scholz SW, Scheet P, Gibbs JR, VanLiere JM, Fung HC, et al. Genotype, haplotype and copy-number variation in worldwide human populations. Nature. 2008;451(7181):998–1003. doi:10.1038/nature06742.
30. Lohmueller KE, Indap AR, Schmidt S, Boyko AR, Hernandez RD, Hubisz MJ, et al. Proportionally more deleterious genetic variation in European than in African populations. Nature. 2008;451(7181):994–7. doi:10.1038/nature06611.
31. Li JZ, Absher DM, Tang H, Southwick AM, Casto AM, Ramachandran S, et al. Worldwide human relationships inferred from genome-wide patterns of variation. Science. 2008;319(5866):1100–4. doi:10.1126/science.1153717.
32. Bryc K, Velez C, Karafet T, Moreno-Estrada A, Reynolds A, Auton A, et al. Colloquium paper: genome-wide patterns of population structure and admixture among Hispanic/Latino populations. Proc Natl Acad Sci U S A. 2010;107 Suppl 2:8954–61. doi:10.1073/pnas.0914618107.
33. Zakharia F, Basu A, Absher D, Assimes TL, Go AS, Hlatky MA, et al. Characterizing the admixed African ancestry of African Americans. Genome Biol. 2009;10(12):R141. doi:10.1186/gb-2009-10-12-r141.
34. Reich D, Patterson N, Campbell D, Tandon A, Mazieres S, Ray N, et al. Reconstructing Native American population history. Nature. 2012;488(7411):370–4. doi:10.1038/nature11258.
35. Genomes Project C, Abecasis GR, Altshuler D, Auton A, Brooks LD, Durbin RM, et al. A map of human genome variation from population-scale sequencing. Nature. 2010;467(7319):1061–73. doi:10.1038/nature09534.
36. Ding L, Wiener H, Abebe T, Altaye M, Go RC, Kercsmar C, et al. Comparison of measures of marker informativeness for ancestry and admixture mapping. BMC Genomics. 2011;12:622. doi:10.1186/1471-2164-12-622.
37. Collins-Schramm HE, Phillips CM, Operario DJ, Lee JS, Weber JL, Hanson RL, et al. Ethnic-difference markers for use in mapping by admixture linkage disequilibrium. Am J Hum Genet. 2002;70(3):737–50. doi:10.1086/339368.
38. Smith MW, O'Brien SJ. Mapping by admixture linkage disequilibrium: advances, limitations and guidelines. Nat Rev Genet. 2005;6(8):623–32. doi:10.1038/nrg1657.
39. Winkler CA, Nelson GW, Smith MW. Admixture mapping comes of age. Annu Rev Genomics Hum Genet. 2010;11:65–89. doi:10.1146/annurev-genom-082509-141523.
40. Stewart C, Kural D, Stromberg MP, Walker JA, Konkel MK, Stutz AM, et al. A comprehensive map of mobile element insertion polymorphisms in humans. PLoS Genet. 2011;7(8):e1002236. doi:10.1371/journal.pgen.1002236.
41. Weir BS, Cockerham CC. Estimating F-Statistics for the Analysis of Population-Structure. Evolution. 1984;38(6):1358–70. doi:10.2307/2408641.
42. Danecek P, Auton A, Abecasis G, Albers CA, Banks E, DePristo MA, et al. The variant call format and VCFtools. Bioinformatics. 2011;27(15):2156–8. doi:10.1093/bioinformatics/btr330.

43. Saitou N, Nei M. The neighbor-joining method: a new method for
     reconstructing phylogenetic trees. Mol Biol Evol. 1987;4(4):406–25.
44. Tamura K, Stecher G, Peterson D, Filipski A, Kumar S. MEGA6: Molecular
     Evolutionary Genetics Analysis version 6.0. Mol Biol Evol. 2013;30(12):2725–9.
     doi:10.1093/molbev/mst197.
45. Cann HM, de Toma C, Cazes L, Legrand MF, Morel V, Piouffre L, et al.
     A human genome diversity cell line panel. Science. 2002;296(5566):261–2.
46. Howie BN, Donnelly P, Marchini J. A flexible and accurate genotype
     imputation method for the next generation of genome-wide association
     studies. PLoS Genet. 2009;5(6):e1000529. doi:10.1371/journal.pgen.1000529.

15

The endonuclease domain of the LINE-1 ORF2 protein can tolerate multiple mutations

Kristine J. Kines, Mark Sokolowski, Dawn L. deHaro, Claiborne M. Christian, Melody Baddoo, Madison E. Smither and Victoria P. Belancio[*]

Abstract

Background: Approximately 17 % of the human genome is comprised of the *Long INterspersed Element*-1 (LINE-1 or L1) retrotransposon, the only currently active autonomous family of retroelements. Though L1 elements have helped to shape mammalian genome evolution over millions of years, L1 activity can also be mutagenic and result in human disease. L1 expression has the potential to contribute to genomic instability via retrotransposition and DNA double-strand breaks (DSBs). Additionally, L1 is responsible for structural genomic variations induced by other transposable elements such as Alu and SVA, which rely on the L1 ORF2 protein for their propagation. Most of the genomic damage associated with L1 activity originates with the endonuclease domain of the ORF2 protein, which nicks the DNA in preparation for target-primed reverse transcription.

Results: Bioinformatic analysis of full-length L1 loci residing in the human genome identified numerous mutations in the amino acid sequence of the ORF2 endonuclease domain. Some of these mutations were found in residues which were predicted to be phosphorylation sites for cellular kinases. We mutated several of these putative phosphorylation sites in the ORF2 endonuclease domain and investigated the effect of these mutations on the function of the full-length ORF2 protein and the endonuclease domain (ENp) alone. Most of the single and multiple point mutations that were tested did not significantly impact expression of the full-length ORF2p, or alter its ability to drive Alu retrotransposition. Similarly, most of those same mutations did not significantly alter expression of ENp, or impair its ability to induce DNA damage and cause toxicity.

Conclusions: Overall, our data demonstrate that the full-length ORF2p or the ENp alone can tolerate several specific single and multiple point mutations in the endonuclease domain without significant impairment of their ability to support Alu mobilization or induce DNA damage, respectively.

Keywords: LINE-1, L1, ORF2, Endonuclease, Mutation, Retrotransposition, Phosphorylation

Background

The *Long INterspersed Element*-1 (LINE-1 or L1) retrotransposon is the only currently active autonomous non-LTR retroelement in the human genome. Most of the approximately 500,000 L1 loci have been truncated or mutated and are incapable of further retrotransposition [1, 2]. L1 is required for the mobilization of non-autonomous retrotransposons such as Alu and SVA elements [3, 4], and

together these retroelements comprise about a third of the human genome [2]. The fully functional L1 element encodes two proteins (ORF1p and ORF2p), both of which are required for its retrotransposition [5]. ORF1p is an RNA-binding protein with nucleic chaperone activity [6, 7] and ORF2p is a multifunctional protein with endonuclease and reverse transcriptase enzymatic activities and a cysteine-rich domain [8, 9]. L1 and L1-driven retroelements amplify through a copy-and-paste mechanism, resulting in *de novo* insertions in the genome. After integration, these retroelements can interfere with gene expression [10, 11] or serve as substrates for non-allelic

* Correspondence: vperepe@tulane.edu
Department of Structural and Cellular Biology, Tulane School of Medicine, Tulane Cancer Center and Tulane Center for Aging, New Orleans, LA 70112, USA

homologous recombination events, which can lead to disease-relevant genomic deletions, inversions or translocations [12–22].

L1 and its Alu and SVA parasites have significantly impacted the human genome for better or worse (reviewed [23]). L1 has had a major role in generating structural variations in the host genome through retrotransposition and post-insertional genomic rearrangements [24]. However, expression of functional L1 loci can potentially be detrimental at the cellular or organismal level. Transient expression of L1 in mammalian cells results in L1 retrotransposition, and the generation of DNA double-strand breaks (DSBs) [25, 26]. This genomic damage can be significant and lead to apoptosis or senescence [27, 28]. Both L1 retrotransposition and L1-induced DSBs result from the activity of the endonuclease domain present in the ORF2 protein, which functions to introduce nicks into the cellular DNA in preparation for L1 integration via target-primed reverse transcription [8, 29]. Mutations within the endonuclease domain have the potential to impair endonuclease activity and thereby diminish subsequent L1-induced genomic damage. Aside from a few critical residues that have been characterized [5, 8, 30, 31], it is not known how well the endonuclease domain can tolerate mutations in terms of maintaining its function.

In this study, we analyzed full-length L1 loci within the human genome to identify naturally occurring mutations within the ORF2 endonuclease domain. Some of these mutated residues are evolutionarily conserved or presumed to be structurally important [8, 31]. Moreover, several of these mutated residues were predicted to be phosphorylation sites for various cellular kinases. Recent studies reported that the phosphorylation of ORF1p is required for L1 retrotransposition [32] and that these and other putative phosphorylation sites within ORF1p may be involved in the regulation of L1 activity by melatonin receptor signaling [33]. These studies raise the question of whether ORF2p could also be post-translationally modified. Focusing on putative phosphorylation sites, as a way to investigate a manageable subset of positions with possible biological relevance to L1 function, we mutated several amino acid residues within the ORF2 endonuclease domain and investigated the effect of these mutations on endonuclease activity. We characterized the ability of the mutated ENp and ORF2p to cause toxicity, induce H2AX phosphorylation (a marker for DNA DSBs [34, 35]), and drive Alu retrotransposition. Our findings demonstrate that all of the tested individual mutations, as well as most of the various combinations of these mutations, are tolerated without significantly impacting the L1 ORF2 endonuclease function, either in the context of the full-length ORF2p or in the ENp alone.

Results

Identification of mutations in the endonuclease domain of full-length human L1 loci

Using L1Base [36], we analyzed the sequences of full-length L1 loci present in the human genome to identify naturally occurring mutations in the ORF2 endonuclease domain. Our search parameters were set to identify L1 loci that contain intact ORF1 and ORF2 sequences (no gaps, premature stops or frameshifts). We identified 134 L1 loci that satisfy these search criteria, the majority of which were L1Ta and L1PA2 families (Additional file 1: Table S1). None of the 134 full-length L1 loci fitting our search criteria had any mutations at amino acid H230 and only one locus contained a mutation at D205 (Fig. 1; Additional file 1: Table S1). These two residues are absolutely critical for ORF2 endonuclease function [8, 31]. The maximum number of mutations found in the endonuclease domain of any of the 134 full-length L1 loci was 11. Aligning the ORF2 protein sequences of the full-length L1 loci extracted from L1Base against the active human L1.3 element revealed that 118 of the 239 amino acids in the endonuclease domain were mutated at least once (Fig. 1; Additional file 2: Figure S1). However, three of these mutations may be specific to the L1PA2 sequence (I15V, A21P, V208L).

The large number of naturally occurring mutations prompted us to narrow our focus on a specific subset of mutated positions within the ORF2 endonuclease. Our bioinformatic analysis identified multiple mutations of serines and threonines in the endonuclease domain, which may be of particular interest as these amino acids are commonly phosphorylated by cellular kinases [37]. We utilized the ELM prediction tool [38] to identify several short linear motifs within the endonuclease domain that were expected to be recognized by serine/threonine kinases. We also used the NetPhos 2.0 prediction program to identify amino acids having a high probability of being phosphorylated [39]. The following amino acids were predicted to be phosphorylated by both programs: S29, S33, S37, S79, T82, S151, S188, T189, T220, T224 and S228 (Additional file 3: Table S2). Additionally, ELM and NetPhos 2.0 identified two very high probability residues just outside of the endonuclease domain (S312 and S335). To evaluate the evolutionary conservation of the putative phosphorylation sites identified by the prediction programs, we aligned the amino acid sequence of the L1 ORF2 endonuclease domain from eight representative species within the Supraprimate clade of mammals (Additional file 4: Figure S2). With respect to the human L1 sequence, S37, S188 and T189 were moderately conserved (present in 50 % or more of the investigated species), while S79, S151 and S228 were highly conserved (present in almost all investigated species). Even the least conserved residues were still shared among the hominid clade of primates (humans, chimpanzees and bonobos).

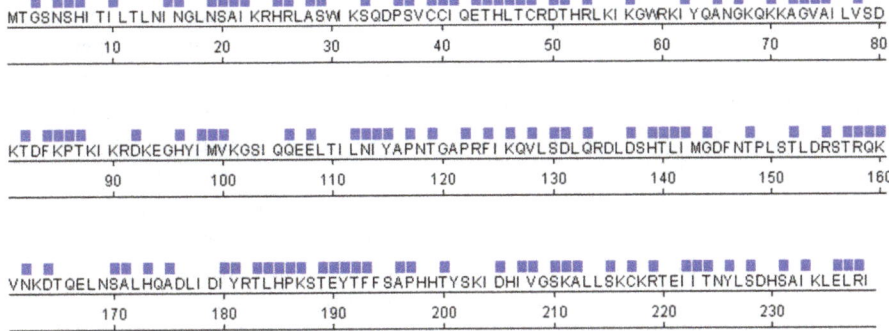

Fig. 1 Mutations in the ORF2 endonuclease domain from full-length L1 loci in the human genome. Bioinformatic analysis using L1Base [36] revealed numerous mutations in the ORF2 endonuclease domains of 134 intact, full-length L1 loci. Positions of mutations relative to the sequence of the L1.3 ORF2 endonuclease domain are indicated by a blue square above the amino acid residue

Mutations of several putative phosphorylation sites in the ORF2 endonuclease domain did not alter its ability to drive Alu retrotransposition, cause toxicity, or impact expression of the full-length ORF2p

As our main interest is in understanding the impact of mutations on L1 endonuclease function, we chose to investigate the functional impact of mutations in putative phosphorylation sites because, if found relevant, these sites could also play a regulatory role. Amino acid residues that were scored by both the ELM and NetPhos prediction programs were selected for mutagenesis (Additional file 3: Table S2). We generated expression plasmids containing codon-optimized human L1 ORF2 sequence with either serine to alanine (S to A) or threonine to alanine (T to A) point mutations in putative phosphorylation sites (Fig. 2). The resulting plasmids contained one (S29A; S33A; S37A; S79A; S188A; S228A), two (S29A/S37A; S79A/T82A; S188A/T189A), three (S29A/S37A/S228A), or four (S29A/S37A/S188A/T189A) point mutations in putative phosphorylation

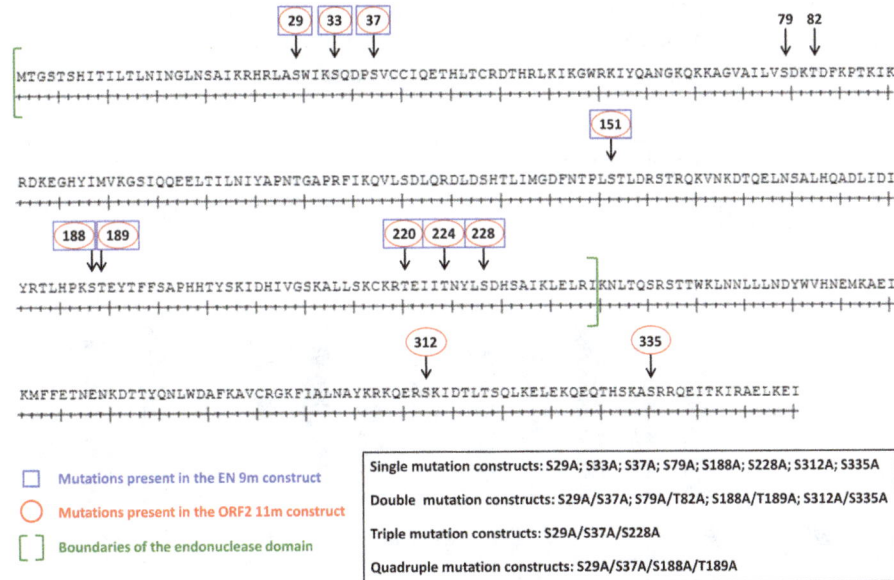

Fig. 2 Schematic representation of the putative phosphorylation sites within ORF2 mutated in this study. Numbered arrows indicate the locations of the putative phosphorylation sites mutated in this study. The boundaries of the ORF2 endonuclease domain are indicated with green brackets. Plasmids encoding for the full-length ORF2p were generated with the following mutations: S29A; S33A; S37A; S79A; S188A; S228A; S29A/S37A; S79A/T82A; S188A/T189A; S29A/S37A/S228A; S29A/S37A/S188A/T189A; S312A; S335A; and S312A/S335A. ORF2 11m contains the following mutations (*red ovals*): S29A/S33A/S37A/S151A/S188A/T189A/T220A/T224A/S228A/S312A/S335A. Plasmids encoding for the endonuclease domain (ENp) alone were generated with the following mutations: S29A; S33A; S37A; S79A; S188A; S228A; S29A/S37A; S79A/T82A; S188A/T189A; S29A/S37A/S228A; and S29A/S37A/S188A/T189A. EN 9m contains the following mutations (*blue boxes*): S29A/S33A/S37A/S151A/S188A/T189A/T220A/T224A/S228A. Plasmids encoding for the full-length L1 containing the following mutations within ORF2 were generated: S29A; S33A; S312A; S335A; and S312A/S335A

sites within the endonuclease domain. The ORF2 11m construct was designed to include mutations in sites predicted to be phosphorylated by kinases in the CMGC group (CDK, MAPK, GSK3 and CLK) [40]. ORF2 11m contains 11 mutations (S29A/S33A/S37A/S151A/S188A/T189A/T220A/T224A/S228A/S312A/S335A); 9 mutations are in the putative phosphorylation sites within the endonuclease domain and the remaining 2 are located between the endonuclease and z-motif region. Because these mutant constructs were tested in transiently transfected HeLa and 293 cells, we used an NGS RNAseq approach to confirm that these cell lines express many cellular kinases (Additional file 5: Table S3) [41].

As an indication of protein function, we investigated the ability of the ORF2 proteins containing the above described mutations to mobilize Alu using a previously described retrotransposition assay [3]. With the exception of the ORF2 11m mutant, all of the full-length mutant ORF2 proteins supported Alu retrotransposition at similar levels to the functional ORF2p (t-test, $P \geq 0.05$) when transiently expressed in HeLa cells (Fig. 3). A significant ~50 % decrease in retrotransposition was observed when Alu mobilization was driven by the ORF2 11m mutant protein (t-test, $P \leq 0.05$).

In addition to genomic damage due to retrotransposition, expression of the ORF2 protein can cause cellular toxicity in a dose-dependent manner when ectopically expressed at high levels [26, 27]. A positive or negative change in ORF2p toxicity may mask or cause subsequent variations in ORF2p-driven Alu mobilization. Using a previously described assay [26], we measured acute

toxicity following transient transfection of the ORF2 putative phosphorylation plasmids to determine if variations in cellular toxicity may contribute to the observed reduction in Alu retrotransposition driven by the ORF2 11m protein (Additional file 6: Figure S3A). For this reason, the same amount of DNA that was transfected in the retrotransposition assay was used for evaluation of potential changes in ORF2p toxicity. Results in Fig. 4a demonstrate that there were no significant differences in toxicity between the functional ORF2p and any of the putative phosphorylation mutant proteins after transient expression in HeLa cells. In contrast to previously reported results [26], we did not observe any toxicity associated with the expression of functional or mutant ORF2 proteins in HeLa cells under our experimental conditions. This discrepancy is likely due to the 20-fold difference in the amount of plasmid DNA transfected per cell between the reported transfection conditions (2 μg plasmid per 100,000 cells in a 6-well plate) and the conditions used here (0.5 μg per 500,000 cells in a T75 flask). Additionally, it was reported that expression of the full-length ORF2p alone was not as efficient in generating γH2AX foci in HeLa cells as was the expression of the full-length L1 [26]. We did observe toxicity after transient expression of the functional ORF2p and putative phosphorylation mutant proteins in 293 cells (Fig. 4b). Consistent with the results obtained in HeLa cells, no significant differences between the toxicity observed after expression of the functional and mutated ORF2 proteins were detected.

To determine whether the significant reduction in Alu retrotransposition driven by the ORF2 11m mutant was

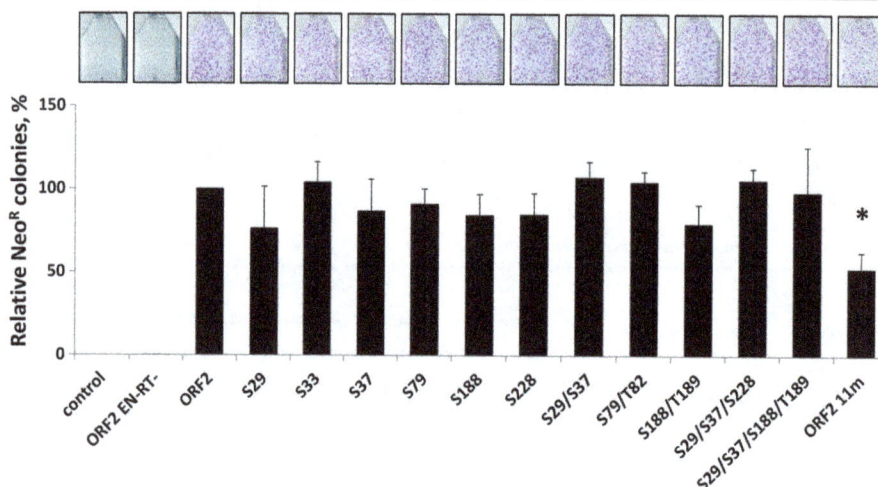

Fig. 3 Alu retrotransposition driven by ORF2 proteins containing mutations in putative phosphorylation sites. ORF2 proteins containing mutations in the indicated putative phosphorylation sites were used to drive Alu retrotransposition in HeLa cells, as previously described [3]. ORF2 is the functional protein and ORF2 EN-RT- is a non-functional protein containing mutations in the endonuclease (D205A) and reverse transcriptase (D702A) domains. Control indicates cells transfected with an empty vector and the Alu retrotransposition reporter plasmid. The graph depicts the relative number of Alu retrotransposition events as represented by Neo^R colonies (Y-axis). Asterisks indicate a statistically significant difference in Alu retrotransposition compared to ORF2 (t-test, $P \leq 0.05$)

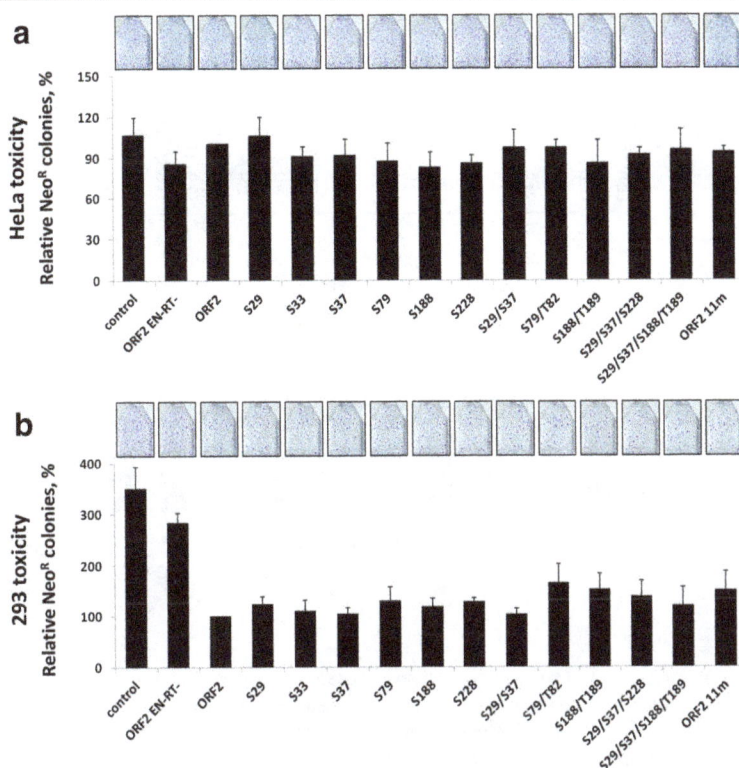

Fig. 4 Acute toxicity assay in HeLa and 293 cells transiently transfected with ORF2 putative phosphorylation mutant plasmids. **a** HeLa cells were cotransfected with a Neo[R] expression vector and the indicated ORF2 putative phosphorylation mutant plasmid. **b** 293 cells were cotransfected with a Neo[R] expression vector and the indicated ORF2 putative phosphorylation mutant plasmid. In both panel **a** and **b** ORF2 is the functional protein and ORF2 EN-RT- is a non-functional protein containing mutations in the endonuclease (D205A) and reverse transcriptase (D702A) domains. Control indicates cells transfected with an empty vector and the Neo[R] expression vector. Colony formation was assayed after 2 weeks under G418 selection (Y-axis) and used as a measure of toxicity as previously described [26, 42]

a result of altered protein expression, we analyzed total protein lysates harvested from HeLa cells transiently transfected with each of the ORF2 mutant plasmids described above. Western blot analysis using antibodies specific to the human L1 ORF2 protein [42, 43] detected ORF2p in the total lysates of transfected HeLa cells (Fig. 5). Quantitation of the relative ORF2p expression levels revealed an approximately 50 % reduction in the steady-state levels of the ORF2 11m protein in comparison to the functional ORF2p. No statistically significant differences in expression were observed between the functional ORF2p and any of the other ORF2 proteins containing mutations in putative phosphorylation sites. The same lysates were also probed with anti-γH2AX antibodies, as histone H2AX is phosphorylated in response to DNA DSBs and can therefore be used as an indication of DNA damage (Additional file 7: Figure S4) [34]. Consistent with our toxicity results in HeLa cells, expression of the functional ORF2p or any of the ORF2 putative phosphorylation mutant proteins generated γH2AX signals that were not significantly different than the background signal observed with the empty vector control or non-functional ORF2 protein (*t*-test, $P \geq 0.05$).

Mutations of several putative phosphorylation sites in the ORF2 endonuclease domain did not alter ENp expression or its ability to induce DNA damage and cause toxicity

Previous in vitro studies have suggested that endonuclease function is repressed in the full-length ORF2 protein [29]. We recently reported that the endonuclease domain of human L1 ORF2p is stable and functional when expressed in mammalian cells [43], which enabled us to characterize the effects of the putative phosphorylation site mutations on the function of the endonuclease domain (ENp) independent of the full-length ORF2p. We generated plasmids containing the sequence of the endonuclease domain (amino acids 1–239) with one (S29A; S33A; S37A; S79A; S188A; S228A), two (S29A/S37A; S79A/T82A; S188A/T189A), three (S29A/S37A/S228A), or four (S29A/S37A/S188A/T189A) point mutations in putative phosphorylation sites (Fig. 2). The EN 9m plasmid includes the following mutations, all of which are also contained in the ORF2 11m construct: S29A/S33A/S37A/S151A/S188A/T189A/T220A/T224A/S228A.

Using a previously reported assay [27, 43], we measured chronic toxicity of these constructs in HeLa cells (Additional file 6: Figure S3B; Fig. 6). With the exception

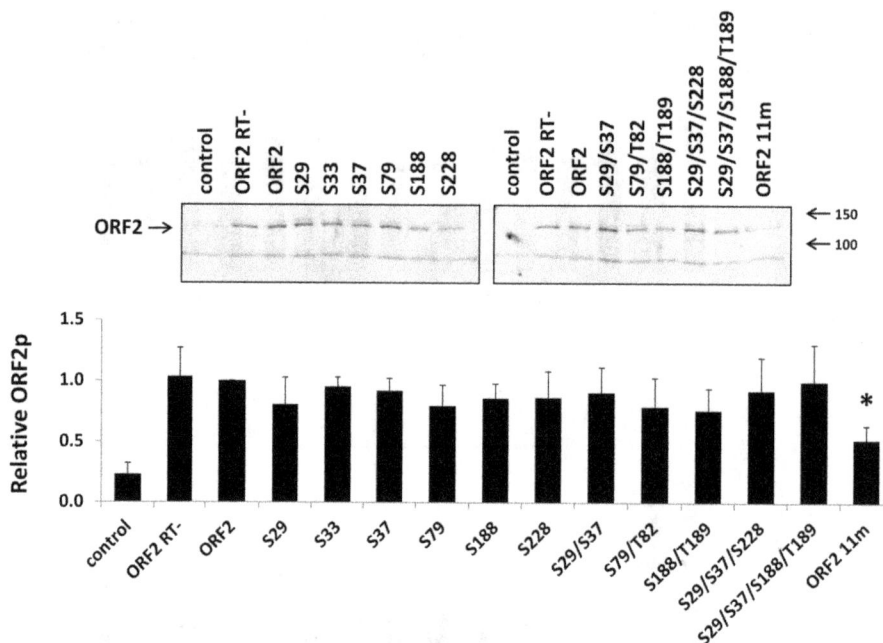

Fig. 5 Expression and detection of ORF2 proteins containing mutations in putative phosphorylation sites. Top panel: Representative western blot analysis of total cell lysates harvested from HeLa cells transfected with the indicated ORF2 putative phosphorylation mutant constructs. ORF2 is the functional protein and ORF2 RT- is a non-functional protein containing a mutation in the reverse transcriptase (D702A) domain. Control lanes indicate cells transfected with an empty vector. Lysates were probed with polyclonal antibodies generated against the human L1 ORF2 protein. Bottom panel: Western blot quantitation. For each sample, the signal detected for ORF2p was normalized to the total protein load. These relative numbers were expressed as a proportion of the relative number detected from the functional ORF2p. Asterisk denotes a significant difference in the steady-state levels relative to the functional ORF2p (t-test, $P \leq 0.05$)

of the EN 9m mutant, all of the mutant endonuclease proteins were as toxic as the functional ENp. Chronic expression of the EN 9m mutant protein resulted in a statistically significant 2.5-fold difference in the relative colony number in comparison to the functional ENp (Fig. 6). Similar results were obtained after transient transfection of HeLa cells with the EN mutant plasmids in an acute toxicity assay (Additional file 6: Figure S3A; Additional file 8: Figure S5).

We have previously reported that expression of the endonuclease domain alone in cultured cells results in a DNA damage response [43]. With the exception of the EN 9m mutant, which was roughly 2-fold higher than the functional ENp, steady-state expression levels were comparable between the functional and mutant proteins (Fig. 7). As previously reported [43], we detected higher steady-state levels of the non-functional ENp (EN-) in comparison to the functional ENp. Similar results were observed with western blot analysis of total protein lysates harvested from 293 cells transiently transfected with the EN expression plasmids (Additional file 9: Figure S6). Western blot analysis detected a γH2AX signal in the total protein lysates from HeLa cells transiently transfected with each of the EN mutant constructs. This result demonstrates that these mutant EN proteins are capable of inducing a DNA damage response (Fig. 7c).

Quantitation of the relative γH2AX signals showed that expression of the functional ENp and the EN putative phosphorylation mutant proteins triggered similar levels of H2AX phosphorylation in HeLa cells (i.e., no statistically significant difference was detected) (Fig. 7c).

Mutations of selected putative phosphorylation sites outside of the endonuclease domain did not alter ORF2p expression or its ability to mobilize Alu, induce DNA damage and cause toxicity

In comparison to the functional ORF2p, expression of the ORF2 11m mutant protein resulted in a reduction in Alu retrotransposition, a decrease in steady-state expression levels and similar levels of toxicity (Figs. 3, 4, and 5). Expression of the EN 9m mutant protein resulted in an increase in steady-state expression levels and less toxicity compared to its functional ENp counterpart (Figs. 6 and 7). The observed differences in toxicity and expression may be due to the presence of the two additional mutations outside of the endonuclease domain in the ORF2 11m protein; alternatively, the nine mutations shared by the ORF2 11m and EN 9m mutants may affect protein function differently in the context of the endonuclease domain alone versus the full-length ORF2p. We generated ORF2 expression plasmids to investigate any independent

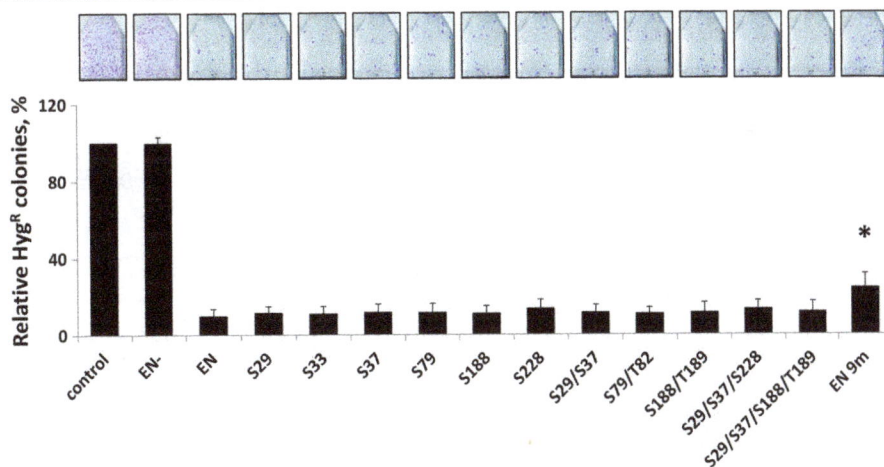

Fig. 6 Chronic expression of EN proteins containing mutations in putative phosphorylation sites causes toxicity. HeLa cells were transfected with a single expression plasmid containing both Hygro^R and the indicated EN putative phosphorylation mutant sequence. Hygromycin selection was maintained for 2 weeks post-transfection, allowing stable expression of ENp throughout the assay. EN is the functional protein and EN- is a non-functional protein containing inactivating mutations (D205A/H230A). Control indicates cells transfected with an empty vector. Colony formation was assayed after 2 weeks under hygromycin (Y-axis) and used as a measure of toxicity as previously described [27, 43]. Asterisks indicate a statistically significant difference in the relative number of Hygro^R colonies compared to EN (t-test, $P \leq 0.05$)

functional effects of the S312A and S335A mutations located outside of the endonuclease domain (Fig. 2).

In contrast to the ORF2 11m protein, ORF2 proteins containing only the S312A, S335A or S312A/S335A mutations supported Alu retrotransposition as efficiently as the functional ORF2p (Fig. 8a). Moreover, there were no significant differences in toxicity after transient expression of these proteins in HeLa and 293 cells (Fig. 8b and c). No statistically significant difference was found between the steady-state expression levels of the functional ORF2p and the ORF2 proteins containing either the individual mutations outside of the endonuclease domain or their combination (Fig. 9).

We also created L1 constructs containing select putative phosphorylation site mutations within the ORF2 sequence, in order to evaluate the effect of these mutations on ORF2p function in the context of the full-length L1. Consistent with the results obtained with the ORF2p expression plasmids, all of the full-length L1 mutants were as efficient as the functional L1 in driving Alu retrotransposition, or their own mobilization, in HeLa cells (Fig. 10).

Discussion

Most L1-induced genomic damage originates with the ORF2 endonuclease, as it initiates endonuclease-dependent retrotransposition events and its activity is implicated in the generation of L1-associated DNA DSBs [8, 26, 29]. The vast majority of full-length L1 loci have become inactive through the accumulation of post-insertional mutations [44]. Given the importance of endonuclease activity to retrotransposition and its relevance to human health, we examined the sequences of full-length L1 loci to find

naturally occurring mutations in the endonuclease domain with the potential to affect its function. Our bioinformatic analysis of 134 full-length human L1 loci revealed that all but one locus have retained the wild-type amino acid at position 205 (D205), a residue that is confirmed to be important for endonuclease function (Additional file 2: Figure S1) [8]. Overall, our bioinformatic analysis identified 118 amino acid positions within the sequence of the endonuclease domain that were mutated relative to the L1.3 sequence. These mutations appear to be randomly distributed throughout the endonuclease sequence (Fig. 1).

Further investigation revealed that 25 % of the 118 mutated positions were serines, threonines, or tyrosines. Combined, serines, threonines and tyrosines comprise about 20 % of the functional endonuclease sequence. More than 50 % of all serine, threonine, or tyrosine residues present in the endonuclease domain were found to be mutated in at least one locus (11 of 20 serines; 15 of 21 threonines; 4 of 7 tyrosines). Most of these mutations did not appear to be due to the presence of CpG dinucleotides within the codons encoding serines and threonines, as there are only two CpG-containing codons (S47, T157) in the human ORF2 endonuclease domain. The underrepresentation of CpG-containing codons is not surprising, given the AT-richness of the L1 coding sequence [45, 46].

Identification of a subset of naturally occurring mutations with potential biological relevance to L1 endonuclease function provided a rationale for investigating the functional effect of mutations at some of these positions (Fig. 2; Additional file 3: Table S2). With the

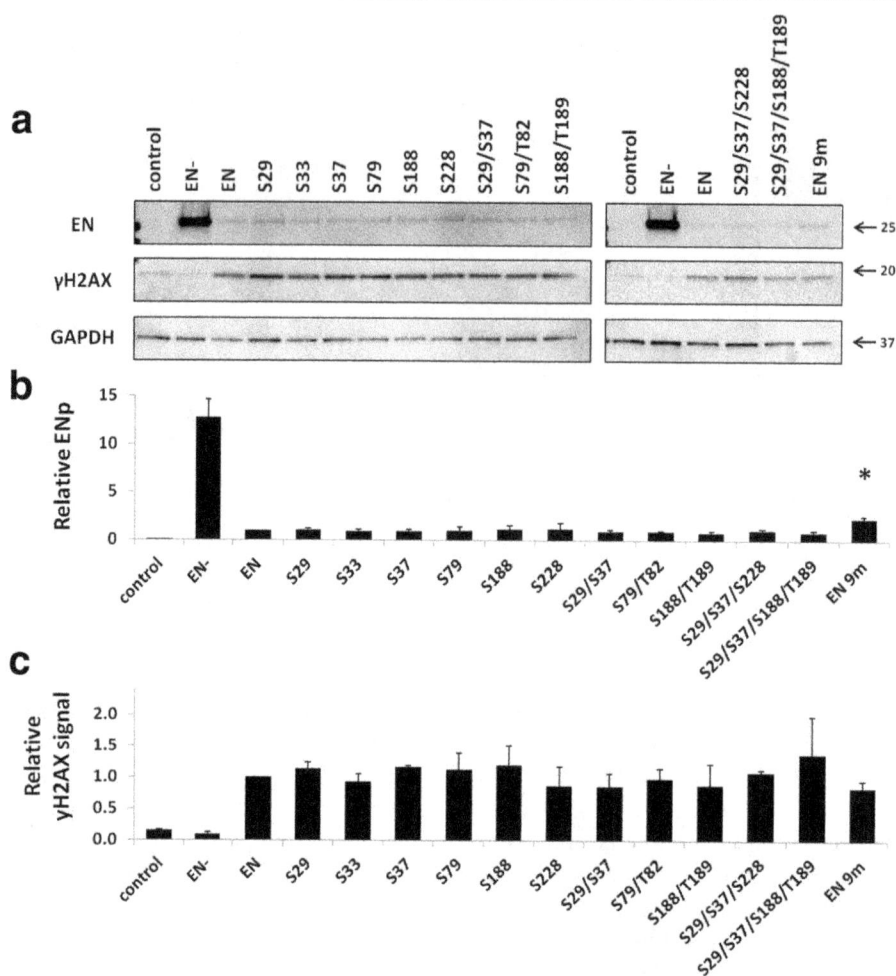

Fig. 7 Expression of EN proteins containing mutations in putative phosphorylation sites can induce DNA damage. **a** Representative western blot analysis of total cell lysates harvested from HeLa cells transiently transfected with the indicated EN putative phosphorylation mutant plasmids. EN is the functional protein and EN- is a non-functional protein containing inactivating mutations (D205A/H230A). Control lanes indicate cells transfected with an empty vector. **a** Lysates were probed with polyclonal antibodies generated against the human L1 ORF2 endonuclease domain [42, 43], top panel; anti-γH2AX antibodies to detect the phosphorylation of histone H2AX in response to DNA damage, middle panel; and anti-GAPDH to serve as a loading control, bottom panel. **b** Western blot quantitation. For each sample, the signal detected for ENp was normalized to the signal detected for GAPDH. These relative numbers were expressed as a proportion of the relative number detected from the functional ENp. Asterisk denotes a significant difference in the steady-state levels relative to the functional ENp (t-test, $P \leq 0.05$). **c** Western blot quantitation. For each sample, the signal detected for γH2AX was normalized to the signal detected for GAPDH. These relative numbers were expressed as a proportion of the relative number detected from the functional ENp

exception of the ORF2 11m mutant, all of the ORF2 proteins with single or multiple point mutations of putative phosphorylation sites behaved similarly to the functional ORF2p in terms of expression and ability to support Alu mobilization. We did not observe any additive effect on the ability of ORF2p to drive Alu retrotransposition as the number of mutations was increased from one to four (Fig. 3). Interestingly, the ORF2 11m protein, which contained 11 point mutations, was still able to drive fairly efficient Alu retrotransposition in HeLa cells, albeit at a 50 % lower level than the functional ORF2p (Fig. 3). This reduction is consistent with the 50 % decrease in the steady-state levels of its

expression relative to the functional ORF2p (Fig. 5). Out of the 11 putative phosphorylation sites that were mutated in the ORF2 11m construct, only 3 sites (S151A, T220A and T224A) were not tested individually or in combination with any of the other 8 mutations for their ability to affect L1 ORF2p expression and retrotransposition. This raises the possibility that the reduction in ORF2 11m protein expression and its ability to drive Alu mobilization may be entirely due to one or more of those three mutations. Alternatively, a combination of all 11 mutations may be responsible for the observed effect. Regardless, the important finding remains that many mutations within the ORF2p endonuclease

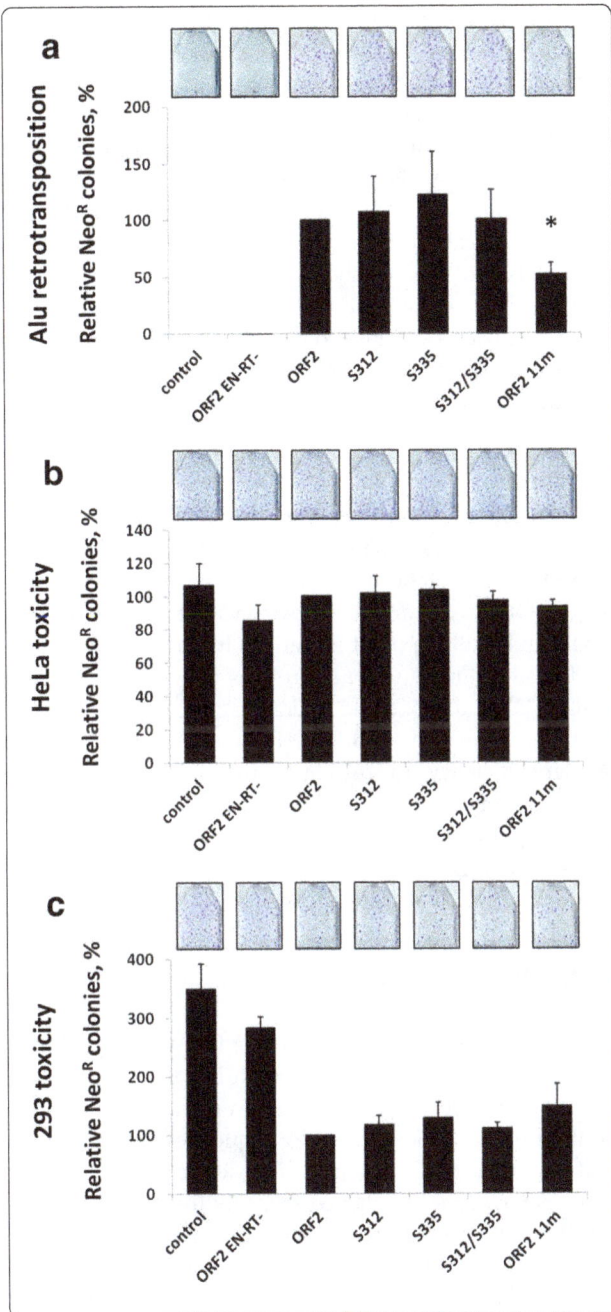

Fig. 8 Analysis of select putative phosphorylation sites outside of the endonuclease domain. **a** Alu retrotransposition: ORF2 proteins containing mutations in the indicated putative phosphorylation sites were used to drive Alu retrotransposition in HeLa cells, as previously described [3]. ORF2 is the functional protein and ORF2 EN-RT- is a non-functional protein containing mutations in the endonuclease (D205A) and reverse transcriptase (D702A) domains. Control indicates cells transfected with an empty vector and the Alu retrotransposition reporter plasmid. The graph depicts the relative number of Alu retrotransposition events as represented by Neo^R colonies (Y-axis). Asterisks indicate a statistically significant difference in Alu retrotransposition compared to ORF2 (t-test, $P \leq 0.05$). **b** Acute toxicity: HeLa cells were cotransfected with a Neo^R expression vector and the indicated ORF2 putative phosphorylation mutant plasmid. **c** Acute toxicity: 293 cells were cotransfected with a Neo^R expression vector and the indicated ORF2 putative phosphorylation mutant plasmid. In both panels **b** and **c**, ORF2 is the functional protein and ORF2 EN-RT- is a non-functional protein containing mutations in the endonuclease (D205A) and reverse transcriptase (D702A) domains. Control indicates cells transfected with an empty vector and the Neo^R expression vector. Colony formation was assayed after 2 weeks under G418 selection (Y-axis) and used as a measure of toxicity as previously described [26, 43]

domain, individually or in various combinations, were tolerated by the enzyme. Despite containing as many as 11 mutations, the Alu retrotransposition potential was only reduced by 50 %.

As with the results from the mutant ORF2 proteins in the retrotransposition assays, we did not observe any additive effect on the ability of ENp to cause toxicity as the number of mutations was increased from one to four (Fig. 6, Additional file 8: Figure S5). The reduction in toxicity observed with the EN 9m mutant in comparison to the functional ENp may suggest that a threshold was crossed at a higher number of mutations. Alternatively, the 3 putative phosphorylation site mutations that were not tested individually (S151A, T220A and T224A) may have been responsible for the observed reduction. Though still highly toxic in comparison to the non-functional ENp, expression of the EN 9m mutant protein was significantly less toxic than the functional ENp in HeLa cells. Additionally, western blot analysis detected significantly higher steady-state levels of the EN 9m protein in comparison to the functional ENp (Fig. 7), consistent with the previously observed increase in expression of non-functional L1 ORF2 and endonuclease proteins relative to their active counterparts [43, 47]. It is particularly interesting that we did not detect any significant differences in endonuclease activity with the EN or ORF2 proteins containing the S228A mutation when compared to the functional proteins. The S228 residue of the ORF2p endonuclease is predicted to be structurally important [8, 31]. This residue is also highly conserved, and it is even present in its ancient evolutionary ancestor APE1 [8, 31]. Additionally, the S228 residue is in close proximity to the critical H230 residue, the mutation of which is known to eliminate endonuclease activity [8, 30, 31].

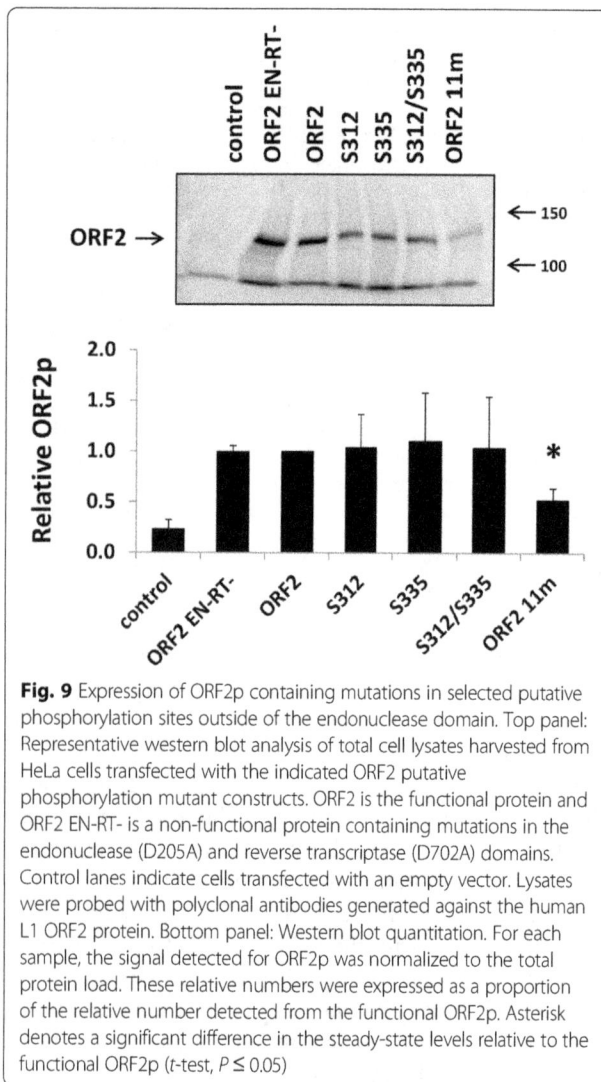

Fig. 9 Expression of ORF2p containing mutations in selected putative phosphorylation sites outside of the endonuclease domain. Top panel: Representative western blot analysis of total cell lysates harvested from HeLa cells transfected with the indicated ORF2 putative phosphorylation mutant constructs. ORF2 is the functional protein and ORF2 EN-RT- is a non-functional protein containing mutations in the endonuclease (D205A) and reverse transcriptase (D702A) domains. Control lanes indicate cells transfected with an empty vector. Lysates were probed with polyclonal antibodies generated against the human L1 ORF2 protein. Bottom panel: Western blot quantitation. For each sample, the signal detected for ORF2p was normalized to the total protein load. These relative numbers were expressed as a proportion of the relative number detected from the functional ORF2p. Asterisk denotes a significant difference in the steady-state levels relative to the functional ORF2p (t-test, $P \leq 0.05$)

Together these results demonstrate that many mutations within the endonuclease domain of ORF2p can be tolerated without substantially impairing endonuclease function and Alu retrotransposition. Although it is plausible that ORF2p mutations may affect L1 and Alu amplification differently, we did not observe any significant variation in the effect of the putative phosphorylation sites mutations tested in both L1 and Alu retrotransposition. The function of the endonuclease domain is predicted to be similar in L1 and Alu retrotransposition, so perhaps a lack of variation is to be expected. The finding that none of the mutations in putative phosphorylation sites that were evaluated in this study eliminated endonuclease activity suggests that, if phosphorylation of these sites occurs, it is not required for endonuclease function in HeLa and 293 cells. However, we cannot rule out the potential impact of these putative phosphorylation sites for ORF2p function in cell types other than the ones tested in this study (HeLa and

293), because cellular kinases often exhibit cell-type-specific expression or activity. All of the putative phosphorylation sites in this study were mutated to alanine, which was a common naturally occurring substitution found in our bioinformatic analysis of the full-length human L1 loci (Additional file 2: Figure S1). Perhaps mutating serine 228 to an alanine was a conservative substitution and did not distort the local protein structure enough to interfere with the function of the neighboring H230. In fact, it is entirely possible that we may have seen different outcomes if any of these sites had been mutated to a different amino acid. It is also worth noting that ORF2p may be phosphorylated at sites other than those investigated in this study.

Conclusions

Our findings demonstrate that the ORF2 endonuclease domain can tolerate many mutations without significantly impacting its function, either in the context of the full-length ORF2p or in the ENp alone. Despite containing single and multiple point mutations in putative phosphorylation sites, the mutant EN proteins were capable of generating DNA damage and toxicity, and the mutant ORF2 proteins were able to drive Alu retrotransposition with similar efficiency to the functional ORF2p. Even the S228A mutation did not significantly alter endonuclease function, despite its proximity to the catalytic H230 residue and its high conservation among mammalian L1s and related phosphohydrolase ancestors.

Methods
Bioinformatic analysis
We searched L1Base to identify mutations in the ORF2 endonuclease domain of full-length human L1 loci [36]. The search criteria were selected to identify L1 loci that contain intact ORF1 and ORF2 sequences (no gaps, premature stops or frameshifts). We found 134 loci fitting these parameters and aligned them by the amino acid sequence of the ORF2 endonuclease domain. All alignments were generated using MegAlign software (DNASTAR v.10.0.1) with human L1.3 as a reference sequence [48]. Alignments using the consensus L1 sequence derived from the 90 intact L1s reported in Brouha et al. [1] resulted in the same findings, data not shown. We also aligned the amino acid sequences of L1 ORF2 endonuclease domains from several orders within the Supraprimate clade of mammals. L1 sequences were obtained from RepBase or GenBank for the following species: human [Homo sapiens, GenBank: L19088.1, [48]]; chimpanzee [Pan troglodytes, GenBank: AY189990.1, [49]]; mouse [Mus musculus domesticus, GenBank: AF081104.1, [50]]; rabbit [Oryctolagus cuniculus, [51]]; rat [Rattus norvegicus, GenBank: U83119.1, [52]]; treeshrew [Tupaia belangeri, [51]]; slow loris [Nycticebus coucang, GenBank: P08548.1, [53]]; and

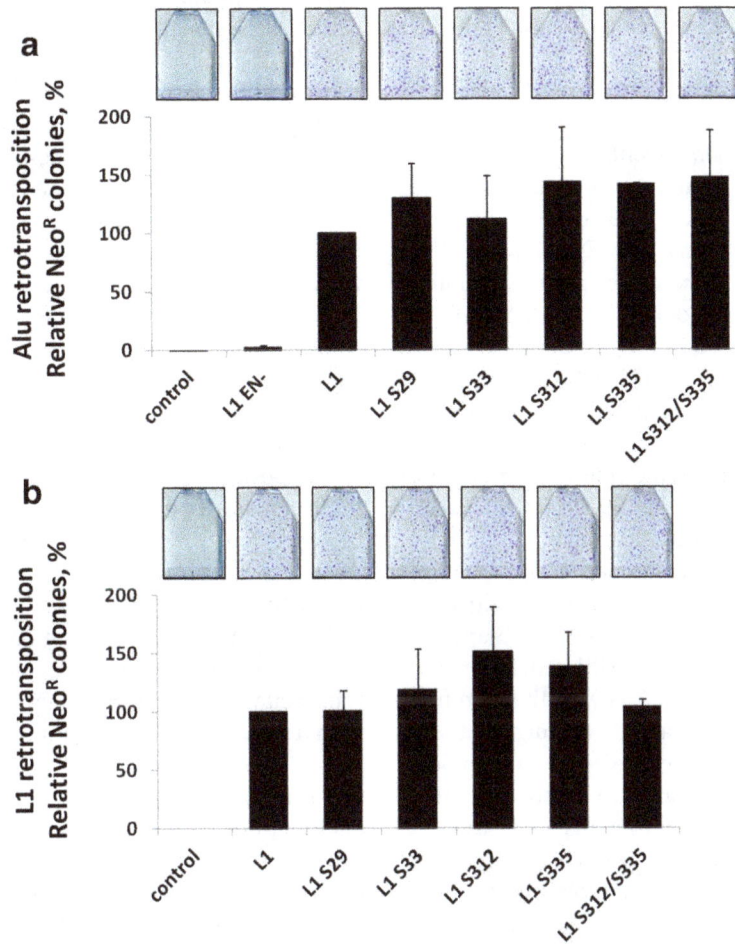

Fig. 10 Retrotransposition driven by full-length L1 elements containing mutations in putative phosphorylation sites within ORF2. **a** Alu retrotransposition: Full-length L1 elements containing the indicated putative phosphorylation mutations within the ORF2 sequence were used to drive Alu retrotransposition in HeLa cells, as previously described [3]. L1 is the functional element and L1 EN- is a non-functional element containing a mutation in the ORF2 endonuclease domain (D205A). Control indicates cells transfected with an empty vector and the Alu retrotransposition reporter plasmid. The graph depicts the relative number of Alu retrotransposition events as represented by Neo^R colonies (Y-axis). **b** L1 retrotransposition: Full-length L1 elements containing the indicated putative phosphorylation mutations within the ORF2 sequence were used in an L1 retrotransposition assay in HeLa cells, as previously described [5, 56]. L1 is the functional element and control indicates cells transfected with an empty plasmid. The graph depicts the relative number of L1 retrotransposition events as represented by Neo^R colonies (Y-axis)

bonobo [*Pan paniscus*, GenBank: AY189988.1, [49]]. Putative protein phosphorylation sites within the human L1 ORF2 endonuclease domain were identified using NetPhos 2.0 and the ELM prediction tool [38, 39].

NGS RNA-seq analysis

RNAseq reads were generated using the Illumina platform and total DNase-treated RNA from HeLa cells (TURBO DNase, Ambion). RNA samples were submitted to the University of Wisconsin Genomics Core for selection of polyadenylated RNAs and TruSeq stranded mRNA library preparation. The raw RNAseq data for the HEK293 cell line were obtained from NCBI's SRA. These files were converted to FASTQ format utilizing fastq-dump:/fastq-dump.2.3.2. The FASTQ files were

aligned to the human genome using RSEM, a package that is used for estimating gene and isoform expression levels from data generated through RNA-Seq [54]. A reference genome for the human genome was prepared using the human genome (version GRCh38) and the *rsem-prepare-reference* command. Each FASTQ file was aligned to this generated reference genome using the *rsem-calculate-expression* command. After these samples were aligned to the human genome, a data matrix was generated utilizing RSEM's EBSEQ using the *rsem-generate-data-matrix* command. A results file was generated from the data matrix using the *rsem-run-ebseq* command. Gene expression from several cellular kinases was analyzed in both cell lines using the genes.results files, generated from the original alignment of each sample to the human genome. The TPM (transcripts per kilobase million) value was obtained by opening genes.results files using Excel and VLOOKUP.

Plasmids

ORF2 putative phosphorylation mutants

Mutations were introduced into a previously reported [26] codon-optimized ORF2 expression plasmid (pBudCE4.1, Invitrogen) using the QuikChange Site-Directed Mutagenesis kit (Stratagene) per the manufacturer's protocol. Plasmids encoding the full-length ORF2 were generated with the following mutations: S29A; S33A; S37A; S79A; S188A; S228A; S29A/S37A; S79A/T82A; S188A/T189A; S29A/S37A/S228A; S29A/S37A/S188A/T189A; S312A; S335A; and S312A/S335A. The ORF2 11m construct contains the following mutations: S29A/S33A/S37A/S151A/S188A/T189A/T220A/T224A/S228A/S312A/S335A. An ORF2 fragment (amino acids 1–348) containing the aforementioned 11m mutations as well as 5'-HindIII and 3'-AflII restriction sites was synthesized by (GenScript). Site-directed mutagenesis was used to introduce HindIII and AflII restriction sites into the ORF2 expression plasmid. The synthesized fragment containing the 11m mutations was digested with the enzymes listed above and cloned into the similarly digested ORF2 expression plasmid to create the ORF2 11m mutant. Site-directed mutagenesis was used to introduce an inactivating D702A reverse transcriptase [9] mutation into the ORF2 expression plasmid to create the ORF2 RT- construct. Site-directed mutagenesis was used to introduce inactivating D205A endonuclease [8] and D702A reverse transcriptase [9] mutations into the ORF2 expression plasmid to create the ORF2 EN-RT-construct. Primer sequences used for site-directed mutagenesis are shown in Additional file 10: Table S4.

EN putative phosphorylation mutants

The codon-optimized endonuclease (EN) and D205A/ H230A endonuclease mutant (EN-) expression plasmids (pcDNA3.1/Hygro+, Invitrogen) were previously reported [43]. The ORF2 putative phosphorylation mutant plasmids were used as PCR templates to create the corresponding endonuclease mutant constructs (amino acids 1–239). Plasmids encoding the ORF2 endonuclease domain were generated with the following mutations: S29A; S33A; S37A; S79A; S188A; S228A; S29A/S37A; S79A/T82A; S188A/T189A; S29A/S37A/S228A; S29A/ S37A/S188A/T189A or EN 9m (S29A/S33A/S37A/ S151A/S188A/T189A/T220A/T224A/S228A). PCR amplification was used to add 5'-NheI and 3'-BamHI restriction sites to the ends of the amplified sequences. The PCR products were subsequently digested with the enzymes listed above and cloned into the similarly digested pcDNA3.1/Hygro+ plasmid.

L1 putative phosphorylation mutants

Mutations were introduced into the ORF2 endonuclease domain of a previously reported [55, 56] codon-optimized L1 expression plasmid and a NeoR-tagged codon-optimized L1 expression plasmid (pBlueScript II, Stratagene) using site-directed mutagenesis as described above. Plasmids encoding the full-length L1 and the full-length L1 tagged with a NeoR reporter cassette were generated with the following mutations: S29A; S33A; S312A; S335A; and S312A/S335A. Primer sequences used for site-directed mutagenesis are shown in Additional file 10: Table S4. The L1 EN- plasmid, a gift from the Deininger laboratory, contains an inactivating D205A [8] mutation in the ORF2 endonuclease domain of the aforementioned L1 expression plasmid.

The NeoR-tagged Alu reporter plasmid used in the retrotransposition assay and the pIRES2-GFP plasmid used in the acute toxicity assay to confer G418 resistance have been previously described [3, 26].

Cell culture

HeLa and 293-FRT cells were cultured as previously described [43, 45]. Cells were seeded 16–18 h prior to transfection and normal growth media was replaced 3 h post-transfection for all experiments.

Retrotransposition assays

Alu retrotransposition assays

Alu retrotransposition experiments were performed as previously described [3]. For ORF2-driven Alu retrotransposition, 500,000 HeLa cells were seeded per T75 flask and co-transfected the following day with 200 ng of the NeoR-tagged Alu reporter plasmid and 200 ng of the ORF2 putative phosphorylation mutant plasmids, using 8 μl of Lipofectamine (Invitrogen) and 4 μl of Plus (Invitrogen). Transfections were performed in duplicate. The experiments shown in Fig. 3 were repeated a minimum of four times and the experiments shown in Fig. 8a were repeated a minimum of three times. For L1-driven Alu retrotransposition, 500,000 HeLa cells were seeded per T75 flask and co-transfected the following day with 200–400 ng of the NeoR-tagged Alu reporter plasmid and 200–400 ng of the L1 putative phosphorylation mutant plasmids, using 8 μl of Lipofectamine and 4 μl of Plus. Experiments shown in Fig. 10a were repeated a minimum of two times. For all Alu retrotransposition experiments, selection medium (400 μg/ml G418) was started 24–48 h post-transfection and maintained for 12–14 days to select for G418-resistant colonies representing Alu retrotransposition events. Colonies were fixed and stained with a crystal violet solution (0.2 % crystal violet, 5 % acetic acid, 2.5 % isopropanol). Statistical significance was evaluated using Student's t-test for samples of equal variance; error bars in figures represent standard deviations.

L1 retrotransposition assays

L1 retrotransposition experiments were performed as previously described [5]. Approximately 500,000 HeLa cells were seeded per T75 flask and co-transfected the following day with 50–800 ng of the Neo^R-tagged L1 putative phosphorylation mutant plasmids and 0–350 ng of empty filler plasmid, using 8 µl of Lipofectamine and 4 µl of Plus. Experiments shown in Fig. 10b were repeated a minimum of two times. Selection medium (400 µg/ml G418) was started 24–48 h post-transfection and maintained for 12–14 days to select for G418-resistant colonies representing L1 retrotransposition events. Colonies were fixed and stained with crystal violet solution as above. Statistical significance was evaluated using Student's t-test for samples of equal variance; error bars in figures represent standard deviations.

Acute toxicity assays

Acute toxicity assay experiments were conducted as previously described, with minor modifications [26, 43]. *ORF2 acute toxicity in HeLa cells:* HeLa cells were seeded at a density of 500,000 cells per T75 flask and transiently co-transfected the following day with 250–400 ng of the ORF2 putative phosphorylation mutant plasmids, 100–250 ng of the pIRES2-GFP plasmid to confer G418 resistance (Neo^R), and 0–150 ng of empty filler plasmid, using 8 µl of Lipofectamine and 4 µl of Plus. Transfections were performed in duplicate. The experiments shown in Figs. 4a and 8b were repeated a minimum of two times. *ORF2 acute toxicity in 293 cells:* 293 cells were seeded at a density of 125,000 cells per T75 flask and transiently co-transfected the following day with 900 ng of the ORF2 putative phosphorylation mutant plasmids and 100 ng of the pIRES2-GFP plasmid to confer G418 resistance (Neo^R), using 8 µl of Lipofectamine and 4 µl of Plus. Transfections were performed in duplicate and the experiments shown in Figs. 4b and 8c were repeated twice. *EN acute toxicity:* 500,000 HeLa cells were seeded per T75 flask and transiently co-transfected the following day with 100 ng of the EN putative phosphorylation mutant plasmids, 150 ng of the Neo^R expression plasmid, and 150 ng of empty filler plasmid, using 8 µl of Lipofectamine and 4 µl of Plus. Transfections were performed in duplicate, the supplemental experiment shown in Additional file 8: Figure S5 was performed once. For all acute toxicity experiments, selection medium (400 µg/ml G418) was added 24–48 h post-transfection and maintained for 12–14 days to select for G418-resistant colonies. Statistical significance was evaluated using Student's t-test for samples of equal variance; error bars in figures represent standard deviations.

Chronic toxicity assays

Chronic toxicity assay experiments were conducted as previously described, with minor modifications [27, 43]. The EN putative phosphorylation mutant sequences were cloned into a plasmid which also expresses a gene for hygromycin resistance. Approximately 500,000 HeLa cells were seeded per T75 flask and co-transfected the following day with 200 ng of the EN putative phosphorylation mutant plasmids and 200 ng of empty filler plasmid, using 8 µl of Lipofectamine and 4 µl of Plus. Transfections were performed in duplicate and the experiments shown in Fig. 6 were repeated four times. Hygromycin selection (220 µg/ml) was initiated 48 h post-transfection and maintained for 12–14 days to allow for constant expression of ENp throughout the duration of the assay. Colonies were fixed and stained as described above. Statistical significance was evaluated using Student's t-test for samples of equal variance; error bars in figures represent standard deviations.

Immunoblot analysis
Transfections

To analyze total protein expression, approximately 400,000 HeLa cells or 1.5 million 293 cells were seeded per T25 flask and transfected the following day with 2 µg of the ORF2 or EN expression plasmids, using 8 µl of Lipofectamine and 4 µl of Plus. Cells were harvested approximately 24 h later and western blots were performed as previously described, with minor modifications [43, 57, 58]. The experiments (transfection and subsequent western blot analysis) shown in Fig. 5 were repeated a minimum of four times, the experiments shown in Fig. 7 were repeated three times, and the experiments shown in Fig. 9 were repeated a minimum of three times. The supplemental experiment shown in Additional file 9: Figure S6 was performed once.

Total protein harvest

Cells were washed once with 1X phosphate buffered saline (PBS) and lysed in 300 µl TLB-sodium dodecyl sulphate (SDS) buffer [50 mM Tris, 150 mM NaCl, 10 mM ethylenediamine-tetraacetic acid (EDTA), 0.5 % SDS, 0.5 % Triton-X, pH 7.2] supplemented with 10 µl/ml each of the Halt protease inhibitor cocktail (Pierce) and Phosphatase inhibitor cocktails 2 and 3 (Sigma). Harvested cells were sonicated three times for 10 s each at 12 W using a Microson XL-2000 sonicator (Misonix). Cell lysates were collected after centrifugation at 14,000 rpm for 15 min at 4 °C. Total protein concentration was calculated using the Bio-Rad Protein Assay.

Western blot

Tris Glycine gels were used for western blot analysis in Figs. 5 and 9. Samples (3.5–15 μg) were boiled for 5 min in denaturing Tris Glycine SDS sample buffer supplemented with β-mercaptoethanol and fractionated on Novex 4 % Tris-Glycine (Invitrogen) gels. Proteins were transferred onto nitrocellulose membranes using the iBlot system (Invitrogen). Membranes were rinsed with PBS-Tween (1x PBS, 0.1 % Tween) and blocked in a mixture containing 5 % non-fat dry milk in 11 ml of PBS-Tween and 4 ml of media collected from NIH-3T3 cells [57]. Membranes were blocked for one hour minimum at room temperature and incubated overnight at 4 °C with custom polyclonal antibodies generated against amino acids 960–973 (NSRWIKDLNVKPKT) of the human L1 ORF2 protein. The ORF2 antibodies were diluted 1:500 in a mixture containing 3 % non-fat dry milk in 11 ml of PBS-Tween and 4 ml of NIH-3T3 media. Membranes were washed and incubated with the secondary HRP-donkey anti-rabbit antibody (Santa Cruz; sc-2317) at a 1:5000 dilution in a mixture containing 3 % non-fat dry milk in 11 ml of PBS-Tween and 4 ml of NIH-3T3 media.

Bis Tris gels were used for western blot analysis in Fig. 7, Additional file 7: Figure S4 and Additional file 9: Figure S6. Samples (3.5–15 μg) were boiled for 5 min in denaturing Laemmli buffer supplemented with β-mercaptoethanol and fractionated on NuPAGE 4–12 % Bis-Tris gels (Invitrogen). Proteins were transferred onto nitrocellulose membranes using the iBlot system (Invitrogen). Membranes were rinsed with PBS-Tween (1x PBS, 0.1 % Tween) and blocked with 5 % non-fat dry milk in PBS-Tween. Membranes were blocked for one hour minimum at room temperature and incubated overnight at 4 °C with polyclonal antibodies generated against the human L1 ORF2 endonuclease domain [42, 43]. Antibodies were diluted in 3 % non-fat dry milk in PBS-Tween as follows: human L1 ORF2 endonuclease domain 1:500; γH2AX (Santa Cruz; sc-101696) 1:100,000; and GAPDH 1:10,000. γH2AX was used as an indicator of DNA damage [34] and GAPDH was used to confirm equal loading of the gel. Membranes were washed and incubated with the secondary antibody, either HRP-donkey anti-goat (Santa Cruz; sc-2020) or HRP-donkey anti-rabbit (Santa Cruz; sc-2317), at a 1:5000 dilution in 3 % milk in PBS-Tween.

All western blots were developed using Clarity Western ECL blotting substrate (Bio-Rad) and the images were captured using a Bio-Rad Gel Doc XR+ imager. The signal intensity of observed bands was quantified before saturation, using Image Lab 4.0.1 software. Statistical significance was evaluated using Student's *t*-test for samples of equal variance. Error bars in figures represent standard deviations.

Additional files

Additional file 1: Table S1. Expanded analysis of mutations in the ORF2 endonuclease domain from full-length human L1 loci. Bioinformatic analysis using L1Base [35] revealed numerous mutations in the ORF2 endonuclease domains of 134 intact, full-length L1 loci. The spreadsheet contains our complete query results from L1Base, including the L1 loci locations and the conservation status of several critical amino acid residues within ORF2. (XLSX 211 kb)

Additional file 2: Figure S1. Alignment of mutations in the ORF2 endonuclease domain from full-length human L1 loci. Bioinformatic analysis using L1Base [35] revealed numerous mutations in the ORF2 endonuclease domains of 134 intact, full-length L1 loci. The ORF2 endonuclease domain sequences (amino acids 1–239) from these 134 L1 loci were aligned using the Clustal W method. The chromosome and L1Base ID number for each loci is listed in the column on the left. Mutations relative to the L1.3 ORF2 endonuclease domain sequence are indicated by the blue square above the amino acid residues. Dots indicate a match to the L1.3 sequence and mutations are denoted by the single letter amino acid code. (PDF 45 kb)

Additional file 3: Table S2. Location of putative phosphorylation sites within the endonuclease domain of L1 ORF2. Putative protein phosphorylation sites within the ORF2 endonuclease domain were identified using the ELM prediction tool [37] and NetPhos 2.0 phosphorylation prediction program [38]. The ELM p-value is a conservative estimate of the probability that an ELM prediction with a given score is a true positive. The significance cut-off is 10e-2. The NetPhos output score is a value in the range of 0.000–1.000 (the higher the score, the higher the confidence of the prediction). (XLSX 10 kb)

Additional file 4: Figure S2. Alignment of L1 ORF2 endonuclease domains from several orders within the Supraprimate clade of mammals. The amino acid sequence of the endonuclease domains of species from various mammalian orders were aligned using the Clustal W method. Residues conserved relative to the human L1 endonuclease sequence are shaded in red. Boxes indicate putative phosphorylation sites of interest. (PNG 13086 kb)

Additional file 5: Table S3. NGS RNAseq analysis of HeLa and 293 cells. RNAseq reads from HeLa and 293 cells were analyzed to confirm gene expression from several kinases and control genes (ACTB, GAPDH). Expression is reported as TPM, transcripts per kilobase million. (XLSX 10 kb)

Additional file 6: Figure S3. Experimental approach for the acute and chronic toxicity assays: A) Acute toxicity assay: Cells are cotransfected with a Neo^R expression vector and either the ORF2 or EN construct. Colony formation was assayed after 2 weeks under G418 selection and used as a measure of toxicity. The full-length ORF2 protein contains an endonuclease (EN), z-motif (z), reverse transcriptase (RT) and Cys-domain (Cys). B) Chronic toxicity assay: The Hygro^R gene is encoded by the same plasmid as the EN gene. Colony formation was assayed after 2 weeks under hygromycin selection and used as a measure of toxicity. (PNG 295 kb)

Additional file 7: Figure S4. Western blot analysis of ORF2 proteins containing mutations in putative phosphorylation sites. Top panel: Representative western blot analysis of total cell lysates harvested from HeLa cells transfected with the indicated ORF2 putative phosphorylation mutant constructs. ORF2 is the functional protein and ORF2 RT- is a non-functional protein containing a mutation in the reverse transcriptase (D702A) domain. Control lanes indicate cells transfected with an empty vector. Lysates were probed with anti-γH2AX antibodies to detect the phosphorylation of histone H2AX in response to DNA damage, top; and anti-GAPDH to serve as a loading control, bottom. Bottom panel: Western blot quantitation. For each sample, the signal detected for γH2AX was normalized to the signal detected for GAPDH. These relative numbers were expressed as a proportion of the relative number detected from the functional ORF2p. (PNG 263 kb)

Additional file 8: Figure S5. Acute toxicity assay in HeLa cells transiently transfected with EN putative phosphorylation mutant plasmids. HeLa cells were cotransfected with a Neo^R expression vector and the indicated EN putative phosphorylation mutant plasmid. EN is the functional protein and EN- is a non-functional protein containing inactivating mutations (D205A/H230A). Control indicates cells transfected with an empty vector and the Neo^R expression vector. Colony formation was assayed after 2 weeks under G418 selection (Y-axis) and used as a measure of toxicity as previously described [26, 42]. (PNG 2274 kb)

Additional file 9: Figure S6. Expression of EN putative phosphorylation site mutant proteins in 293 cells generates DNA damage. Representative western blot analysis of total cell lysates harvested from 293 cells transiently transfected with the indicated EN putative phosphorylation mutant plasmids. EN is the functional protein and EN- is a non-functional protein containing inactivating mutations (D205A/H230A). Control lanes indicate cells transfected with an empty vector. Lysates were probed with polyclonal antibodies generated against the human L1 ORF2 endonuclease domain [41, 42]; anti-γH2AX antibodies to detect the phosphorylation of histone H2AX in response to DNA damage; and anti-GAPDH antibodies to serve as a loading control. (PNG 280 kb)

Additional file 10: Table S4. Sequences of primers used in this study for site-directed mutagenesis. (XLSX 9 kb)

Abbreviations
APE1: apurinic/apyrimidic endonuclease 1; DNA DSB: DNA double-strand break; EN: N-terminal endonuclease; HRP: horseradish peroxidase; kDa: kilodalton; L1 or LINE-1: long interspersed element-1; ORFs: open reading frames; PBS: phosphate buffered saline; RT: reverse transcriptase; SVA: SINE-VNTR-Alu elements.

Competing interests
The authors declare that they have no competing interests.

Authors' contributions
VPB and KJK conceived the idea; KJK, MS, DLD, CMC, MES, MB and VPB designed and performed experiments, analyzed collected data, and wrote the manuscript. All authors read and approved the final manuscript.

Acknowledgements
This work was supported in part by the Louisiana State Board of Regents Graduate Research Fellowship to MS; Life Extension Foundation to VPB; National Institutes of Health [P20GM103518] to VPB; and Kay Yow Cancer Fund to VPB. We would like to thank Joseph Combs for his assistance in making the ORF2 S79 and ORF2 S79/T82 plasmids. We would like to thank Prescott Deininger and Geraldine Servant for the use of the L1 EN- plasmid.

References
1. Brouha B, Schustak J, Badge RM, Lutz-Prigge S, Farley AH, Moran JV, Kazazian HH. Hot L1s account for the bulk of retrotransposition in the human population. Proc Natl Acad Sci U S A. 2003;100(9):5280–5.
2. Lander ES, Linton LM, Birren B, Nusbaum C, Zody MC, Baldwin J, Devon K, Dewar K, Doyle M, FitzHugh W, et al. Initial sequencing and analysis of the human genome. Nature. 2001;409(6822):860–921.
3. Dewannieux M, Esnault C, Heidmann T. LINE-mediated retrotransposition of marked Alu sequences. Nat Genet. 2003;35(1):41–8.
4. Ostertag EM, Goodier JL, Zhang Y, Kazazian HH. SVA elements are nonautonomous retrotransposons that cause disease in humans. Am J Hum Genet. 2003;73(6):1444–51.
5. Moran JV, Holmes SE, Naas TP, DeBerardinis RJ, Boeke JD, Kazazian HH. High frequency retrotransposition in cultured mammalian cells. Cell. 1996;87(5):917–27.
6. Kolosha VO, Martin SL. In vitro properties of the first ORF protein from mouse LINE-1 support its role in ribonucleoprotein particle formation during retrotransposition. Proc Natl Acad Sci U S A. 1997;94(19):10155–60.
7. Martin SL, Bushman FD. Nucleic acid chaperone activity of the ORF1 protein from the mouse LINE-1 retrotransposon. Mol Cell Biol. 2001;21(2):467–75.
8. Feng Q, Moran JV, Kazazian HH, Boeke JD. Human L1 retrotransposon encodes a conserved endonuclease required for retrotransposition. Cell. 1996;87(5):905–16.
9. Mathias SL, Scott AF, Kazazian HH, Boeke JD, Gabriel A. Reverse transcriptase encoded by a human transposable element. Science. 1991;254(5039):1808–10.
10. Wheelan SJ, Aizawa Y, Han JS, Boeke JD. Gene-breaking: a new paradigm for human retrotransposon-mediated gene evolution. Genome Res. 2005; 15(8):1073–8.
11. Nigumann P, Redik K, Mätlik K, Speek M. Many human genes are transcribed from the antisense promoter of L1 retrotransposon. Genomics. 2002;79(5):628–34.
12. Han K, Lee J, Meyer TJ, Remedios P, Goodwin L, Batzer MA. L1 recombination-associated deletions generate human genomic variation. Proc Natl Acad Sci U S A. 2008;105(49):19366–71.
13. Elliott B, Richardson C, Jasin M. Chromosomal translocation mechanisms at intronic alu elements in mammalian cells. Mol Cell. 2005;17(6):885–94.
14. Han K, Sen SK, Wang J, Callinan PA, Lee J, Cordaux R, Liang P, Batzer MA. Genomic rearrangements by LINE-1 insertion-mediated deletion in the human and chimpanzee lineages. Nucleic Acids Res. 2005;33(13):4040–52.
15. Morales ME, White TB, Streva VA, DeFreece CB, Hedges DJ, Deininger PL. The contribution of alu elements to mutagenic DNA double-strand break repair. PLoS Genet. 2015;11(3):e1005016.
16. Symer DE, Connelly C, Szak ST, Caputo EM, Cost GJ, Parmigiani G, Boeke JD. Human l1 retrotransposition is associated with genetic instability in vivo. Cell. 2002;110(3):327–38.
17. Lin C, Yang L, Tanasa B, Hutt K, Ju BG, Ohgi K, Zhang J, Rose DW, Fu XD, Glass CK, et al. Nuclear receptor-induced chromosomal proximity and DNA breaks underlie specific translocations in cancer. Cell. 2009;139(6):1069–83.
18. Robberecht C, Voet T, Zamani Esteki M, Nowakowska BA, Vermeesch JR. Nonallelic homologous recombination between retrotransposable elements is a driver of de novo unbalanced translocations. Genome Res. 2013;23(3):411–8.
19. Gilbert N, Lutz-Prigge S, Moran JV. Genomic deletions created upon LINE-1 retrotransposition. Cell. 2002;110(3):315–25.
20. Gilbert N, Lutz S, Morrish TA, Moran JV. Multiple fates of L1 retrotransposition intermediates in cultured human cells. Mol Cell Biol. 2005; 25(17):7780–95.
21. Kines KJ, Belancio VP. Expressing genes do not forget their LINEs: transposable elements and gene expression. Front Biosci (Landmark Ed). 2012;17:1329–44.
22. Belancio VP, Hedges DJ, Deininger P. LINE-1 RNA splicing and influences on mammalian gene expression. Nucleic Acids Res. 2006;34(5):1512–21.
23. Belancio VP, Hedges DJ, Deininger P. Mammalian non-LTR retrotransposons: for better or worse, in sickness and in health. Genome Res. 2008;18(3):343–58.
24. Kazazian HH. Mobile elements: drivers of genome evolution. Science. 2004; 303(5664):1626–32.
25. Ostertag EM, Madison BB, Kano H. Mutagenesis in rodents using the L1 retrotransposon. Genome Biol. 2007;8 Suppl 1:S16.
26. Gasior SL, Wakeman TP, Xu B, Deininger PL. The human LINE-1 retrotransposon creates DNA double-strand breaks. J Mol Biol. 2006;357(5):1383–93.
27. Wallace NA, Belancio VP, Deininger PL. L1 mobile element expression causes multiple types of toxicity. Gene. 2008;419(1–2):75–81.
28. Belancio, V.P.; Roy-Engel, A.M.; Pochampally, R.R.; Deininger, P. Somatic expression of LINE-1 elements in human tissues. Nucleic Acids Res. 2010; 38(12):3909–3922.
29. Cost GJ, Feng Q, Jacquier A, Boeke JD. Human L1 element target-primed reverse transcription in vitro. EMBO J. 2002;21(21):5899–910.
30. Morrish TA, Gilbert N, Myers JS, Vincent BJ, Stamato TD, Taccioli GE, Batzer MA, Moran JV. DNA repair mediated by endonuclease-independent LINE-1 retrotransposition. Nat Genet. 2002;31(2):159–65.
31. Weichenrieder O, Repanas K, Perrakis A. Crystal structure of the targeting endonuclease of the human LINE-1 retrotransposition. Structure. 2004;12(6):975–86.
32. Cook PR, Jones CE, Furano AV. Phosphorylation of ORF1p is required for L1 retrotransposition. Proc Natl Acad Sci U S A. 2015;112(14):4298–303.
33. deHaro D, Kines KJ, Sokolowski M, Dauchy RT, Streva VA, Hill SM, Hanifin JP, Brainard GC, Blask DE, Belancio VP. Regulation of L1 expression and retrotransposition by melatonin and its receptor: implications for cancer risk associated with light exposure at night. Nucleic Acids Res. 2014;42(12):7694–707.
34. Rogakou EP, Pilch DR, Orr AH, Ivanova VS, Bonner WM. DNA double-stranded breaks induce histone H2AX phosphorylation on serine 139. J Biol Chem. 1998;273(10):5858–68.
35. Burma S, Chen BP, Murphy M, Kurimasa A, Chen DJ. ATM phosphorylates histone H2AX in response to DNA double-strand breaks. J Biol Chem. 2001; 276(45):42462–7.
36. Penzkofer T, Dandekar T, Zemojtel T. L1Base: from functional annotation to prediction of active LINE-1 elements. Nucleic Acids Res. 2005;33(Database issue):D498–500.
37. Edelman AM, Blumenthal DK, Krebs EG. Protein serine/threonine kinases. Annu Rev Biochem. 1987;56:567–613.
38. Dinkel H, Van Roey K, Michael S, Davey NE, Weatheritt RJ, Born D, Speck T, Krüger D, Grebnev G, Kuban M et al. The eukaryotic linear motif resource ELM: 10 years and counting. Nucleic Acids Res. 2014;42(Database issue):D259–66.

39. Blom N, Gammeltoft S, Brunak S. Sequence and structure-based prediction of eukaryotic protein phosphorylation sites. J Mol Biol. 1999;294(5):1351–62.

40. Varjosalo M, Keskitalo S, Van Drogen A, Nurkkala H, Vichalkovski A, Aebersold R, Gstaiger M. The protein interaction landscape of the human CMGC kinase group. Cell Rep. 2013;3(4):1306–20.

41. Dauchy RT, Dauchy EM, Mao L, Belancio VP, Hill SM, Blask DE. A new apparatus and surgical technique for the dual perfusion of human tumor xenografts in situ in nude rats. Comp Med. 2012;62(2):99–108.

42. Ergün S, Buschmann C, Heukeshoven J, Dammann K, Schnieders F, Lauke H, Chalajour F, Kilic N, Strätling WH, Schumann GG. Cell type-specific expression of LINE-1 open reading frames 1 and 2 in fetal and adult human tissues. J Biol Chem. 2004;279(26):27753–63.

43. Kines KJ, Sokolowski M, deHaro DL, Christian CM, Belancio VP. Potential for genomic instability associated with retrotranspositionally-incompetent L1 loci. Nucleic Acids Res. 2014;42(16):10488–502.

44. Sassaman DM, Dombroski BA, Moran JV, Kimberland ML, Naas TP, DeBerardinis RJ, Gabriel A, Swergold GD, Kazazian HH. Many human L1 elements are capable of retrotransposition. Nat Genet. 1997;16(1):37–43.

45. Perepelitsa-Belancio V, Deininger P. RNA truncation by premature polyadenylation attenuates human mobile element activity. Nat Genet. 2003;35(4):363–6.

46. Han JS, Boeke JD. A highly active synthetic mammalian retrotransposon. Nature. 2004;429(6989):314–8.

47. Goodier JL, Ostertag EM, Engleka KA, Seleme MC, Kazazian HH. A potential role for the nucleolus in L1 retrotransposition. Hum Mol Genet. 2004;13(10):1041–8.

48. Dombroski BA, Scott AF, Kazazian HH. Two additional potential retrotransposons isolated from a human L1 subfamily that contains an active retrotransposable element. Proc Natl Acad Sci U S A. 1993;90(14):6513–7.

49. Mathews LM, Chi SY, Greenberg N, Ovchinnikov I, Swergold GD. Large differences between LINE-1 amplification rates in the human and chimpanzee lineages. Am J Hum Genet. 2003;72(3):739–48.

50. DeBerardinis RJ, Goodier JL, Ostertag EM, Kazazian HH. Rapid amplification of a retrotransposon subfamily is evolving the mouse genome. Nat Genet. 1998;20(3):288–90.

51. Jurka J, Kapitonov VV, Pavlicek A, Klonowski P, Kohany O, Walichiewicz J. Repbase Update, a database of eukaryotic repetitive elements. Cytogenet Genome Res. 2005;110(1–4):462–7.

52. Ilves H, Kahre O, Speek M. Translation of the rat LINE bicistronic RNAs in vitro involves ribosomal reinitiation instead of frameshifting. Mol Cell Biol. 1992;12(9):4242–8.

53. Hattori M, Kuhara S, Takenaka O, Sakaki Y. L1 family of repetitive DNA sequences in primates may be derived from a sequence encoding a reverse transcriptase-related protein. Nature. 1986;321(6070):625–8.

54. Li B, Dewey CN. RSEM: accurate transcript quantification from RNA-Seq data with or without a reference genome. BMC Bioinformatics. 2011;12:323.

55. Wagstaff BJ, Kroutter EN, Derbes RS, Belancio VP, Roy-Engel AM. Molecular reconstruction of extinct LINE-1 elements and their interaction with nonautonomous elements. Mol Biol Evol. 2013;30(1):88–99.

56. Wagstaff BJ, Barnerssoi M, Roy-Engel AM. Evolutionary conservation of the functional modularity of primate and murine LINE-1 elements. PLoS One. 2011;6(5):e19672.

57. Sokolowski M, DeFreece CB, Servant G, Kines KJ, deHaro DL, Belancio VP. Development of a monoclonal antibody specific to the endonuclease domain of the human LINE-1 ORF2 protein. Mob DNA. 2014;5(1):29.

58. Sokolowski M, DeHaro D, Christian CM, Kines KJ, Belancio VP. Characterization of L1 ORF1p self-interaction and cellular localization using a mammalian two-hybrid system. PLoS One. 2013;8(12):e82021.

16

HERV-K(HML-2) *rec* and *np9* transcripts not restricted to disease but present in many normal human tissues

Katja Schmitt[1,4], Kristina Heyne[2], Klaus Roemer[2], Eckart Meese[1] and Jens Mayer[1,3*]

Abstract

Background: Human endogenous retroviruses of the HERV-K(HML-2) group have been associated with the development of tumor diseases. Various HERV-K(HML-2) loci encode retrovirus-like proteins, and expression of such proteins is upregulated in certain tumor types. HERV-K(HML-2)-encoded Rec and Np9 proteins interact with functionally important cellular proteins and may contribute to tumor development. Though, the biological role of HERV-K(HML-2) transcription and encoded proteins in health and disease is less understood. We therefore investigated transcription specifically of HERV-K(HML-2) *rec* and *np9* mRNAs in a panel of normal human tissues.

Results: We obtained evidence for *rec* and *np9* mRNA being present in all examined 16 normal tissue types. A total of 18 different HERV-K(HML-2) loci were identified as generating *rec* or *np9* mRNA, among them loci not present in the human reference genome and several of the loci harboring open reading frames for Rec or Np9 proteins. Our analysis identified additional alternative splicing events of HERV-K(HML-2) transcripts, some of them encoding variant Rec/Np9 proteins. We also identified a second HERV-K(HML-2) locus formed by L1-mediated retrotransposition that is likewise transcribed in various human tissues.

Conclusions: HERV-K(HML-2) *rec* and *np9* transcripts from different HERV-K(HML-2) loci appear to be present in various normal human tissues. It is conceivable that Rec and Np9 proteins and variants of those proteins are part of the proteome of normal human tissues and thus various cell types. Transcription of HERV-K(HML-2) may thus also have functional relevance in normal human cell physiology.

Keywords: Human endogenous retrovirus, Provirus, Transcription, Splicing, HERV-K Rec protein, HERV-K Np9 protein, Retrotransposition, L1 element

Background

Human endogenous retroviruses (HERVs) stem from ancient germ line infections by exogenous retroviruses. About 8% of the human genome mass consists of retroviral sequences in *sensu stricto* and sequences with retroviral portions. There are about 40 phylogenetically distinct HERV groups documenting germ line integration, that is, provirus formations by different ancient exogenous retroviruses millions of years ago. Re-infections and intracellular amplifications often increased numbers of proviruses per HERV group for limited evolutionary time periods following initial integration events. Most HERV groups no longer encode former retroviral proteins due to long time presence in the genome and thus accumulation of nonsense mutations including smaller and larger indels. Some retroviral proteins, in particular Envelope (Env), have been conserved during evolution to contribute important Env-mediated functions such as fusion of cell membranes [1-4].

The so-called HERV-K(HML-2) group (in short, HML-2) includes a number of evolutionarily young proviruses, some of which formed in the human lineage after the evolutionary split of human from chimpanzee about 6 million years ago. Especially the young HML-2 loci often harbor open reading frames (ORFs) for retroviral proteins such as Gag, Protease, Polymerase, and

* Correspondence: jens.mayer@uks.eu
[1]Institute of Human Genetics, Center of Human and Molecular Biology, Medical Faculty, University of Saarland, 66424 Homburg/Saar, Germany
[3]Center of Human and Molecular Biology, University of Saarland, 66424 Homburg/Saar, Germany
Full list of author information is available at the end of the article

Envelope. Analyses of HML-2 proviral transcripts had identified typical retroviral splicing events generating an *env* mRNA and a sub-spliced *env* mRNA, originally named *cORF* and later re-named *rec*, with most of the envelope coding sequence removed. Historically, HML-2 proviruses have been divided into type 2 loci, the transcripts of which can be sub-spliced to *rec* mRNA, and type 1 loci that lack a characteristic 292-bp sequence located about 50 bp into the *env* coding sequence [5,6]. Lack of the 292-bp sequence in type 1 loci impairs sub-splicing of *env* mRNA to *rec* mRNA because of lack of the *rec* splice donor (SD) site located within the deleted region. Instead, a SD site just upstream of the 292-bp deletion is now employed in combination with a splice acceptor (SA) site located at the 3′ end of *env* that is the same SA for splicing of transcripts from type 1 and type 2 loci. Such spliced transcripts derived from HML-2 type 1 loci have been named *np9* [7,8] (Figure 1).

Clinical relevance of HML-2 transcription and proteins has been investigated in the context of various human diseases. Especially germ cell tumors (GCT) display strongly upregulated HML-2 transcription and expression of HML-2 proteins already in early stages of tumor development. GCT patients display strong antibody titers against HML-2 Gag and Env proteins at the time of tumor detection (reviewed in ref. [2]).

Both *rec* and *np9* mRNA can encode proteins with potentially important cellular functions that may be relevant to disease development. HML-2 Rec protein is basically a functional homologue of HIV$_{Rev}$ protein [9-13]. Nude mice transgenic for Rec protein develop lesions reminiscent of testicular carcinoma *in situ* [14]. Rec protein was shown to interact with several functionally relevant cellular proteins such as promyelocytic zinc finger protein (PLZF), testicular zinc finger protein (TZFP), Staufen-1, and human small glutamine-rich

tetratricopeptide repeat protein (hSGT). Np9 protein was shown to interact with PLZF and ligand of Numb protein X (LNX). All of those interactions may have important cellular consequences depending on cellular context [15-20].

Several recent studies have identified a number of HML-2 loci transcribed in various disease as well as normal conditions by assigning specifically generated HML-2 cDNA sequences to genomic HML-2 loci employing characteristic sequence differences between HML-2 loci, with transcription patterns varying considerably between conditions (for instance, see [21-26]). Several of the transcribed HML-2 loci can, in principle, encode *rec* or *np9* mRNA. We have previously analyzed HML-2 loci specifically for coding capacity for *rec* mRNA and protein by analysis of HML-2 locus sequences for features required for *rec* mRNA splicing and presence of a Rec ORF within predicted mRNA sequences. We also had identified a number of HML-2 loci generating *rec* mRNA by means of cDNA sequence assignments to genomic HML-2 loci [21,27].

The role(s) of HML-2 Rec and Np9 proteins in human biology is still little understood. As various HML-2 loci are also transcribed in normal human tissue types, it is conceivable that HML-2 proteins also exert biological functions apart from potential roles in disease development. To contribute to a better understanding of a potential biological relevance of HML-2 Rec and Np9, we investigated presence of *rec* and *np9* mRNA in a collection of normal human tissue types and identified HML-2 loci generating *rec* and *np9* mRNA. We also identified additional HML-2 loci not present in the human reference genome sequence and additional splicing variants of HML-2 transcripts potentially encoding HML-2 protein variants in the course of our studies.

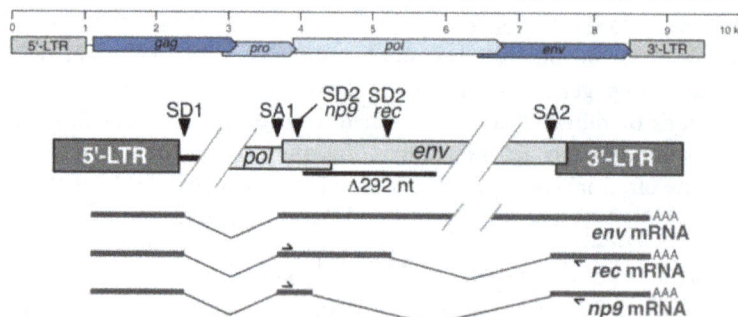

Figure 1 Schematic of HERV-K(HML-2) provirus and splicing of *rec* and *np9* mRNAs. Irrelevant proviral regions are omitted. *rec* mRNA is generated by a second splicing event of *env* mRNA removing most of the *env* gene region. Both *rec* and *np9* mRNA utilize the same splice acceptor site just upstream of the 3′LTR. Because SD2 for *rec* mRNA is located within the 292-bp sequence missing in HML-2 type 1 proviruses, an alternative *np9*-specific SD2 is used instead also resulting in a translation frameshift when spliced to exon 3. Location of PCR primers used for amplification of *rec* and *np9* mRNA/cDNA is indicated. Note that the lower provirus map is not drawn to scale. LTR, long terminal repeat.

Results

Identification of HERV-K(HML-2) *rec* and *np9* mRNA in normal human tissues

Recent findings indicated transcription of HERV-K (HML-2) loci in various human cell and tissue types. Several proteins encoded by some HML-2 loci are considered to be involved in the development of some diseases, among them HML-2 Rec and Np9 proteins in the development of certain tumor types. Carrying on the identification of transcribed HERV and especially HML-2 loci in various disease and normal conditions, we were now interested in whether there are *rec* and *np9* mRNAs in normal human tissues and, if so, from which HML-2 loci those *rec* and *np9* mRNAs were generated.

To identify *rec* and *np9* mRNA, we made use of a multiple tissue cDNA panel that included cDNAs from 16 different tissue types (15 actual tissues and peripheral blood leukocytes (PBL), henceforth all designated as 'tissues' for the sake of simplicity). We amplified *rec* and *np9* mRNA-derived cDNA by using PCR primers located within exons 2 and 3 of HML-2 proviral full-length transcripts (Figure 1) and considering sequence variations between HML-2 loci within PCR primer binding regions to compensate for potentially suboptimal amplification of respective cDNAs.

PCR products of *ca.* 580 bp indicative of *rec* mRNA could be amplified from all 16 tissue cDNAs, with amplification from cDNA from PBL resulting in only a faint band after gel electrophoresis. PCR products of *ca.* 360 bp indicative of *np9* mRNA could be amplified from all 16 tissue cDNAs as well, with amplification from cDNA from PBL producing a relatively strong PCR product (Figure 2). PCR products of *ca.* 1 kb amplified from liver and testis cDNAs and *ca.* 2.2 kb amplified from spleen and thymus cDNAs were not further regarded in this study. Taken together, *rec* and *np9* mRNA appeared to be present in all tissue types examined in this study.

Identification of *rec* and *np9* mRNA encoding HERV-K (HML-2) loci

We then identified HML-2 loci having generated those *rec* and *np9* mRNAs. To do so, we cloned *rec* and *np9* mRNA representing PCR products and sequenced inserts from randomly selected plasmid clones. We then assigned resulting cDNA sequences, on average 41 (min. 29, max. 46) per tissue type, to specific HML-2 type 1 and type 2 loci in the human reference genome sequence by means of characteristic sequence differences between the various HML-2 loci. Despite the rather short-sized PCR products, thus short cDNA sequences (excluding primer regions), there was a sufficient number of sequence differences between relevant exon regions of HML-2 loci for unambiguously assigning generated *rec* and *np9* cDNA sequences to loci. For *rec* mRNA-derived cDNAs, only two HML-2 loci in chromosome 1 were identical in sequence for the regarded exon regions, and respective *rec* transcripts could thus, in principle, not be assigned to either one of them (Additional file 1: Figure S1).

We identified *rec* transcripts originating from, in total, nine different genomic HML-2 loci. Some tissue types (lung and colon) appeared to contain *rec* mRNA from up to five different HML-2 loci, while in kidney tissue, *rec* transcripts originated from only one HML-2 locus. Other tissue types displayed intermediate numbers of transcribed *rec* mRNA coding loci. *rec* transcripts from two HML-2 loci in chromosome 2q32.1 and 5q15 were found in 15 and 13, respectively, of the examined tissues (Table 1). As we will describe below, several of those loci are special with regard to *rec* mRNA.

We identified *np9* transcripts originating from, in total, seven different HML-2 loci present in the human reference genome sequence. As for *rec* mRNA, *np9* mRNA originated from variable numbers of HML-2 loci depending on tissue type. *np9* mRNA transcripts from HML-2 loci in chromosomes 1q22 and 3q12.3 were found in 14 and 13, respectively, of the examined tissues (Table 1). Two additional *np9* mRNA encoding HML-2

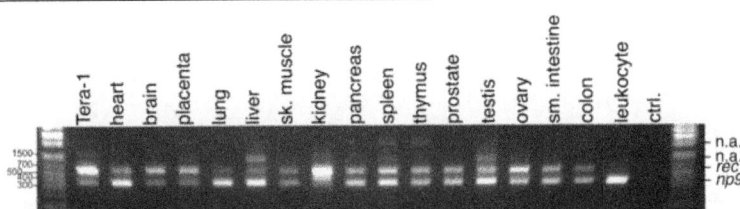

Figure 2 Gel electrophoretic separation of *rec*- and *np9*-specific PCR products amplified from a panel of cDNAs. Generated from 15 normal human tissues and peripheral blood leukocytes. Expected product sizes were *ca.* 580 and 360 bp for *rec* and *np9*, respectively. cDNA generated from GCT-derived Tera-1 cell line, known to strongly express HML-2 both on the RNA and protein level (for instance, see ref. [28]), served as a positive control. 'ctrl.' indicates a PCR control reaction without template DNA. A faint band representing *rec*-specific PCR product from PBL is not properly reproduced. 'n.a.' indicates additional PCR products of approx. 1 kb in the liver and testis and approx. 2.2 kb in the spleen and the thymus not further investigated in this study.

Table 1 Summary of absolute numbers of *np9* and *rec* representing cDNA sequences assignable to HERV-K(HML-2) loci[a]

Type 1/2 loci	HGNC	Band	Heart	Heart tot. RNA	Brain	Brain tot. RNA	Placenta	Lung	Liver	Skeletal muscle	Kidney	Pancreas	Spleen	Thymus	Prostate	Testis	Ovary	Small intestine	Colon	Colon tot. RNA	Leukocyte	ORF	mRNA
chr1:75615359-75621731	*ERVK-1*	1p31.1					1					1										Y	
chr1:153863081-153872260	*ERVK-7*	1q22	1	*7*			11	15	21	2	1	1	22	11	4	7		10	8	*6*	24	Y	
chr1:158927199-158936430	*ERVK-18*	1q23.3												3			1		1		2		
chr3:102893427-102902549	*ERVK-5*	3q12.3	1	*10*	6	*1*		7	3	1	2	1	1	2	4	7		3	5	*12*		Y	
chr3:114225814-114234972	*ERVK-3*	3q13.2													1							Y	*np9*
chr4:166136289-166143518	-	4q32.3														1							
chr22:17306187-17315361	*ERVK-24*	22q11.21			1																	Y	
HERV-K111			16	*6*	10	*5*	2		2	5	6	1		2	2		1	3	4	*11*			
Venter locus			1	*13*	1		6	2		3	4	15			5		13	5	3				
Not assignable *np9*-like			15	*25*	7	*3*	4		2	12	4	5	2	4	12	3	8	5	4	*18*			
chr2:187093879-187095344	*ERVK-30*	2q32.1	10	*13*	2	*2*	6	5	4	4		13	13	12	3	1	5	4	8	*28*	17		
chr3:127091992-127101129	*ERVK-4*	3q21.2				*1*												1		*1*		Y	
chr5:30522516-30531962	-	5p13.3													1							Y	
chr5:92818136-92819668	*ERVK-31*	5q15	2	*8*	13	*89*	10	11	1			4	7	1	11	7	10	9	7	*12*			
chr6:78483381-78492802	*ERVK-9*	6q14.1						1	15													Y	*rec*
chr7:4588583-4606557	*ERVK-6*	7p22.1						1							1				2			Y	
chr10:101570559-101577735	*ERVK-17*	10q24.2					1		8			1				3			1			Y	
chr12:57007509-57016965	*ERVK-21*	12q14.1		*3*			2	1			20			1			1	1	1			Y	
chr19:32820338-32829201	*ERVK-27*	19q12			2																	Y	
Total # of sequences			46	*85*	42	*101*	43	43	41	42	37	42	45	36	44	29	39	41	44	*88*	43		

[a]Normal tissue types from which cDNA sequences were generated are indicated on the top. Values in italics indicate results from experiments starting from total RNA instead of pre-made cDNA specifically for heart, brain, and colon tissue (see paper text). HML-2 loci including chromosomal coordinates in the hg18 human reference genome sequence are indicated in the left-most column, followed by HUGO Gene Nomenclature Committee (HGNC)-assigned locus designations [29] and location in chromosomal bands. Total numbers of analyzed cDNA sequences per tissue are indicated at the bottom. *np9*-like cDNA sequences unambiguously assignable to recently described HERV-K111 and Venter locus sequences not present in the human reference genome sequence (see text for details) are given separately. *np9*-like cDNA sequences not assignable to any of the listed or other HML-2 type 1 locus sequences are also given separately. Presence of ORFs for Np9 or Rec protein ('Y') in predicted mRNA sequences derived from the various HML-2 loci is indicated. See Figure 5 for actual protein sequences.

loci, from which transcripts were identified in many tissues and which are not present in the human reference genome sequence, are described below.

Taken together, our results indicate that *rec* or *np9* mRNA originated from at least 18 different HML-2 loci in various normal human tissue types.

An additional HML-2 locus formed by L1-mediated retrotransposition of *rec* mRNA

We have previously reported a HML-2 locus located in chromosome 2q32.1 that was formed by L1-mediated retrotransposition of a *rec* mRNA [21]. In the present study, that locus was found to be transcribed in almost all examined normal human tissues (Table 1). We now identified an additional HML-2 locus located in chromosome 5q15 (hg18; chr5: 92818136–92819668), also transcribed in almost all of the investigated tissues (Table 1), that very likely was also formed by L1-mediated retrotransposition of a *rec* mRNA. The 1.5-kb-long locus is flanked by target site duplications (5'-TTAAAAATGT-3') typical of an L1 target site consensus sequence [30] with a poly-A tail and a poly-A signal located (more) upstream of the locus' 3' end. Apart from an approximately 250-bp 5' truncation of the retrotransposed mRNA, portions typically missing from a full-length proviral HML-2 sequence are those likewise not present in a *rec* mRNA and boundaries of missing sequence portions coincide with known splice donor and acceptor sites of *rec* mRNA. Also, those sites are basically identical with the ones of the retrotransposed locus in chromosome 2q32.1. The locus in chromosome 5q15 is evolutionarily old as it is also present in the homologous genome regions of chimpanzee, gorilla, orangutan, and gibbon, but missing in rhesus, baboon (data not shown), and the common marmoset (Figure 3). The latter, as a new world primate, is lacking HML-2 homologous sequences entirely. The locus in chromosome 5q15 does not encode a Rec(–like) protein due to a stop codon 20 triplets into the coding sequence.

True transcription of two retrotransposed HML-2 loci

We employed for amplification of *rec* and *np9* transcripts a pre-made panel of cDNAs. As opposed to true splicing events, the two retrotransposed HML-2 loci in chromosomes 2q32.1 and 5q15 would produce identically sized PCR products when amplified from mRNA/cDNA or genomic DNA. We were therefore concerned that amplified PCR products assignable to those two loci were due to traces of genomic DNA present in the pre-made cDNAs, thus conceivably a false indication of those two HML-2 loci being transcribed in the examined tissues. To investigate this further, we generated cDNA from commercially available total RNA from three normal human tissues, specifically heart, brain, and colon, following own previously established protocols for rigorous DNA removal and including strict controls for DNA contamination [21]. We assigned, on average, 91 cDNA sequences from each of the three tissue RNAs to genomic HML-2 loci. Overall, when taking higher numbers of cDNA sequences generated per tissue into account, we obtained very similar numbers regarding transcribed HML-2 loci when compared to the pre-made cDNA panel, especially regarding cDNA sequences assignable to the loci in chromosomes 2q32.1 and 5q15 (Table 1). It thus appears that the two retrotransposed HML-2 loci are truly transcribed in quite a number of normal human tissue types.

In accord with an HUGO Gene Nomenclature Committee initiative [29], the two HML-2 loci in chromosomes 2q32.1 and 5q15 were designated *ERVK-30* and *ERVK-31*, respectively.

Transcription of HML-2 loci not present in the human reference genome sequence

Generated cDNA sequences regularly included *np9* mRNA-like sequences that could not be assigned to HML-2 type 1 loci in the human reference genome sequence. At first glance, one population of sequences was most similar to the *ERVK-5* locus in chromosome 3q12.3, yet almost all those sequences uniformly displayed eight different nucleotide positions to that locus. Further analysis provided evidence that about half of those cDNA sequences were very likely transcribed from recently reported HERV-K(HML-2) loci not present in the human reference genome sequence. Specifically, out of 280 cDNA sequences deemed unassignable to HML-2 loci in the human reference genome, a subset of 76 cDNA sequences were identical to the recently reported HERV-K111 sequence (GenBank acc. no. GU476554; ref. [32]) along the comparable proviral regions and thus were presumably transcribed from the HERV-K111 locus. Another subset of 71 cDNA sequences were identical to a recently reported, 4214-bp-long sequence entry consisting of a partial HERV-K(HML-2) type 1 locus (GenBank acc. no. ABBA01159463; ref. [33]) (DNA donor: J. Craig Venter; henceforth named 'Venter locus') and were thus most likely transcribed from the Venter locus (Additional file 1: Figure S2). Another subset of 133 unassignable sequences was neither identical to HERV-K111 nor the Venter locus.

A fourth cDNA sequence population was most similar to the *ERVK-18* locus in human chromosome 1q23.3, yet uniformly displayed 25 differing nucleotides to that locus. That sequence population harbored an additional 245 nt compared to the amplified *np9* cDNA-derived PCR product. The difference in length was due to a SD signal located 252 nt downstream within the *env* gene region and a different SA2 located 7 nt downstream from the canonical *rec/np9* SA2 (see below). Also, those sequences lacked the 292-bp sequence discriminating HML-2 type 1 and type 2 loci, so that they cannot be interpreted as *rec*-like mRNA transcribed from a HML-2 type 2 locus (Figure 4). The sequences displayed between 2- and 5-nt differences along the comparable 567 nt of cDNA sequence to the Venter locus and to HERV-K111. It is conceivable that those 245-bp longer cDNA sequences, compared to *np9* mRNA, represent alternatively spliced transcripts from HML-2 type 1 loci, potentially from sequence alleles of HERV-K111, several of which have been reported recently [34], or an allele of the Venter locus, or other hitherto unknown sequence or presence/absence alleles of HML-2 loci, several of which were partially described recently [35] (see also the Discussion section).

All of the abovementioned sequences appeared to have been spliced differently from *np9* mRNA. Specifically, the SA site of intron 2 (removing most of the envelope coding region) was located 7 bp downstream from the canonical *rec* and *np9* mRNA SA2 site [7,8]. This was most likely due to both HERV-K111 and the Venter locus harboring a mutated SA2 (5'-TGTT*A*GTCTG-3' → 5'-TGTT*GG*TCTG-3') and a SA signal located 7 bp downstream (5'-CTGC*A*GGTGT-3') being used instead (Figure 4).

Taken together, based on detected sequence similarities, our analysis provided evidence for HERV-K111 and the Venter locus being transcribed and encoding *np9*-like mRNA in various normal human tissue types. It was

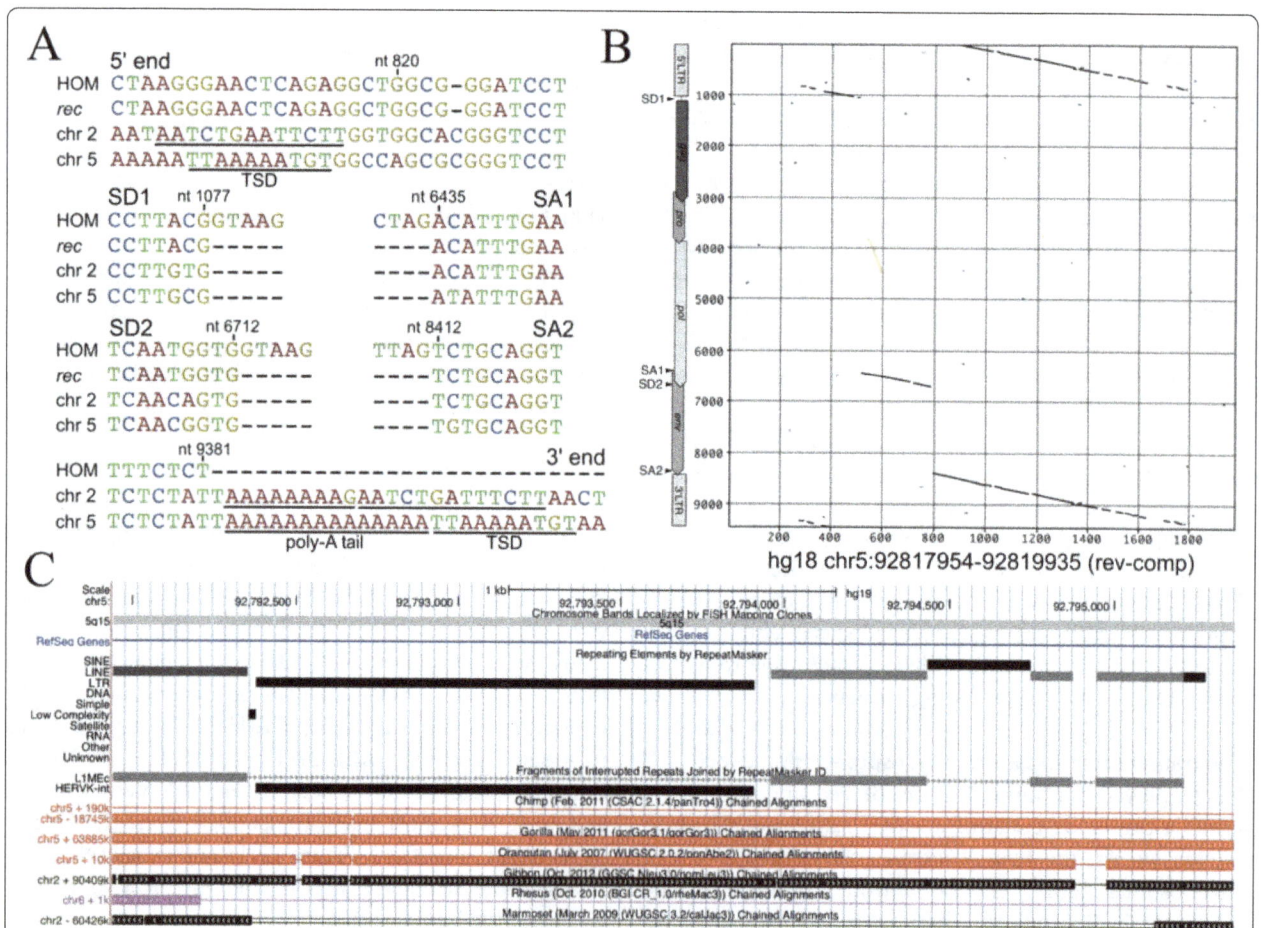

Figure 3 A HML-2 locus in human chromosome 5q15 once formed by L1-mediated retrotransposition of *rec* mRNA. (A) Multiple sequence alignments of relevant regions of the HERV-K(HML-2.HOM) proviral sequence (GenBank acc. no. AF074086; ref. [31]), a *rec* mRNA sequence (GenBank acc. no. X72790; ref. [8]), a recently described retrotransposed *rec* mRNA in human chromosome 2q32.1 [21], and the HML-2 locus in human chromosome 5q15 are shown, depicting target site duplications, *rec*-typical intron/exon boundaries and poly-A tails immediately upstream of the 3'TSD of the chromosome 2q32.1 and 5q15 loci. See ref. [21] for a more detailed description of the chromosome 2q32.1 locus. **(B)** A dot plot matrix comparison of the chromosome 5q15 locus sequence (reverse complemented) with the HERV-K(HML-2.HOM) reference sequence is shown with proviral regions indicated for the latter, likewise depicting the *rec*-typical proviral regions present for that HML-2 locus. **(C)** Presence of the HML-2 locus in the homologous regions of human chromosome 5q15 in the genomes of chimpanzee, gorilla, orangutan, gibbon, and lack in rhesus monkey (the homologous HML-2 locus also lacking in baboon is not shown). Marmosets, as new world monkeys, generally lack HERV-K(HML-2) homologous sequences. LTR, long terminal repeat; SA, splice acceptor; SD, splice donor; TSD, target site duplication.

unclear whether one or several alleles of those two loci or another HML-2 type 1 locus encodes an alternatively spliced, 245-bp longer mRNA.

Coding capacity of *rec* and *np9* mRNAs

We analyzed whether transcribed *rec* and *np9*(–like) mRNAs also have potential to encode Rec and Np9 proteins. We have previously analyzed the capacity of genomic HML-2 type 2 loci to encode Rec protein and have identified a number of HML-2 loci encoding *rec* mRNA [27]. We now analyzed in a similar fashion the capacity of HML-2 type 1 loci to potentially encode Np9 protein by (i) presence of canonical SD and SA sites, (ii) and an ORF for Np9 protein within the predicted mRNA sequence. We identified a total of 12 HML-2 type 1 loci to potentially produce a spliced mRNA and to harbor an ORF for Np9 protein as previously reported in size [7]. Several of the resulting Np9 proteins displayed amino acid differences compared among each other (Figure 5). For instance, a Np9 protein potentially encoded by a locus on chromosome 1 (hg18: 205875079–205879259) would harbor a deletion of three amino acids and additional amino acid differences overlapping with previously reported nuclear localization and LNX protein interaction domains [27]. Other HML-2 type 1 loci could potentially only encode Np9-like proteins about ten or more amino acids shorter than full-length Np9, or being similar to Np9 only within the N-terminal third of a (shorter) protein. This is also the case for proteins potentially encoded by HERV-K111 and the Venter locus that are identical with the canonical Np9 protein sequence only for the N-terminal 15 aa (Figure 5). Notably, several of the potentially protein encoding HML-2 type 1 loci were found transcribed in various normal human tissues.

Similar to our study from a decade ago [27], we re-analyzed for the present study HML-2 type 2 loci potentially encoding Rec protein by examining sequence features required for the splicing of *rec* mRNA and the translation of a Rec protein. Our re-analyses identified ORFs for Rec protein presumably in up to 19 HML-2 type 2 loci, among them the two polymorphic HERV-K113 and HERV-K115 loci [36]. Two other HML-2 loci on human chromosome 4 (hg18: 4029946–4039536 and 9268677–9278272) may potentially encode much longer, Rec-like proteins if they were transcribed. Several other loci only harbor (very) short Rec protein ORFs. The potential full-length Rec proteins display various amino acid differences when compared among each other (Figure 5). As for Np9, several of the potentially Rec protein encoding loci were found transcribed in various normal human tissues.

Discussion

Several lines of evidence suggest biological significance of HERV-K(HML-2)-encoded proteins Rec and Np9 in the development of tumor diseases, for instance, germ cell tumors and melanoma (for reviews, see [2,37]). However, biological roles of those proteins still need to be investigated in much more detail. Since transcripts from several HML-2 loci have been identified in non-tumor and even normal tissues, it is also conceivable that Rec and Np9 exert biological roles in normal human tissues. We therefore investigated in this study whether there are *rec* and *np9* mRNA transcripts also in normal human tissues. Our strategy for identification of such transcripts involved the amplification of a PCR product encompassing the second and third exons of *rec* and *np9* mRNA. Sequence differences within primer binding regions were also considered to potentially demonstrate transcripts from sequence-diverged HML-2 loci (see also ref. [21]). Contrary to previous studies (for instance, see [21,38]), this study does not allow for estimating relative transcript levels for different HML-2 loci as *rec* and *np9* mRNA representing PCR products were isolated and cloned in a combined fashion thus potentially falsifying relative frequencies of *rec* and *np9* mRNA-derived sequences and transcribed loci. It is also conceivable that full-length transcripts from some HML-2 loci are spliced more efficiently to *rec* or *np9* mRNA than those from other loci. Nevertheless, higher numbers of cDNA sequence derived from particular loci may hint towards higher transcript levels or more efficient splicing of full-length transcripts from those loci. Also, amplified cDNA sequences do not fully document the structure of the actual *rec* and *np9* mRNA sequences as we amplified exons 2 and 3 encompassing intron 2, disregarding exon 1 and intron 1 of *rec* and *np9* mRNAs. It is therefore, in principle, possible that for some HML-2 loci transcript, regions outside of the examined regions are spliced in some non-canonical way, though Rec and Np9 protein coding regions appear to be spliced properly (see also below).

Independent of that, *rec* and *np9* mRNA appear to be present in quite a number of human tissue types as indicated by respective PCR products amplified from 16 normal human tissues investigated in this study.

Our assignments of *rec* and *np9* cDNA sequences to HML-2 loci also indicate that at least 18 different HML-2 loci can be transcribed in normal human tissues and very likely encode *rec* or *np9* mRNAs. Additional transcribed loci encoding *rec* or *np9* may be identified, and tissue-specific patterns of such *rec* or *np9* mRNA encoding loci may be identified when much higher numbers of cDNA sequences than in this study are assigned to HML-2 loci.

Our assignment of cDNA sequences to HML-2 loci also lends support to peripheral blood leukocytes not being the sole source of *rec* and *np9* mRNAs in the various tissues as all tissues would then have displayed a PBL-typical basic pattern of transcribed HML-2 loci. However, HML-2 loci identified as transcribed and

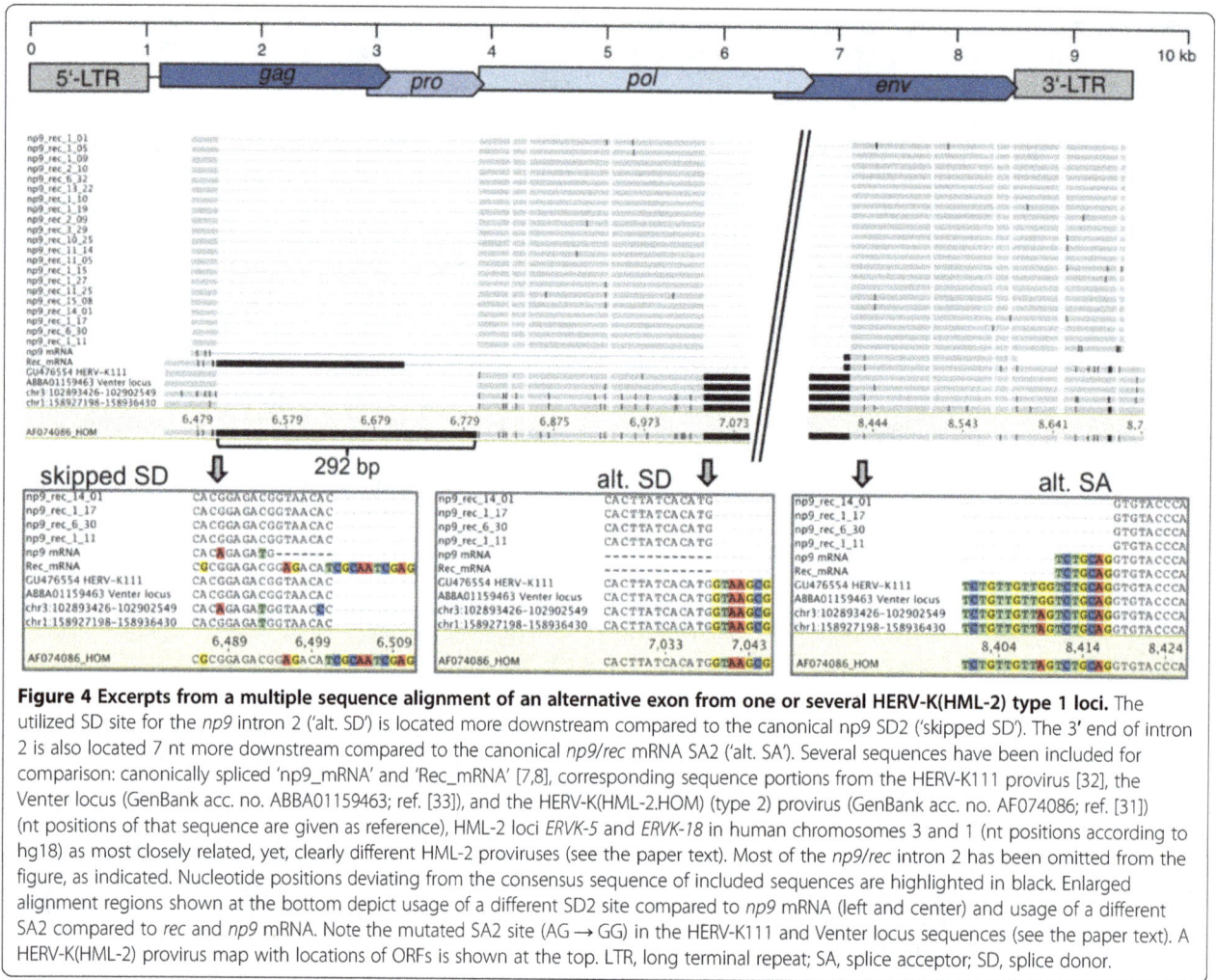

Figure 4 Excerpts from a multiple sequence alignment of an alternative exon from one or several HERV-K(HML-2) type 1 loci. The utilized SD site for the *np9* intron 2 ('alt. SD') is located more downstream compared to the canonical np9 SD2 ('skipped SD'). The 3' end of intron 2 is also located 7 nt more downstream compared to the canonical *np9/rec* mRNA SA2 ('alt. SA'). Several sequences have been included for comparison: canonically spliced 'np9_mRNA' and 'Rec_mRNA' [7,8], corresponding sequence portions from the HERV-K111 provirus [32], the Venter locus (GenBank acc. no. ABBA01159463; ref. [33]), and the HERV-K(HML-2.HOM) (type 2) provirus (GenBank acc. no. AF074086; ref. [31]) (nt positions of that sequence are given as reference), HML-2 loci *ERVK-5* and *ERVK-18* in human chromosomes 3 and 1 (nt positions according to hg18) as most closely related, yet, clearly different HML-2 proviruses (see the paper text). Most of the *np9/rec* intron 2 has been omitted from the figure, as indicated. Nucleotide positions deviating from the consensus sequence of included sequences are highlighted in black. Enlarged alignment regions shown at the bottom depict usage of a different SD2 site compared to *np9* mRNA (left and center) and usage of a different SA2 compared to *rec* and *np9* mRNA. Note the mutated SA2 site (AG → GG) in the HERV-K111 and Venter locus sequences (see the paper text). A HERV-K(HML-2) provirus map with locations of ORFs is shown at the top. LTR, long terminal repeat; SA, splice acceptor; SD, splice donor.

encoding *rec* and *np9* mRNAs in PBL are quite different from locus patterns observed for the various tissues. As tissues are always composed of various cell types, it remains to be investigated which cell types in the regarded tissues actually produce *rec* and *np9* mRNAs. Different scenarios are conceivable. For instance, some HML-2 loci may be transcribed and produce a subset of *rec* and/ or *np9* mRNAs in some cell types within a particular tissue, while other cell types within that tissue contribute a different mRNA subset because of other HML-2 loci being transcribed in those cells. The proportions of HML-2 loci transcribed in different cell types could differ considerably between tissue types. Eventually, cell type-specific transcription patterns will have to be established.

As for Rec and Np9 protein levels, it is currently not known how efficiently *rec* and *np9* mRNAs are translated into respective proteins, how stable those proteins are in normal human cells, and whether there is protein translated in one or the other tissue/cell type at all.

Cell culture-based experiments demonstrated a relatively stable HML-2 Rec protein with a half-life >8 h [39]. Np9 protein seems much shorter-lived in cell culture experiments [18,19]. Nevertheless, little appears to be known about the expression of Rec and Np9 proteins in normal and diseased conditions. Rec protein expression was reported in some melanoma tissue samples, but not in melanocytes or normal lymph nodes [40,41] and in normal synovial, rheumatoid arthritis, and osteoarthritis specimens [42]. Np9 protein was identified in EBV-positive Raji cells that are derived from a Burkitt's lymphoma, and in an EBV-transformed human lymphoblastoid cell line, IB4 [43]. The detection of Rec and Np9 proteins was accompanied in those studies by detection of *rec* and *np9* mRNAs. Therefore, *rec* and *np9* mRNAs present in normal human tissues imply presence of Rec and Np9 protein in normal human tissues. However, specially designed studies on Rec and Np9 protein levels, half-life, cellular distribution, and so on, in normal human cells will be required. Well-suited Rec- and Np9-specific antibodies

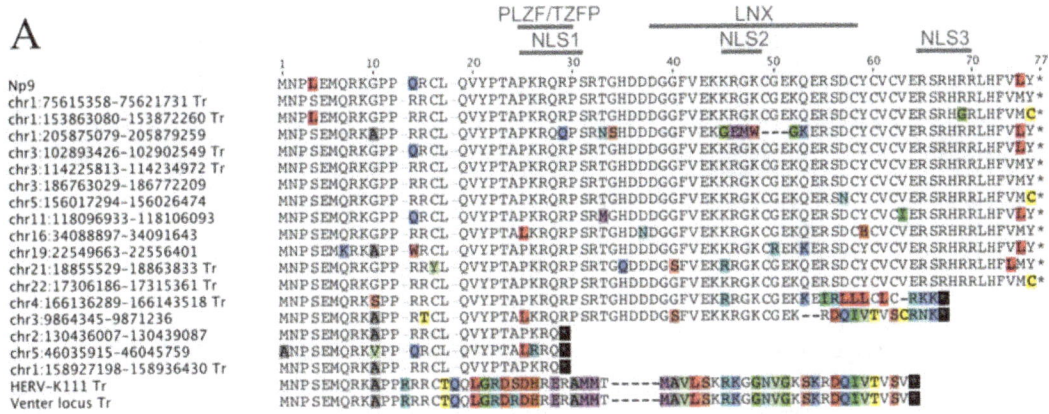

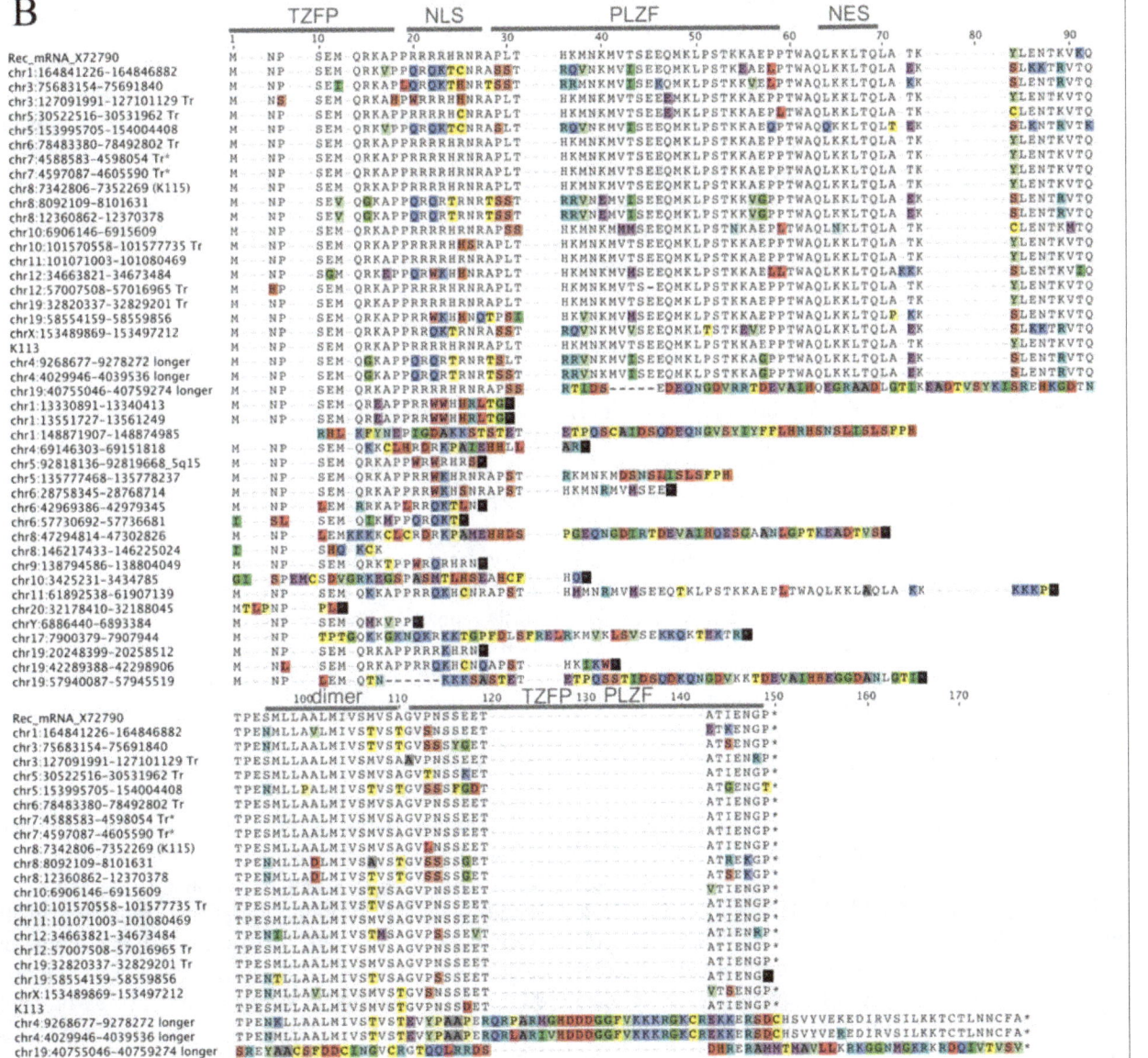

Figure 5 (See legend on next page.)

appear crucial for such protein studies including Western blot and immunohistochemistry and immunocytochemistry for examination of tissue and cellular distributions, respectively. It seems unclear whether current Rec and Np9 antibodies will be fully suited for such studies especially when considering that several HML-2 protein variants with amino acid sequences and protein sizes very similar to Rec and Np9 proteins have been described recently (for instance, see [21,39]) and in this study.

Our analysis of transcribed HML-2 loci furthermore identified two loci (designated *ERVK-30* and *ERVK-31*), both once formed by retrotransposition of *rec* mRNA by L1 machinery, as transcribed in normal human tissues. Locus *ERVK-30*, located in chromosome 2q32.1, has already been described before [21]. The present study identified another such locus, *ERVK-31*, located in chromosome 5q15, that is due to L1-mediated retrotransposition as it displays typical hallmarks of that process and is identical regarding exon-intron junctions compared to the *ERVK-30* locus in 2q32.1 and *rec* mRNA. The *ERVK-31* locus in chromosome 5q15, located central within a ~57-kb intron of the *NR2F1 antisense RNA 1* (*NR2F1-AS1*) gene producing a non-coding RNA, is about as evolutionarily old as the *ERVK-30* locus in 2q32.1; none of the two loci is present in the homologous regions of the rhesus monkey and baboon genomes but both homologous loci are present in the genomes of subsequent primate species.

Evidence for both loci being transcribed was not due to artifactual amplification from contaminating DNA potentially still present in employed cDNA tissue panels, as demonstrated by our control experiments from total RNA from three selected normal tissues. Sequence data from, for instance, the ENCODE project provide additional support for the retrotransposed *rec* mRNA loci *ERVK-30*/2q32.1 and *ERVK-31*/5q15 being transcribed. Numerous single-pass and paired-read RNA-seq reads generated from various cell lines and normal cell types were mapped to the two loci's sequence portions (data not shown; ref. [44]). We thus describe here the second instance of a HML-2 *rec* mRNA that was retrotransposed by L1 machinery and is now transcribed by a hitherto unknown promoter active in many human tissues. Contrary to the retrotransposed *rec* mRNA

locus in chromosome 2q32.1, the locus in chromosome 5q15 does not appear to encode a Rec-like protein as it harbors a stop mutation about 20 triplets and a frameshift about 63 triplets into the Rec coding sequence.

Our study presumably identified in various human tissues transcripts from two recently described HML-2 loci that are not present in the human reference genome sequence. One population of *np9* cDNA sequences was identical to the recently described HERV-K111 provirus [32]; another population of *np9* cDNA sequences was identical to sequence portions in GenBank acc. no. ABBA01159463, a sequence identified in the genome of J. Craig Venter, that consists of HML-2 *pol*, *env*, and 3′ long terminal repeat (LTR) sequence portions, starting at *ca.* nt 5000 relative to the HERV-K(HML-2.HOM) proviral sequence (GenBank acc. no. AF074086.2; ref. [31]). Notably, transcripts from both loci employed a SA2 site located 7 nt downstream from the canonical SA2 site due to a mutation within that site. Transcripts assignable to the HERV-K111 provirus were identified in our study in all but four of the investigated normal tissue types including lack of detection of HERV-K111 transcript in PBL. The HERV-K111 provirus was previously reported to be specifically active during HIV infection [32].

Transcripts from the 'Venter locus' were identified in 11 of the investigated tissues. Since several of the employed tissue cDNA panels were pools from higher numbers of donors, it seems less likely that lack of transcripts in several tissues is due to a polymorphic presence/absence status of the HERV-K111 and the Venter locus in respective tissues. In any case, more detailed analyses will be required to characterize especially the status, genome location, and exact sequence of the Venter locus. We note in this context that numerous sequence variants of HERV-K111 were recently reported [34]. The Venter locus sequence displays 35-nt differences to HERV-K111 along 4.178 kb of comparable sequence and only 1-nt difference along the amplified cDNA sequence portion. While the sequence of the Venter locus is not identical to any of the reported HERV-K111 variants (data not shown), it is nevertheless conceivable that the Venter locus is, in fact, an unde-scribed variant of HERV-K111.

Additional analyses will also be required to identify the HML-2 source locus (or loci) producing an alterna-tively spliced, HML-2 type 1 locus-derived mRNA that utilizes a splice donor site located 252 nt downstream from the canonical *np9* mRNA's SD2 and thus resulting in a longer mRNA. The isolated cDNA sequences dis-play an open reading frame of 408 nt, starting at nt 41 of the cDNA sequences (nt 6805 relative to the HERV-K (HML-2.HOM) sequence, GenBank acc. no. AF074086), encoding a 135 aa long protein identical to HML-2 Env along the N-terminal 80 aa. Since our employed forward PCR primer overlapped with the start codon of the Env, Rec, and Np9 ORFs, we currently do not have informa-tion as to whether the source locus of the alternative splice variant has a start codon identical to the start codon of Env/Rec/Np9 proteins. That is, the ORF may extend further upstream from the start codon at nt 41 of the available cDNA sequence up to the Env/Rec/Np9 canonical start codon.

In this context, we also obtained cDNA sequence evidence for splicing of transcripts from a previously characterized HML-2 locus in human chromosome 10 very likely formed by reverse transcription and integra-tion of a HML-2 transcript lacking *env* gene portions [21]. Testis- and colon-derived cDNA sequences in-cluded one sequence each lacking a 334-nt sequence compared to full-length cDNA sequences from that chromosome 10 locus. The missing sequence portion was compatible with that sequence being a spliced out intron. The resulting protein from that splice variant would encode a chimeric protein consisting of Rec and Np9 portions (Additional file 1: Figure S3).

Conclusions

Our study demonstrates HERV-K(HML-2) *rec* and *np9* transcripts from various HML-2 loci in various normal human tissues. Among them are Rec or Np9 protein coding loci and it is thus conceivable that Rec and Np9 proteins might be present in normal human tissues. Rec/ Np9/Env-like proteins potentially encoded by retrotran-sposed HERV-K(HML-2) loci identified recently and in this study may also be present in various normal tissues. Besides potential roles in disease development, it seems worthwhile hypothesizing that various HERV-K(HML-2)-encoded proteins exert biological functions also in normal human tissues. Better knowledge of specific HERV-K(HML-2) loci encoding *rec* and *np9* mRNAs in the various human tissue-composing cell types and knowledge of amounts of Rec and Np9 protein present in those cell types will likely contribute to a better un-derstanding of those proteins' functions under normal cellular conditions.

Methods

Multiple tissue cDNA panel and tissue total RNAs

We utilized the Human MTC™ Panel I and II (Clontech/ Takara Bio, Otsu, Japan). The two panels included nor-malized, first-strand cDNA preparations from RNA from, in total, 15 different normal human tissues and peripheral blood leukocytes. cDNAs for each tissue con-sisted of pools from 3 to 98 Caucasian individuals and 550 male/female Caucasians in the case of peripheral blood leukocytes. Lung and liver derived cDNAs were from one male Caucasian each.

We also utilized commercially available total RNA from normal human heart, brain, and colon tissues (Clontech/Takara Bio; catalogue numbers 636532; 636530; 636553).

Amplification of *np9* and *rec* PCR products, cloning of PCR products, plasmid preparation, sequencing

We amplified *rec* and *np9* mRNA representing PCR products from tissue cDNA panels employing forward primers rec-np9-for-1: 5'-ATG AAC CCA TCA GAG ATG CAA-3'; rec-np9-for-2: 5'-ATG AAT CCA TCA GAG ATG CAA-3'; rec-np9-for-3: 5'-GCG AAC CCT TCA GAG ATG CAA-3'; rec-np9-for-4: 5'-ATG AAC CCA TCG GAG ATG AAA-3'; that were combined in a ratio of 85/5/5/5, and reverse primers rec-np9-rev-1: 5'-AGC ATC TGT TTA ACA AAG CA-3'; rec-np9-rev-2: 5'-AGC ATG TTT AAC AAA GCA-3' 5% combined in a ratio of 95/5. The various primer variants considered sequence differences of HERV-K(HML-2) loci within primer binding regions. PCR products were amplified using standard conditions with AmpliTaq Gold (Applied Biosystems/Life Technologies, Carlsbad, CA, USA) DNA polymerase and the following PCR program: 12 min 95°C; 35 cycles: 50 s 95°C, 50 s 58°C, 30 s 72°C, and final elong-ation 10 min 72°C. PCR products were separated by agar-ose gel electrophoresis; *np9* and *rec* representing PCR products were purified from gels using NucleoSpin Gel and PCR Clean-Up Kit (Macherey-Nagel GmbH & Co. KG, Düren, Germany). Products were cloned into pCR II-TOPO (Invitrogen/Life Technologies) and transformed into *Escherichia coli* DH-5α cells. Plasmid DNA from randomly selected bacterial colonies was purified and subjected to Sanger sequencing (see below).

We amplified *rec* and *np9* mRNA representing cDNA from total RNA from heart, brain, and colon tissues by RT-PCR following a previously established procedure [21]. PCR products were cloned into pGEM T-Easy (Promega GmbH, Mannheim, Germany). Plasmid DNA from randomly selected bacterial colonies was prepared as described before [32].

cDNA inserts were sequenced using vector-specific T7 primer and an Applied Biosystems 3730 DNA-Analyzer (Seq-IT GmbH, Kaiserslautern, Germany). Sequence qualities were verified by eye, and poor quality sequence reads were excluded from further analysis.

Chromosomal assignment of cDNA sequences

We assigned *rec* and *np9* representing cDNA sequences to specific HML-2 loci by sequence comparisons essentially as described before [21,31]. Sequences that could be unambiguously assigned to a HML-2 locus in the human reference genome sequence with less than three mismatches were considered for analysis. Sequences not matching a locus in the human reference genome sequence are described in the main text. For assignment of *np9* cDNA sequences, we omitted the rather short (23 nt when excluding primer binding region) and thus little informative exon 2 of *np9* mRNA (SA1 - SD2) because the remaining sequence portions provided a sufficient number of sequence differences between (type 1) loci (see Additional file 1: Figure S1).

Analysis of HML-2 loci for *np9* mRNA coding capacity

Similar to a recent analysis of HML-2 *rec* coding capacity [27], we examined in a multiple alignment of genomic HML-2 sequences, plus polymorphic HML-2 proviruses not present in the human reference genome sequence, features required for *np9* mRNA splicing and Np9 protein coding capacity, specifically presence of 5′ and 3′LTRs, splice donor and acceptor sites, and open reading frames as previously described [7].

Availability of supporting data

LN624403 and LN624404 are accession numbers of cDNA sequences assignable to two HML-2 loci in human chromosomes 2q32.1 and 5q15 reported in this study and previously [21]. Accession numbers LN680257 to LN680271 are 245 bp longer, *np9*-like cDNA sequences. Sanger sequence reads generated in the course of this study have been deposited at the European Nucleotide Archive (study accession number PRJEB8273).

Competing interests
The authors declare that they have no competing interests.

Authors' contributions
KS, KH, and JM performed the experiments and the additional analyses. KR, EM, and JM conceived the study. JM wrote the paper. All authors read and approved the final manuscript.

Acknowledgements
This study was supported by grant number Ma2298/8-1 provided by the Deutsche Forschungsgemeinschaft to JM.

Author details
[1]Institute of Human Genetics, Center of Human and Molecular Biology, Medical Faculty, University of Saarland, 66424 Homburg/Saar, Germany. [2]José Carreras Research Center, Medical Faculty, University of Saarland, 66424 Homburg/Saar, Germany. [3]Center of Human and Molecular Biology, University of Saarland, 66424 Homburg/Saar, Germany. [4]Sanofi-Aventis Deutschland GmbH, Industriepark Hoechst, K703, Elisabeth Kuhn Street, Frankfurt/Main 65926, Germany.

References

1. Kurth R, Bannert N. Beneficial and detrimental effects of human endogenous retroviruses. Int J Cancer. 2010;126(2):306–14.
2. Ruprecht K, Mayer J, Sauter M, Roemer K, Mueller-Lantzsch N. Endogenous retroviruses: endogenous retroviruses and cancer. Cell Mol Life Sci. 2008;65(21):3366–82.
3. Stoye JP. Studies of endogenous retroviruses reveal a continuing evolutionary saga. Nat Rev Microbiol. 2012;10(6):395–406.
4. Dupressoir A, Lavialle C, Heidmann T. From ancestral infectious retroviruses to bona fide cellular genes: role of the captured syncytins in placentation. Placenta. 2012;33(9):663–71.
5. Ono M. Molecular cloning and long terminal repeat sequences of human endogenous retrovirus genes related to types A and B retrovirus genes. J Virol. 1986;58(3):937–44.
6. Ono M, Yasunaga T, Miyata T, Ushikubo H. Nucleotide sequence of human endogenous retrovirus genome related to the mouse mammary tumor virus genome. J Virol. 1986;60(2):589–98.
7. Armbruester V, Sauter M, Krautkraemer E, Meese E, Kleiman A, Best B. A novel gene from the human endogenous retrovirus K expressed in transformed cells. Clin Cancer Res. 2002;8(6):1800–7.
8. Löwer R, Tönjes RR, Korbmacher C, Kurth R, Löwer J. Identification of a Rev-related protein by analysis of spliced transcripts of the human endogenous retroviruses HTDV/HERV-K. J Virol. 1995;69(1):141–9.
9. Magin-Lachmann C, Hahn S, Strobel H, Held U, Löwer J, Löwer R. Rec (formerly Corf) function requires interaction with a complex, folded RNA structure within its responsive element rather than binding to a discrete specific binding site. J Virol. 2001;75(21):10359–71.
10. Yang J, Bogerd H, Le SY, Cullen BR. The human endogenous retrovirus K Rev response element coincides with a predicted RNA folding region. RNA. 2000;6(11):1551–64.
11. Magin C, Hesse J, Löwer J, Löwer R. Corf, the Rev/Rex homologue of HTDV/HERV-K, encodes an arginine-rich nuclear localization signal that exerts a trans-dominant phenotype when mutated. Virology. 2000;274(1):11–6.
12. Yang J, Bogerd HP, Peng S, Wiegand H, Truant R, Cullen BR. An ancient family of human endogenous retroviruses encodes a functional homolog of the HIV-1 Rev protein. Proc Natl Acad Sci U S A. 1999;96(23):13404–8.
13. Magin C, Löwer R, Löwer J. cORF and RcRE, the Rev/Rex and RRE/RxRE homologues of the human endogenous retrovirus family HTDV/HERV-K. J Virol. 1999;73(11):9496–507.
14. Galli UM, Sauter M, Lecher B, Maurer S, Herbst H, Roemer K, et al. Human endogenous retrovirus rec interferes with germ cell development in mice and may cause carcinoma *in situ*, the predecessor lesion of germ cell tumors. Oncogene. 2005;24(19):3223–8.
15. Hanke K, Chudak C, Kurth R, Bannert N. The Rec protein of HERV-K(HML-2) upregulates androgen receptor activity by binding to the human small glutamine-rich tetratricopeptide repeat protein (hSGT). Int J Cancer. 2013;132(3):556–67.
16. Hanke K, Hohn O, Liedgens L, Fiddeke K, Wamara J, Kurth R, et al. Staufen-1 interacts with the human endogenous retrovirus family HERV-K(HML-2) rec and gag proteins and increases virion production. J Virol. 2013;87(20):11019–30.
17. Kaufmann S, Sauter M, Schmitt M, Baumert B, Best B, Boese A, et al. Human endogenous retrovirus protein Rec interacts with the testicular zinc-finger protein and androgen receptor. J Gen Virol. 2010;91(Pt 6):1494–502.
18. Denne M, Sauter M, Armbruester V, Licht JD, Roemer K, Mueller-Lantzsch N. Physical and functional interactions of human endogenous retrovirus proteins Np9 and rec with the promyelocytic leukemia zinc finger protein. J Virol. 2007;81(11):5607–16.
19. Armbruester V, Sauter M, Roemer K, Best B, Hahn S, Nty A, et al. Np9 protein of human endogenous retrovirus K interacts with ligand of numb protein X. J Virol. 2004;78(19):10310–9.
20. Boese A, Sauter M, Galli U, Best B, Herbst H, Mayer J, et al. Human endogenous retrovirus protein cORF supports cell transformation and associates with the promyelocytic leukemia zinc finger protein. Oncogene. 2000;19(38):4328–36.
21. Schmitt K, Reichrath J, Roesch A, Meese E, Mayer J. Transcriptional profiling of human endogenous retrovirus group HERV-K(HML-2) loci in melanoma. Genome Biol Evol. 2013;5(2):307–28.
22. Ruprecht K, Ferreira H, Flockerzi A, Wahl S, Sauter M, Mayer J, et al. Human

endogenous retrovirus family HERV-K(HML-2) RNA transcripts are selectively packaged into retroviral particles produced by the human germ cell tumor line Tera-1 and originate mainly from a provirus on chromosome 22q11.21. J Virol. 2008;82(20):10008–16.

23. Flockerzi A, Ruggieri A, Frank O, Sauter M, Maldener E, Kopper B, et al. Expression patterns of transcribed human endogenous retrovirus HERV-K (HML-2) loci in human tissues and the need for a HERV Transcriptome Project. BMC Genomics. 2008;9(1):354.

24. Agoni L, Guha C, Lenz J. Detection of human endogenous retrovirus K (HERV-K) transcripts in human prostate cancer cell lines. Front Oncol. 2013;3:180.

25. Fuchs NV, Loewer S, Daley GQ, Izsvak Z, Lower J, Lower R. Human endogenous retrovirus K (HML-2) RNA and protein expression is a marker for human embryonic and induced pluripotent stem cells. Retrovirology. 2013;10:115.

26. Sugimoto J, Matsuura N, Kinjo Y, Takasu N, Oda T, Jinno Y. Transcriptionally active HERV-K genes: identification, isolation, and chromosomal mapping. Genomics. 2001;72(2):137–44.

27. Mayer J, Ehlhardt S, Seifert M, Sauter M, Muller-Lantzsch N, Mehraein Y, et al. Human endogenous retrovirus HERV-K(HML-2) proviruses with Rec protein coding capacity and transcriptional activity. Virology. 2004;322(1):190–8.

28. Sauter M, Schommer S, Kremmer E, Remberger K, Dölken G, Lemm I, et al. Human endogenous retrovirus K10: expression of Gag protein and detection of antibodies in patients with seminomas. J Virol. 1995;69(1):414–21.

29. Mayer J, Blomberg J, Seal RL. A revised nomenclature for transcribed human endogenous retroviral loci. Mob DNA. 2011;2(1):7.

30. Jurka J. Sequence patterns indicate an enzymatic involvement in integration of mammalian retroposons. Proc Natl Acad Sci U S A. 1997;94(5):1872–7.

31. Mayer J, Sauter M, Racz A, Scherer D, Mueller-Lantzsch N, Meese E. An almost-intact human endogenous retrovirus K on human chromosome 7. Nat Genet. 1999;21(3):257–8.

32. Contreras-Galindo R, Kaplan MH, Contreras-Galindo AC, Gonzalez-Hernandez MJ, Ferlenghi I, Giusti F, et al. Characterization of human endogenous retroviral elements in the blood of HIV-1-infected individuals. J Virol. 2012;86(1):262–76.

33. Levy S, Sutton G, Ng PC, Feuk L, Halpern AL, Walenz BP, et al. The diploid genome sequence of an individual human. PLoS Biol. 2007;5(10):e254.

34. Contreras-Galindo R, Kaplan MH, He S, Contreras-Galindo AC, Gonzalez-Hernandez MJ, Kappes F, et al. HIV infection reveals widespread expansion of novel centromeric human endogenous retroviruses. Genome Res. 2013;23(9):1505–13.

35. Marchi E, Kanapin A, Magiorkinis G, Belshaw R. Unfixed endogenous retroviral insertions in the human population. J Virol. 2014;88(17):9529–37.

36. Turner G, Barbulescu M, Su M, Jensen-Seaman MI, Kidd KK, Lenz J. Insertional polymorphisms of full-length endogenous retroviruses in humans. Curr Biol. 2001;11(19):1531–5.

37. Hohn O, Hanke K, Bannert N. HERV-K(HML-2), the best preserved family of HERVs: endogenization, expression, and implications in health and disease. Front Oncol. 2013;3:246.

38. Schmitt K, Richter C, Backes C, Meese E, Ruprecht K, Mayer J. Comprehensive analysis of human endogenous retrovirus group HERV-W locus transcription in multiple sclerosis brain lesions by high-throughput amplicon sequencing. J Virol. 2013;87(24):13837–52.

39. Ruggieri A, Maldener E, Sauter M, Mueller-Lantzsch N, Meese E, Fackler OT, et al. Human endogenous retrovirus HERV-K(HML-2) encodes a stable signal peptide with biological properties distinct from Rec. Retrovirology. 2009;6:17.

40. Büscher K, Hahn S, Hofmann M, Trefzer U, Ozel M, Sterry W, et al. Expression of the human endogenous retrovirus-K transmembrane envelope, Rec and Np9 proteins in melanomas and melanoma cell lines. Melanoma Res. 2006;16(3):223–34.

41. Muster T, Waltenberger A, Grassauer A, Hirschl S, Caucig P, Romirer I, et al. An endogenous retrovirus derived from human melanoma cells. Cancer Res. 2003;63(24):8735–41.

42. Ehlhardt S, Seifert M, Schneider J, Ojak A, Zang KD, Mehraein Y. Human endogenous retrovirus HERV-K(HML-2) Rec expression and transcriptional activities in normal and rheumatoid arthritis synovia. J Rheumatol. 2006;33(1):16–23.

43. Gross H, Barth S, Pfuhl T, Willnecker V, Spurk A, Gurtsevitch V, et al. The NP9 protein encoded by the human endogenous retrovirus HERV-K(HML-2) negatively regulates gene activation of the Epstein-Barr virus nuclear antigen 2 (EBNA2). Int J Cancer. 2011;129(5):1105–15.

44. Rosenbloom KR, Sloan CA, Malladi VS, Dreszer TR, Learned K, Kirkup VM, et al. ENCODE data in the UCSC genome browser: year 5 update. Nucleic Acids Res. 2013;41(Database issue):D56–63.

Permissions

The contributors of this book come from diverse backgrounds, making this book a truly international effort. This book will bring forth new frontiers with its revolutionizing research information and detailed analysis of the nascent developments around the world.

We would like to thank all the contributing authors for lending their expertise to make the book truly unique. They have played a crucial role in the development of this book. Without their invaluable contributions this book wouldn't have been possible. They have made vital efforts to compile up to date information on the varied aspects of this subject to make this book a valuable addition to the collection of many professionals and students.

This book was conceptualized with the vision of imparting up-to-date information and advanced data in this field. To ensure the same, a matchless editorial board was set up. Every individual on the board went through rigorous rounds of assessment to prove their worth. After which they invested a large part of their time researching and compiling the most relevant data for our readers.

The editorial board has been involved in producing this book since its inception. They have spent rigorous hours researching and exploring the diverse topics which have resulted in the successful publishing of this book. They have passed on their knowledge of decades through this book. To expedite this challenging task, the publisher supported the team at every step. A small team of assistant editors was also appointed to further simplify the editing procedure and attain best results for the readers.

Apart from the editorial board, the designing team has also invested a significant amount of their time in understanding the subject and creating the most relevant covers. They scrutinized every image to scout for the most suitable representation of the subject and create an appropriate cover for the book.

The publishing team has been an ardent support to the editorial, designing and production team. Their endless efforts to recruit the best for this project, has resulted in the accomplishment of this book. They are a veteran in the field of academics and their pool of knowledge is as vast as their experience in printing. Their expertise and guidance has proved useful at every step. Their uncompromising quality standards have made this book an exceptional effort. Their encouragement from time to time has been an inspiration for everyone.

The publisher and the editorial board hope that this book will prove to be a valuable piece of knowledge for researchers, students, practitioners and scholars across the globe.

List of Contributors

Travis B White and Prescott L Deininger
Tulane Cancer Center, 1430 Tulane Avenue, New Orleans, LA 70112, USA

Vincent A Streva
Tulane Cancer Center, 1430 Tulane Avenue, New Orleans, LA 70112, USA
Tulane University, Tulane Cancer Center and the Department of Epidemiology, 1430 Tulane Ave, New Orleans, LA 70112, USA

Joshua Fenrich
Bio-Rad Laboratories, 750 Alfred Nobel Drive, Hercules, CA, 94547, USA

Adam M McCoy
Bio-Rad Laboratories, 750 Alfred Nobel Drive, Hercules, CA, 94547, USA

Jui Wan Loh
Department of Genetics, The State University of New Jersey, Piscataway 08854, NJ, USA

Hongseok Ha and Jinchuan Xing
Department of Genetics, The State University of New Jersey, Piscataway 08854, NJ, USA
Human Genetic Institute of New Jersey, Rutgers, The State University of New Jersey, Piscataway 08854, NJ, USA

Patricia E. Carreira, Adam D. Ewing, Stephanie N. Schauer, Allister C. Fagg, Daniel J. Gerhardt, Santiago Morell, Michaela Kindlova, Patricia Gerdes and Sandra R. Richardson
Mater Research Institute - University of Queensland, TRI Building, Woolloongabba, QLD 4102, Australia

Kyle R. Upton
Mater Research Institute - University of Queensland, TRI Building, Woolloongabba, QLD 4102, Australia
School of Chemistry and Molecular Biosciences, University of Queensland, Brisbane, QLD 4072, Australia

Geoffrey J. Faulkner
Mater Research Institute - University of Queensland, TRI Building, Woolloongabba, QLD 4102, Australia
Queensland Brain Institute, University of Queensland, Brisbane, QLD 4072, Australia

Guibo Li and Jun Wang
BGI-Shenzhen, Shenzhen 518083, China
Department of Biology and the Novo Nordisk Foundation Center for Basic Metabolic Research, University of Copenhagen, Copenhagen 1599, Denmark

Bo Li
BGI-Shenzhen, Shenzhen 518083, China

Paul M. Brennan
Edinburgh Cancer Research Centre, IGMM, University of Edinburgh, Edinburgh EH42XR, UK

Pragathi Achanta
Molecular Biology and Genetics, Johns Hopkins University School of Medicine, Baltimore, MD, USA

Reema Sharma and Wan Rou Yang
Department of Pathology, Johns Hopkins University School of Medicine, Miller Research Building (MRB) Room 447, 733 North Broadway, Baltimore, MD 21205, USA

Jared P. Steranka and Kathleen H. Burns
Department of Pathology, Johns Hopkins University School of Medicine, Miller Research Building (MRB) Room 447, 733 North Broadway, Baltimore, MD 21205, USA
McKusick-Nathans Institute of Genetic Medicine, Johns Hopkins University School of Medicine, Miller Research Building (MRB) Room 447, 733 North Broadway, Baltimore, MD 21205, USA

Nemanja Rodić
Department of Pathology, Johns Hopkins University School of Medicine, Miller Research Building (MRB) Room 447, 733 North Broadway, Baltimore, MD 21205, USA

Cheng Ran Lisa Huang
Department of Pathology, Johns Hopkins University School of Medicine, Miller Research Building (MRB) Room 447, 733 North Broadway, Baltimore, MD 21205, USA

Anna M. Schneider
Department of Pathology, Johns Hopkins University School of Medicine, Miller Research Building (MRB) Room 447, 733 North Broadway, Baltimore, MD 21205, USA

Sisi Ma
Center for Health Informatics and Bioinformatics, New York University Langone Medical Center, New York, NY, USA

Zuojian Tang, Mark Grivainis and David Fenyö
Center for Health Informatics and Bioinformatics, New York University Langone Medical Center, New York, NY, USA
Institute for Systems Genetics, New York University Langone Medical Center, ACLSW Room 503, 430 East 29th Street, New York, NY 10016, USA

Jef D. Boeke
Institute for Systems Genetics, New York University Langone Medical Center, ACLSW Room 503, 430 East 29th Street, New York, NY 10016, USA
McKusick-Nathans Institute of Genetic Medicine, 733 North Broadway, Miller Research Building Room 469, Baltimore, MD 21205, USA
High Throughput (HiT) Biology Center, 733 North Broadway, Miller Research Building Room 469, Baltimore, MD 21205, USA
Department of Molecular Biology and Genetics, The Johns Hopkins University School of Medicine, Baltimore, MD, USA

Gary L. Gallia and Gregory J. Riggins
Department of Neurosurgery, Johns Hopkins University School of Medicine, Baltimore, MD

Alfredo Quinones-Hinojosa
Department of Neurosurgery, Johns Hopkins University School of Medicine, Baltimore, MD
Mayo Clinic, Jacksonville, FL, USA

John G. Zampella
Department of Dermatology, Johns Hopkins University School of Medicine, 733 North Broadway, Miller Research Building Room 469, Baltimore, MD 21205, USA

Nemanja Rodić, Wan Rou Yang and Toby C. Cornish
Department of Pathology, Johns Hopkins University School of Medicine, 733 North Broadway, Miller Research Building Room 469, Baltimore, MD 21205, USA

Cheng Ran Lisa Huang, Jane Welch and Veena P. Gnanakkan
McKusick-Nathans Institute of Genetic Medicine, 733 North Broadway, Miller Research Building Room 469, Baltimore, MD 21205, USA

Kathleen H. Burns
Department of Pathology, Johns Hopkins University School of Medicine, 733 North Broadway, Miller Research Building Room 469, Baltimore, MD 21205, USA
McKusick-Nathans Institute of Genetic Medicine, 733 North Broadway, Miller Research Building Room 469, Baltimore, MD 21205, USA
High Throughput (HiT) Biology Center, 733 North Broadway, Miller Research Building Room 469, Baltimore, MD 21205, USA
The Sidney Kimmel Comprehensive Cancer Center, Johns Hopkins University School of Medicine, 733 North Broadway, Miller Research Building Room 469, Baltimore, MD 21205, USA

Pamela R. Cook
Laboratory of Cell and Molecular Biology, National Institute of Diabetes and Digestive and Kidney Diseases, National Institutes of Health, 8 Center Drive, Bethesda, MD 20892, USA

G. Travis Tabor
National Institute of Child Health and Human Development, National Institutes of Health, 35 Convent Drive, Bethesda, MD 20892, USA

Roger Horton
Max Planck Institute for molecular Genetics, Ihnestr. 63-73, 14195 Berlin, Germany

Karin Moelling
Max Planck Institute for molecular Genetics, Ihnestr. 63-73, 14195 Berlin, Germany
Institute of Medical Microbiology, University of Zurich, Gloriastr. 32, 8006 Zurich, Switzerland

Felix Broecker
Max Planck Institute for molecular Genetics, Ihnestr. 63-73, 14195 Berlin, Germany
Institute of Medical Microbiology, University of Zurich, Gloriastr. 32, 8006 Zurich, Switzerland
Max Planck Institute of Colloids and Interfaces, Am Mühlenberg 1, 14424 Potsdam, Germany

Alexandra Franz
Max Planck Institute for molecular Genetics, Ihnestr. 63-73, 14195 Berlin, Germany
University of Zurich, Institute of Molecular Life Sciences, Winterthurerstr. 190, 8057 Zurich, Switzerland

Michal-Ruth Schweiger
Max Planck Institute for molecular Genetics, Ihnestr. 63-73, 14195 Berlin, Germany
Functional Epigenomics, CCG, Cologne University Hospital, University of Cologne, Weyertal 115b, 50931 Cologne, Germany

Hans Lehrach
Max Planck Institute for molecular Genetics, Ihnestr. 63-73, 14195 Berlin, Germany
Dahlem Centre for Genome Research and Medical Systems Biology, Fabeckstr. 60-62, 14195 Berlin, Germany

Jochen Heinrich
Institute of Medical Microbiology, University of Zurich, Gloriastr. 32, 8006 Zurich, Switzerland

Lindsay M. Payer
Department of Pathology, Johns Hopkins University School of Medicine, Miller Research Building (MRB) Room 447, 733 North Broadway, Baltimore, MD 21205, USA

Maria S. Kryatova Jared P. Steranka and Kathleen H. Burns
Department of Pathology, Johns Hopkins University School of Medicine, Miller Research Building (MRB) Room 447, 733 North Broadway, Baltimore, MD 21205, USA
McKusick-Nathans Institute of Genetic Medicine, Johns Hopkins University School of Medicine, Miller Research Building (MRB) Room 447, 733 North Broadway, Baltimore, MD 21205, USA

Manoj Kannan
Department of Biological Sciences, Birla Institute of Technology and Science Pilani, Pilani 333031, Rajasthan, India
Laboratory of Immunobiology, Mouse Cancer Genetics Program and Basic Research Laboratory, Center for Cancer Research, National Cancer Institute, Frederick, MD 21702, USA

Teresa L. Sullivan
Laboratory of Immunobiology, Mouse Cancer Genetics Program and Basic Research Laboratory, Center for Cancer Research, National Cancer Institute, Frederick, MD 21702, USA

Jingfeng Li
Laboratory of Immunobiology, Mouse Cancer Genetics Program and Basic Research Laboratory, Center for Cancer Research, National Cancer Institute, Frederick, MD 21702, USA
Department of Cancer Biology and Genetics, The Ohio State University, Columbus, OH, USA
Department of Internal Medicine, The Ohio State University, Columbus, OH, USA

Pawan Kumar Tiwary
Laboratory of Immunobiology, Mouse Cancer Genetics Program and Basic Research Laboratory, Center for Cancer Research, National Cancer Institute, Frederick, MD 21702, USA

David E. Symer
Laboratory of Immunobiology, Mouse Cancer Genetics Program and Basic Research Laboratory, Center for Cancer Research, National Cancer Institute, Frederick, MD 21702, USA
Department of Cancer Biology and Genetics, The Ohio State University, Columbus, OH, USA
Human Cancer Genetics Program, and Department of Biomedical Informatics, The Ohio State University, Columbus, OH, USA
Human Cancer Genetics Program, Department of Cancer Biology and Genetics, and Department of Biomedical Informatics, The Ohio State University, Tzagournis Research Facility, Room 440, 420 West 12th Ave, Columbus, OH 43210, USA

Sarah E. Fritz
Biomedical Sciences Graduate Program, The Ohio State University, Columbus, OH, USA
Present Address: National Heart, Lung and Blood Institute, National Institutes of Health, Bethesda, MD, USA

Kathryn E. Husarek
Biomedical Sciences Graduate Program, The Ohio State University, Columbus, OH, USA

Jonathan C. Sanford
Biomedical Sciences Graduate Program, The Ohio State University, Columbus, OH, USA

Wenfeng An
Department of Molecular Biology and Genetics, The Johns Hopkins University School of Medicine, Baltimore, MD, USA

Julie Feusier, David J. Witherspoon, W. Scott Watkins, Clément Goubert, Thomas A. Sasani and Lynn B. Jorde
Department of Human Genetics, University of Utah School of Medicine, Salt Lake City, UT, USA

Nicole Grandi, Marta Cadeddu, Maria Paola Pisano and Francesca Esposito
Department of Life and Environmental Sciences, University of Cagliari, Cagliari, Italy

Enzo Tramontano
Department of Life and Environmental Sciences, University of Cagliari, Cagliari, Italy
Istituto di Ricerca Genetica e Biomedica, Consiglio Nazionale delle Ricerche (CNR), Monserrato, Cagliari, Italy

Jonas Blomberg
Department of Medical Sciences, Uppsala University, Uppsala, Sweden

Geraldine Servant
Tulane University, Tulane Cancer Center and the Department of Epidemiology, 1430 Tulane Ave, New Orleans, LA 70112, USA

Prescott L. Deininger
Tulane University, Tulane Cancer Center and the Department of Epidemiology, 1430 Tulane Ave, New Orleans, LA 70112, USA
Tulane Cancer Center, SL66, Tulane University Health Sciences Center, 1430 Tulane Ave., New Orleans, LA 70112, USA

Debpali Sur, Raj Kishor Kustwar, Savita Budania and Prabhat K. Mandal
Department of Biotechnology, IIT Roorkee, Roorkee, Uttarakhand, India

Vijay Yadav
School of Environmental Sciences, Jawaharlal Nehru University, New Delhi, India

Anita Mahadevan and S. K. Shankar
Human Brain Tissue Repository (HBTR), Neurobiology Research Centre, NIMHANS, Bangalore 560 029, India

Dustin C. Hancks
Department of Human Genetics, University of Utah, Salt Lake City, UT, USA

Lavanya Rishishwar and I. King Jordan
School of Biology, Georgia Institute of Technology, 310 Ferst Drive, Atlanta, GA 30332-0230, USA.
PanAmerican Bioinformatics Institute, Cali, Valle del Cauca, Colombia
BIOS Centro de Bioinformática y Biología Computacional, Manizales, Caldas, Colombia

Carlos E. Tellez Villa
PanAmerican Bioinformatics Institute, Cali, Valle del Cauca, Colombia
Escuela de Ingeniería de Sistemas y Computación, Universidad del Valle, Santiago de Cali, Colombia

Kristine J. Kines, Mark Sokolowski, Dawn L. deHaro, Claiborne M. Christian, Melody Baddoo, Madison E. Smither and Victoria P. Belancio
Department of Structural and Cellular Biology, Tulane School of Medicine, Tulane Cancer Center and Tulane Center for Aging, New Orleans, LA 70112, USA

Eckart Meese
Institute of Human Genetics, Center of Human and Molecular Biology, Medical Faculty, University of Saarland, 66424 Homburg/Saar, Germany

Jens Mayer
Institute of Human Genetics, Center of Human and Molecular Biology, Medical Faculty, University of Saarland, 66424 Homburg/Saar, Germany
Center of Human and Molecular Biology, University of Saarland, 66424 Homburg/Saar, Germany

Katja Schmitt
Institute of Human Genetics, Center of Human and Molecular Biology, Medical Faculty, University of Saarland, 66424 Homburg/Saar, Germany
Sanofi-Aventis Deutschland GmbH, Industriepark Hoechst, K703, Elisabeth Kuhn Street, Frankfurt/Main 65926, Germany

Kristina Heyne and Klaus Roemer
José Carreras Research Center, Medical Faculty, University of Saarland, 66424 Homburg/Saar, Germany

Index

www.ingramcontent.com/pod-product-compliance
Lightning Source LLC
Chambersburg PA
CBHW082042190326
41458CB00010B/3438